ENCYCLOPAEDIA OF MATHEMATICS AND ITS APPLICATIONS

EDITED BY G.-C. ROTA

Volume 31

Functional equations in several variables

ENCYCLOPAEDIA OF MATHEMATICS AND ITS APPLICATIONS

Functional equations in several variables

with applications to mathematics, information theory and to the natural and social sciences

J. ACZÉL
University of Waterloo, Ontario, Canada

J. DHOMBRES
Université de Nantes, France

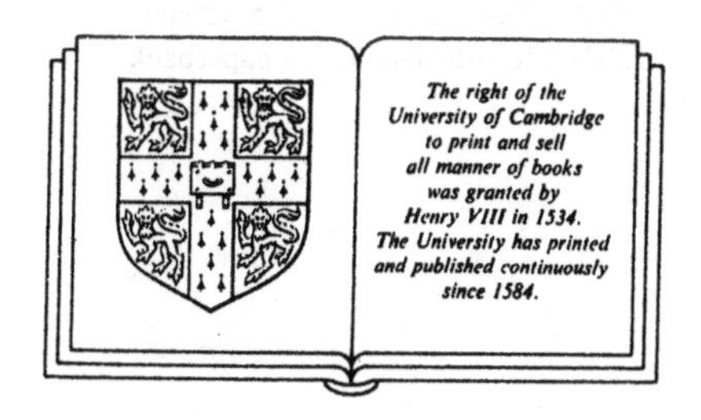
The right of the
University of Cambridge
to print and sell
all manner of books
was granted by
Henry VIII in 1534.
The University has printed
and published continuously
since 1584.

CAMBRIDGE UNIVERSITY PRESS
Cambridge
New York New Rochelle
Melbourne Sydney

CAMBRIDGE UNIVERSITY PRESS
Cambridge, New York, Melbourne, Madrid, Cape Town, Singapore, São Paulo

Cambridge University Press
The Edinburgh Building, Cambridge CB2 8RU, UK

Published in the United States of America by Cambridge University Press, New York

www.cambridge.org
Information on this title: www.cambridge.org/9780521352765

First published 1989
Reprinted 1991
This digitally printed version 2008

A catalogue record for this publication is available from the British Library

Library of Congress Cataloguing in Publication data

Functional equations in several variables J. Aczél and
J. Dhombres.
p. cm. – (Encyclopedia of mathematics and its applications)
Bibliography: p.
Includes indexes.
ISBN 0 521 35276 2
1. Functional equations. 2. Functions of several real variables.
I. Dhombres, Jean G. II. Title. III. Series.
QA431.A334 1989
515.8′4 - - dcl9 87-38107
CIP

ISBN 978-0-521-35276-5 hardback
ISBN 978-0-521-06389-0 paperback

CONTENTS

Dedicated to
Beckie, Cathy, Julie, Pascale, Robbie, Sylvestre and Thomas

PREFACE

Functional equations are equations in which the unknown (or unknowns) are functions (but we shall not cover such a large domain). As the title shows, we deal in this book with 'functional equations in several variables'. This does not mean that we consider only equations in which the unknown functions are of several variables (in other words, multiplace functions) but, rather, that there are several variables *in the equation.* Cauchy's equation,

$$f(x+y)=f(x)+f(y),$$

containing *two* variables x, y and one unknown function f of *one* variable, is a classical example. On the other hand, iterative equations, differential, difference–differential, difference, integral and similar equations contain (in most cases) just one variable, if the unknown function is of one variable (a one-place function). We will deal here with *functional equations in which the number of variables is greater than the number of places in the unknown function* (or, if there is more than one unknown function, greater than the number of places in the unknown function with the smallest number of places).

There are several thousand works on functional equations, even in this restricted sense, and it is impossible to summarize their contents in a book of this size. So we have tried to make the *bibliography* encyclopaedic (although not complete) – it is organized by years. The reader is encouraged to refer to as many of the works listed in the bibliography as possible. (Some are quite elementary.) An amazing number of rediscoveries can be spotted in this way.

We felt it might be useful to the reader for us to include a historical chapter at the end of the book (Chapter 21). This chapter can be read *first* in order to get a bird's eye view (which makes a longer introduction unnecessary), ignoring or checking the few references to other sections in it. But, also, we think that this last chapter can be read again with profit after the first twenty chapters have been studied.

From this historical chapter, the reader will see that a somewhat general theory of functional equations in several variables is a relatively recent development. It can also be seen that, from their very beginnings, functional equations arose from applications, were developed mostly for the sake of applications and, indeed, were applied quite intensively as soon as they were developed. Such a course of development is not typical in all theoretical aspects of mathematics. A characteristic example of this process can be found in the first, introductory, chapter which is devoted to the parallelogram law for vector addition.

This brings us to the natural question of motivation: *To what kind of problems can functional equations be applied* (and not only within mathematics). G.H. Hardy said that mathematicians are 'makers of patterns'. They often construct new notions, partly based on older ones, which may come from mathematics or from the natural, behavioural and social sciences, for instance from physics or economics. Thus the notion of affine vectors evolved in order to represent forces, the (homogeneous) linear function to describe proportionality, the logarithmic function to transform geometric sequences into algebraic ones, and the trigonometric functions to determine unknown parts of triangles from known ones. The next task of the mathematician is to describe the properties of these new objects, that is, to establish their relation to other objects in order to include them in a systematic and orderly way into an existing or new theory. (This is important because it is convenient to organize the notions in a pattern ruled by the same structure, with unity in the style of proofs.) The vectors, for instance, form the foundation of linear algebra, the trigonometric functions that of trigonometry. After stating the properties of the new objects (or at least some of them), what next?

If we are lucky, the properties deduced from the definitions of mathematical objects can be quite numerous, even too rich. According to the principle of economy of reasoning and to a legitimate desire for elegance, it is desirable to check whether the newly introduced objects are the *only ones* which have some of *the most important properties* that we have just established. It is here where functional equations may enter. One takes these properties as points of departure and tries to determine all objects satisfying them and, in particular, to find conditions under which there is unicity. This is done in the framework of the classical structures of mathematics, be they algebraic, order-theoretic, topological, or simply the real axis which carries all these structures. This is the fundamental procedure which motivates functional equations. We see that it is axiomatic in nature. But we see also that the axioms are not arbitrarily chosen, we do not generalize just for the sake of generalization. In fact, one prefers to take as points of departure those which are considered the most useful for applications. (Also practical concerns determine the procedure since we want to attain uniqueness.) Thus there is a feedback from the applications to the theory.

This is not all. After such a uniqueness result is established, for instance that concerning the equation of the cosine (d'Alembert's equation), one also tries to deduce (using the equation, not the solutions) the classical properties of the cosine directly. One thus arrives at an elegant way of developing trigonometry which, with some changes, can also be used to introduce and develop elliptic and, especially, hyperbolic trigonometry. This procedure was followed (with varying degrees of clarity) by mathematicians in different ages, such as Oresme, Euler, d'Alembert, Abel, Cauchy, Lobachevskiĭ, Darboux, Picard, etc. Another famous example, outside the framework of our book, is the introduction of the gamma function through its functional equation, as developed by Artin.

As in all parts of mathematics, different authors' viewpoints and levels of discussion of functional equations are different. Also, sometimes the methods of reduction of one functional equation to another or the proof of their equivalence or discussion and/or reduction of regularity conditions, extension of domains, etc. give rise to further intrinsic developments. This is a further sign of the increasing maturity of this field of mathematics. On the other hand, the characterization of functions by their equations, as described above, involves many branches of pure and applied mathematics in the development of the theory of functional equations.

Our general intention in this book is to give the reader at least a general impression of what this subject is about by focusing on a relatively small number of examples, chosen with particular regard to applications but without neglecting theory. This explains why most technical connections among the different chapters are minimal, so that every chapter may constitute a unit of study in itself. Indeed, the reader can start with any chapter (and even many sections) with only occasional backward references. (Here the subject index at the end of the book will prove helpful.) An obvious exception is Chapter 2 which deals with Cauchy's equation and is intended to provide the reader with a certain number of definitions, techniques and results which will be used throughout the book (such as domain, extension of solutions, general solutions, regular solutions, etc.). However, from a more general mathematical point of view, connections between chapters can be found, and such connections can be used, for instance in both undergraduate and graduate courses or seminars. So can the several chapters and sections (or parts of them) devoted to connections with other parts of mathematics and to applications in and outside of mathematics. We give some examples below:

Combinatorics, probability and information theory	3; 5.4; 12; 16.4
Economics, decision making and mean values	7.3; 15; 17; 20
Almost periodic functions and harmonic analysis	5.3; 7.2; 8; 10.5

Functional analysis, operator theory and Heaviside functions	6; 10; 11; 19.4; 19.5; 19.6
Geometry, nomography and physics	1; 4.2; 5.2; 6; 8; 9.1; 11.4; 18; 19.4
Groups, groupoids and semigroups	4.4; 5.3; 16.2; 16.3; 19.1; 19.2; 19.3
Trigonometric functions and number theory	1; 8; 13; 14; 16.3; 16.4; 16.5; 16.6.

(There are several more items related to these and other applications in the 'exercises and further results' at the end of different chapters.)

We also indicate some longer sequences which can be used for individual courses. The fundamental Chapter 2 should be added to each of them, and, clearly, some paragraphs of the following chapters and sections can be omitted if they do not fit into the subject of the course:

- 3; 4.1; 4.2; 4.4; 9.1; 11.2; 16.3; 16.5; 16.6; 19.1; 19.2; 20
- 5.1; 5.2; 5.3; 8; 10.5; 12; 13; 14
- 5.2; 6; 9.2; 11.3; 11.4; 14; 16.3; 16.4; 16.5; 16.6
- 10.4; 15; 16.4; 17; 18; 19.3
- 3; 6; 7; 10.2; 10.3; 11.2; 16.1; 16.2
- 5.1; 5.4; 6; 7.2; 10.1; 10.2; 19.4; 19.5; 19.6
- 1; 4.3; 4.4; 5.2; 7.2; 8; 11.1; 12; 13; 15; 16.2; 17; 18; 19.1; 19.2.

The pedagogical purpose of the book and the wish to at least indicate some further results and directions of research has led us to include more than 400 items as 'exercises and further results'. Some can be solved by a direct application of the material explained in the chapter in which the exercise appears; these exercises serve the usual aims of practice and of further applications of the results and proofs. Others are 'further results': extensions of the results described in the text to more general (algebraic, topological) settings and (more or less) related theorems. We sometimes give details of such generalizations and further results. Even where we do not generalize, the reader may want to try to prove such generalizations. Some 'exercises and further results' go deeper in the theory (and the most difficult ones are marked with a *). For some selected exercises and results we provide the reader with hints (which we hope won't become hindrances!), but we do not give hints to exercises which we consider to be easy or to results which are preceded by easily accessible references. In a large number of cases, these references are to the original contributions to the solution of the problem; in some other cases we preferred to give more modern or more easily accessible literature. On the other hand, in the main text, we have tried to be accurate concerning priorities, except for general background references for which we again wanted to quote easily accessible sources.

Now we have to say a few words about the prerequisites which are necessary for a fruitful use of the book. We do try to avoid all unnecessary complications by limiting ourselves to simpler cases and we think that most chapters in the book can be read at the sophomore level, after mastering calculus, general and linear algebra, and perhaps basic Lebesgue theory of integration in $\mathbb{R}$ or $\mathbb{R}^n$. Incidentally, a previous monograph by one of the authors (Aczél 1966*c*) provides a more elementary approach to some parts of this book and also some further topics. Some applications require more specialized knowledge which we indicate there (some Banach algebra techniques for example, duality in functional analysis, and some measure theory). We try to explain at least the results we use. But, generally, no previous knowledge of the field to which functional equations are applied is required. This is true, for instance, in information theory, number theory, mean values, consensus allocations, geometric objects, almost periodic functions, relativity theory, etc.

While we have tried to 'homogenize' the presentation to some degree, the attentive reader will notice which parts were written by which author, due to their different styles (and mother tongues). Hopefully, this is not too terrible.

Acknowledgements

Che Tat Ng has suggested several essential improvements and shortcuts. Gord Sinnamon helped checking the galley proofs. Richard Atkins and Gord Sinnamon helped finishing the subject index. Susan Aczél has compiled the author index and worked long and hard to put the bibliography together. Linda Gregory and Brenda Law have typed (in several versions), Dong Nguyen, Richard Atkins, Nicole Brillouët-Belluot and, last but not least, Bert Schweizer have read and corrected the manuscript and suggested improvements. Our warmest thanks are due to all of them and to colleagues, including anonymous referees, who have helped us with their comments. Support by the Natural Sciences and Engineering Research Council of Canada and the France-Canada Scientific Cooperation Agreement is gratefully acknowledged. We express also our thanks to the publishing and printing houses for their conscientious work.

Let us add that we welcome comments on the book or on any aspect of the subject.

János Aczél
Jean Dhombres

FURTHER INFORMATION

References, numbering. The references in the bibliography are listed by years, within the same year alphabetically according to the authors' names, and works of the same author(s) in the same year are distinguished by letters. References are quoted in the text with the authors' names, the years, and, where necessary, with distinguishing letters. Theorems, propositions, lemmata, and corollaries are numbered consecutively in each chapter (so that Lemma 1 may be followed by Theorem 2 and that by Corollary 3) and formulas are separately numbered, also consecutively, within the chapters. In the same chapter they are referred to by these numbers, while in the references to another chapter the number of that chapter is attached, for example Theorem 2.3 or formula (3.5). (Some longer chapters are subdivided into sections but numbering of formulas, etc., is within chapters, not sections.) As usual, easier theorems are called 'Propositions'.

Exercises and further results at the end of the chapters are numbered (separately) and quoted in a similar way. Those with stars (*) are thought (by the authors) to be more difficult.

An *Author Index* and a *Subject Index* are, as usual, at the end. *Notations* are mostly standard, but we also summarize less standard notations at the end of the book.

1

Axiomatic motivation of vector addition

In this book we try to emphasize applications as motivation for functional equations. Of course we cannot present them in all exact details (as we try to do in the treatment of functional equations themselves). This first chapter is meant to give such a motivation for the Cauchy and d'Alembert equations. The reader who needs no convincing of their importance (or has any difficulty with Chapter 1) may proceed directly to Chapter 2.

One of the first problems to which functional equations were ever applied, and for whose sake functional equations were first solved, was the so called 'parallelogram of forces' or, in more modern language, the axiomatic motivation of the customary rule for addition of vectors (d'Alembert 1769; Poisson 1804, 1811; Picard 1928, pp. 4–17; Aczél 1966*c*, pp. 120–4, 1969*a*, pp. 7–11, 1976*a* are a few relevant references). Here we give a somewhat more satisfactory treatment, based on slightly weaker assumptions.

We start with the space V of all 3-tuples $\mathbf{p}=(x_1,x_2,x_3)$, $x_i\in\mathbb{R}$. This space is equipped with the euclidean distance

$$d(\mathbf{p},\mathbf{q})=((x_1-y_1)^2+(x_2-y_2)^2+(x_3-y_3)^2)^{1/2}$$

between $\mathbf{p}$ and $\mathbf{q}=(y_1,y_2,y_3)$. Our purpose is to characterize the additive law of such 3-tuples by geometric conditions.

Our fundamental tools will then be rotations on V, which are bijective transformations from V onto itself preserving $\mathbf{0}=(0,0,0)$ and distances in V. We recall that a rotation is completely determined by an axis through $\mathbf{0}$ and an angle, which is a real number modulo 2π. Given any two 3-tuples, $\mathbf{p}$ and $\mathbf{q}$, having the same positive distance from the origin, there exists one and only one rotation mapping $\mathbf{p}$ into $\mathbf{q}$. The angle of this rotation, say Θ, is the angle of the two 3-tuples, which can be determined by the analytic formula

$$\cos\Theta=\frac{x_1y_1+x_2y_2+x_3y_3}{(x_1^2+x_2^2+x_3^2)^{1/2}(y_1^2+y_2^2+y_3^2)^{1/2}},$$

where $\Theta \in [0, \pi]$ or $\Theta \in [\pi, 2\pi]$ according to the orientation chosen on V (for example by choosing $(1, 0, 0)$, $(0, 1, 0)$ and $(0, 0, 1)$ to be a directed basis). The first will be called the smaller, the second the larger angle of the two vectors.

For the sake of simplicity, we call the 3-tuple

$$\mathbf{p} = (x_1, x_2, x_3)$$

the vector $\mathbf{p}$. We say that two vectors $\mathbf{p} = (x_1, x_2, x_3)$ and $\mathbf{q} = (y_1, y_2, y_3)$ have the same (or opposite) direction if there exist real numbers λ, μ of the same (respectively, opposite) sign such that

$$\lambda x_i = \mu y_i, \quad i = 1, 2, 3.$$

We allow that one (but not both) of λ and μ be 0. For convenience we call $\mathbf{p}$ and $\mathbf{q}$ of the same direction in this case too. Naturally, the length of a vector $\mathbf{p}$ is its euclidean norm

$$|\mathbf{p}| = (x_1^2 + x_2^2 + x_3^2)^{1/2}.$$

With this in mind, we can state our theorem:

Theorem 1. *If, and only if, a binary operation* $\circ$ *on the set* V *of all vectors in the oriented three dimensional euclidean space*

(i) *is invariant under (really, covariant with) the rotations of the space, that is, the result of the operation (the 'resultant vector') undergoes the same rotation as the two factors ('components'),*
(ii) *is commutative and associative,*
(iii) *for two vectors of the same direction,* $\circ$ *reduces to arithmetic addition (the resultant vector also points in that direction and its length is the sum of the lengths of the components), and*
(iv) *the resultant of two vectors of equal length depends continuously upon their angle,*

then the operation $\circ$ *is the usual vector addition (the resultant is obtained by the 'parallelogram rule').*

Remark. Strictly speaking, suppositions like (i) and (iii) are themselves functional equations. However, for this introductory chapter we prefer the above verbal descriptions.

Proof. The most important consequence of (i) and of the commutativity (ii) is that *the resultant of two vectors of equal lengths lies in the plane spanned by them,* more exactly, *in the direction of the bisector of their angle* (using this direction as an axis for a rotation). Drawn from their intersection, the resultant could still lie either inside the smaller or the larger angle formed by the two vectors.

Condition (iii) assures that the zero-vector is a unit under $\circ$ $(\mathbf{p} \circ \mathbf{0} = \mathbf{0} \circ \mathbf{p} = \mathbf{p})$, and then again (i) shows that in $(V, \circ)$ *a vector* $\mathbf{p}$ *and a vector* $-\mathbf{p}$ *of the same length but opposite direction are inverses*: $(-\mathbf{p})\circ\mathbf{p}=\mathbf{p}\circ(-\mathbf{p})=\mathbf{0}$. Thus, with (ii), we have

(v) $(V, \circ)$ *is an abelian group with* $\mathbf{0}$ *as unit and* $-\mathbf{p}$ *as the inverse of* $\mathbf{p}$.

By (iii), the resultant of two vectors of the same length and of the same direction (angle 0) lies in the same direction, that is, in the direction of the bisector of their smaller angle (0). If there existed an angle of two vectors of the same length for which the resultant would lie in the bisector of their larger angle (Figure 1) then, by the continuity (iv), there would exist an angle, smaller than π, under which two vectors of the same length would have the zero-vector as resultant. But these would then be inverses to each other in $(V, \circ)$, while we have just seen that the inverse vectors always form angles π. (By (v) there exists just one inverse.) This contradiction shows that the resultant of two vectors of equal length always lies inside their smaller angle (and is the zero-vector if and only if the two vectors of equal and non-zero length have the angle π).

So we have determined the direction (in the bisector of their smaller angle) of the resultant of two vectors of equal length x. Keeping their angle fixed,

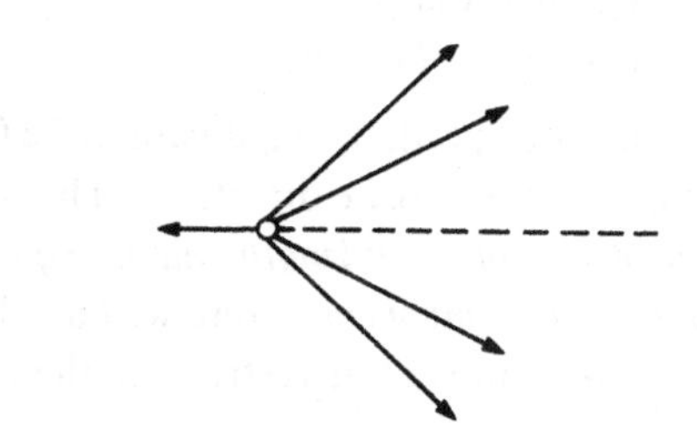
Fig. 1

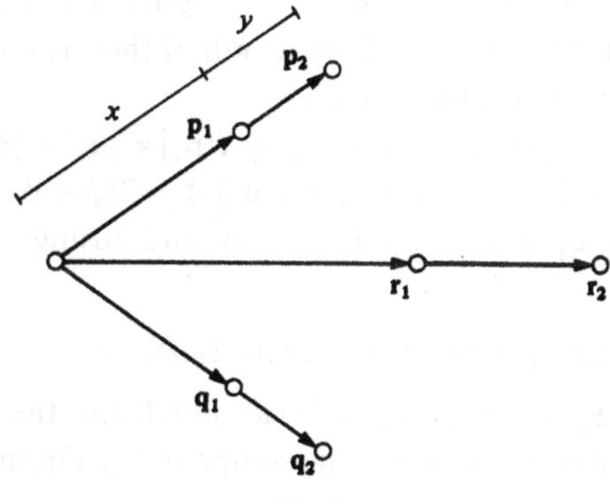

Fig. 2

we now determine its *length*, which we will denote by $g(x)$. This function g is defined on the set of all *nonnegative* numbers and is, in view of the above, nonnegative. Also (see Figure 2), if the vectors $\mathbf{p}_1, \mathbf{p}_2$ have the same direction, and also the vectors $\mathbf{q}_1, \mathbf{q}_2$, and ($|\mathbf{p}|$ denoting the length of the vector $\mathbf{p}$)

$$|\mathbf{p}_1| = |\mathbf{q}_1| = x, \quad |\mathbf{p}_2| = |\mathbf{q}_2| = y,$$

then, by (iii),

$$|\mathbf{p}_1 \circ \mathbf{p}_2| = |\mathbf{q}_1 \circ \mathbf{q}_2| = x + y$$

and, in view of the definition of g,

$$|\mathbf{r}_1| = |\mathbf{p}_1 \circ \mathbf{q}_1| = g(x), \quad |\mathbf{r}_2| = |\mathbf{p}_2 \circ \mathbf{q}_2| = g(y), \quad |(\mathbf{p}_1 \circ \mathbf{p}_2) \circ (\mathbf{q}_1 \circ \mathbf{q}_2)| = g(x+y).$$

But also, by (ii) and (iii), since $(\mathbf{p}_1 \circ \mathbf{q}_1)$ and $(\mathbf{p}_2 \circ \mathbf{q}_2)$ are collinear,

$$|(\mathbf{p}_1 \circ \mathbf{p}_2) \circ (\mathbf{q}_1 \circ \mathbf{q}_2)| = |(\mathbf{p}_1 \circ \mathbf{q}_1) \circ (\mathbf{p}_2 \circ \mathbf{q}_2)| = g(x) + g(y), \tag{1}$$

and so

$$g(x+y) = g(x) + g(y) \quad \text{for all nonnegative } x, y. \tag{2}$$

As mentioned in the introduction, this is Cauchy's equation (Cauchy 1821). Furthermore, as pointed out above,

$$g(x) \geqslant 0 \quad \text{for all nonnegative } x. \tag{3}$$

We will prove in Chapter 2 that (2) and (3) can hold if, and only if, there exists a nonnegative constant c such that

$$g(x) = cx \quad (x \geqslant 0).$$

The constant c has to be nonnegative, but, by (v), it cannot be 0 except when the two vectors of equal length have opposite directions. This shows that *the length of the resultant of two vectors of equal length and of angle different from π is proportional to the length of the components.* Here we have kept the angle of the two vectors constant. If we remove this restriction, then c will depend upon that angle.

We now determine the dependence of the length of the resultant of two vectors of equal length on their angle ($\neq \pi$). Because of what we have just proved, it is enough to take two vectors of unit length ('*unit vectors*'). Denote their angle, for convenience, by 2ϕ and the length of their resultant by $2f(\phi)$ (this is really our c, now dependent upon ϕ).

Take four unit vectors (Figure 3) $\mathbf{p}_1, \mathbf{p}_2, \mathbf{q}_1, \mathbf{q}_2$ ($|\mathbf{p}_1| = |\mathbf{p}_2| = |\mathbf{q}_1| = |\mathbf{q}_2| = 1$) with angles $(\mathbf{p}_1, \mathbf{p}_2)\measuredangle = 2\psi = (\mathbf{q}_1, \mathbf{q}_2)\measuredangle$, $(\mathbf{p}_1, \mathbf{q}_1)\measuredangle = 2(\phi + \psi)$, $(\mathbf{p}_2, \mathbf{q}_2)\measuredangle = 2(\phi - \psi)$. Then $(\mathbf{p}_1 \circ \mathbf{p}_2, \mathbf{q}_1 \circ \mathbf{q}_2)\measuredangle = (\mathbf{p}, \mathbf{q})\measuredangle = 2\phi$ and (using (i)) $|\mathbf{p}_1 \circ \mathbf{p}_2| = |\mathbf{q}_1 \circ \mathbf{q}_2| = 2f(\psi)$. Also,

$$|(\mathbf{p}_1 \circ \mathbf{p}_2) \circ (\mathbf{q}_1 \circ \mathbf{q}_2)| = 2f(\psi) 2f(\phi) \tag{4}$$

(because the length of $\mathbf{p}_1 \circ \mathbf{p}_2$ or $\mathbf{q}_1 \circ \mathbf{q}_2$ is $2f(\psi)$, not 1 and the length of the resultant is proportional to the length of the components). On the other hand,

$$|\mathbf{r}_1| = |\mathbf{p}_1 \circ \mathbf{q}_1| = 2f(\phi + \psi), \quad |\mathbf{r}_2| = |\mathbf{p}_2 \circ \mathbf{q}_2| = 2f(\phi - \psi)$$

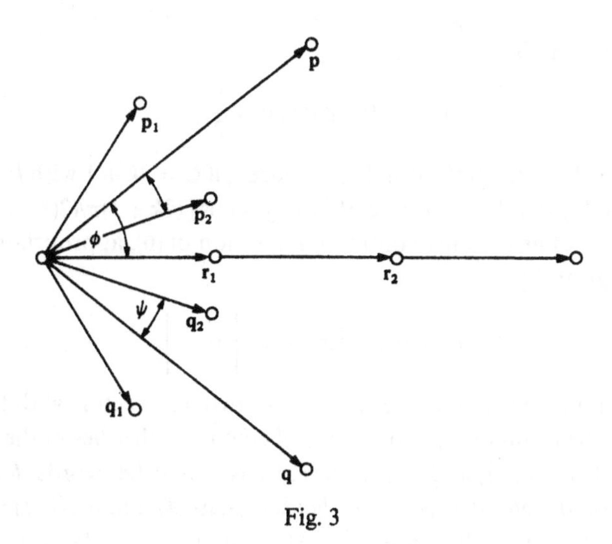

Fig. 3

and, by (4), (ii), (iii), and since $\mathbf{p}_1, \mathbf{q}_1$ and $\mathbf{p}_2, \mathbf{q}_2$ have a common bisector,

$$\begin{aligned}4f(\phi)f(\psi) &= |(\mathbf{p}_1 \circ \mathbf{p}_2)\circ(\mathbf{q}_1 \circ \mathbf{q}_2)| = |(\mathbf{p}_1 \circ \mathbf{q}_1)\circ(\mathbf{p}_2 \circ \mathbf{q}_2)| \\ &= |\mathbf{r}| = |\mathbf{r}_1 \circ \mathbf{r}_2| = 2f(\phi+\psi) + 2f(\phi-\psi),\end{aligned} \tag{5}$$

or

$$f(\phi+\psi) + f(\phi-\psi) = 2f(\phi)f(\psi) \quad \text{whenever } 0 \leqslant \psi \leqslant \phi \leqslant \frac{\pi}{4}. \tag{6}$$

This is d'Alembert's functional equation.

By (iv), the function f, defined above on $[0, \pi/2]$, is *continuous*. As we will see in Chapter 8, a continuous function f satisfies (6) if (and only if) it has one of the following forms:

$$f(\phi) = 0, \quad \text{or } f(\phi) = \cosh C\phi, \quad \text{or } f(\phi) = \cos C\phi \tag{7}$$

for all $\phi \in [0, \pi/2]$ where C is a (real) constant. We may take $C \geqslant 0$, because (7) gives the same result for negative as for positive C. (We may also extend, if we wish, (6) and (7) to all real ϕ and ψ.) As we have seen, for vectors of equal lengths and the same or opposite directions,

$$f(0) = 1 \quad \text{and} \quad f\left(\frac{\pi}{2}\right) = 0$$

respectively. These exclude the first and second solutions enumerated (since cosh is nowhere 0) in (7) and specify

$$C = 2k+1, \quad \text{where } k \text{ is a nonnegative integer}, \tag{8}$$

in the third solution. But, as we have seen, by (v) the only vector which gives the zero vector as resultant with a given unit vector is the unit vector of opposite direction. So

$$f(\phi) \neq 0 \quad \text{if } 0 \leqslant \phi < \frac{\pi}{2}. \tag{9}$$

Therefore $C = 1$ in the third formula (7). Indeed, if $C = 2k + 1$ with $k > 0$, then $0 < \pi/[2(2k+1)] < \pi/2$ and $f[\pi/(2(2k+1))] = \cos[(2k+1)\pi/(2(2k+1))] = 0$, contrary to (9). Thus the only continuous solution of (6), appropriate for our purposes, is given by

$$f(\phi) = \cos\phi \quad \text{for all } \phi \in \left[0, \frac{\pi}{2}\right].$$

In conclusion, the length of the resultant of two unit vectors with the angle 2ϕ is $2\cos\phi$, the resultant of two vectors of equal length x lies in the bisector of their angle ($2\phi \in [0, \pi]$) and its length is $2x\cos\phi$. In other words, *for vectors of equal length, the parallelogram rule holds* (Figure 4), that is, if $|\mathbf{p}| = |\mathbf{q}| = x$ and $(\mathbf{p}, \mathbf{q}) \measuredangle = 2\phi \in [0, \pi]$, then $(\mathbf{p}, \mathbf{p} \circ \mathbf{q}) \measuredangle = (\mathbf{p} \circ \mathbf{q}, \mathbf{q}) \measuredangle = \phi$ and $|\mathbf{p} \circ \mathbf{q}| = 2x\cos\phi$.

The generalization of this result to vectors of *unequal* length can now be carried out by purely geometric considerations which we include here for completeness's sake.

First we consider *perpendicular* vectors $\mathbf{p}, \mathbf{q}$ (Figure 5). Take the diagonals of the rectangle spanned by $\mathbf{p}$ and $\mathbf{q}$ and draw parallels to one at the end points of $\mathbf{p}$ and $\mathbf{q}$, and at their common origin O, to the other. Thus we get two isosceles triangles. Denoting their sides, with the appropriate orientation (as in Figure 5), by the vectors $\mathbf{p}_1, \mathbf{p}_2$ and $\mathbf{q}_1, \mathbf{q}_2$ (in addition to $\mathbf{p}$ and $\mathbf{q}$), we have by our previous results

$$\mathbf{p}_1 \circ \mathbf{p}_2 = \mathbf{p}, \quad \mathbf{q}_1 \circ \mathbf{q}_2 = \mathbf{q}, \quad \mathbf{p}_1 \circ \mathbf{q}_1 = \mathbf{0} \quad \text{(the zero vector)}$$

and, by obvious geometric facts (segments cut from parallel lines by parallel lines are equal), $\mathbf{p}_2 = \mathbf{q}_2 = \frac{1}{2}\mathbf{r}$, where $\mathbf{r}$ is the vector of the diagonal coming

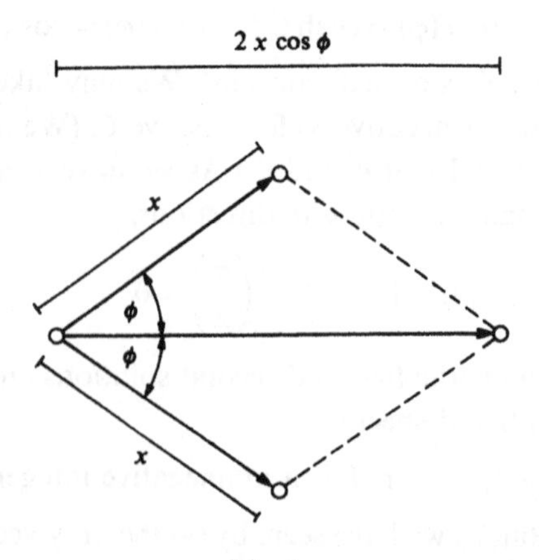

Fig. 4

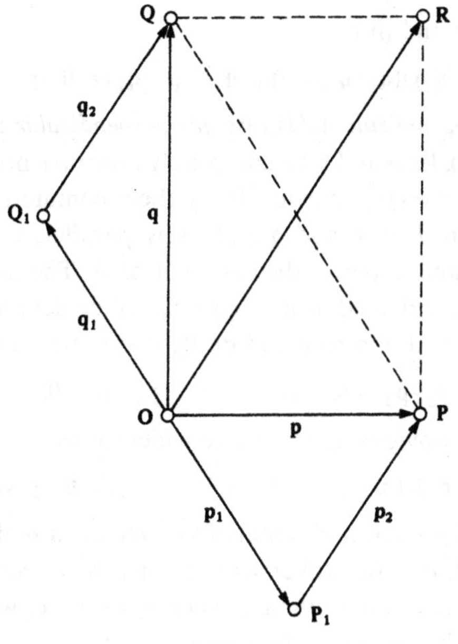

Fig. 5

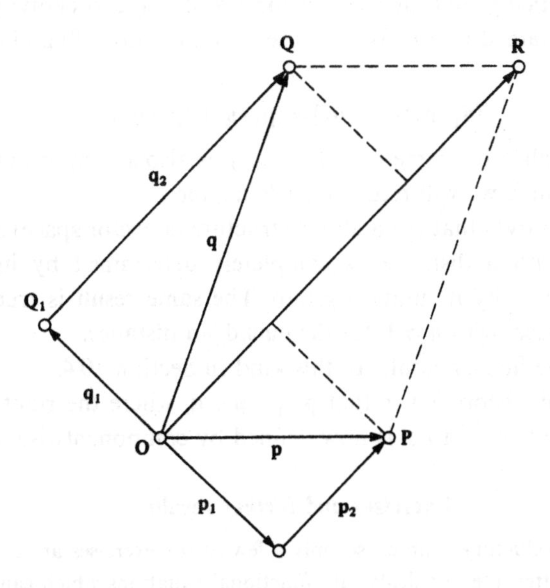

Fig. 6

from $\mathbf{0}$. Thus, cf. (ii) and (iii),

$$\mathbf{p}\circ\mathbf{q}=(\mathbf{p}_1\circ\mathbf{p}_2)\circ(\mathbf{q}_1\circ\mathbf{q}_2)=(\mathbf{p}_1\circ\mathbf{q}_1)\circ(\mathbf{p}_2\circ\mathbf{q}_2)=\mathbf{0}\circ(\mathbf{p}_2\circ\mathbf{q}_2)=\mathbf{r}, \tag{10}$$

that is, *the parallelogram rule holds also for perpendicular vectors.*

Finally (Figure 6), let $\mathbf{p}$ and $\mathbf{q}$ be completely *arbitrary* nonparallel vectors. Take the diagonal $\mathbf{r}=\overrightarrow{OR}$, starting from their common origin $\mathbf{0}$, of the parallelogram spanned by $\mathbf{p}$ and $\mathbf{q}$ and draw parallels to it from the end points of $\mathbf{p}$ and $\mathbf{q}$ and a perpendicular to it at $\mathbf{0}$. The sides of these two rectangular triangles, oriented as in Figure 6, will be denoted by the vectors $\mathbf{p}_1, \mathbf{p}_2$ and $\mathbf{q}_1, \mathbf{q}_2$ (in addition to $\mathbf{p}$ and $\mathbf{q}$). By our previous results,

$$\mathbf{p}_1\circ\mathbf{p}_2=\mathbf{p},\quad \mathbf{q}_1\circ\mathbf{q}_2=\mathbf{q},\quad \mathbf{p}_1\circ\mathbf{q}_1=\mathbf{0},$$

and by (v), (iii), and obvious geometric considerations,

$$\mathbf{p}\circ\mathbf{q}=(\mathbf{p}_1\circ\mathbf{p}_2)\circ(\mathbf{q}_1\circ\mathbf{q}_2)=(\mathbf{p}_1\circ\mathbf{q}_1)\circ(\mathbf{p}_2\circ\mathbf{q}_2)=\mathbf{0}\circ(\mathbf{p}_2\circ\mathbf{q}_2)=\mathbf{r}. \tag{11}$$

The only case still not covered is that of two vectors $\mathbf{p}, \mathbf{q}$ of different lengths and opposite directions ((iii) takes care of any two vectors of the same direction). Let $\mathbf{p}$ be, say, longer than $\mathbf{q}$. Then $\mathbf{p}=\mathbf{r}\circ(-\mathbf{q})$ where $\mathbf{r}$'s length is $|\mathbf{p}|-|\mathbf{q}|$ and it has the same direction as $\mathbf{p}$. By (v),

$$\mathbf{p}\circ\mathbf{q}=[\mathbf{r}\circ(-\mathbf{q})]\circ\mathbf{q}=\mathbf{r}\circ[(-\mathbf{q})\circ\mathbf{q}]=\mathbf{r},$$

and so the parallelogram rule determines the resultant $\mathbf{p}\circ\mathbf{q}$ of two arbitrary vectors $\mathbf{p}, \mathbf{q}$. This concludes the proof of Theorem 1 (under acceptance of the regular solutions of (2) and (6)).

It is worth noticing that, in most of the proof, the associativity and commutativity contained in (ii) has been used – see (1), (5), (10) and (11) – in the form

$$(\mathbf{p}_1\circ\mathbf{p}_2)\circ(\mathbf{q}_1\circ\mathbf{q}_2)=(\mathbf{p}_1\circ\mathbf{q}_1)\circ(\mathbf{p}_2\circ\mathbf{q}_2). \tag{12}$$

This identity, called bisymmetry or *mediality*, is also an important functional equation, to which we will return later (Chapter 17).

Theorem 1 proves that the additive structure (a vector space) of the space V, equipped with a distance, is completely determined by its group of isometries, that is, by its unitary group. The same result is true for any n dimensional space, with $n\geqslant 3$, for the euclidean distance.

We shall give further results of this kind in Section 10.4.

The result of Theorem 1 is that $\mathbf{p}\circ\mathbf{q}=\mathbf{p}+\mathbf{q}$, where the right hand side, the 'sum' of the two vectors, is now defined by componentwise addition.

Exercises and further results

This was an introductory chapter, so only a few of the exercises are closely related to its subject matter; the rest deals with functional equations which can be handled directly, without use of the methods discussed in the rest of this book.

(For Exercises 1–3, cf. Aczél 1966*c*, pp. 24–5, 27–30.)

1. Show that the general solution $f:\mathbb{R}\to\mathbb{R}$ of
$$f(x+y)+f(x-y)=2f(x)\cos y \quad (x, y\in\mathbb{R})$$
is given by $f(x)=a\cos x+b\sin x$ $(x\in\mathbb{R})$, where a, b are arbitrary real constants. (As usual, the set of all real numbers is denoted by $\mathbb{R}$.)
2. Let E be a euclidean space of dimension $n>2$. Let $f:E\times E\to\mathbb{R}$ be invariant under any linear unitary transformation T of E ($Tx=A\cdot x$ where A is a matrix with determinant 1), that is, $f(Tx, Ty)=f(x, y)$ for x, y in E. Suppose also that f is linear in the first variable [$f(\alpha x+\beta z, y)=\alpha f(x, y)+\beta f(z, y)$ for all $\alpha, \beta\in\mathbb{R}$, $x, y, z\in E$]. Show that f is, up to a scalar multiple, the scalar product on E.
3. Let $\mathbb{R}^3$ be the usual euclidean space with three coordinates. Let $f:\mathbb{R}^3\times\mathbb{R}^3\to\mathbb{R}^3$ be such that, for any rotation T, $f(Tx, Ty)=T(f(x, y))$; $x, y\in\mathbb{R}^3$. Suppose that f is linear in the first variable. Show that f is, up to a scalar multiple, the vector product on $\mathbb{R}^3$.
4. (Aczél 1966*c*, pp. 223–4.) Find the general solution of the functional equation
$$F(x, y)+F(y, z)=F(x, z),$$
where x, y and z belong to a nonempty set S and $F:S\times S\to G$, where $(G, +)$ is a group.
5. Find the general solution $F:\mathbb{R}^2\to\mathbb{R}$ of the functional equation
$$F(x, y)F(y, z)=F(x, z), \quad (x, y, z\in\mathbb{R}).$$
6. (Moszner 1965.) Find the general solution $F:\mathbb{R}^2\to\mathbb{R}$ of the equation in Exercise 5 if we suppose only that the equation in the previous exercise holds for ordered triples $x\leqslant y\leqslant z$.
7. Show that, among $f:\mathbb{R}\to\mathbb{R}$, only constant functions satisfy
$$f\left(\frac{x+y}{2}\right)^2=f(x+y)f(x-y), \quad (x, y\in\mathbb{R}).$$
8. For any $x\in\mathbb{R}$, one defines the (leibnizian) *dilogarithm* by
$$\mathrm{Li}(x)=-\int_0^x\frac{\ln 1-t}{t}dt$$
and, for $x\neq 0$, $x\neq 1$,
$$M(x)=\mathrm{Li}(x)+\tfrac{1}{2}\ln|x(1-x)|.$$
Show that the following functional equation holds for $x\neq 1$, $y\neq 1$:
$$M(xy)=M(x)+M(y)+M\left(\frac{x}{x-1}(1-y)\right)$$
$$+M\left(\frac{y}{y-1}(1-x)\right)-2\lambda(x+y+xy)M(2),$$
where λ is the characteristic function of $[1, \infty[$ ($\lambda(x)=1$ if $x\geqslant 1$, $\lambda(x)=0$ if $x<1$).
9. (Aczél 1956*b*; Samuelson 1974; Christian 1983.) By forming the difference quotient and using the continuity and convexity of a^x, show that the derivative of a^x exists

everywhere and is proportional to a^x:

$$\frac{da^x}{dx} = a^x l(a).$$

Give a geometric motivation to

$$l(a^t) = tl(a) \quad \text{for all real numbers } t.$$

Show that there exists an exponential function whose derivative at 0 is 1. Call its base e. Then show that

$$l(s) = \log_e s = \ln s \quad \text{for all } s > 0.$$

(The above can serve as a definition of the natural logarithm.)

10. (Aczél 1966c, p. 26.) Determine all solutions $f:\mathbb{R} \to \mathbb{R}$ of

$$f(x+y) + f(x-y) = f(x)(y+2) - y(x^2 - 2y).$$

11. Suppose that $f:\mathbb{R}_+ \to \mathbb{R}_+$ ($\mathbb{R}_+$ is the set of positive numbers) is continuous, strictly decreasing and satisfies

$$f(x+y) + f(f(x) + f(y)) = f[f(x + f(y)) + f(y + f(x))] \quad \text{for all } x, y \in \mathbb{R}_+.$$

Prove that $f(x) = f^{-1}(x)$ for all $x \in \mathbb{R}_+$ (f^{-1} is the inverse function of f).

2

Cauchy's equation. Hamel basis

The purpose of this chapter is to introduce the main notions for functional equations: domain of the equation, solution and extension of a solution, regularity conditions, and the general solution of a functional equation. We use the example of the Cauchy equation on the real axis of which we give the solutions under different regularity conditions, and also the general solution.

2.1 General considerations, extensions, and regular solutions

We first consider *Cauchy's equation* (Cauchy 1821):

$$f(x+y)=f(x)+f(y) \quad \textit{for all } (x,y)\in\mathbb{R}^2. \tag{1}$$

(As usual, $\mathbb{R}$ is the set of all reals, $\mathbb{R}_+$ the set of positive numbers, and $\bar{\mathbb{R}}_+$ that of all nonnegative numbers, all equipped with the usual topology; and, for any set S, the cartesian square is $S^2 := \{(x,y) \mid x\in S, y\in S\}$.) If we compare (1) to (1.2) we notice a difference. There the equation was supposed to hold only for *nonnegative* x, y, here for all *real* x, y. The set of all values of the variables, on which the equation is supposed to hold, is called the *domain of the equation* (not to be confused with the domain(s) of the unknown function(s); the domain of (1) is $\mathbb{R}^2$, while the domain of f is $\mathbb{R}$). A function satisfying a functional equation on a given domain is called a *solution* of the equation on that domain. It is sometimes (but not always, see equations (3.1), (3.18), and Chapter 7) possible to *extend* a solution of an equation from a more restricted domain to a wider one. For (1.2), that is, for

$$g(x+y)=g(x)+g(y) \quad \text{for all } (x,y)\in\bar{\mathbb{R}}_+^2, \tag{2}$$

this means the following (cf. Aczél–Erdös 1965; Aczél–Baker–Djoković–Kannappan–Radó 1971; Grząślewicz–Sikorski, 1979, for general algebraic structures):

Theorem 1. *To every solution* $g:\bar{\mathbb{R}}_+ \to \mathbb{R}$ *of* (2) *there exists a solution* $f:\mathbb{R}\to\mathbb{R}$ *of* (1), *for which*

$$f(x) = g(x) \quad \textit{for all } x \in \bar{\mathbb{R}}_+. \tag{3}$$

Proof. We construct this f, that is, we extend g to all of $\mathbb{R}$ by the definition

$$f(s-t) = g(s) - g(t) \quad \text{for all } (s,t) \in \bar{\mathbb{R}}_+^2. \tag{4}$$

This definition is unambiguous (and so is indeed a definition) because, whenever $s - t = u - v$, where s, t, u and v are in $\bar{\mathbb{R}}_+$, that is, whenever $s + v = u + t$, then, by (2),

$$g(s) + g(v) = g(s+v) = g(u+t) = g(u) + g(t),$$

so

$$g(s) - g(t) = g(u) - g(v),$$

as asserted. Also (3) holds: if $s - t = u \in \bar{\mathbb{R}}_+$, then $s = u + t$ where s, t and u belong to $\bar{\mathbb{R}}_+$ and

$$g(s) = g(u+t) = g(u) + g(t), \tag{5}$$

that is,

$$f(u) = f(s-t) = g(s) - g(t) = g(u)$$

for all $u \in \bar{\mathbb{R}}_+$. (One could also prove first $g(0) = 0$ from (2), and deduce (3) from (4).) We have still to prove that f defined by (4) satisfies equation (1). Indeed, if $x = s - t$, $y = u - v$, and so $x + y = (s+u) - (t+v)$ where s, t, u, and v belong to $\bar{\mathbb{R}}_+$, then, by (4),

$$f(x) = g(s) - g(t), \quad f(y) = g(u) - g(v)$$

and, by (2),

$$f(x+y) = g(s+u) - g(t+v) = g(s) + g(u) - g(t) - g(v) = f(x) + f(y).$$

This concludes the proof of Theorem 1.

A similar theorem (with a similar proof) holds if $\bar{\mathbb{R}}_+$ is replaced by $\mathbb{R}_+$ and, in general, if $\mathbb{R}$ is replaced by an arbitrary abelian group G generated by a subsemigroup S (a semigroup is a set with an associative operation defined on it) and if $\bar{\mathbb{R}}_+$ (or $\mathbb{R}_+$) is replaced by S. Also, the values of the unknown function f may be in an arbitrary abelian group rather than in $\mathbb{R}$ (cf. Aczél–Baker–Djoković–Kannappan–Radó 1971).

A solution $f:S\to\mathbb{R}$ of a Cauchy equation for all x, y in a groupoid (a set S equipped with an addition $+$) is called *additive* on S. If $S = \mathbb{R}$ (or, later, $S = \mathbb{R}^n$), then we call the solution an additive function. The same terminology will be kept if f is no longer a numerical function but takes its values in a group G or groupoid ($f:S\to G$).

Now we look at some direct consequences of (1).

Lemma 2. *If a function f satisfies* (1), *then it also satisfies*

$$f\left(\sum_{k=1}^{n} r_k x_k\right) = \sum_{k=1}^{n} r_k f(x_k) \quad \textit{for all real } x_k \textit{ and for all rational } r_k$$
$$(k = 1, 2, \ldots, n; n \geqslant 1). \tag{6}$$

Proof. Putting into (1) $y = 0$ or $y = -x$, one gets

$$f(0) = 0 \tag{7}$$

and

$$f(-x) = -f(x), \tag{8}$$

respectively. Also,

$$f(x_1 + x_2 + \cdots + x_n) = f(x_1) + f(x_2) + \cdots + f(x_n) \quad \text{for all real } x_k$$
$$(k = 1, 2, \ldots, n; n \geqslant 1) \tag{9}$$

can easily be proved from (1) by induction. Choosing $x_1 = x_2 = \cdots = x_n = x$ in (9), one gets

$$f(nx) = nf(x) \quad \text{for all real } x \text{ and all positive integers } n. \tag{10}$$

Now let $r = m/n$ be a positive rational number (m, n positive integers) and t an arbitrary real number. Then, for $x = rt = (m/n)t$, we have $nx = mt$ and, by (10),

$$nf(x) = f(nx) = f(mt) = mf(t),$$

or

$$f(rt) = f(x) = \frac{m}{n} f(t) = rf(t),$$

for all real t and all positive rationals r. By (8) and (7) the same holds for negative rationals and 0, so that

$$f(rt) = rf(t) \quad \text{for all rational } r \text{ and all real } t. \tag{11}$$

Combined with (9), we see that (6) holds and the Lemma 2 is proved.

From (11), with $t = 1$, $f(1) = c$, we get

$$f(r) = rf(1) = cr \quad \text{for all } r \in \mathbb{Q}. \tag{12}$$

(As usual, we denote the set of all rational numbers by $\mathbb{Q}$.) We may wonder whether this can be extended to all *reals*. (It certainly can if f is continuous.) In Chapter 1 we mentioned that under certain circumstances the solutions of (1) (or of (2)) are given by

$$f(x) = cx \quad \text{for all real } x, \tag{13}$$

where c is a real constant (it was nonnegative under the condition (1.3), and there $x \in \bar{\mathbb{R}}_+$). Certainly, all functions of the form (13) satisfy (1). The following theorem shows that solutions of (1) which are *not* of the form (13) have to be pretty 'weird' (cf. Hamel 1905; San Juan 1946; Wilansky 1967; Hewitt–Zuckerman 1969).

The *graph of a function* f on a set S is the set

$$G = \{(x, y) | y = f(x), x \in S\}.$$

In our case, $S = \mathbb{R}$ and $f(S) = \{f(x) | x \in S\} \subset \mathbb{R}$, so the graph is a subset of $\mathbb{R}^2$:

$$G = \{(x, y) | y = f(x), x \in \mathbb{R}\}. \tag{14}$$

Theorem 3. *The graph of each solution of* (1), *which is not of the form* (13), *is everywhere dense in the plane* $\mathbb{R}^2$.

Proof. Take $x_1 \neq 0$. If f is *not* of the form (13), then there exists a nonzero real number x_2 such that

$$\frac{f(x_1)}{x_1} \neq \frac{f(x_2)}{x_2} \tag{15}$$

(or else let $x_2 = x$ run through the nonzero reals and denote $c = f(x_1)/x_1$, in order to get (13) for $x \neq 0$ and also, by (7), for $x = 0$). In other words,

$$\begin{vmatrix} x_1 & f(x_1) \\ x_2 & f(x_2) \end{vmatrix} \neq 0,$$

so that the vectors $\mathbf{p}_1 = (x_1, f(x_1))$ and $\mathbf{p}_2 = (x_2, f(x_2))$ are linearly independent and thus *span* the whole plane $\mathbb{R}^2$. This means that for *any* plane vector $\mathbf{p}$ there exist reals ρ_1, ρ_2 such that

$$\mathbf{p} = \rho_1 \mathbf{p}_1 + \rho_2 \mathbf{p}_2$$

(by definition, $\rho\mathbf{p} = \rho(x, y) = (\rho x, \rho y)$ and $\mathbf{q}_1 + \mathbf{q}_2 = (x_1, y_1) + (x_2, y_2) = (x_1 + y_1, x_2 + y_2)$). If we allow only *rational* r_1, r_2, then, by their appropriate choice, with $r_1\mathbf{p}_1 + r_2\mathbf{p}_2$ we can get only *arbitrarily close* to any given plane vector $\mathbf{p}$ (since the set of (pairs of) rationals is dense in the set of (pairs of) reals). Now

$$\begin{aligned} r_1\mathbf{p}_1 + r_2\mathbf{p}_2 &= r_1(x_1, f(x_1)) + r_2(x_2, f(x_2)) \\ &= (r_1x_1 + r_2x_2, r_1f(x_1) + r_2f(x_2)) \\ &= (r_1x_1 + r_2x_2, f(r_1x_1 + r_2x_2)) \end{aligned}$$

(the last equation coming from (6)). Thus for $x_1(\neq 0)$ and $x_2(\neq 0)$ satisfying (15), the set

$$G_{1,2} = \{(x, y) | y = f(x), x = r_1x_1 + r_2x_2; (r_1, r_2) \in \mathbb{Q}^2\}$$

is everywhere dense in $\mathbb{R}^2$. Since, see (14), $G_{1,2} \subseteq G$, the graph G is also dense in $\mathbb{R}^2$ and Theorem 3 is proved.

Since we will use Theorem 3 to prove that (1.2) and (1.3) imply $f(x) = cx$, we have to emphasize that we did not fall into a circuitous reasoning. In the above proof, vector addition is not an operation axiomatically defined by (i)–(iv) in Theorem 1.1, but is simply and *a priori* defined coordinatewise by

$$\mathbf{q}_1 + \mathbf{q}_2 = (x_1, y_1) + (x_2, y_2) = (x_1 + x_2, y_1 + y_2)$$

(which is, of course, the ordinary vector addition).

While, according to Theorem 3, the graph of every solution of (1), which is not of the form (13), is everywhere dense in $\mathbb{R}^2$, Jones (1942*a*) has shown examples where this graph is, nevertheless, connected.

Theorem 3 shows that a solution of (1), which is not of the form (13), is globally irregular. Local irregularity of such an f can also be obtained, as is easy to deduce by localization from Theorem 3.

Corollary 4. *For a solution of* (1), *which is not of the form* (13), *the image of any interval* $]a, b[$ $(a < b)$ *is dense in* $\mathbb{R}$.

On the other hand, as this may indicate, some regularity for a solution of (1) implies that the solution is of the form (13).

Corollary 5. *If a function f satisfies* (1) *and is continuous at a point or monotonic or bounded from one side on an interval of positive length, then there exists a constant c such that*

$$f(x) = cx \quad \textit{for all real } x. \tag{13}$$

Indeed, see Figure 7, under any of these conditions the graph of f would

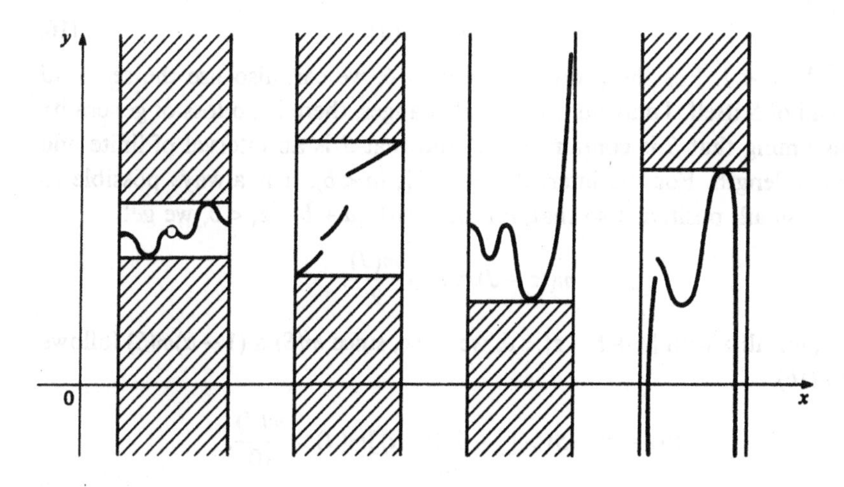

Fig. 7

avoid the regions shadowed on that diagram. Thus it could not be everywhere dense in $\mathbb{R}^2$ and, therefore by Theorem 3, f has to be of the form (13).

Conversely, as mentioned before, functions of the form (13) always satisfy (1) whatever the constant c is, but the monotonicity or boundedness restrictions may restrict c. For instance, if f is nonnegative (bounded from below by 0 on $\bar{\mathbb{R}}_+$ or on a subinterval of $\bar{\mathbb{R}}_+$), then $c \geqslant 0$.

We now generalize Corollary 5 by showing that all functions satisfying (1), which are bounded from one side on a set of positive measure, are given by (13). Thus all solutions which are not of this form are not bounded on any set of positive measure. We will need some basic facts of the Lebesgue theory, in particular the following result of the Steinhaus (1920) type; we quote his result here in a slightly altered form and use a variant of the usual proof (see e.g. Halmos 1950, p. 68), due to Ng (unpublished).

Theorem 6. *If $S \subset \mathbb{R}$ has positive Lebesgue measure, then the set*

$$S + S := \{x + y \mid x \in S, y \in S\},$$

the sum of S and S, contains an interval of positive length.

Proof. It is equivalent to prove that there exists an interval I (of positive length) such that, for all z in I, the intersection of the set S with $z - S$ is nonempty (where $z - S$ denotes the set of all $z - y$ when y runs through S). A way of proving that this intersection is not empty is to show that its Lebesgue measure is positive. In other words, we show that $m((z-S) \cap S) > 0$.

We will show in Lemma 7 that we are able to find an interval J, of positive length, for which the intersection of J with S is almost as rich as J, more exactly, such that

$$m(S \cap J) \geqslant \alpha m(J), \tag{16}$$

with $\frac{4}{5} < \alpha < 1$. For the proof of Theorem 6, we can also use the set $S \cap J$ instead of S itself. To avoid notational changes, there is no loss of generality in assuming that S is contained in J and that J is an interval of finite and positive length. For the interval $J =]a, b[$, $(a < b)$, it is always possible to find a (small) positive δ so that, for all z with $|a + b - z| < \delta$, we get

$$m((z-J) \cap J) > \frac{m(J)}{2}.$$

Then, for all z with $|a + b - z| < \delta$, we have, since $m(S) \leqslant (1-\alpha)m(S)$ follows from (16),

$$m((z-S) \cap S) > [\tfrac{1}{2} - 2(1-\alpha)]m(J) > \frac{m(J)}{10}.$$

This shows that $(x - S) \cap S \neq \varnothing$ for all z in the interval $]a + b - \delta, a + b + \delta[$.

It only remains to be shown that there exists an interval J for which (16) holds. We prove this in a separate lemma (see e.g. Halmos 1950, p. 68).

Lemma 7. *If $S \subseteq \mathbb{R}$ has positive Lebesgue measure and if $0 < \alpha < 1$, then there exists an interval of positive length J such that*

$$m(S \cap J) \geqslant \alpha m(J). \tag{16}$$

Proof. Without loss of generality, by 'trimming down' S, we may suppose that $m(S)$ is finite $(0 < m(S) < \alpha)$. The regularity of the Lebesgue measure yields an open subset U such that $S \subseteq U$ and $\alpha m(U) \leqslant m(S)$ $(\leqslant m(U))$. But this open subset of $\mathbb{R}$ is the countable union of disjoint intervals J_n of positive length. Using the σ-additivity of the Lebesgue measure, we get

$$\alpha m(U) = \alpha \sum_{n=1}^{\infty} m(J_n) \leqslant \sum_{n=1}^{\infty} m(S \cap J_n) = m(S).$$

Therefore there exists at least one integer $n_0(\geqslant 1)$ for which

$$\alpha m(J_{n_0}) \leqslant m(S \cap J_{n_0}).$$

So (16) is satisfied with $J = J_{n_0}$, which concludes the proof of Lemma 7 and also of Theorem 6.

The following theorem was proved by Ostrowski (1929) and again by Kestelman (1946).

Theorem 8. *There exists a constant c such that*

$$f(x) = cx \quad \textit{for all real } x \tag{13}$$

if and only if $f: \mathbb{R} \to \mathbb{R}$ satisfies (1) *and is bounded from above (or below) on a set of positive measure.*

Proof. Let S be a set of positive Lebesgue measure on which f is bounded from above, that is,

$$f(x) \leqslant M \quad \text{for all } x \in S. \tag{17}$$

From this and from (1), $f(x+y) = f(x) + f(y) \leqslant 2M$ for all $x, y \in S$ follows, that is, $f(z) \leqslant 2M$ for all $z \in S + S$. Theorem 6 shows that $S + S$ contains an interval of positive length. Therefore f is bounded from above on this interval. Corollary 5 yields the form (13) for f.

Notes. See Exercise 7.15 for another proof of Theorem 6 valid in a more general setting. In Section 9.2 we will prove converse theorems, characterizing sets E for which a solution of (1), bounded from one side on E, is always continuous and so of the form (13).

A nice consequence of these results is the measurability part of the following theorem, proved by Fréchet (1913), Sierpinski (1920*a*), and Banach (1920).

Other authors have found new proofs since then. See, for example, Alexiewicz–Orlicz (1945).

Corollary 9. *If a function* $g:\bar{\mathbb{R}}_+ \to \mathbb{R}$ *satisfies* (2) *and is continuous at a point or monotonic or Lebesgue measurable or bounded from one side on a set of positive measure, then there exists a constant c such that*

$$g(x) = cx \quad \textit{for all } x \geqslant 0. \tag{18}$$

In particular, if (2) *and* $g(x) \geqslant 0$ *are supposed, then* (18) *holds and* $c \geqslant 0$.

The last statement was the first assertion which we have used in Chapter 1.

For the result under the measurability condition, we just recall that a Lebesgue measurable function is bounded on some subset of positive Lebesgue measure.

These results show that it is of interest to find *all* solutions of a given functional equation within a certain *class of admissible functions* (e.g. functions continuous at a point or bounded on an interval or measurable). The set of all such solutions is called the *general solution of the equation in that class.* (In the most general case, *any* function mapping a given domain into a given set is admissible.) Just as functions are often given by formulas yielding all their values by substituting for the variables all elements of their domains, general solutions are often given by *formulas* which give all solutions (in the class of admissible functions) by specifying the *constants* or other arbitrary elements (e.g. functions) in these formulas. So, for instance, the general nonnegative solution of (2) is given by (18) with $c \geqslant 0$, and the general solution of (1), continuous at a point, is given by (13).

2.2 General solutions

Let x_0 and y_0 be two given real numbers. Except if $x_0 = 0$, $y_0 \neq 0$ there always exists at least one $f:\mathbb{R} \to \mathbb{R}$ satisfying the Cauchy equation (1) and such that

$$f(x_0) = y_0. \tag{19}$$

If $x_0 \neq 0$, just take $f(x) = (y_0/x_0)x$, but if $x_0 = 0$, then y_0 has to be zero.

Clearly a *continuous* solution of (1) is completely determined by (19). However, we will see that the discontinuous solutions of (1) are not uniquely determined by (19). Replacing a singleton $\{x_0\}$ by a subset E, a natural question is to find those subsets E of $\mathbb{R}$ on which the values of a solution of (1) can be arbitrarily chosen. We therefore look for subsets E of $\mathbb{R}$ such that for any $g:E \to \mathbb{R}$ there exists a *unique* f satisfying (1) and coinciding with g on E. For this purpose, we need Hamel bases. Hamel (1905), using the axiom of choice, proved that *there exists a subset H of* $\mathbb{R}$ *such that every real number x can be expressed in a unique way* (up to 0 coefficients) *in the form*

$$x = \sum_{k=1}^{n} r_k h_k \quad \text{where } h_k \in H \quad \text{and } r_k \text{ are rational} \quad (k = 1, 2, \ldots, n). \tag{20}$$

The set H is a Hamel basis, and the formula (20) the *Hamel expansion* of x with *coefficients* r_k $(k = 1, 2, \ldots, n)$. Notice that the sum in (20) has only a finite number of terms, but the number n and the choice of the r_k and h_k $(k = 1, 2, \ldots, n)$ depend, of course, upon x. A Hamel basis is just a basis, in the sense of linear algebra, for the space $\mathbb{R}$ viewed as a linear space over the rational field $\mathbb{Q}$.

Now we proceed to prove that the subsets E, which we are looking for, are the Hamel bases H. First we prove the following:

Theorem 10. *The general solution* $f:\mathbb{R} \to \mathbb{R}$ *of* (1) *is given with the aid of the Hamel expansion* (20), *by choosing the values of* f *on* H *arbitrarily and defining*

$$f(x) = f\left(\sum_{k=1}^{n} r_k h_k\right) = \sum_{k=1}^{n} r_k f(h_k). \tag{21}$$

Proof. The above definition is unambiguous because 0 coefficients do not change either (20) or (21). In particular, if $x = h \in H$, then (21) just gives $f(h) = f(h)$, so it does not restrict the arbitrary choice of f on H. That *every solution of* (1) *can be written in the form* (21) is an immediate consequence of (20) and (6).

Conversely, all $f:\mathbb{R} \to \mathbb{R}$ defined in this way satisfy (1). In order to prove this, take two arbitrary real numbers with the Hamel expansions

$$x = \sum_{k=1}^{m} r_k h_k \quad \text{and} \quad y = \sum_{k=1}^{m} s_k h_k \quad (h_k \in H, \text{ and } r_k, s_k \text{ rational}; \ k = 1, 2, \ldots, m).$$

(By adding a sufficient number of terms with 0 coefficients, we can always have the same Hamel basis elements and the same number of terms in the Hamel expansions of x and y.) Then

$$x + y = \sum_{k=1}^{m} (r_k + s_k) h_k \quad (h_k \in H, r_k + s_k \text{ rational}; \ k = 1, 2, \ldots, m)$$

is *a* Hamel expansion of $x + y$ on the basis H, and by the *uniqueness* of Hamel expansions *it is, up to* 0 *coefficients, the only Hamel expansion.* Thus, by (20),

$$f(x) = \sum_{k=1}^{m} r_k f(h_k), \quad f(y) = \sum_{k=1}^{m} s_k f(h_k) \quad \text{and} \quad f(x + y) = \sum_{k=1}^{m} (r_k + s_k) f(h_k),$$

and so (1) is indeed satisfied, which concludes the proof of Theorem 10.

Theorems 1 and 10 have the following obvious consequence:

Corollary 11. *The general solution* $g:\bar{\mathbb{R}}_+ \to \mathbb{R}$ *of* (2) *is given by taking an arbitrary solution of* (1) *as constructed in Theorem* 10 *and restricting it to* $\bar{\mathbb{R}}_+$.

As we have seen, Theorem 3 shows that, *if* (1) has solutions which are *not* of the form (13), *then* their graphs are everywhere dense in the plane. This does *not* prove that such solutions exist (even if the following statement is true: 'if there exist angels, then they have wings', this does not prove that there are indeed angels around). Theorem 10 and Corollary 11 show, however, that *equations* (1) *or* (2) *have solutions which are not of the forms* (13) *and* (18) *respectively*. In order to find such solutions, it is enough to choose the values of f on two Hamel basis elements h_1, h_2 such that $f(h_1)/h_1 \neq f(h_2)/h_2$ (cf. (15)). By Corollary 5, such solutions are nowhere continuous or monotonic. By Corollary 9 they are also not locally Lebesgue measurable, not bounded from either side on any set of positive measure and, by Theorem 3, their graphs are everywhere dense.

Proposition 12. *A function defined arbitrarily on a subset E of* $\mathbb{R}$ *can be extended uniquely to an additive function* (*to a solution of* (1) *on* $\mathbb{R}$) *if and only if E is a Hamel basis.*

Proof. Let H be a Hamel basis and $g:H \to \mathbb{R}$. Theorem 10 states that there exists a solution f of (1) with $f = g$ on H. Let h be a solution of (1) and $h = g$ on H. Using (21), we see that, for any x in $\mathbb{R}$,

$$h(x) = \sum_{k=1}^{n} r_k h(x_k) = \sum_{k=1}^{n} r_k g(x_k) = \sum_{k=1}^{n} r_k f(x_k) = f(x) \quad (x_k \in H, r_k \in \mathbb{Q}),$$

which proves the uniqueness of f.

Conversely let E be a subset of $\mathbb{R}$ having the property that a function defined arbitrarily on E can be extended uniquely to an additive function. First we prove that E is independent over $\mathbb{Q}$. Were this not the case, then there would exist a nontrivial relation $\sum_{k=1}^{n} r_k x_k = 0$ where $x_k \in E$ ($x_i \neq x_j$ if $i \neq j$), $r_k \in \mathbb{Q}$ ($k = 1, 2, \ldots, n$), and at least one $r_k \neq 0$. Then, by (6) and (7), any additive f must satisfy $\sum_{k=1}^{n} r_k f(x_k) = 0$ and thus the values $f(x_k)$ ($k = 1, \ldots, n$) cannot be all arbitrary. In other words, the restriction g of an additive f to E cannot be any function defined on E. This contradicts our hypothesis that any function g defined on E permits (at least) one additive extension.

Second, we show that E spans $\mathbb{R}$ (as a linear space over $\mathbb{Q}$). Were this not the case, the independent set E could be extended to a Hamel basis H, strictly containing E. Then any $g:E \to \mathbb{R}$ has more than one extension to H, say g_1 and g_2, with $g_1 \neq g_2$, but $g_1(x) = g_2(x)$ for all x in E. Theorem 10 shows that there exist additive functions f_1 and f_2 such that $f_i(y) = g_i(y)$ ($i = 1, 2$), for all y in H. Therefore f_1 and f_2 are two distinct additive functions extending the given function $g:E \to \mathbb{R}$. This contradicts the hypothesis that g allows, at most, one such additive extension.

Thus we have proved that E is independent over $\mathbb{Q}$ and spans $\mathbb{R}$ as a linear space over $\mathbb{Q}$. In other words, E is a Hamel basis.

We obtain as a by-product the following result.

Corollary 13. *A subset E of $\mathbb{R}$ contains a Hamel basis if, and only if, any f satisfying* (1) *and equal to zero on E is identically equal to zero.*

From Corollary 13 and Corollary 5 of Theorem 3, we deduce that any interval of positive length, or even any subset of $\mathbb{R}$ of positive Lebesgue measure, contains a Hamel basis. However, there exist sets with zero Lebesgue measure which contain a Hamel basis. Let us construct such a set.

The *Cantor set* C is the set of all t in $[0, 1]$ whose base 3 expansion contains no 1s:

$$t = \sum_{i=1}^{\infty} \frac{\varepsilon_i}{3^i}. \quad \text{where } \varepsilon_i = 0 \text{ or } \varepsilon_i = 2.$$

First, we show that the Cantor set contains a Hamel basis. Choose any x in $[0, 1]$ and consider an expansion of x in base 3.

$$x = \sum_{i=1}^{\infty} \frac{x_i}{3^i}, \quad \text{where } x_i = 0, 1 \text{ or } 2. \tag{22}$$

If $x_i = 0$ or $x_i = 2$ we define $y_i = z_i = x_i$. If $x_i = 1$, we define $y_i = 2$, and $z_i = 0$ if i is even and $y_i = 0$, $z_i = 2$ if i is odd. Therefore we have

$$y = \sum_{i=1}^{\infty} \frac{y_i}{3^i} \quad \text{and} \quad z = \sum_{i=1}^{\infty} \frac{z_i}{3^i}.$$

Both y and z are in the Cantor set C and our construction yields

$$x = \frac{y+z}{2}. \tag{23}$$

Let $f:\mathbb{R} \to \mathbb{R}$ be an additive function equal to zero on C. From (23) and Lemma 2 we deduce that f is zero on $[0, 1]$, thus zero everywhere. Corollary 13 shows that C contains a Hamel basis.

Secondly, it is well known that the Cantor set C is of Lebesgue measure zero. We prove this here for completeness' sake by use of another step-by-step construction of the Cantor set. Let E_1 be the subset of all t in $[0, 1]$, such that, for at least one expansion (22) of t, the first digit t_1 is different from 1. More generally, let E_n be the subset of all t in E_{n-1} such that, for at least one expansion (22) of t, the nth coordinate t_n is different from 1. The sets $E_1 \supset E_2 \supset \cdots \supset E_n \supset \cdots$ are closed subsets of $[0, 1]$. The Lebesgue measure of E_n is $\frac{2}{3}$ of the Lebesgue measure of E_{n-1}. The Cantor set C is the intersection of all E_n, $n \geqslant 1$. Therefore its measure is less than $(\frac{2}{3})^n$ for all $n \geqslant 1$, which means it is zero.

Exercises and further results

1. (E.g. Aczél 1966*c*, pp. 46–7.) Let n be a given integer ($n \geqslant 3$). Characterize those functions $f:\mathbb{R} \to \mathbb{R}$ for which $x_i \in \mathbb{R}$ ($i = 1, \ldots, n$) and $x_1 + \cdots + x_n = 0$ imply $f(x_1) + \cdots + f(x_n) = 0$.
2. Determine all solutions $f:\mathbb{R} \to \mathbb{R}$ of the equation
$$f(x+y) + f(x-y) = 2f(x) \quad (x, y \in \mathbb{R})$$
by reducing it to Cauchy's equation. Show that the same reduction also works when $f:G \to H$ and the equation holds for all x, y in G where G is any group and H a 2-cancellative abelian group (2-*cancellativity* means that $2a = 2b \Rightarrow a = b$).
3. Find the general solution $f:\mathbb{R} \to \mathbb{R}$ of Lobachevskiĭ's functional equation
$$f(x+y)f(x-y) = f(x)^2 \quad (x, y \in \mathbb{R}).$$
Determine all continuous solutions on $\mathbb{R}$.
4. Find the general solution $f:\mathbb{R} \to \mathbb{R}$ of the functional equation
$$f\left(\frac{x+y}{2}\right)^2 = f(x)f(y) \quad (x, y \in \mathbb{R}).$$

(For Exercises 5–7, cf. Aczél 1967; Aczél–Fischer 1968; Dhombres 1979, pp. 1.11–1.15.)

5. Let $f:\mathbb{R} \to \mathbb{R}$ be an additive function ($f(x+y) = f(x) + f(y)$ for all $x, y \in \mathbb{R}$). Suppose that $m > 1$ is a given integer and that $f(x^m) = f(x)^m$ holds for all $x > 0$ such that $f(x) > 0$. Does it follow that f is continuous, and what is its general form?
6. Same question as in the previous exercise but with the equation $f(x^m) = f(x)^m$ valid for all $x \neq 0$ such that $f(x) \neq 0$ and, when m is an integer (positive or negative), different from 1 and 0.
7. Same question as in Exercise 5, but with the equation $|f(x^m)| = f(x)^m$ valid for all $x \neq 0$, $f(x) \neq 0$, m being an even integer different from 0.
8. Let $f:\mathbb{R} \to \mathbb{R}$ be an additive function and $g:\bar{\mathbb{R}}_+ \to \bar{\mathbb{R}}_+$ continuous and not constant. Suppose that
$$f(g(x)) \geqslant g(|f(x)|) \quad \text{for all } x > 0.$$
Does it follow that f is continuous?
9. (Volkmann 1982.) Let $g:\bar{\mathbb{R}}_+ \to \bar{\mathbb{R}}_+$ be a continuous and strictly increasing function such that $g(0) < 0$, $g(1) < 1$. Prove that the identity is the only function $f:\mathbb{R} \to \mathbb{R}$ such that $f(1) = 1$,
$$f(x+y) \geqslant f(x) + f(y) \quad (x, y \in \mathbb{R})$$
and
$$f(g(x)) \geqslant g(|f(x)|) \quad (x \in \bar{\mathbb{R}}_+).$$
10. (Kurepa 1964*a*, Jurkat 1965.) Let $f:\mathbb{R} \to \mathbb{R}$ be an additive function and suppose that $f(1/x) = f(x)/x^2$ for all $x \neq 0$. Show that f is continuous.

11*. (E.g. Zariski–Samuel 1958, pp. 120–4. For a simple proof see Segal 1969.) Prove that there exists a not identically zero additive function f on $\mathbb{R}$ satisfying

$$f(xy) = xf(y) + yf(x) \quad \text{for all } x, y \in \mathbb{R}.$$

(Such functions are called (nontrivial) *derivations* on $\mathbb{R}$.)

12. Show that there exists a discontinuous additive $f: \mathbb{R} \to \mathbb{R}$ for which $f(1/x) = -f(x)/x^2$ for all $x \neq 0$ in $\mathbb{R}$.

(For Exercises 13–15, cf. Kannappan-Kurepa 1970.)

13*. Let $f: \mathbb{R} \to \mathbb{R}$ be an additive function. Suppose that

$$f(x^n) = ax^{n-m}f(x^m) \quad \text{for all } x \in \mathbb{R} \backslash \{0\},$$

where a is a real constant and n and m integers such that $n \neq m$. Show that f is continuous. Moreover, if $a \neq 1$ then f is identically zero if $n \neq m$.

14*. Let $f, g: \mathbb{R} \to \mathbb{R}$ be two additive functions, not identically zero. Suppose there exists a continuous function $\alpha: \mathbb{R} \backslash \{a, b\} \to \mathbb{R}$, where $a \neq b$ are real constants, such that

$$f((x-a)^{-1}) = \alpha(x)g((x-b)^{-1}) \quad (x \in \mathbb{R} \backslash \{a, b\}.)$$

Show that f and g are *linearly dependent*, that is, there exist real constants A, B, not both 0, such that $Af(x) + Bg(x) = 0$ for all $x \in \mathbb{R}$. Show, moreover, that f and g are continuous if and only if $\alpha(x) = c[(x-b)/(x-a)]$ where c is a constant.

15*. Let $f, g: \mathbb{R} \to \mathbb{R}$ be two additive functions and $\alpha: \mathbb{R} \to \mathbb{R}$ be a continuous function. Suppose there exists real constants a, b $(a \neq b)$ such that

$$f(x-a) = \alpha(x)g(x-b) \quad (x \in \mathbb{R}).$$

Show that f and g are linearly dependent.

(For Exercises 16 and 17, cf. Grząślewicz 1978*b*.)

16*. Let $f, g: \mathbb{R} \to \mathbb{R}$ be additive functions. A derivation d is an additive function such that $d(xy) = xd(y) + yd(x)$ for all $x, y \in \mathbb{R}$ (Exercise 11). Show that $f(x)^2 = x^4 g(1/x)^2$ for all $x \in \mathbb{R} \backslash \{0\}$ if and only if either $g + f$ is continuous and $g - f$ is a derivation or $g + f$ is a derivation and $g - f$ is continuous.

17. Determine all additive functions $f: \mathbb{R} \to \mathbb{R}$, which are not identically zero and satisfy

$$f(x) = x^{n+1} f(1/x)^n \quad \text{where } n \text{ is an integer.}$$

18*. Let $f: \mathbb{R} \to \mathbb{R}$ be an additive function. Let I be a proper real interval and $\phi: I \to \mathbb{R}$ a continuous function. Prove or disprove the following: if $f(x)f(\phi(x)) \geqslant a > 0$ for all x in I, then f is continuous.

19. (Aczél 1972*b*.) Reduce the following functional equation to the Cauchy equation:

$$f\left(\frac{x+y}{2} - (xy)^{1/2}\right) + f\left(\frac{x+y}{2} + (xy)^{1/2}\right) = f(x) + f(y) \quad (x, y \in \mathbb{R}_+).$$

20. (Benz 1980*b*.) Show that the affine functions $(f(x) = ax + b)$ are the only continuous functions $f: \mathbb{R} \to$ such that

$$f(xy) + f(x+y) = f(xy+x) + f(y) \quad (x, y \in \mathbb{R})$$

21*. Prove or disprove that two additive functions $f, g: \mathbb{R} \to \mathbb{R}$ are linearly dependent (cf. Exercise 14) if and only if either $f(x)g(x) \geqslant 0$ for all $x \in \mathbb{R}$ or $f(x)g(x) \leqslant 0$ for all $x \in \mathbb{R}$.

22. Let E be a bounded closed subset of $\mathbb{R}$ with positive Lebesgue measure. There exists an open subset $\mathcal{O}$ such that $\mathcal{O}$ contains E and that the Lebesgue measure of $\mathcal{O} \backslash E$ is strictly less than half of the Lebesgue measure of E. Take $\delta = \inf |z - y|$, where $z \in E$ and y is the complement of $\mathcal{O}$. Show that $]-\delta, +\delta[$ is contained in $E - E$. Deduce from this another proof for Theorem 8.
23*. (Sierpiński 1920*a*.) Prove the following result and use it to give another proof of Theorem 8: Let two subsets A and B of $\mathbb{R}$ be of positive Lebesgue measure. There exist a point in A and a point in B whose distance is a rational number.
24*. (Blanuša 1970; Daróczy 1971; Światak 1971*a*.) Let $f: \mathbb{R} \to \mathbb{R}$ satisfy

$$f(x + y - xy) + f(xy) = f(x) + f(y) \quad (x, y \in \mathbb{R}).$$

Show that there exists an additive function $g: \mathbb{R} \to \mathbb{R}$, and a constant $a \in \mathbb{R}$, such that

$$f(x) = g(x) + a.$$

25. Prove that every measurable subset E of a Hamel basis H in $\mathbb{R}$ is of Lebesgue measure zero.
26. Suppose $f: \mathbb{R} \to \mathbb{R}$ is additive and bounded from one side on the Cantor set. Show that it is additive everywhere on $\mathbb{R}$. (Theorem 8 does not apply here since the Cantor set is of Lebesgue measure 0).
27. Show that there exists an additive $f: \mathbb{R} \to \mathbb{R}$ such that $f(\mathbb{R})$ is countably infinite.
28. Give an example of an additive $f: \mathbb{R} \to \mathbb{R}$ whose graph is *not* connected.

3

Three further Cauchy equations. An application to information theory

Cauchy's equation on the real axis $\mathbb{R}$ is the equation of a homomorphism for the additive group structure of $\mathbb{R}$. In the present chapter we deal with three more equations from $\mathbb{R}$ into $\mathbb{R}$ which are deduced from homomorphisms depending on the additive or multiplicative structure on $\mathbb{R}$. The relation of these three equations to the additive Cauchy equation is shown and this yields the general as well as the regular solutions of all three equations. An application to information theory is given.

Cauchy's logarithmic equation

$$g(xy) = g(x) + g(y) \quad \text{for all } (x, y) \in \mathbb{R}_+^2 \tag{1}$$

can easily be reduced to Cauchy's equation (2.1) by a convenient change of variables and functions. Indeed if one chooses

$$x = e^u, \quad y = e^v, \quad \text{that is,} \quad u = \log x, \quad v = \log y,$$

then one has associated *positive* x, y *to all real* u, v and, at the same time, one has associated *real* u, v *to all positive* x, y. Introducing a function f by

$$f(u) = g(e^u), \quad (u \in \mathbb{R}), \tag{2}$$

the equation (1) goes over into

$$f(u + v) = f(u) + f(v), \quad \text{for all real } u, v, \tag{3}$$

that is, into Cauchy's equation (2.1). By (2),

$$g(x) = f(\log x) \quad \text{for all positive } x. \tag{4}$$

Since all steps of this reduction are reversible, we have proved the following (for this and other results in this chapter see, e.g., Cauchy 1821; Aczél 1966c, pp. 39–41).

Theorem 1. *The general solution* $g:\mathbb{R}_+ \to \mathbb{R}$ *of* (1) *is given by* (4), *where* f *is an arbitrary solution of* (3).

In view of our results in Chapter 2, we also have the following:

Corollary 2. *The general solution of* (1) *in the classes of functions continuous at a point or measurable or bounded, from one side, on a set of positive measure is given by*

$$g(x) = c \log x \quad (x \in \mathbb{R}_+), \tag{5}$$

where c *is an arbitrary real constant.*

Equation (1) is an example of an equation whose solutions *cannot be extended even to* $\bar{\mathbb{R}}_+$ *from* $\mathbb{R}_+$. Indeed, *the equation*

$$\bar{g}(xy) = \bar{g}(x) + \bar{g}(y) \quad \text{for all } \textit{nonnegative } x, y \tag{6}$$

has only the identically 0 function as a solution, as one can see immediately by putting $y = 0$ into (6):

$$\bar{g}(0) = \bar{g}(x) + \bar{g}(0), \quad \text{that is, } \bar{g}(x) = 0 \quad \textit{for all nonnegative } x.$$

So, for instance, *the solution of* (1), *given by*

$$g(x) = \log x \quad (x \in \mathbb{R}_+),$$

cannot be extended to a solution of (6) *on* $\bar{\mathbb{R}}_+$.

This shows that $g(0)$ cannot be *defined* (except as $g(0) = \infty$ or $g(0) = -\infty$) so that (1) remain satisfied, if g is not identically zero.

Nevertheless, the extension is possible to

$$(\mathbb{R}^*)^2 = \{(x, y) \mid x, y \textit{ real } xy \neq 0\},$$

that is, the most general solutions $h:\mathbb{R}^* \to \mathbb{R}$ ($\mathbb{R}^*$ being the set of all nonzero reals) of the equation

$$h(xy) = h(x) + h(y) \quad \text{for all real nonzero } x, y \tag{7}$$

can be found and are extensions of the solutions of (1).

Indeed, (7) implies that the *restriction* of h to $\mathbb{R}_+$ satisfies (1), that is,

$$h(x) = f(\log x) \quad \text{for positive } x, \text{ where } f \text{ satisfies (3).} \tag{8}$$

Now, $h(1) = 0$ is a consequence of (7) or (8) (put $x = 1$). By substituting $x = y = -1$ into (7) we get

$$2h(-1) = h(1) = 0 \quad \text{or} \quad h(-1) = 0.$$

Further, with $x = |t|$, $y = \operatorname{sign} t$ ($t \neq 0$), equation (7) gives $h(t) = h(|t| \operatorname{sign} t) = h(|t|) + h(\operatorname{sign} t) = h(|t|)$, or $h(t) = h(|t|)$ for all $t \neq 0$. Note that, by definition, $\operatorname{sign} t = 1$ if $t > 0$, $\operatorname{sign} t = -1$ if $t < 0$ and $\operatorname{sign} 0 = 0$, but here the sign function is applied only to $t \neq 0$.

Compared to (8), we get the following.

Theorem 3. *The general solution* $h:\mathbb{R}^* \to \mathbb{R}$ *of* (7) *is given by*

$$h(x) = f(\log|x|) \quad \textit{for all real } x \neq 0, \tag{9}$$

where f *is an arbitrary solution of* (3). *In particular, the general solution of* (7) *in the class of functions bounded, from one side, on a set of positive measure is given by*

$$h(x) = c\log|x| \quad (x \in \mathbb{R}^*), \tag{10}$$

where c *is an arbitrary real constant.*

Of course, g in (4) or (5) is the restriction to $\mathbb{R}_+$ of h in (9) and (10) respectively.

On the other hand, the domain of Cauchy's logarithmic equation can be further restricted. Take, for instance, the equation

$$I(pq) = I(p) + I(q) \quad \text{for all } p \in]0,1], \quad q \in]0,1]. \tag{11}$$

Setting

$$p = e^{-x},\ q = e^{-y} \quad (x \geqslant 0,\ y \geqslant 0), \quad g(x) = I(e^{-x}),$$

it is transformed into

$$g(x+y) = g(x) + g(y) \quad \textit{for all nonnegative } x, y, \tag{12}$$

that is, into (2.2). So we have the following (cf. also Corollary 2.5).

Proposition 4. *The general solution* $I:]0,1] \to \mathbb{R}$ *of* (11) *is given by*

$$I(p) = g(-\log p) \quad (p \in]0,1]),$$

where g *is an arbitrary solution of* (12). *In particular, the general nonnegative solution of* (11) *is given by*

$I(p) = -c\log p \quad (p \in]0,1])$, *where* c *is an arbitrary nonnegative constant.* (13)

This has an immediate application in the elements of information theory (Luce 1960; Rényi 1960; Aczél–Daróczy 1975, pp. 2–5).

It is usual to assume that *the amount of information furnished by an event depends* (only) *upon the* (nonzero) *probability of the event, is nonnegative, and that the amount of information obtained from two independent events is the sum of the amounts of information obtained from the single events.* If the amount of information belonging to the probability p is denoted by $I(p)$, then the last condition means exactly that equation (11) holds and the preceding condition means that I is *nonnegative.* So, by Proposition 4, *the amount of information yielded by an event of probability p is given by* (13). Usually a unit (one bit) is chosen by fixing $I(\frac{1}{2}) = 1$. Then $c = 1/\log 2$ and

$$I(p) = -\frac{\log p}{\log 2} = -\log_2 p$$

is the information yielded by an event of probability p.

In Section 16.4 we will see that the entropies which are of principal importance in information theory and its applications are *mean values* of the $I(p)$'s with the p's as weights.

Next we consider *Cauchy's exponential equation*

$$h(x+y) = h(x)h(y) \tag{14}$$

in three cases:

$$x \in \mathbb{R}, \quad y \in \mathbb{R}, \tag{15}$$

$$x \in \mathbb{R}_+, \quad y \in \mathbb{R}_+, \tag{16}$$

and

$$x \in \bar{\mathbb{R}}_+, \quad y \in \bar{\mathbb{R}}_+. \tag{17}$$

There are some complications if h may be 0 at certain points. *In cases* (15) *and* (16) *we will prove that a function h satisfying* (14) *is either everywhere or nowhere* 0.

The proof is simplest in the case (15). If there exists an x_1 at which $h(x_1) = 0$, then in (14) we set $x = x_1$, $y = t - x_1$ and get

$$h(t) = h(x_1)h(t - x_1) = 0$$

for all $t \in \mathbb{R}$, as asserted. In case (16) this reasoning proves $h(t) = 0$ *only for* $t \geqslant x_1$. Now take an arbitrary $x \in]0, x_1[$. For sufficiently large integers n, $nx > x_1$, so $h(nx) = 0$. On the other hand, just as (2.10) has followed from (2.1), we have $0 = h(nx) = h(x)^n$ by induction, and thus again $h(x) = 0$ for all $x \in \mathbb{R}_+$.

In the case (17), the above argument proves $h(x) = 0$ for $x > 0$ but not for $x = 0$. Equation (14) gives, for $x = y = 0$, $h(0) = h(0)^2$, so *either* $h(0) = 0$ *or* $h(0) = 1$. Indeed the function h given by

$$h(x) = \begin{cases} 0 & \text{for } x > 0 \\ 1 & \text{for } x = 0 \end{cases} \tag{18}$$

satisfies (14) on the domain (17) (but h, defined by $h(x) = 0$ for $x \neq 0$, $h(0) = 1$, does *not* satisfy (14) on (15)). Also, of course,

$$h(x) = 0 \quad (\text{for all } x \in \mathbb{R} \quad \text{or} \quad x \in \mathbb{R}_+ \quad \text{or} \quad x \in \bar{\mathbb{R}}_+) \tag{19}$$

satisfies (14) on the domains (15), (16) and (17), respectively. *For all other solutions h is nowhere* 0. Equation (14) with $x = y = t/2$ also gives

$$h(t) = h\left(\frac{t}{2}\right)^2 \geqslant 0,$$

so, except for the solutions (18) and (19), we have everywhere

$$h(t) > 0. \tag{20}$$

When we know this, then we may take logarithms of both sides of (14) and, with

$$f(x) = \log h(x),$$

get

$$f(x+y)=f(x)+f(y), \tag{21}$$

that is, Cauchy's equation. So we have proved the following.

Theorem 5. *The general solutions of* (14) *on the domains* (15), (16) *or* (17) *are given by* (19) *and by*

$$h(x)=e^{f(x)},$$

where f satisfies (21) *on the same domains; in the case* (17) *we have the additional solution* (18), *but no others. In particular, the general solutions of* (14) *in the class of functions bounded from above on an interval or a set of positive measure (or in the class of measurable functions), are given by* (19), *by*

$$h(x)=e^{cx} \quad (\textit{for } x\in\mathbb{R}, \quad x\in\mathbb{R}_+ \quad \textit{or} \quad x\in\bar{\mathbb{R}}_+) \tag{22}$$

and, in the case (17), *by* (18). *Instead of boundedness from above, boundedness (on a set of positive measure) from below, by a positive bound, may also be supposed. This supposition, however, excludes* (18) *and* (19).

The restriction to positive lower bounds is there because (see (18), (19) and (20)) 0 is a lower bound for *every* solution of (14). In particular,

$$h(x)=e^{f(x)},$$

where f is a solution of Cauchy's equation which is *not* of the form $f(x)=cx$, gives a solution of (14) satisfying $h(x)>0$ but which has *neither* of the forms (18), (19) or (22). We have seen in Theorem 2.10 and Corollary 2.11 that such solutions exist. Also, remember that only logarithms of *positive* numbers are finite real numbers.

The solution (18) of (14) on the domain (17) is again an example which *cannot be extended from* $\bar{\mathbb{R}}_+$ to $\mathbb{R}$. Indeed, this function is bounded from above (and continuous) on $\mathbb{R}_+$, so its extension would have the same properties (on $\mathbb{R}_+$). But all solutions of (14) on (15), bounded from above on an interval, have one of the forms (19) or (22), and (18) is *not* the restriction of either of them to $\bar{\mathbb{R}}_+$.

Finally, we consider *Cauchy's power equation*

$$g(xy)=g(x)g(y), \tag{23}$$

first in case (16),

$$x\in\mathbb{R}_+, \quad y\in\mathbb{R}_+. \tag{24}$$

By writing

$$x=e^u, \quad y=e^v, \quad h(u)=g(e^u),$$

we immediately get

$$h(u+v)=h(u)h(v) \quad \text{for all real } u, v,$$

that is equation (14) on the domain (15). From Theorem 5 we have the following:

Proposition 6. *The general solutions* $g:\mathbb{R}_+ \to \mathbb{R}$ *of* (23) *for* (24) *are given by*

$$g(x) = 0 \quad \text{for all } x > 0 \tag{25}$$

and by

$$g(x) = \exp(f(\log x)) \quad \text{for all } x > 0, \tag{26}$$

where f is an arbitrary solution of Cauchy's equation:

$$f(u+v) = f(u) + f(v) \quad \text{for all real } u, v.$$

In particular, the general solutions of (23) *for* (24), *in the class of functions continuous at a point or bounded (from above) on a set of positive measure, are given by*

$$g(x) = 0 \quad \text{and} \quad g(x) = x^c \quad \text{for all } x > 0, \quad (c \text{ an arbitrary constant}). \tag{27}$$

Next we consider (23) for the domain

$$x \in \bar{\mathbb{R}}_+, \quad y \in \bar{\mathbb{R}}_+. \tag{28}$$

By putting $y = 0$ into (23), we get

$$g(0)[1 - g(x)] = 0,$$

that is, either $g(0) = 0$ or

$$g(x) = 1 \quad \text{for all } x \geqslant 0. \tag{29}$$

This gives the following:

Corollary 7. *We get the general solutions of* (23) *for* (28) *under the same conditions as in Proposition 6, if we supplement the solutions in* (25), (26) *and* (27) *by* $g(0) = 0$ *and take the one additional solution* (29).

Finally, we solve equation (23) for the domain

$$x \in \mathbb{R}, \quad y \in \mathbb{R}. \tag{30}$$

Putting $y = 1$ into (23) we get

$$g(x)[1 - g(1)] = 0,$$

that is, either $g(1) = 1$ or

$$g(x) = 0 \quad \text{for all real } x. \tag{31}$$

The function g, given by (31), indeed satisfies (23) for (30) but we will disregard it in the next portion of our arguments (it will reappear just before stating Corollary 8). Thus, $g(1) = 1$ and, putting $x = y = -1$ into (23), we have

$$1 = g(1) = g(-1)^2,$$

that is, either

$$g(-1) = 1 \tag{32}$$

or

$$g(-1) = -1. \tag{33}$$

In both cases we put $x = |t|$, $y = \operatorname{sign} t$ $(t \neq 0)$ into (23) and get

$$g(t) = g(|t|)g(\operatorname{sign} t) \quad \text{for } t \neq 0,$$

that is, in cases (32) and (33),

$$g(t) = g(|t|) \quad \text{for all } t \neq 0 \tag{34}$$

or

$$g(t) = g(|t|) \operatorname{sign} t \quad \text{for all } t \neq 0 \tag{35}$$

respectively. The solution (31) satisfies both (34) and (35). So we conclude the following.

Corollary 8. *We get the general solutions of* (23) *for* (30) *under the same conditions as in Proposition* 6, *if we extend all solutions given in Corollary* 7 *either by* (34) *or by* (35).

(It is easy to check that all functions described in Corollaries 7 and 8 indeed satisfy (23) for (28) and (30) respectively.) Notice that here the extensions of the solutions (26) (or (27)) from (24) or (28) to (30) *exist but are not unique.* The result in Corollary 8 is even simpler if we restrict the domain of g to $\mathbb{R}^* = \mathbb{R}\backslash\{0\}$ and the domain of (23) to $(\mathbb{R}^*)^2$.

Corollary 9. *The general solutions* $g: \mathbb{R}^* \to \mathbb{R}$ *of* (23) *for* $(\mathbb{R}^*)^2$, *in the class of functions continuous at a point, are given by*

$$g(x) = 0, \quad g(x) = |x|^c \quad \textit{and} \quad g(x) = |x|^c \operatorname{sign} x \quad \textit{for all } x \in \mathbb{R}^*, \tag{36}$$

where c is an arbitrary real constant.

Exercises and further results

1. Take the following definition of the exponential function

$$e^x = \lim_{n \to \infty} \left(1 + \frac{x}{n}\right)^n.$$

Using the binomial theorem, show that e^x satisfies Cauchy's exponential equation (14).

(For Exercises 2 and 3 see Elliott 1978.)

2. Let $\alpha, \beta; f, g$ be real valued functions defined on $\mathbb{R}_+$. Suppose that $\beta(x) \neq 0$ for all $x \in \mathbb{R}_+$ and that the asymptotic relation

$$\lim_{x \to \infty} (\alpha(x^y) - f(y)\alpha(x))/\beta(x) = g(y) \in \mathbb{R} \quad (y \in \mathbb{R}_+)$$

holds. Determine f and g and find an asymptotic relation between α and β (which does not contain f and g).

3. Let $g:\mathbb{R}_+ \to \mathbb{R}^* = \mathbb{R}\setminus\{0\}$ be such that

$$\lim_{x\to\infty} \frac{g(x^y)}{g(x)} = f(y)$$

exists for all $y\in\mathbb{R}_+$. Find a functional equation satisfied by f.

4. (Thielman 1949*b*.) Prove that the general continuous solutions $f:\bar{\mathbb{R}}_+ \to \mathbb{R}$ of the functional equation

$$f((x^n + y^n)^{1/n}) = f(x) + f(y)$$

is given by $f(x) = cx^n$, where c is an arbitrary real constant.

(For Exercises 5 and 6 see Dhombres 1979, pp. 2.2–2.8.)

5. Find all $f:\mathbb{R}\to\mathbb{R}$, continuous or bounded on an interval (or on a set of positive measure), such that

$$f((x^2 + y^2)^{1/2}) = f(x)f(y) \quad (x, y\in\mathbb{R}).$$

6*. Show that all continuous nonidentically zero functions $f:\mathbb{R}\to\mathbb{C}$ ($\mathbb{C}$ is the set of complex numbers) satisfying Gauss's functional equation

$$f((x^2 + y^2)^{1/2}) = f(x)f(y) \quad (x, y\in\mathbb{R}),$$

are of the form

$$f(x) = \exp(ax^2) \quad (x\in\mathbb{R}),$$

where a is a complex constant. (This gives a characterization of the normal distribution.)

7. (E.g. Aczél 1966*c*, pp. 107–9.) Prove that all functions $f:\mathbb{R}\to\mathbb{R}$ such that:

   (a) f is of class C^1 (has everywhere a continuous derivative);
   (b) $\int_{-\infty}^{\infty} f(x)\mathrm{d}x = 1$;
   (c) for every choice of x_1, x_2 and x_3, the function $g(x) = f(x_1 - x)f(x_2 - x)f(x_3 - x)$ has its maximum at $x = (x_1 + x_2 + x_3)/3$ are given by

$$f(x) = 1/((2\pi)^{1/2}\sigma)\exp(-x^2/2\sigma^2).$$

   (This is another characterization of the normal distribution given by Gauss.)

8. (Aczél 1966*c*, pp. 64–5.) Prove that the general solution $f:\mathbb{R}\to\mathbb{R}$ of the functional equation

$$f(x+y) = a^{xy}f(x)f(y) \quad (x, y\in\mathbb{R}),$$

where $a > 0$ is a constant, is given by

$$f(x) = a^{x^2/2}\, g(x),$$

where g is an arbitrary solution of Cauchy's exponential equation.

9*. (Ward 1970*a*, *b*; Reich 1971.) Prove that the general nonconstant solution $f:\mathbb{R}\to\mathbb{R}$ of the functional equation

$$f\left(\frac{x+y}{x-y}\right) = \frac{f(x)+f(y)}{f(x)-f(y)}$$

(for all real x, y such that $f(x) \neq f(y)$) is $f(x) = x$, by showing that every solution of this equation satisfies Cauchy's power equation (23).

10. (Moszner 1980.) Suppose that $f,g:\mathbb{R}\to\mathbb{C}$ have the same set C of zeros, and that C is nonempty, symmetric and does not contain 0. Prove that all solutions of

$$f(x)f(y)f(-x-y)=g(x)g(y)g(-x-y)\quad (x,y\in\mathbb{R})$$

are given by g arbitrary (with the above restriction on its set C of zeros) and

$$f(x)=\begin{cases}0 & \text{if } x\in C\\ aH(x)g(x) & \text{if } x\notin C,\end{cases}$$

where H is a nonzero function satisfying Cauchy's exponential equation and $a^3=1$.

11. Find the general solution $f:\mathbb{R}\to\mathbb{R}$ of the functional equation

$$f(x+y)=f(x)+f(y)+\lambda f(x)f(y)\quad (x,y\in\mathbb{R}),$$

where λ is a real constant.

4

Generalizations of Cauchy's equations to several multiplace vector and matrix functions. An application to geometric objects

The present chapter gives the general as well as the regular solutions of Cauchy's equation for multiplace functions, that is, when the domain of definition is in $\mathbb{R}^n$. In this context, we give an application to the theory of geometric objects and obtain a characterization of the determinant of a matrix. We continue the chapter with Pexider equations, which are generalizations of Cauchy equations and show how just one equation can determine several unknown functions, which is a typical feature of functional equations. We take this opportunity to show how it is possible to generalize the results from $\mathbb{R}^n$ to quite general structures, such as semigroups, a generalization which may prepare for further studies in the analysis of partial differential equations, for example.

4.1 Multiplace and vector functions

The analogue of Cauchy's equation (2.1) for n-place functions $f:\mathbb{R}^n \to \mathbb{R}$ is

$$f(x_1 + y_1, x_2 + y_2, \ldots, x_n + y_n) = f(x_1, x_2, \ldots, x_n) + f(y_1, y_2, \ldots, y_n)$$
$$(x_k \in \mathbb{R}, y_k \in \mathbb{R}; k = 1, 2, \ldots, n). \quad (1)$$

If we write

$$f_1(t_1) = f(t_1, 0, \ldots, 0), \ldots, f_k(t_k) = f(0, \ldots, 0, t_k, 0, \ldots, 0), \ldots,$$
$$f_n(t_n) = f(0, \ldots, 0, t_n),$$

then we get, by repeated application of (1),

$$\begin{aligned} f(t_1, t_2, \ldots, t_n) &= f(t_1 + 0, 0 + t_2, 0 + t_3, \ldots, 0 + t_n) \\ &= f(t_1, 0, \ldots, 0) + f(0, t_2, t_3, \ldots, t_n) \\ &= f_1(t_1) + f(0 + 0, t_2 + 0, 0 + t_3, \ldots, 0 + t_n) \\ &= f_1(t_1) + f_2(t_2) + f(0, 0, t_3, \ldots, t_n) = \cdots \\ &= f_1(t_1) + f_2(t_2) + \cdots + f_{n-1}(t_{n-1}) + f(0, \ldots, 0, t_n), \end{aligned}$$

or

$$f(t_1,t_2,\ldots,t_n)=\sum_{k=1}^{n} f_k(t_k) \quad \textit{for all } (t_1,t_2,\ldots,t_n)\in\mathbb{R}^n. \tag{2}$$

It is obvious from (1) (put $x_j=y_j=0$ for all $j\neq k$ and $x_k=x, y_k=y$) that

$$f_k(x+y)=f_k(x)+f_k(y) \quad \text{for all } x\in\mathbb{R},\quad y\in\mathbb{R} \quad (k=1,2,\ldots,n). \tag{3}$$

Conversely, any function of the form (2), where the f_k $(k=1,2,\ldots,n)$ satisfy (3), is a solution of (1). So we have proved the following:

Proposition 1. *The general solution* $f:\mathbb{R}^n\to\mathbb{R}$ *of* (1) *is given by* (2), *where the* f_k $(k=1,2,\ldots,n)$ *are arbitrary solutions of Cauchy's equation* (3).

Using Corollary 2.5, it is easy to find the general solution of (1) in the class of functions continuous at a point. If f is continuous at the point $(a_1,a_2,\ldots,a_n)$, then the functions

$$x_k\mapsto f(a_1,\ldots,a_{k-1},x_k,a_{k+1},\ldots,a_n) \quad (k=1,2,\ldots,n) \tag{4}$$

are also continuous at a_k. But, by (2),

$$\begin{aligned} f(a_1,\ldots,a_{k-1},x_k,a_{k+1},\ldots,a_n)&=f_k(x_k)+f_1(a_1)+\cdots\\ &\quad+f_{k-1}(a_{k-1})+f_{k+1}(a_{k+1})+\cdots+f_n(a_n), \end{aligned}$$

that is, f_k has to be continuous at a_k. So, in view of (3), Corollary 2.5 gives

$$f_k(x_k)=c_kx_k \quad \text{for all } x_k\in\mathbb{R} \quad (k=1,2,\ldots,n), \tag{5}$$

with real constants c_k, and we have proved the following:

Corollary 2. *The general solution of* (1) *in the class of functions continuous at a point is given by*

$$f(x_1,x_2,\ldots,x_n)=\sum_{k=1}^{n} c_kx_k. \tag{6}$$

As can be seen from the proof, *it is enough that the functions* (4) *be continuous at* a_k $(k=1,2,\ldots,n)$. Of course, *the same solutions could be found under other, weaker, regularity conditions*, similar to those in Chapter 2, *for instance boundedness from one side on a set of positive measure (in* $\mathbb{R}^n$*)*.

We postpone to Section 9.2 the study of converse theorems, that is, the characterization of subsets E of $\mathbb{R}^n$ so that a solution of (1), bounded above on E, is necessarily continuous and so of the form (6).

Let us take now the function $\mathbf{f}:\mathbb{R}^n\to\mathbb{R}^m$. For this

$$\mathbf{f}(\mathbf{x}+\mathbf{y})=\mathbf{f}(\mathbf{x})+\mathbf{f}(\mathbf{y}) \quad (\mathbf{x},\mathbf{y}\in\mathbb{R}^n) \tag{7}$$

would be the analogue of Cauchy's equation. Breaking up (7) into components we get m equations

$$f_j(x_1+y_1, x_2+y_2, \ldots, x_n+y_n) = f_j(x_1, x_2, \ldots, x_n) + f_j(y_1, y_2, \ldots, y_n)$$
$$(j = 1, 2, \ldots, m;\ x_k \in \mathbb{R},\ y_k \in \mathbb{R};\ k = 1, 2, \ldots, n) \tag{8}$$

of the form (1), so we can easily find the general solution of (7). We just formulate the result for continuous vector–vector functions.

Corollary 3. *The general solution of* (7) *in the class of functions* $\mathbf{f}:\mathbb{R}^n \to \mathbb{R}^m$, *continuous at a point, is given by*

$$\mathbf{f}(\mathbf{x}) = \mathbf{C}\cdot\mathbf{x} \quad (\mathbf{x} \in \mathbb{R}^n), \tag{9}$$

where $\mathbf{C}$ *is a constant m by n matrix and '·' denotes multiplication of a vector by a matrix.*

Indeed, combination of (8), (1) and (6) gives

$$f_j(x_1, x_2, \ldots, x_n) = \sum_{k=1}^{n} c_{jk} x_k \quad (j = 1, 2, \ldots, m),$$

which is equivalent to (9).

In accordance with Corollary 3 ($m = 1$), Corollary 2 states that that the *general solution* of $f:\mathbb{R}^n \to \mathbb{R}$ of

$$f(\mathbf{x}+\mathbf{y}) = f(\mathbf{x}) + f(\mathbf{y}) \quad (\mathbf{x} \in \mathbb{R}^n, \mathbf{y} \in \mathbb{R}^n) \tag{10}$$

in the class of functions continuous at a point is given by

$$f(\mathbf{x}) = \mathbf{c}\cdot\mathbf{x},$$

where $\mathbf{c}$ *is an arbitrary constant vector and '·' denotes the inner product of two vectors.*

If, on the other hand, $n = 1$, then *the general solution* $f:\mathbb{R} \to \mathbb{R}^m$ *of*

$$\mathbf{f}(x+y) = \mathbf{f}(x) + \mathbf{f}(y) \quad (x \in \mathbb{R}, y \in \mathbb{R})$$

in the class of functions continuous at a point is, of course (cf. (5)), *given by*

$$\mathbf{f}(x) = x\mathbf{c}$$

where $\mathbf{c}$ *is an arbitrary constant vector, multiplied by the scalar x.*

Another way of stating Corollary 3 is to say that, if M and N are real normed vector spaces of finite dimension, $f:N \to M$ is continuous at a point and

$$f(x+y) = f(x) + f(y) \quad \text{for all } x, y \in N, \tag{11}$$

then f is a *linear operator*, that is, it satisfies, in addition to (11), also

$$f(\lambda x) = \lambda f(x) \quad \text{for all } x \in N, \lambda \in \mathbb{R}. \tag{12}$$

It is not difficult to see that the result remains true when the dimensions of M and/or N are infinite and, more generally, when M and N are real linear Hausdorff spaces. Of course, f is then a continuous linear operator and if, in particular, M and N are real normed spaces, then f is a *bounded linear operator*, that is,

$$\|f\|_{N,M} := \sup_{\substack{x\in N \\ \|x\|_N=1}} \|f(x)\|_M < \infty$$

holds in addition to (11) and (12). We show that the linearity (11), (12) may be obtained from weaker assumptions than continuity (in the following five paragraphs the rudiments of functional analysis and of measure theory are used).

Let N and M be two real normed vector spaces. Let E be a nonempty subset of N such that there exists an x_0 in E and, for any straight line D in N passing through x_0, the intersection $D\cap E$ has a positive measure (D being a straight line, we can identify it with $\mathbb{R}$ and $D\cap E$ with a subset of $\mathbb{R}$). We shall then say that E has a *positive linear Lebesgue measure.*

A function $f:N\to M$, which is bounded on a subset E of positive linear Lebesgue measure (that is, there exists $A>0$ so that, for all x in E, $\|f(x)\| \leqslant A$) *and which satisfies*

$$f(x+y) = f(x) + f(y) \quad (x\in N, y\in N), \tag{11}$$

also satisfies

$$f(\lambda x) = \lambda f(x) \quad (x\in N, \lambda\in\mathbb{R}). \tag{12}$$

To prove this assertion, using (11) and a convenient translation, we may suppose without loss of generality that x_0 is the origin of N. Let ϕ be a real, continuous linear form on M (an element of the topological dual M^* of M). Let x be any nonzero vector of N. Consider $g(\lambda) = \phi\circ f(\lambda x) = \phi[f(\lambda x)]$ (here $\circ$ is the operation of composition of functions). Thus $g:\mathbb{R}\to\mathbb{R}$ satisfies the Cauchy equation (2.1) and is bounded (in absolute value) on a set of positive Lebesgue measure. Therefore

$$g(\lambda) = \lambda g(1) = \lambda(\phi\circ f(x)) = \phi(\lambda f(x)).$$

Thus, for every ϕ in M^* and for every λ in $\mathbb{R}$,

$$\phi(f(\lambda x)) = \phi(\lambda f(x)).$$

As M^* separates points of M, we conclude that (12) is true. (Notice that M^* has here a meaning different from $M\setminus\{0\}$.) Instead of assuming that f is bounded on E, we could have only supposed that, for any straight line D in N passing through x_0, there exists a constant $A(D)$ such that, for all x in $D\cap E$, $\|f(x)\| \leqslant A(D)$.

A warning is necessary. Since the norm on N plays no role in the above discussion, an additive $f:N\to M$ bounded on a subset E of positive linear Lebesgue measure is not necessarily a bounded linear operator. As a counterexample, let N be a real linear space equipped with two nonequivalent norms, say $\|\cdot\|_1$ and $\|\cdot\|_2$ and suppose, e.g., that $\sup_{\|x\|_2=1}\|x\|_1=\infty$. Let N_1 and N_2 be the corresponding normed spaces. Let E be a ray in the unit open ball of N_2. This set E is clearly of positive linear Lebesgue measure. Then the identity map $N_2 \to N_1$ is additive, bounded on E and linear as well. However, this linear map is not bounded as the two norms are nonequivalent (N_1 has to be of infinite dimension, cf. Exercise 19).

Generalizing further (cf. e.g., Rudin 1987, pp. 95–115), it is not difficult to obtain similar results for locally convex real linear spaces instead of just normed real spaces. The situation appears to be different for normed linear spaces over a valued complete nondiscrete field K (instead of $\mathbb{R}$) as the topological dual may reduce to zero.

Let us proceed to other functional equations of the Cauchy type.

We have no difficulty with Cauchy's exponential equation for $h: \mathbb{R}^n \to \mathbb{R}$,

$$h(\mathbf{x}+\mathbf{y}) = h(\mathbf{x})h(\mathbf{y}) \quad (\mathbf{x}, \mathbf{y} \in \mathbb{R}^n). \tag{13}$$

One shows, as in Chapter 3 (right after (3.17)), that either

$$h(\mathbf{x}) = 0 \quad (\mathbf{x} \in \mathbb{R}^n), \tag{14}$$

or h is everywhere positive. In the latter case one can take logarithms on both sides of (13).

Proposition 4. *The general solutions* $h: \mathbb{R}^n \to \mathbb{R}$ *of* (13) *are given by* (14) *and by*

$$h(\mathbf{x}) = \exp f(\mathbf{x}) \quad (x \in \mathbb{R}^n), \tag{15}$$

where f is an arbitrary solution of (10). *In particular, the general solutions of* (13), *in the classes of functions continuous at a point or, say, bounded from above or measurable on a set of positive measure, are given by* (14) *and by*

$$h(\mathbf{x}) = e^{\mathbf{c}\cdot\mathbf{x}}, \quad \textit{where } \mathbf{c} \in \mathbb{R}^n \textit{ is an arbitrary constant vector.} \tag{16}$$

We get results similar to those above if we replace $\mathbb{R}^n$ by $\mathbb{R}^n_+$. We will come back to (13) for complex *valued* functions in Chapter 5.

4.2 A matrix functional equation and a characterization of densities in the theory of geometric objects

Of course, Cauchy type equations can also be further generalized. For instance, those containing multiplication, e.g. (3.23), can be generalized to matrices since matrices form a ring (under addition, matrices behave like vectors, so (7) would give nothing new for matrices). Let us give an example. The equation ('det' stands for the determinant)

$$g(XY) = g(X)g(Y) \quad (\det XY \neq 0) \tag{17}$$

for real (scalar) valued functions of square matrices has important applications in the *theory of geometric objects* (cf. Aczél–Gołąb 1960, pp. 43–7.)

A homogeneous linear purely differential geometric object of first class with one component is a real number depending on the coordinates in the n dimensional space in such a way that under regular (that is, differentiable and locally one to one) changes of coordinates

$$\tilde{x}_k = \phi_k(x_1, x_2, \ldots, x_n) \quad (k = 1, 2, \ldots, n)$$

the value d of the object in the old coordinate system changes into

$$\tilde{d} = g(X)d \tag{18}$$

in the new system, where g is a real valued function of the Jacobian matrix

$$X = \left[\frac{\partial \tilde{x}_k}{\partial x_j}\right] \quad (j, k = 1, 2, \ldots, n; \det X \neq 0).$$

Of course, we may change again to a third system,

$$\tilde{\tilde{x}}_l = \psi_l(\tilde{x}_1, \tilde{x}_2, \ldots, \tilde{x}_n) \quad (l = 1, 2, \ldots, n),$$

of coordinates with the Jacobian matrix

$$Y = \left[\frac{\partial \tilde{\tilde{x}}_l}{\partial \tilde{x}_k}\right] \quad (k, l = 1, 2, \ldots, n; \det Y \neq 0),$$

and arrive at a new value $\tilde{\tilde{d}}$ of the object. By applying the rule of derivation for composite functions of several variables, we get

$$\left[\frac{\partial \tilde{\tilde{x}}_l}{\partial x_j}\right] = \left[\sum_{k=1}^{n} \frac{\partial \tilde{\tilde{x}}_l}{\partial \tilde{x}_k} \frac{\partial \tilde{x}_k}{\partial x_j}\right] = Y \cdot X, \tag{19}$$

where $\cdot$ denotes multiplication of matrices.

So, from (18) and (19), we have

$$\tilde{\tilde{d}} = g(Y \cdot X)d \quad \text{and also } \tilde{\tilde{d}} = g(Y)\tilde{d} = g(Y)g(X)d.$$

Thus (except for the trivial case $d \equiv 0$), we get the functional equation

$$g(Y \cdot X) = g(Y)g(X),$$

which is the same as (17). In order to solve (17), we will first use the facts that (omitting the dots for multiplication)

$$g(YX) = g(XY) \quad \text{and} \quad g(XYZ) = g(XZY), \text{ etc.,} \tag{20}$$

follows from (17), because g is scalar valued. All we will need from matrix theory are the following concepts and results (see, e.g., Gantmacher 1960, pp. 273–6):

Every nonsingular matrix X can be represented as the product of a hermitian and a unitary matrix ($X = HU$). All hermitian and also all unitary matrices are similar to diagonal matrices.

Remember that a matrix X is *nonsingular* if $\det X \neq 0$, $\det X$ being the determinant of the matrix X (cf. (17)). Two matrices X and Z are *similar* if there exists a nonsingular matrix T, with inverse T^{-1}, such that

$$X = T^{-1}ZT.$$

Similar matrices represent the same linear operation in different bases. A matrix is *diagonal* if 0 stands at all places except in the main diagonal. The unit matrix I is an example of a diagonal matrix. In I there are all 1s in the diagonal. Of course $T^{-1}T = TT^{-1} = I$. So, by (20),

$$g(X) = g(T^{-1}ZT) = g(T^{-1}TZ) = g(Z), \tag{21}$$

that is, the (scalar) *value of g, satisfying* (17), *is the same for any two similar matrices.* A diagonal matrix can be written as a product of matrices, each of which differs in at most one element of the main diagonal from the unit matrix:

$$\begin{bmatrix} d_1 & 0 & \cdots & 0 \\ 0 & d_2 & & 0 \\ \vdots & & & \\ 0 & & & d_n \end{bmatrix} = \begin{bmatrix} d_1 & 0 & \cdots & 0 \\ 0 & 1 & & 0 \\ \vdots & & & \\ 0 & 0 & & 1 \end{bmatrix} \cdot \begin{bmatrix} 1 & 0 & \cdots & 0 \\ 0 & d_2 & & 0 \\ \vdots & & & \\ 0 & 0 & & 1 \end{bmatrix} \cdots \begin{bmatrix} 1 & 0 & \cdots & 0 \\ 0 & 1 & & 0 \\ \vdots & & & \\ 0 & 0 & & d_n \end{bmatrix}.$$

It is easy to see that each diagonal matrix on the right hand side (for instance the second) is similar to one where it is the first element of the diagonal (all other elements of the diagonal being 0). The product of such matrices is again a matrix with only the first element possibly different from that of the unit matrix. Of course, in all these transformations, the value of the determinant does not change.

We then introduce the notation

$$f(t) = g\left(\begin{bmatrix} t & 0 & \cdots & 0 \\ 0 & 1 & & 0 \\ \vdots & & & \\ 0 & 0 & & 1 \end{bmatrix}\right)$$

It is clear that from (17) that

$$f(tu) = f(t)f(u) \quad \text{for all } t \neq 0, \quad u \neq 0, \tag{22}$$

and that the value of g at a diagonal matrix, with $d_1, d_2, \ldots, d_n$ in the diagonal, is $f(d_1)f(d_2)\cdots f(d_n)$. By (22), this last expression is equal to $f(d_1 d_2 \cdots d_n)$. As $d_1 d_2 \cdots d_n$ is the value of the determinant of the diagonal matrix, we have proved for such diagonal matrices X:

$$g(X) = f(\det X) \quad (\det X \neq 0). \tag{23}$$

But using the decomposition $X = HU$, (17) and (21), we see that this relation holds for any nonsingular matrix X.

So (cf. (3.23) and Corollary 3.9), we have proved the following (the above proof essentially follows that of Hosszú 1960):

Theorem 5. *The general real valued solution of* (17) *is given by* (23) *where f is an arbitrary solution of Cauchy's power equation* (22).

Corollary 6. *The general, not identically* 0, *real valued solutions of* (17), *continuous at a point, are given by*

$$g(X) = |\det X|^c \quad \textit{and} \quad g(X) = |\det X|^c \operatorname{sign}(\det X), \tag{24}$$

where c is an arbitrary real constant.

Corollary 7. *The only nonzero homogeneous linear purely differential geometric objects, depending continuously on the nonsingular and differentiable transformations of coordinates, are objects which transform according to* (18), *g being one of the functions given by* (24), *while c is a constant.*

The geometric objects which are transformed according to (18) with the second g in (24) are called '*(ordinary) densities*'; those with the first g in (24) are called '*Weyl densities*'. In both cases c is 'the weight of the density'.

We have solved (17) under the restriction that both X and Y are nonsingular matrices because that is what the application requires. (Equation (17) is a conditional equation, cf. Chapter 6.) However, we may allow, for completeness sake, X and Y to be singular in the functional equation (17), then Theorem 5 remains valid by removing the condition $tu \neq 0$ in equation (22).

To prove this assertion, let X be a singular matrix. There exists a matrix Y, similar to X, whose first column consists of zeros. Let D_k be a diagonal matrix with all diagonal entries equal to 1, except the kth entry which is zero. We clearly get $Y = YD_1$. If we prove that $g(D_1) = 0$, then we will get

$$g(X) = g(Y) = g(Y)g(D_1) = 0.$$

But $\det X = 0$. Therefore, if $f(0) = 0$, then $g(X) = f(\det X)$, which is (23). So, equation (22) will hold for $tu = 0$ as well. We will prove that $g(D_1) = 0$ except if $f(t) \equiv 1$. (If $f(t) \not\equiv 1$, we also have $f(0) = 0$ from (22).)

Indeed, let us compute $g(D_1)$. From (17), with no restriction on X or Y, we deduce either $g(0) = 0$ or $g(X) \equiv 1$. The first case leads to $g(D_1) \cdots g(D_n) = 0$ as $D_1 D_2 \cdots D_n$ is the zero matrix. But all matrices D_k are similar. Therefore $g(D_1) = g(D_2) = \cdots = g(D_n) = 0$, as asserted. In the second case, where $g \equiv 1$, we get $f \equiv 1$, which satisfies both (22) and (23) without restriction.

Theorem 5 and its version which does not exclude singular matrices do not depend upon the particular field $\mathbb{R}$, but are true for any field F, not necessarily commutative, as shown by Djoković (1970). We denote the semigroup of n by n matrices over F by $M_n(F)$.

Theorem 8. *Let F be a field. If $g: M_n(F) \to F$ satisfies, for all X, Y in $M_n(F)$,*

$$g(XY) = g(X)g(Y),$$

then there exists $f: F \to F$ *such that* $g(X) = f(\det X)$ *and* f *satisfies, for all* t, u *in* F,

$$f(tu) = f(t)f(u).$$

If singular matrices are excluded, the same result holds with $tu \neq 0$ *in the last equation.*

4.3 Pexider equations

It is a remarkable feature of functional equations, which distinguishes them from most other (differential, difference, integral, etc.) equations that *one* equation can determine *several* unknown functions. The classical example for this is a generalization of Cauchy's equation, namely the Pexider equation (Pexider 1903):

$$k(x + y) = g(x) + h(y),$$

where g, h, k are *all* unknown functions. Similar generalizations are possible for the other three Cauchy equations, which were treated in Chapter 3 (cf. Corollary 11 below). We will consider this equation immediately in a general case which contains both scalar (real or complex), vector and matrix functions. Let $(N, +)$ be a *groupoid* (a set with an operation $+: N \times N \to N$) with a neutral element 0 ($x + 0 = x = 0 + x$ for all $x \in N$), and let $(G, +)$ be a (additively written, but not necessarily commutative) group. We look for the solutions g, h, k: $N \to G$ of

$$k(x + y) = g(x) + h(y) \quad (x, y \in N). \tag{25}$$

Substituting first $x = 0$, then $y = 0$ into (25) and writing $a = g(0)$, $b = h(0)$, we get

$$h(y) = -a + k(y), \quad \text{and} \quad g(x) = k(x) - b \tag{26}$$

respectively. Putting these into (25) we have

$$k(x + y) = k(x) - b - a + k(y). \tag{27}$$

If we subtract a on the left and b on the right from both sides of (27) and introduce the new function $f: N \to G$ by

$$f(z) = -a + k(z) - b, \tag{28}$$

we obtain

$$f(x + y) = f(x) + f(y) \quad \text{for all } x, y \in N.$$

This looks like Cauchy's equation and just says that f is a homomorphism from $(N, +)$ into $(G, +)$ (just as the solutions of Cauchy's equation (7) are homomorphisms of $(\mathbb{R}^n, +)$ into $(\mathbb{R}^m, +)$ where $+$ is the usual addition of vectors). Taking (26) and (28) into consideration also, we have proved the following.

Theorem 9. *Let $(N, +)$ be a groupoid with a neutral element and $(G, +)$ a group. The general solutions g, h, k: $N \to G$ of (25) are given by*

$$g(x) = a + f(x), \quad h(y) = f(y) + b, \quad k(z) = a + f(z) + b \quad \text{for all } x, y, z \in N, \qquad (29)$$

where a and b are arbitrary constant elements of G and f is an arbitrary homomorphism from N into G.

(It is easy to check that the functions given by (29) always satisfy (25), whichever constants a, b and whichever homomorphism $f: N \to G$ we choose.) The general solutions $\mathbf{g}, \mathbf{h}, \mathbf{k}: \mathbb{R}^n \to \mathbb{R}^m$ of

$$\mathbf{k}(\mathbf{x} + \mathbf{y}) = \mathbf{g}(\mathbf{x}) + \mathbf{h}(\mathbf{y}) \quad (\mathbf{x}, \mathbf{y} \in \mathbb{R}^n) \qquad (30)$$

are readily constructed by aid of (8), Proposition 1 and Theorem 2.10. As to the general continuous solutions, from Corollary 3 we get the following:

Corollary 10. *The general solutions of (30) in the class of functions $\mathbf{g}, \mathbf{h}, \mathbf{k}: \mathbb{R}^n \to \mathbb{R}^m$, where at least one of $\mathbf{g}, \mathbf{h}, \mathbf{k}$ is continuous at a point, are given by*

$$\mathbf{g}(\mathbf{x}) = \mathbf{C} \cdot \mathbf{x} + \mathbf{a}, \quad \mathbf{h}(\mathbf{y}) = \mathbf{C} \cdot \mathbf{y} + \mathbf{b}, \quad \mathbf{k}(\mathbf{z}) = \mathbf{C} \cdot \mathbf{z} + \mathbf{a} + \mathbf{b} \quad (\mathbf{x}, \mathbf{y}, \mathbf{z} \in \mathbb{R}^n),$$

where $\mathbf{a}$, $\mathbf{b}$ are arbitrary constants in $\mathbb{R}^m$ and $\mathbf{C}$ is an arbitrary constant $m \times n$ matrix.

We can extend Theorem 8 in a similar way:

Corollary 11. *Let F be a field (not necessarily commutative) and $M_n(F)$ the multiplicative groupoid (monoid) of $n \times n$ matrices over F. The general solutions $g, h, k: M_n(F) \to F$ of*

$$k(XY) = g(X)h(Y) \quad (X, Y \in M_n(F)) \qquad (31)$$

are given by

$$g(Z) = k(Z) = 0 \quad (Z \in M_n(F)), \quad h \text{ arbitrary}, \qquad (32)$$

$$h(Z) = k(Z) = 0 \quad (Z \in M_n(F)), \quad g \text{ arbitrary}, \qquad (33)$$

and by

$$g(Z) = af(\det Z), \quad h(Z) = f(\det Z)b, \quad k(Z) = af(\det Z)b, \qquad (34)$$

where $a, b \in F \setminus \{0\}$ are arbitrary constants and f an arbitrary homomorphism from $(F, \cdot)$ into $(F, \cdot)$.

Proof. The monoid (semigroup with neutral element) $M_n(F)$ is, under multiplication, certainly a groupoid with neutral element (the unit matrix I). However, only the nonzero elements of F form a group under multiplication. The proof of Theorem 9 also works in this situation if $g(I) \neq 0$, $h(I) \neq 0$. So, in this case, we have the solution (34) which indeed satisfies (31) if f is a homomorphism. If, say, $g(I)=0$ then, from (31), $k(Y)=k(IY)=g(I)h(Y)=0$ for all $Y\in M_n(F)$. But then, again from (31), $0=g(X)h(Y)$ and, since fields have no zero-divisors, either $g(X)\equiv 0$ or $h(Y)\equiv 0$. The same holds when $h(I)=0$. If k and g (or h) are identically zero, then h (respectively g) can be arbitrary and (31) is still satisfied. These give (32) and (33), respectively, and the Corollary 11 is proved.

We could again restrict ourselves, for instance, to real or nonsingular matrices.

It is also readily seen that, e.g., Proposition 1 remains true for $f:N_1\times N_2\times\cdots\times N_n\to G$, where $N_1,N_2,\ldots,N_n$ are groupoids with neutral elements and G an abelian group. Some further 'tricks' are needed if these contain no neutral element 0, see Aczél 1964*b*, 1975; Taylor 1980; Kuczma 1972; and Szymiczek 1973.

We first show that a result similar to Theorem 9 also holds for abelian semigroups which do not necessarily contain a neutral element (Aczél 1964*b*).

Theorem 12. *Let* $(S,+)$ *be a commutative semigroup and* $(G,\cdot)$ *a group. The general solutions* $g,h,k:S\to G$ *of*

$$k(x+y)=g(x)h(y)\quad \text{for all } x,y\in S \tag{35}$$

are given by

$$g(x)=af(x),\quad h(y)=f(y)b,\quad k(t)=af(t)b\quad (x,y\in S,\, t\in S+S), \tag{36}$$

where $a,b\in G$ *are arbitrary constants and* f *is an arbitrary homomorphism from* S *into* G.

(In (35), as usual, since there is no danger of misunderstanding, we have omitted the dot $\cdot$ for brevity. In (36), again, $S+S=\{x+y\,|\,x\in S, y\in S\}$.)

Proof. By commutativity,

$$g(x)h(y)=k(x+y)=k(y+x)=g(y)h(x).$$

We fix $y=y_0$ and write

$$\alpha=g(y_0),\quad \beta=h(y_0) \tag{37}$$

in order to get

$$g(x)=\alpha h(x)\beta^{-1}. \tag{38}$$

Substitution of this into (35) gives

$$k(x+y)=\alpha h(x)\beta^{-1}h(y),$$

and, since

$$k[(x+y)+y_0]=k[x+(y+y_0)],$$

we now have

$$\alpha h(x+y)\beta^{-1}h(y_0) = \alpha h(x)\beta^{-1}h(y+y_0).$$

In view of (37), and since G is a group, we get, with $e(y) = \beta^{-1}h(y+y_0)$,

$$h(x+y) = h(x)e(y). \tag{39}$$

Again by commutativity,

$$h(x)e(y_0) = h(y_0)e(x),$$

or, with (37) and $\gamma = e(y_0)^{-1}$,

$$h(x) = \beta e(x)\gamma. \tag{40}$$

Putting (40) back into (39) gives

$$e(x+y) = e(x)\gamma e(y)\gamma^{-1} \tag{41}$$

and associativity again yields

$$e(x+y)\gamma e(z)\gamma^{-1} = e[(x+y)+z] = e[x+(y+z)] = e(x)\gamma e(y+z)\gamma^{-1},$$

or

$$e(x)\gamma e(y)e(z) = e(x)\gamma e(y)\gamma e(z)\gamma^{-1}.$$

Cancelling $e(x)\gamma e(y)$ we get $e(z) = \gamma e(z)\gamma^{-1}$, that is,

$$e(z)\gamma = \gamma e(z), \tag{42}$$

so that (41) reduces to

$$e(x+y) = e(x)e(y), \tag{43}$$

that is, e is a homomorphism from $(S, +)$ into $(G, \cdot)$. By (40), (38), (35), (42), and (43), we get

$$h(x) = \beta e(x)\gamma,$$
$$g(x) = \alpha\beta\gamma e(x)\beta^{-1},$$

and

$$k(x+y) = g(x)h(y) = \alpha\beta\gamma e(x)e(y)\gamma = \alpha\beta\gamma e(x+y)\gamma.$$

We will be better off, however, with the new constants $a = \alpha\beta\gamma\beta^{-1}$, $b = \beta\gamma$ and with

$$f(x) := \beta e(x)\beta^{-1}.$$

This transform f of e is also a homomorphism

$$f(x+y) = f(x)f(y) \quad (x, y \in S), \tag{44}$$

and we now have

$$g(x) = a\beta e(x)\beta^{-1} = af(x) \quad (x \in S),$$
$$h(y) = \beta e(y)\beta^{-1}b = f(y)b \quad (x \in S),$$

and

$$k(z) = a\beta e(z)\beta^{-1}b = af(z)b \quad (z \in S+S).$$

This is (36) and these g, h, k always satisfy (35) if (44) holds, so Theorem 12 is proved.

4.4 Cauchy-type equations on semigroups

Now we come to similar generalizations of Proposition 1. Obviously, the proof of Proposition 1 remains true for additive functions (homomorphisms) $f: N_1 \times N_2 \times \cdots \times N_n \to S$, where N_k $(k = 1, \ldots, n)$ are groupoids with neutral elements, and S is a commutative semigroup. The following somewhat more sophisticated result is due to Kuczma (1972):

Theorem 13. *Let $S_1, S_2, \ldots, S_n$ be commutative semigroups, and G an abelian group. The mapping f is a homomorphism from $S_1 \times S_2 \times \cdots \times S_n$ into G, that is,*

$$f(x_1 + y_1, x_2 + y_2, \ldots, x_n + y_n) = f(x_1, x_2, \ldots, x_n) + f(y_1, y_2, \ldots, y_n)$$
$$(x_j, y_j \in S_j; j = 1, 2, \ldots, n), \tag{45}$$

if, and only if, there exist homomorphisms $f_j: S_j \to G$ $(j = 1, 2, \ldots, n)$ such that

$$f(x_1, x_2, \ldots, x_n) = \sum_{j=1}^{n} f_j(x_j) \quad (x_j \in S_j; j = 1, 2, \ldots, n). \tag{46}$$

Proof. Let $c_j \in S_j$ $(j = 1, 2, \ldots, n)$ be constants. We define

$$f_j(x_j) = f(c_1, \ldots, c_{j-1}, x_j + c_j, c_{j+1}, \ldots, c_n) - f(c_1, c_2, \ldots, c_n).$$

By applying (45) repeatedly, we get

$$\begin{aligned} f(x_1, x_2, \ldots, x_n) + nf(c_1, c_2, \ldots, c_n) &= f(x_1 + nc_1, x_2 + nc_2, \ldots, x_n + nc_n) \\ &= f(x_1 + c_1, c_2, \ldots, c_n) \\ &\quad + f(c_1, x_2 + c_2, c_3, \ldots, c_n) \\ &\quad + \cdots + f(c_1, \ldots, c_{n-1}, x_n + c_n) \\ &= \sum_{j=1}^{n} f_j(x_j) + nf(c_1, c_2, \ldots, c_n). \end{aligned}$$

Thus we have proved (46). We prove that the f_j's are homomorphisms similarly, by repeated application of (45):

$$\begin{aligned} f_j(x_j + y_j) + 2f(c_1, c_2, \ldots, c_n) &= f(c_1, \ldots, c_{j-1}, x_j + y_j + c_j, c_{j+1}, \ldots, c_n) \\ &\quad + f(c_1, c_2, \ldots, c_n) \\ &= f(c_1, \ldots, c_{j-1}, x_j + c_j, c_{j+1}, \ldots, c_n) \\ &\quad + f(c_1, \ldots, c_{j-1}, y_j + c_j, c_{j+1}, \ldots, c_n) \\ &= f_j(x_j) + f_j(y_j) + 2f(c_1, c_2, \ldots, c_n). \end{aligned}$$

Since (46) satisfies (45) whenever the $f_j: S_j \to G$ are homomorphisms, we have proved Theorem 13.

We may note that in Theorem 13, as compared to Theorem 12, the range is supposed to be in a *commutative* group. However, the condition that G be commutative can be eliminated from the 'only if' part of Theorem 13, and it is even sufficient to suppose that the S_j have nonempty centers (Exercise 3; Szymiczek 1973).

We have generalized Cauchy's power equation (Theorems 5 and 8) and Cauchy's exponential equation (Proposition 4) to the case of scalar valued functions of square matrices or of vectors in $\mathbb{R}^n$. In the theory of differential equations, important generalizations of Cauchy's exponential equation are needed (see, e.g., Hille–Phillips 1957, pp. 388–528; Kato 1976, pp. 365–426). First the values of the function may belong to the linear space of bounded operators over a normed space. Second, the variable could be in a semigroup rather than in $\mathbb{R}$ or $\mathbb{R}^n$. (In the remainder of this section, the rudiments of operator theory are used.) Let us agree on the notation.

Let N be a real Banach space and $L(N)$ be the real normed space of all bounded linear operators from N into N. Let $(S, +)$ be a semigroup. We look for functions $f: S \to L(N)$ such that

$$f(x+y) = f(x) \circ f(y) \quad (x, y \in S). \tag{47}$$

Here $\circ$ denotes the composition of operators of $L(N)$. The image $f(S)$ is a semigroup of bounded linear operators on N.

For example, take $S = \bar{\mathbb{R}}_+$ under the usual addition and let C be a given element of $L(N)$. As N is a Banach space, so is the space $L(N)$ for the usual norm of bounded operators over N. Therefore the following series is convergent on $L(N)$:

$$f(x) = I + xC + \frac{x^2}{2} C^2 + \cdots + \frac{x^n}{n!} C^n + \cdots \quad (x \geqslant 0), \tag{48}$$

where I is the identity operator on N. The series (48) defines

$$f(x) = e^{xC}. \tag{49}$$

Then (49) is a solution of (47) for all x, y in $\bar{\mathbb{R}}_+$ as is proved by the usual manipulation of convergent series. However, the general solution of (47) is not of the form (49). An extensive literature has been devoted to the problem of finding regularity conditions for a solution f of (47) so that it reduces to (48). In this case, the operator C is called the ***infinitesimal generator*** of the semigroup $f(\bar{\mathbb{R}}_+)$. We quote here Hille–Phillips 1957 and Kato 1976 as classical textbooks on the subject. We shall only mention two typical results.

Theorem 14. *A function* $f: \bar{\mathbb{R}}_+ \to L(N)$ *satisfying the functional equation*

$$f(x+y) = f(x) \circ f(y) \quad (x, y \geqslant 0) \tag{50}$$

is of the form (49) *for some* C *in* $L(N)$ *if (and only if)* $f(0) = I$, *the identity operator, and* $\lim_{x \to 0} \| f(x) - f(0) \| = 0$.

Let $f:\bar{\mathbb{R}}_+ \to L(N)$ be a solution of (50) for which $f(0)=I$. Let n be a vector in N such that the following limit exists:

$$D_n = \lim_{x\to 0+} \frac{f(x)n - n}{x}. \qquad (51)$$

The set $\mathscr{D}$ of all n in N for which the limit exists is a linear subspace of N on which $n \mapsto D_n$ is a linear operator D. In general, $\mathscr{D} \neq N$ and D is not bounded on $\mathscr{D}$. However, one gets the following:

Theorem 15. *$\mathscr{D} = N$ and D is bounded if and only if the conditions of Theorem 14 apply. In this case $D = C$.*

Conversely, let D be a linear operator mapping $\mathscr{D}$, a subspace of N, into N. An interesting question is to decide when there exists a solution f of (50), with $f(0)=I$, and for which (51) holds with $D_n = D(n)$.

If $\mathscr{D} = N$ and if D is bounded, $f(x) = e^{xD}$ provides a solution for which f is continuous. There exist deeper results when $\mathscr{D}$ is dense in N and when D is a closed linear operator. For the existence of a solution f of (50), with $f(0)=I$ and with $\lim_{x\to 0} f(x)n - n = 0$ for all n in N such that (51) holds, a necessary and sufficient condition is the existence of real numbers M and w such that, for every real $\lambda > w$, $(\lambda I - D)^{-1}$ exists and satisfies the inequalities

$$\|(\lambda I - D)^{-n}\| \leqslant \frac{M}{(\lambda - w)^n} \quad (n = 1, 2, \ldots).$$

Exercises and further results

1. Find the general solution of the functional equation $f:\mathbb{R}\to\mathbb{R}$:
$$f(x+y) + f(z) = f(x) + f(y+z) \quad (x, y, z \in \mathbb{R}).$$
2. Find the general solutions $f, g, h, k: N \to G$ of the equation
$$f(x+y) + g(z) = h(x) + k(y+z) \quad (x, y, z \in N),$$
where N is a groupoid with a neutral element and G is an abelian group.
3. (Szymiczek 1973.) Let $S_1, S_2, \ldots, S_n$ be semigroups with nonempty centres and G a group. Prove that, if (45) holds, then there exist homomorphisms $f_j: S_j \to G$ such that f can be written in the form (46).

(For Exercises 4–5, see Dhombres 1984c.)

4. Determine the general continuous solutions $f, g: \mathbb{R}\to\mathbb{R}$ of
$$f(x)f(y) - f(xy) = g(x+y) \quad (x, y \in \mathbb{R}),$$
with $g(1) = 1$, $g(0) = 0$.

5*. Let K be a commutative field of characteristic different from 2. Find all functions $f: K \to K$ such that $f(x)f(y) - f(xy)$ depends only upon $x + y$.

6*. Let f map a group into an abelian group A *divisible by* 2 (that is, for every $a \in A$, the equation $x + x = A$ has a unique solution). Are the two functional equations
$$2(f(x) + f(y)) = f(xy) + f(yx)$$

and

$$f(xy) = f(x) + f(y)$$

equivalent?

7. Let $\phi: \mathbb{R}^n_+ \to \mathbb{R}_+$ and $f: \mathbb{R}^n_+ \to \mathbb{R}$ be such that, for all $\alpha \in \mathbb{R}^n_+$,

$$\lim_{\|x\| \to \infty} \phi(x)[f(x+\alpha) - f(x)] = g(\alpha)$$

exists, is finite and is not identically zero ($\|\cdot\|$ is any norm on $\mathbb{R}^n$). Suppose that ϕ and f are Lebesgue measurable. Determine the general form of g.

(For Exercises 8, 9, see Aczél 1966c, pp. 66–72.)

8. Determine all solutions, continuous at a point (or bounded on a set of positive measure) of the equation

$$f(\alpha x + \beta y + \gamma) = Af(x) + Bf(y) + C \quad (x, y \in \mathbb{R}),$$

($\alpha, \beta, \gamma, A, B, C$ are given constants, $\alpha\beta AB \neq 0$). For what $\alpha, \beta, \gamma, A, B, C$ do there exist nonconstant continuous solutions?

9*. Determine all solutions of

$$f(\alpha x + y) = Af(x) + f(y) \quad (x, y \in \mathbb{R}),$$

where $\alpha A \neq 0$. (Notice that no regularity condition was supposed here.) For what α, A do there exist not identically zero solutions.

10. (Jurkat–Shawyer 1981; Aczél 1983a.) Let A and B be two linear summation methods for real sequences, that is, each maps sequences $\{a_n\}$, $\{b_n\}$ of real numbers into real numbers $\lim_A \{a_n\}$ and $\lim_B \{b_n\}$, respectively, and $\lim_X \{\alpha c_n + \beta d_n\} = \alpha \lim_X \{c_n\} + \beta \lim_X \{d_n\}$ ($X = A$ or B; $\alpha, \beta \in \mathbb{R}$). Let $\{1\} = \{1, 1, \ldots\}$, $\{(-1)^n\} = \{1, -1, 1, -1, \ldots\}$ and $\lim_A \{1\} = \phi$, $\lim_B \{1\} = \psi$, $\lim_A \{(-1)^n\} = \theta$, $\lim_B \{(-1)^n\} = \omega$ with $\theta \neq \pm\phi$, $\omega \neq \pm\psi$. Suppose that $f: \mathbb{R} \to \mathbb{R}$ is bounded on a set of positive measure and that, whenever $\lim_A x_n$ exists, so does $\lim_B f(x_n)$ and

$$\lim_B f(x_n) = f(\lim_A x_n).$$

Prove that $f(x) = \gamma x + \delta$ ($x \in \mathbb{R}$) for some real constants γ and δ. On the other hand, if $f(x) = x$, or $f(x) = 1$ are solutions, then $\phi = \psi$, $\theta = \omega$ and $\psi = 1$ respectively.

11. (Aczél 1966c, pp. 353–5.) Let S be an arbitrary set with at least three elements a, b, c and $F, G, H: S^2 \to M$ where M is the semigroup of nth order matrices over a commutative field. Suppose $\det G(a, y) \det H(b, c) \neq 0$ for all $y \in S$ and

$$F(x, z) = G(x, y)H(y, z) \quad \text{for all } x, y, z \in S.$$

Determine the general form of F, G, H.

(For Exercises 12 and 13, see Aczél–Haruki–McKiernan–Saković 1968.)

12. Find all continuous functions $f: \mathbb{R}^2 \to \mathbb{R}$ for which the following functional equation holds:

$$f(x+t, y+t) + f(x+t, y-t) + f(x-t, y+t) + f(x-t, y-t) = 4f(x, y)$$
$$(x, y, t \in \mathbb{R}).$$

13. Find all continuous functions $f: \mathbb{R}^2 \to \mathbb{R}$ for which the following functional

equation holds:

$$f(x+t,y)+f(x-t,y)+f(x,y+t)+f(x,y-t)=4f(x,y) \quad (x,y,t\in\mathbb{R}).$$

14. (Haruki 1969.) Find all continuous functions $f:\mathbb{R}^3\to\mathbb{R}$ for which the following functional equation holds

$$\begin{aligned} &f(x+t,y+t,z+t)+f(x+t,y+t,z-t)+f(x+t,y-t,z+t)\\ &\quad+f(x+t,y-t,z-t)+f(x-t,y+t,z+t)+f(x-t,y+t,z-t)\\ &\quad+f(x-t,y-t,z+t)+f(x-t,y-t,z-t)=8f(x,y,z) \quad (x,y,z,t\in\mathbb{R}). \end{aligned}$$

(For Exercises 15 and 16, see Sweet 1981.)

15*. Take an integer $n\geq 1$. Let G be an abelian group divisible by 2 and F be a real vector space. Let $f:G^n\to F$ and $\mathbf{x}=(x_1,\ldots,x_n)$ $(x_i\in G)$. For every t and x_i $(i=1,2,\ldots,n)$ in G define

$$P_i^t f(\mathbf{x})=f(x_1,\ldots,x_i-t,\ldots,x_n)+f(x_1,\ldots,x_i+t,\ldots,x_n).$$

Consider the two following functional equations:

$$O_n:\quad \left(\sum_{i=1}^{n} P_i^t\right)f(\mathbf{x})=2nf(\mathbf{x}) \quad (\mathbf{x}\in G^n, t\in G),$$

$$C_n:\quad \left(\prod_{i=1}^{n} P_i^t\right)f(\mathbf{x})=2^n f(\mathbf{x}) \quad (\mathbf{x}\in G^n, t\in G).$$

Is every solution of O_n a solution of C_n?

16*. With the same notation as in Exercise 15, examine whether, at least for $n\leq 4$, every solution of C_n is a solution of O_n.

17. Find all $f:I\to\mathbb{R}$ (where $I\subset\mathbb{R}$ is an interval) for which there exists a continuous function $g:I\to\mathbb{R}$ such that

$$f(x)-f(y)=(x-y)g\left(\frac{x+y}{2}\right)$$

for all $x,y\in I$. Give an interpretation of the result as a characterization of the parabola.

18*. (Sh. Haruki 1979; Aczél 1985*a*.) Let F be a commutative field of characteristic different from 2. Determine all functions $f,g:F\to F$ which satisfy

$$f(x)-f(y)=(x-y)g(x+y)$$

for all $x,y\in F$. (Of course, no regularity of g, or of f, is supposed here.)

19. Let N and M be real normed spaces, N of finite dimension, and E a subset of N of positive linear Lebesgue measure. Prove or disprove that every solution $f:N\to M$ of (11), bounded on E, is a bounded linear operator.

(For Exercises 20, 21 see Daróczy 1986.)

20. Let $(G,\cdot)$ be a commutative semigroup and let $F:G^2\to\mathbb{R}_+$ be such that

$$F(xy,z)F(x,y)=F(x,yz)F(y,z) \quad (x,y,z\in G)$$

and

$$F(x,y)=\tfrac{1}{2}(F(x,x)+F(y,y)) \quad (x,y\in G).$$

Prove or disprove that, for $g:G\to\mathbb{R}_+$ defined by $g(x)=F(x,x)$,

$$g(xy)(g(x)+g(y))-g(x)g(y)=b\quad (x,y\in G)$$

holds for some nonnegative constant b.

21. Let G be a commutative semigroup. Find the general solution of

$$f(xy)(f(x)+f(y))-f(x)f(y)=1\quad (x,y\in G).$$

(For Exercises 22–4 see Dhombres 1984c.)

22. Find all solutions $f:\mathbb{R}\to\mathbb{R}$ of

$$f(x+y)+f(x)+f(y)=f(xy)+f(x)f(y).$$

23. Find all solutions $f:\mathbb{R}\to\mathbb{R}$ of

$$f(x+y)+f(xy)=f(x)+f(y)+f(x)f(y).$$

24*. Let F be a field. Find all $f:F\to F$ such that $f(x)f(y)-f(xy)$ depends only upon $x+y$ for all $x,y\in F$ (cf. Exercise 5).

25. Let $K_{r,c}$ be the field of real numbers which can be constructed with ruler and compass, i.e., $K_{r,c}$ is the set of real numbers which are algebraic over $\mathbb{Q}$ and whose degree is a power of 2. Let $K^1_{r,c}=K_{r,c}\cap[1,\infty[$. Solve the following Cauchy equations

(i) $f(x+y)=f(x)+f(y)$, (ii) $f(x+y)=f(x)\cdot f(y)$,
(iii) $f(x\cdot y)=f(x)+f(y)$, (iv) $f(xy)=f(x)\cdot f(y)$,

where f is a continuous function from $K^1_{r,c}$ into $K^1_{r,c}$.

26*. (Jessen–Karpf–Thorup 1968, cf. also Aczél–Daróczy 1975, pp. 98–100.) Prove or disprove that the general symmetric ($F(x,y)=F(y,x)$) solution $F:\mathbb{R}^2\to\mathbb{R}$ of

$$F(x,y)+F(x+y,z)=F(y,z)+F(x,y+z)\quad (x,y,z\in\mathbb{R}).$$

is given by

$$F(x,y)=f(x+y)-f(x)-f(y),$$

where $f:\mathbb{R}^2\to\mathbb{R}$ is an arbitrary function. To what groups more general than the additive group of $\mathbb{R}$ can you generalize this result?

27. (Fenyö 1985.) Determine the general solution $g:\mathbb{R}^2\to\mathbb{R}$ of the system of functional equations

$$g(x,z)-g(x+y,z)=g(x,y)-g(x+z,y),$$
$$g(x,y)+g(x,z)-g(x,y+z)=g(y,x)+g(y,z)-g(y,x+z)\quad (x,y,z\in\mathbb{R}).$$

5

Cauchy's equations for complex functions. Applications to harmonic analysis and to information measures

It is quite natural to investigate the various Cauchy equations on classical fields, in particular on the complex field $\mathbb{C}$. In this chapter, we give the general as well as the regular solutions of the Cauchy equations on $\mathbb{C}$. The three applications that we present in this chapter show on one hand that it is important to consider not only the real but also the complex field, on the other that the general solution – and not only regular solutions – play an important role in applications. The first example is about field endomorphisms and shows that $\mathbb{R}$ and $\mathbb{C}$ behave in a totally different manner. The second example uses the general solution of a complex valued Cauchy equation to characterize a class of functions which plays an important role in harmonic analysis. And the last example, with recursive entropies, brings us back to information theory. At the same time, it shows how an integrability assumption on the solutions of some functional equations is enough to yield differentiability properties.

5.1 Cauchy's equation and the exponential equation for complex functions

As a linear space over the reals, the complex field $\mathbb{C}$ is just the linear space $\mathbb{R}^2$, and as such the generalization of Cauchy's equation to $\mathbb{C}$ is plainly a rewriting of Corollary 4.3. However, it is worthwhile explaining some details, using the fact that $\mathbb{C}$ is also a field.

An illustration of the case $n=1, m=2$ of Corollary 4.3 is given by *complex valued functions of a real variable*, that is, $f:\mathbb{R}\to\mathbb{C}$. Then

$$f(x)=f_1(x)+if_2(x) \tag{1}$$

and

$$f(x+y)=f(x)+f(y) \quad (x\in\mathbb{R}, y\in\mathbb{R}) \tag{2}$$

means

$$f_k(x+y) = f_k(x) + f_k(y) \quad (k=1,2). \tag{3}$$

Proposition 1. *The general solution $f:\mathbb{R}\to\mathbb{C}$ of (2) is given by (1) with (3) and, in particular, the general solution $f:\mathbb{R}\to\mathbb{C}$ of (2), in the class of functions continuous at a point or measurable on an interval, is given by*

$$f(x) = (c_1 + ic_2)x = cx, \tag{4}$$

where c is an arbitrary complex constant.

Similarly, the case $n=m=2$ of Corollary 4.3 leads to *complex valued functions of a complex variable*, that is, $f:\mathbb{C}\to\mathbb{C}$. Then

$$x = x_1 + ix_2, \quad f(x) = f_1(x) + if_2(x) \tag{5}$$

and for

$$f(x+y) = f(x) + f(y) \quad (x\in\mathbb{C}, y\in\mathbb{C}), \tag{6}$$

we get, by Corollary 4.3, under supposition of the continuity of f at a point, the solutions

$$f(x) = c_{11}x_1 + c_{12}x_2 + i(c_{21}x_1 + c_{22}x_2).$$

But $x = x_1 + ix_2$, $\bar{x} = x_1 - ix_2$ (the conjugate of x), so

$$f(x) = (c_{11} + ic_{21})\frac{x+\bar{x}}{2} + (c_{22} - ic_{12})\frac{x-\bar{x}}{2} = ax + b\bar{x},$$

where

$$a = \frac{c_{11}+c_{22}}{2} + i\frac{c_{21}-c_{12}}{2}, \quad b = \frac{c_{11}-c_{22}}{2} + i\frac{c_{21}+c_{12}}{2},$$

and, since

$$f(x) = ax + b\bar{x} \quad (x\in\mathbb{C}) \tag{7}$$

satisfies (6) for arbitrary complex constants a, b, we have the following:

Proposition 2. *The general solutions $f:\mathbb{R}\to\mathbb{C}$ and $f:\mathbb{C}\to\mathbb{C}$ of (2) and (6), in the class of functions continuous at a point or measurable on an interval, are given by (4) and (7), respectively, where a, b, and c are arbitrary complex constants.*

Not all solutions of the form (7) are differentiable (in the complex domain), however, since $x\mapsto\bar{x}$ is not a differentiable complex function. *The general differentiable solution $f:\mathbb{C}\to\mathbb{C}$ of (6) is given by*

$$f(x) = ax \quad (x\in\mathbb{C}),$$

where a is an arbitrary complex constant.

Let us proceed now to the determination of the complex valued solutions $f:\mathbb{R}\to\mathbb{C}$ and $f:\mathbb{C}\to\mathbb{C}$ of Cauchy's exponential equation

$$h(x+y)=h(x)h(y) \quad (x,y\in\mathbb{R}) \tag{8}$$

or

$$h(x+y)=h(x)h(y) \quad (x,y\in\mathbb{C}). \tag{9}$$

For such solutions, more properties of complex numbers are needed (properties of $\mathbb{C}$ as a field).

As with every complex number, we can write h (in both cases) in the exponential form

$$h(x)=r(x)\exp(i\phi(x)), \tag{10}$$

where $r(x)\geqslant 0$ is the absolute value of $h(x)$ and $\phi(x)$ is also real valued, but determined modulo 2π if $r(x)\neq 0$. Most technical difficulties are caused by this lack of uniqueness. We consider (8) first, so $x\in\mathbb{R}$ in (10). Substituting (10) into (8), we get

$$r(x+y)\exp(i\phi(x+y))=r(x)r(y)\exp(i[\phi(x)+\phi(y)]).$$

The absolute values of both sides are equal, so

$$r(x+y)=r(x)r(y). \tag{11}$$

As we have seen in Chapter 3, either $r=0$ (thus also $h=0$) everywhere, or r is nowhere 0. In the latter case,

$$\exp(i\phi(x+y))=\exp(i[\phi(x)+\phi(y)]) \quad (x,y\in\mathbb{R}) \tag{12}$$

follows.

Now, the functional equation (11) is exactly (3.14), this time in the case (3.15). By Theorem 3.5, we have

$$r(x)=0 \quad (x\in\mathbb{R}) \quad \text{or} \quad r(x)=e^{f(x)} \quad (x\in\mathbb{R}),$$

where f is an arbitrary solution of Cauchy's equation (2.1). As to (12), it is equivalent to

$$\phi(x+y)\equiv\phi(x)+\phi(y) \pmod{2\pi} \quad \text{for all } x;y\in\mathbb{R}, \tag{13}$$

since the (complex) exponential function is periodic with period $2\pi i$ and thus ϕ is only determined up to a multiple of 2π. The general solution of (13) can be constructed in a similar way to that of (2.1) in Theorem 2.10 (van der Corput 1940*a*, *b*; Vietoris 1944). Anyway, we have the following:

Theorem 3. *The general solutions $h:\mathbb{R}\to\mathbb{C}$ of* (8) *are given by*

$$h(x)=0 \quad (x\in\mathbb{R}) \quad \textit{and} \quad h(x)=\exp(f(x)+i\phi(x)),$$

where $f:\mathbb{R}\to\mathbb{R}$ is an arbitrary solution of Cauchy's equation and ϕ is an arbitrary solution of (13).

The determination of the regular, say measurable, solutions is not so simple. We shall apply a very useful technique for obtaining very regular solutions when only integrability or measurability is assumed and which uses only basic facts on measurable and integrable functions. Of course, those of (11) are taken care of by Theorem 3.5 since, with h, r is also measurable. So we get

$$r(x) \equiv 0 \quad \text{or} \quad r(x) = e^{cx} \quad \text{for all } x \in \mathbb{R} \quad (c \in \mathbb{R} \text{ an arbitrary constant}). \tag{14}$$

The first solution gives $h(x) = 0$ everywhere. We now deal with the second. Let us look at (12). If h is measurable on an interval then so is $g: \mathbb{R} \to \mathbb{C}$, defined by

$$g(x) = \exp(i\phi(x)), \tag{15}$$

because $g(x) = e^{-cx}h(x)$. Moreover, since

$$|g(x)| = 1, \quad (x \in \mathbb{R}), \tag{16}$$

g is also bounded and thus (Lebesgue) integrable on the same interval (which can be supposed to be finite). With (15), equation (12) goes over into

$$g(x + y) = g(x)g(y) \quad (x, y \in \mathbb{R}). \tag{17}$$

It is obvious from (17) that g is integrable on *any* finite interval.

There exists a number $b \in \mathbb{R}$ such that

$$a = \int_0^b g(x)\,dx \neq 0. \tag{18}$$

Indeed, if, for every $t \in \mathbb{R}_+$, we had

$$\int_0^t g(y)\,dy = 0,$$

then $g(x) = 0$ almost everywhere on $\mathbb{R}_+$, due to a Lebesgue theorem (see, e.g., Rudin 1987, p. 30), in contradiction to (16). Integrate both sides of (17) with respect to y over $[0, b]$. Then by (18) we get, with the change of variables $t = x + y$,

$$g(x) = \frac{1}{a}\int_0^b g(x + y)\,dy = \frac{1}{a}\int_x^{x+b} g(t)\,dt = \frac{1}{a}\int_0^{x+b} g(t)\,dt - \frac{1}{a}\int_0^x g(t)\,\mathrm{d}t. \tag{19}$$

Since the (Lebesgue) integral as function of the upper limit is continuous, so the right hand side of (19) is continuous and so is the left hand side, that is, g is continuous. But the integral of a continuous function as a function of the upper limit is differentiable, thus the right hand side of (19) as a function of x is differentiable and thus g is differentiable too. Differentiate (17) with respect to y and put $y = 0$ in order to get

$$g'(x) = \delta g(x) \quad (x \in \mathbb{R}),$$

where $\delta = g'(0)$. The integration of this differential equation gives

$$g(x) = \gamma e^{\delta x} \tag{20}$$

where γ, δ are complex constants. Substitution of (20) into (17) gives either $\gamma = 0$, which contradicts (16), or $\gamma = 1$ and, by (16), $\delta = di$, where d is a real number, that is,

$$g(x) = e^{idx} \quad (x \in \mathbb{R}).$$

Putting this and (14) into (10), we obtain the following:

Theorem 4. *The general solutions $h: \mathbb{R} \to \mathbb{C}$ of* (8) *in the class of functions measurable on an interval are given by*

$$h(x) = 0 \quad (x \in \mathbb{R}) \quad \text{or} \quad h(x) = \exp((c + id)x) = e^{\alpha x} \quad (x \in \mathbb{R}),$$

where α is an arbitrary complex number.

All the above results would *remain essentially unchanged, if we took $\mathbb{R}_+$ instead of* $\mathbb{R}$.

We also notice that as distinct from the real valued case, Theorem 3.5, for complex valued solutions of (8) boundedness (cf. (16)), alone does not imply continuity.

It is perhaps worthwhile restating the result we obtained for (13). Suppose $\phi: \mathbb{R} \to \mathbb{R}$ is (Lebesgue) measurable and satisfies (13). Then there exists a real constant d such that, for all x in $\mathbb{R}$,

$$\phi(x) \equiv dx \pmod{2\pi}.$$

The proof is the one we already gave using $g(x) = e^{i\phi(x)}$ and noting that g is measurable if ϕ is measurable.

If ϕ were supposed continuous on $\mathbb{R}$, then we would get

$$\phi(x) = dx + 2n\pi$$

for a real constant d and some integer n (because a continuous integer valued function on $\mathbb{R}$ is constant).

Now we consider (9). In view of (10), we again have (11) and (12), this time for all $x, y \in \mathbb{C}$. Again, under addition the complex numbers behave like two dimensional vectors, so we have, by Propositions 4.1 and 4.4,

$$\begin{cases} r(x) = r(x_1 + ix_2) = r(x_1)r(ix_2), \\ e^{i\phi(x)} = g(x) = g(x_1 + ix_2) = g(x_1)g(ix_2). \end{cases}$$

The functions $x_1 \mapsto g(x_1)$, $x_2 \mapsto g(ix_2)$ have the same properties (16) and (17). So we have the following:

Theorem 5. *The general solutions $h: \mathbb{C} \to \mathbb{C}$ of* (9) *are given by*

$$\begin{aligned} h(x) = 0 \quad \text{and} \quad h(x) &= h(x_1 + ix_2) \\ &= \exp[f_1(x_1) + f_2(x_2) + i\phi_1(x_1) + i\phi_2(x_2)], \end{aligned}$$

where f_1 and f_2 are arbitrary solutions of Cauchy's equation (2.1) *and ϕ_1, ϕ_2*

are arbitrary real valued solutions of (13). *In particular, the general measurable solutions* $h:\mathbb{C}\to\mathbb{C}$ *of* (9) *are given by*

$$h(x)=0 \quad \textit{and} \quad h(x)=h(x_1+ix_2)=\exp(\gamma x_1+\delta x_2)=\exp(\alpha x+\beta\bar{x})$$

where α, β *(and* γ, δ*) are arbitrary constant complex numbers.*

5.2 Endomorphisms of the real and complex fields

The 'Pexider-type generalization' of the results in the previous section is again immediate. However, we would be wrong if we thought that, since both $\mathbb{R}$ and $\mathbb{C}$ are fields, $\mathbb{R}$ and $\mathbb{C}$ behave in the same way in regard to Cauchy's functional equations. We know that the application $x\mapsto\bar{x}$ from a complex number onto its conjugate is a nontrivial ($f(x)\not\equiv x, f(x)\not\equiv 0$) automorphism of the complex field. There are no such automorphisms or even endomorphisms for $\mathbb{R}$ (Darboux 1880). An *endomorphism* is a homomorphism of a structure into itself, just as an *automorphism* is an isomorphism of the structure onto itself. The identity and the identically zero mapping are *trivial endomorphisms.*

Proposition 6. *All field endomorphisms from* $\mathbb{R}$ *into* $\mathbb{R}$ *are trivial.*

Proof. A field endomorphism from $\mathbb{R}$ into $\mathbb{R}$ is a function $f:\mathbb{R}\to\mathbb{R}$ for which both of the following functional equations hold for every x, y in $\mathbb{R}$:

$$f(x+y)=f(x)+f(y), \tag{21}$$

$$f(xy)=f(x)f(y). \tag{22}$$

(If $f(x)\not\equiv 0$ then (22) implies $f(1)=1$ and $f(x^{-1})=f(x)^{-1}$.) Equation (22) provides us with $f(x^2)=f(x)^2$, that is, $f(t)\geqslant 0$ for all $t\geqslant 0$. So, by Corollary 2.5, there exists an $a\geqslant 0$ such that

$$f(x)=ax \quad \text{for all } x\in\mathbb{R}.$$

But then (22) implies either $a=1$ and so $f(x)\equiv x$ or $a=0$ and so $f(x)\equiv 0$. These are the trivial endomorphisms of $\mathbb{R}$ and Proposition 6 is proved.

In other words, the identity is the only automorphism (that is, bijective endomorphism) on $\mathbb{R}$.

The situation is different for $\mathbb{C}$. There even exist very weird automorphisms of $\mathbb{C}$, as we shall now see. To construct an endomorphism $\mathbb{C}\to\mathbb{C}$ we have to construct a set which behaves like a Hamel basis for $\mathbb{C}$, both for the additive and the multiplicative operations. Somewhat more is required in order to get a bijection. We will use algebra and set theory a little more here than in other parts of the book.

We prove the existence of a subset T of $\mathbb{C}$ with the following properties:

(A) For any $n \geqslant 1$, any polynomial of n variables with rational coefficients, which is equal to zero for n different elements of T, is identically zero.
(B) Every complex number is an algebraic number over the field generated (in $\mathbb{C}$) by the set T (that is, for every complex number z there exists a not identically zero polynomial P, with coefficients in the field $R(T)$ generated by T, such that $P(z)=0$).
(C) There exist at least two points x_1 and x_2 in T such that $x_1 \neq x_2$ and $x_1 \neq \bar{x}_2$.

We first consider nonempty subsets S of $\mathbb{C}$ having properties (A) and (C) and the following property (compare it to (A)):

(D) No element z of S is an algebraic number over the field generated in $\mathbb{C}$ by all other elements of S.

There exist x_1, x_2 in $\mathbb{C}$ such that $S_0=\{x_1, x_2\}$ has the properties (A), (C) and (D) (e.g. $x_1=1$, $x_2=\pi i$). So S_0 itself is also a set S as described above. We now order, by inclusion, the family $\mathscr{S}$ of all subsets S containing S_0. Applying Hausdorff's maximality theorem, there exists a maximal totally ordered subset $\mathscr{S}^1$ of $\mathscr{S}$. Then define T to be the union of all subsets S^1 of $\mathscr{S}^1$. This T clearly has the properties (C) and (D). Property (A) involves only a finite number of elements of T each time. Therefore, as these elements belong to some S^1 of $\mathscr{S}^1$, property (A) is clearly satisfied for T. Let us now prove that (B) is also satisfied for T. Were this not so, there would exist a z_0 in $\mathscr{C}$ and, for every nonidentically zero polynomial P with coefficients in $R(T)$, we would get $P(z_0) \neq 0$. Consider the set

$$S = T \cup \{z_0\}.$$

It is not difficult to see directly that S satisfies the properties (A), (C) and (D). But this result contradicts the maximality of the family $\mathscr{S}^1$ with respect to these properties. Therefore we have proved by contradiction that T has the property (B).

Let $f: T \to T$ be a bijection so that $f(x_1)=x_2$ where x_1 and x_2 are two distinct elements of T such that $x_1 \neq \bar{x}_2$ (property (C)). Let $z \in R(T)$. There exist two polynomials P and Q, in several variables, with rational coefficients, such that

$$z = P(z_1, \ldots, z_n)/Q(z_1, \ldots, z_n) \text{ where } z_1, \ldots, z_n \text{ are different elements of } T. \tag{23}$$

We define

$$g(z) = \frac{P(f(z_1), \ldots, f(z_n))}{Q(f(z_1), \ldots, f(z_n))}. \tag{24}$$

Using property (A), we notice that g is then unambiguously defined by (24) from (23) as a function on $R(T)$, taking its values in $R(T)$. Naturally, g extends

f to $R(T)$. Moreover, g is a ring endomorphism of $R(T)$, in the sense that g satisfies (21) and (22), as can be seen from (23) and (24). The function g is an injection (see (24) and (A)). For any $z' = P(z'_1, \ldots, z'_n)/Q(z'_1, \ldots, z'_n)$ in $R(T)$, define $z = P(f^{-1}(z'_1), \ldots, f^{-1}(z'_n))/Q(f^{-1}(z'_1), \ldots, f^{-1}(z'_n))$. We notice that $g(z) = z'$. In other words, g is a surjection onto $R(T)$. To summarize, g is an automorphism of $R(T)$.

Let $h: H \to I$ be an isomorphism on a subfield H of $\mathbb{C}$ containing $R(T)$, extending the automorphism g. The family $\mathscr{H}$ of all such extensions h of g is partially ordered by extension (inclusion). Using Hausdorff's maximality theorem there exists a maximal totally ordered subfamily $\mathscr{H}'$ of $\mathscr{H}$. Define j to be the union of $\mathscr{H}'$ and J the union of the subfields contained in $\mathscr{H}'$ as domains. Then j is an extension of g, isomorphically mapping J onto a subfield $J' \subset \mathbb{C}$.

Let us prove by contradiction that $J = \mathbb{C}$. If not, let z_0 be in $\mathbb{C} \backslash J$. Due to the property (B) and to $R(T) \subset J$, this z_0 is algebraic over J. Let $P(z) = z^m + a_1 z^{m-1} + \cdots + a_m$ be the unique irreducible monic polynomial with coefficients in J such that $P(z_0) = 0$. Choose z' in $J' \subseteq \mathbb{C}$ such that

$$0 = z'^m + j(a_1) z'^{m-1} + \cdots + j(a_m).$$

For an x in $\mathbb{C}$ with $x = Q(z_0) = \sum_{k=0}^{n} \alpha_k z_0^k$, where the α_k's belong to J, we define

$$h(x) = \sum_{k=0}^{n} j(\alpha_k)(z')^k.$$

If Q_1 and Q_2 are two polynomials with coefficients in J such that $Q_1(z_0) = Q_2(z_0)$, then $Q_1 - Q_2$ is a multiple of P (since P is irreducible). This shows that h is unambiguously defined. Moreover, h is a homomorphism into $\mathbb{C}$, defined over the field J'' generated by J and z_0. It extends j. Therefore h is an isomorphism from J'' onto a subfield of $\mathbb{C}$. But the existence of h contradicts the maximality of j. We have thus proved $J = \mathbb{C}$.

Now, every polynomial with coefficients in J' must have all its roots in J', since J' is the isomorphic image of $J = \mathbb{C}$ and since $\mathbb{C}$ has this property (fundamental theorem of algebra). However, J' contains $g(R(T)) = R(T)$. Using property (B), we deduce that $J' = \mathbb{C}$. In other words, j is an automorphism of $\mathbb{C}$.

This automorphism is not the identity as $j(x_1) = g(x_1) = f(x_1) = x_2$, and is not the conjugate mapping as $x_2 \neq \bar{x}_1$. Using Proposition 2, we see that the automorphism j is nowhere continuous. We have proved the following:

Theorem 7. *There exists a nowhere continuous automorphism of* $\mathbb{C}$.

Theorem 7 was conjectured by C. Segre (1889*a, b*), almost proved by Lebesgue (1907*a*) and completely established by B. Segre (1947) and Kestelman (1951).

5.3 Bohr groups

The general solutions of (8), for $f:\mathbb{R}\to\mathbb{C}$, play an interesting role in harmonic analysis. Here we shall see an example of the Bohr group. The rudiments of harmonic analysis will be used in this section.

A trigonometric polynomial f, not identically zero, is a complex valued function defined for all real values by

$$f(x)=\sum_{j=1}^{n} c_j \exp(i\lambda_j x) \tag{25}$$

where $c_1, c_2, \ldots, c_n$ are nonzero complex numbers; $\lambda_1, \lambda_2, \ldots, \lambda_n$ are different real numbers and n is a positive integer.

It should be noticed that the representation (25) is unique. For the sake of simplicity, let us introduce a convenient classical notation, the uniform norm:

$$\|f\|_\infty = \sup_{x\in\mathbb{R}} |f(x)|.$$

From (25) we deduce that

$$\|f\|_\infty \leqslant \sum_{j=1}^{n} |c_j|,$$

that is, with this norm all trigonometric polynomials are bounded.

Let $\mathscr{P}$ be the complex linear space of all trigonometric polynomials (including the zero polynomial). It is clear that the infinite dimensional space $\mathscr{P}$ is an algebra because the pointwise product of two trigonometric polynomials is also a trigonometric polynomial. In order to get a better understanding of the structure of $\mathscr{P}$, it is convenient, as usual in functional analysis, to consider the set $\mathscr{B}$ of all functionals $\phi:\mathscr{P}\to\mathbb{C}$ with the following properties:

(a)	$\phi(\lambda f+\mu g)=\lambda\phi(f)+\mu\phi(g)$	for all λ, μ in $\mathbb{C}$ and for all f, g in $\mathscr{P}$,
(b)	$\phi(fg)=\phi(f)\phi(g)$	for all f, g in $\mathscr{P}$,
(c)	$\phi(1)=1$	(1 is the constant function identically equal to 1),
(d)	$|\phi(f)|\leqslant \|f\|_\infty$	for all f in $\mathscr{P}$.

(cf., e.g., Rudin 1973, pp. 268–75, 1987, pp. 364–65). The set $\mathscr{B}$, the *spectrum* of $\mathscr{P}$, is not empty and even contains a copy of the set $\mathbb{R}$ of all real numbers. Let x be any real number and define $\hat{x}$ by

$$\hat{x}(f)=f(x).$$

Clearly, $\hat{x}$ belongs to $\mathscr{B}$. Moreover, for two different x_1, x_2, the $\hat{x}_1$ and $\hat{x}_2$ are different functionals over $\mathscr{P}$.

However, the set $\mathscr{B}$ contains more elements than this simple copy of $\mathbb{R}$. To see this, we shall characterize $\mathscr{B}$. We introduce a notation. By $e_\lambda(\lambda\in\mathbb{R})$

we denote the complex exponential:

$$e_\lambda(x) = e^{i\lambda x}.$$

Let ϕ be an element of $\mathscr{B}$ and consider the action of ϕ over all e_λ where λ runs through $\mathbb{R}$. We define $\Phi:\mathbb{R}\to\mathbb{C}$ by

$$\Phi(\lambda) = \phi(e_\lambda).$$

Using $e_\lambda \cdot e_\mu = e_{\lambda+\mu}$ and (b), we deduce

$$\Phi(\lambda+\mu) = \Phi(\lambda)\Phi(\mu) \quad (\lambda, \mu \in \mathbb{R}), \tag{26}$$

which amounts to (8) where Φ is again complex valued.

From (d) we get information concerning the bound for Φ,

$$|\Phi(\lambda)| \leqslant 1 \quad (\lambda \in \mathbb{R}),$$

and from (c), $\Phi(0) = 1$. We may sharpen this bound using (26) with $\lambda = -\mu$. It gives $\Phi(-\lambda)\Phi(\lambda) = 1$. Therefore, for all λ in $\mathbb{R}$,

$$|\Phi(\lambda)| = 1. \tag{27}$$

(Also $\Phi(-\lambda) = \overline{\Phi(\lambda)}$ and $\phi(\bar{f}) = \overline{\phi(f)}$ are easily proved.)

Conversely, let $\Phi:\mathbb{R}\to\mathbb{C}$ satisfy both (26) and (27). We shall build a ϕ from this Φ. We define ϕ on the space $\mathscr{P}$ in view of (25) by

$$\phi(f) = \sum_{j=1}^{n} c_j \Phi(\lambda_j) \quad \text{for } f = \sum_{j=1}^{n} c_j e_{\lambda_j}$$

and $\phi(0) = 0$. This ϕ is a well defined linear functional on $\mathscr{P}$. We also prove a multiplication property, first with any e_μ and any e_λ, using (26):

$$\begin{aligned}\phi(e_\mu \cdot e_\lambda) = \phi(e_{\mu+\lambda}) &= \Phi(\mu+\lambda) \\ &= \Phi(\mu)\Phi(\lambda) \\ &= \phi(e_\mu)\phi(e_\lambda).\end{aligned}$$

Then, by the linearity of ϕ over $\mathscr{P}$, we deduce that, for all f in $\mathscr{P}$ and any e_λ,

$$\phi(fe_\lambda) = \phi(f)\phi(e_\lambda).$$

Therefore, again by the linearity of ϕ over $\mathscr{P}$, we get for all f, g in $\mathscr{P}$,

$$\phi(fg) = \phi(f)\phi(g).$$

Clearly, also $\phi(1) = 1$. So we have proved (a), (b) and (c) for our ϕ. It remains to establish (d). We shall make use of a result from number theory:

Theorem 8. *Let $z_1, z_2, \ldots, z_n$ be n complex numbers of modulus one. Let $\mu_1, \mu_2, \ldots, \mu_n$ be n real and distinct numbers. Suppose that, for any n integers $d_1, \ldots, d_n$ satisfying*

$$\sum_{j=1}^{n} d_j \mu_j = 0,$$

we have the equation

$$\prod_{j=1}^{n} z_j^{d_j} = 1.$$

Then, for any $\varepsilon > 0$, *there exists a real number* x *such that*

$$|e_{\mu_j}(x) - z_j| \leqslant \varepsilon \quad j = 1, 2, \ldots, n.$$

Proof. Define $f(t) = 1 + \sum_{i=1}^{n} \bar{z}_j e_{\mu_j}(t)$ and let p be a positive integer. We may also write $f(t)^p$ as a trigonometric polynomial

$$f(t)^p = \sum_{l=1}^{N} c_l e_{\lambda_l}(t),$$

where the λ_l are distinct. In order to compute the coefficients c_l, we may use the following easy evaluation of the limit of an integral mean value:

$$\lim_{T \to \infty} \frac{1}{T} \int_0^T f(x)^p e_\lambda(x)\, dx = \begin{cases} 0 & \text{if } \lambda \neq \lambda_l, \quad l = 1, \ldots, N \\ c_l & \text{if } \lambda = \lambda_l \quad \text{for some } l = 1, \ldots, N. \end{cases}$$

Our purpose is to find a good majorization for f. We clearly have

$$|f(x)| \leqslant n + 1 \quad (x \in \mathbb{R}).$$

So, if for some nonnegative α (possibly only for $\alpha = 0$),

$$|f(x)| \leqslant (n+1)e^{-\alpha},$$

then

$$\left| \lim_{T \to \infty} \frac{1}{T} \int_0^T f(x)^p e_\lambda(x) dx \right| \leqslant (n+1)^p e^{-p\alpha}.$$

Therefrom we deduce the inequality

$$|c_l| \leqslant (n+1)^p e^{-p\alpha} \quad \text{for } l = 1, 2, \ldots, N.$$

Due to the multinomial theorem we have

$$(1 + x_1 + \cdots + x_n)^p = \sum_{0 \leqslant k_1 + \cdots + k_n \leqslant p} a_{k_1, \ldots, k_n} x_1^{k_1} \cdots x_n^{k_n},$$

where the k_i's are nonnegative integers and $a_{k_1, \ldots, k_n} > 0$. Also, as one sees by substituting $x_1 = x_2 = \cdots = x_n = 1$,

$$\sum_{0 \leqslant k_1 + \cdots + k_n \leqslant p} a_{k_1, \ldots, k_n} = (n+1)^p$$

holds. Thus another expression is obtained for $f(t)^p$:

$$f(t)^p = \sum_{0 \leqslant k_1 + \cdots + k_n \leqslant p} a_{k_1, \ldots, k_n} e_k(t) \overline{\prod_{j=1}^{n} z_j^{k_j}},$$

where

$$k = \sum_{j=1}^{n} k_j \mu_j.$$

If $\sum_{j=1}^{n} k_j \mu_j = \sum_{j=1}^{n} k'_j \mu_j$, the hypotheses in Theorem 8 on the μ_j's and the z_j's imply

$$\overline{\prod_{j=1}^{n} z_j^{k_j}} = \overline{\prod_{j=1}^{n} z_j^{k'_j}}.$$

From this we deduce an equality between the coefficients c_l and some sums of the $a_{k_1,\dots,k_n}$:

$$|c_l| = \sum_{\substack{0 \leqslant k_1 + \cdots + k_n \leqslant p \\ \sum_{j=1}^{n} k_j \mu_j = \lambda_l}} a_{k_1,\dots,k_n}.$$

Thus

$$\sum_{\substack{0 \leqslant k_1 + \cdots + k_n \leqslant p \\ \sum_{j=1}^{n} k_j \mu_j = \lambda_l}} a_{k_1,\dots,k_n} \leqslant (n+1)^p e^{-p\alpha}.$$

As

$$(n+1)^p = \sum_{l=1}^{N} \left(\sum_{\substack{0 \leqslant k_1 + \cdots + k_n \leqslant p \\ \sum_{j=1}^{n} k_j \mu_j = \lambda_l}} a_{k_1,\dots,k_n} \right),$$

we obtain

$$e^{p\alpha} \leqslant N.$$

The integer N measures the number of distinct values of $\sum_{j=1}^{n} k_j \mu_j$ for integers k_j such that $0 \leqslant k_1 + \cdots + k_n \leqslant p$. The maximum value for N is obtained when the μ_j are linearly independent over $\mathbb{Q}$. In this case, as each k_j has $(p+1)$ possible values, we get

$$N = (p+1)^n.$$

Thus, in general,

$$e^{p\alpha} \leqslant (p+1)^n.$$

The integer p is variable. Using the comparison of an exponential and a power, we have that $\alpha = 0$. So

$$\sup_{t \in \mathbb{R}} |f(t)| = \sup_{t \in \mathbb{R}} \left| 1 + \sum_{j=1}^{n} \bar{z}_j e_{\mu_j}(t) \right| = n + 1.$$

But $|\bar{z}_j e_{\mu_j}(t)| = 1$ and

$$|f(t)| = \left| 1 + \sum_{j=1}^{n} \bar{z}_j e_{\mu_j}(t) \right| \leqslant 1 + \sum_{j=1}^{n} |\bar{z}_j| \, |e_{\mu_j}(t)| = n + 1$$

with equality exactly when all terms are 1. Therefore $|f(t)|$ can be close to its supremum only when each term in $1 + \sum_{j=1}^{n} \bar{z}_j e_{\mu_j}(t)$ is close to 1 and thus $f(t)$ is close to $n+1$. In other words, for every $\varepsilon > 0$, there exists an $x \in \mathbb{R}$ such that

$$|f(x) - (n+1)| \leqslant \varepsilon.$$

Thus

$$\left|\sum_{j=1}^{n}(\bar{z}_j e_{\mu_j}(x)-1)\right|\leqslant\varepsilon.$$

But $|\bar{z}_j e_{\mu_j}(x)|=1$, and so

$$\left|\sum_{j=1}^{n}(\bar{z}_j e_{\mu_j}(x)-1)\right|\geqslant|\bar{z}_j e_{\mu_j}(x)-1|.$$

As $|z_j|=1$, we finally obtain, for all $j=1,\ldots,n$,

$$|e_{\mu_j}(x)-z_j|\leqslant\varepsilon.$$

This concludes the proof of Theorem 8.

Remark. The above proof follows an idea of Bohr (1934). Theorem 8, with linearly independent μ_j's over $\mathbb{Z}$, was first proved by Kronecker (1884). There are other proofs in the geometrical or arithmetical vein. See Hardy and Wright (1979, pp. 375–93) for two more proofs and for further references. Cf. also Venkov (1970, pp. 34–6) and Favard (1933, pp. 18–21).

We shall apply Theorem 8 with $z_j=\Phi(\lambda_j)$ $(j=1,\ldots,n)$ and with the real values $\mu_1=\lambda_1,\ \mu_2=\lambda_2,\ldots,\mu_n=\lambda_n$. From the equation

$$\sum_{j=1}^{n}d_j\lambda_j=0,\quad \text{with integers } d_1,d_2,\ldots,d_n,$$

we immediately get

$$1=\Phi(0)=\prod_{j=1}^{n}\Phi(\lambda_j)^{d_j}=\prod_{j=1}^{n}z_j^{d_j}.$$

Therefore the condition in Theorem 8 is satisfied. Thus, for any $\varepsilon>0$, we can find an x in $\mathbb{R}$ such that

$$|e_{\lambda_j}(x)-\Phi(\lambda_j)|\leqslant\varepsilon\quad (j=1,\ldots,n).$$

In other words,

$$|\phi(f)|\leqslant\sum_{j=1}^{n}|c_j|\varepsilon+\left|\sum_{j=1}^{n}c_j e_{\lambda_j}(x)\right|,$$

or

$$|\phi(f)|\leqslant\|f\|_\infty+\varepsilon\sum_{j=1}^{n}|c_j|.$$

But, as ε is arbitrary, this inequality leads exactly to (d). We have thus proved the following:

Proposition 9. *A complex valued linear functional ϕ on $\mathscr{P}$ belongs to $\mathscr{B}$ (that is, satisfies* (a), (b), (c) *and* (d)) *if and only if $\Phi(\lambda)=\phi(e_\lambda)$ satisfies both* (26) *and* (27).

If we take $\hat{x}$, an element of $\mathscr{B}$, obtained from an element x of $\mathbb{R}$ by $\hat{x}(f) = f(x)$, we notice that the associated Φ is given by

$$\Phi(\lambda) = e^{i\lambda x} \quad (x \in \mathbb{R});$$

which is a continuous function on $\mathbb{R}$.

However, we have already investigated the existence of discontinuous solutions of both (26) and (27). We take $\Phi(\lambda) = e^{ih(\lambda)}$ where $h: \mathbb{R} \to \mathbb{R}$ is a discontinuous solution of Cauchy's functional equation on $\mathbb{R}$ (cf. Theorem 3). And these discontinuous solutions, as proved in Proposition 9, still lead to elements of $\mathscr{B}$. Thus the copy of $\mathbb{R}$, mentioned above, does not exhaust $\mathscr{B}$.

Now we shall prove that $\mathscr{B}$ can be equipped with an operation $*$, for which $\mathscr{B}$ is an abelian group, and such that $*$, restricted to the injected image of $\mathbb{R}$ in $\mathscr{B}$, is a copy of the ordinary addition on $\mathbb{R}$.

Let ϕ_1 and ϕ_2 be in $\mathscr{B}$ and define the action of $\phi_1 * \phi_2$ first on e_λ, for any λ in $\mathbb{R}$:

$$\phi_1 * \phi_2(e_\lambda) = \Phi_1(\lambda)\Phi_2(\lambda), \tag{28}$$

where $\Phi_1(\lambda) = \phi_1(e_\lambda)$, $\Phi_2(\lambda) = \phi_2(e_\lambda)$. We then notice that the mapping $\lambda \mapsto \phi_1 * \phi_2(e_\lambda)$ satisfies both (26) and (27). Proposition 9 proves that such a mapping leads to an element of $\mathscr{B}$, an element which we denote by $\phi_1 * \phi_2$.

Using equation (28), we easily deduce that $*$ is an associative and commutative binary operation on $\mathscr{B}$. If ϕ in $\mathscr{B}$ and $\hat{0}$ is the element in $\mathscr{B}$ coming from $x = 0$ in $\mathbb{R}$, we compute

$$\phi * \hat{0} = \hat{0} * \phi = \phi,$$

as $\hat{0}(e_\lambda) = 1$ for all λ. In the same way, for any ϕ in $\mathscr{B}$, consider the mapping $\lambda \mapsto \overline{\phi(e_\lambda)}$. It gives rise, using Proposition 9, to an element ϕ_1 in $\mathscr{B}$. This element satisfies

$$\phi_1 * \phi = \phi * \phi_1 = \hat{0}.$$

Thus $(\mathscr{B}, *)$ is an *abelian group.*

Now let x, y be elements of $\mathbb{R}$. Let us consider $\hat{x} * \hat{y}$. We get

$$\begin{aligned} \hat{x} * \hat{y}(e_\lambda) &= e_\lambda(x)e_\lambda(y) \\ &= e_\lambda(x + y) \\ &= \widehat{x + y}(e_\lambda). \end{aligned}$$

In other words, $\hat{x} * \hat{y} = \widehat{x + y}$ and so $*$ extends $+$ to $\mathscr{B}$; $(\mathscr{B}, *)$ is called the Bohr group of $\mathbb{R}$. A function $\Phi: \mathbb{R} \to \mathbb{C}$ satisfying (26) and (27) is called a *character* of the group $\mathbb{R}$. These are the elements of the Bohr group of $\mathbb{R}$.

In order to go further, some topology is needed here. It can be proved that there exists some appropriate topology on $\mathscr{B}$ with respect to which $(\mathscr{B}, *)$ is a *compact topological abelian group.* As we will see in Section 10.5, there exists a continuous isomorphism from $\mathbb{R}$ onto a dense subset of $(\mathscr{B}, *)$.

It should be noted that the set $\mathscr{P}$ is not closed under the uniform norm.

Its closure constitutes an algebra $\mathscr{A}$ of continuous complex valued functions, the algebra of so-called *Bohr almost periodic functions*. (The set $\mathscr{B}$ will have on $\mathscr{A}$ the same properties (a), (b), (c) and (d) as it has on $\mathscr{P}$.) Cf. Weil 1940, pp. 130–5; Rudin 1962, pp. 30–2; Hewitt–Ross 1963, pp. 245–61; and Section 10.5 for more details and references.

The procedure of constructing a compact group (the Bohr group) from $\mathbb{R}$ using all characters of $\mathbb{R}$ is quite general and can be applied to any locally compact abelian group, using its characters, that is, functions satisfying (26) and (27). This technique is typical in abstract harmonic analysis.

5.4 Recursive entropies

We now return to the method we applied in proving Theorem 4. We used the hypothesis that a solution of a functional equation is integrable, in order to deduce its differentiability, and then proceeded more easily to the determination of the general (integrable) solution. This method can be widely applied (see, e.g., Kac 1937; Aczél 1960*b*), as will also be seen in Chapter 14. As a further example, we shall determine here the integrable solutions of a functional equation which arises from the axiomatic theory of information measures.

About the amount of information to be expected from a complete system of events (say all possible outcomes of an experiment), with probabilities p_1, $p_2, \ldots, p_n$, which we suppose to be positive and which satisfy

$$\sum_{k=1}^{n} p_k = 1 \quad (k = 1, 2, \ldots, n; n \geqslant 2),$$

of course, it is traditional to assume the following (and more). The amount of information (as after Proposition 3.4) *depends only upon the probabilities.* We denote this amount by $H_n(p_1, p_2, \ldots, p_n)$ and suppose that H_n is a symmetric function (the information does not depend upon the labelling of the different possible outcomes of the experiment). We also suppose the following.

If an event (outcome) is split into two events, the amount of information for the new experiment can be determined in the following way from that of the original one:

$$\begin{aligned} H_n(p_1, p_2, p_3, \ldots, p_n) &= H_{n-1}(p_1 + p_2, p_3, \ldots, p_n) \\ &\quad + (p_1 + p_2) H_2\left(\frac{p_1}{p_1 + p_2}, \frac{p_2}{p_1 + p_2}\right) \\ &\quad \left(p_k > 0, \sum_{k=1}^{n} p_k = 1; n = 3, 4, \ldots\right). \end{aligned} \tag{29}$$

This supposition is called *recursivity* for obvious reasons (H_n can be determined for all $n \geqslant 2$, if H_2 is known). We prove the following (for somewhat stronger theorems see Tverberg 1958 and Aczél–Daróczy 1975, pp. 84–93).

Theorem 10. *The general solution of* (29) *in the class of functions, where* H_3 *is symmetric and* $x \mapsto H_2(1-x, x)$ *is integrable on every closed subinterval of* $]0,1[$, *is given by*

$$H_n(p_1, p_2, \ldots, p_n) = c \sum_{k=1}^{n} p_k \log p_k, \quad \textit{whenever} \sum_{k=1}^{n} p_k = 1$$
$$(p_k > 0; k = 1, 2, \ldots, n; n = 2, 3, \ldots). \tag{30}$$

In (30), c is an arbitrary constant. With the traditional choice of unit (commonly called a 'bit'),

$$H_2(\tfrac{1}{2}, \tfrac{1}{2}) = 1,$$

one gets $c = -1/\log 2$ or

$$H_n(p_1, p_2, \ldots, p_n) = -\sum_{k=1}^{n} p_k \log_2 p_k \tag{31}$$

the *Shannon entropy*.

We have supposed $p_k > 0$ at the beginning because $\log p_k$ is not defined for $p_k = 0$ and neither is the last term in (29) for $p_1 = p_2 = 0$. However, a similar theorem is true if we allow $p_k = 0$ and define $0 \log 0 = 0$, and $0\, H_2(\tfrac{0}{0}, \tfrac{0}{0}) = 0$.

Proof. Define a function $f:]0,1[\to \mathbb{R}$ by

$$f(x) = H_2(1-x, x) \quad (x \in]0,1[). \tag{32}$$

In the above statement of Theorem 10, f is supposed to be *integrable on every closed subinterval of* $]0,1[$ (that is, locally integrable on $]0,1[$). By the symmetry of H_3, and by (29) for $n = 3$, we see that H_2 is symmetric too, so (32) gives

$$f(x) = f(1-x) \quad \textit{for all } x \in]0,1[. \tag{33}$$

Again by the symmetry of H_3, by (32) and by (29) for $n = 3$, we also get

$$\begin{aligned} f(x) + (1-x) f\left(\frac{y}{1-x}\right) &= H_2(1-x, x) + (1-x) H_2\left(\frac{1-x-y}{1-x}, \frac{y}{1-x}\right) \\ &= H_3(1-x-y, y, x) = H_3(1-x-y, x, y) \\ &= H_2(1-y, y) + (1-y) H_2\left(\frac{1-x-y}{1-y}, \frac{x}{1-y}\right) \\ &= f(y) + (1-y) f\left(\frac{x}{1-y}\right), \end{aligned}$$

that is,

$$f(x)+(1-x)f\left(\frac{y}{1-x}\right)=f(y)+(1-y)f\left(\frac{x}{1-y}\right) \quad \text{whenever } x, y, x+y\in]0,1[. \tag{34}$$

Equation (34) is the *fundamental equation of information.*

Choose x arbitrarily in $]0,1[$ and let α, β be such that

$$0<\alpha<\beta<1-x \quad (<1). \tag{35}$$

Then, whenever $y\in[\alpha,\beta]\subset]0,1[$, we also have

$$0<\frac{x}{1-\alpha}\leqslant\frac{x}{1-y}\leqslant\frac{x}{1-\beta}<1 \quad \text{and} \quad 0<\frac{\alpha}{1-x}\leqslant\frac{y}{1-x}\leqslant\frac{\beta}{1-x}<1. \tag{36}$$

By supposition, f is integrable between the limits in (35) and (36). Integrating (34) with respect to y from α to β yields

$$\begin{aligned}(\beta-\alpha)f(x)=\int_\alpha^\beta f(x)\,dy&=\int_\alpha^\beta f(y)\,dy+\int_\alpha^\beta(1-y)f\left(\frac{x}{1-y}\right)dy\\ -(1-x)\int_\alpha^\beta f\left(\frac{y}{1-x}\right)dy&=\int_\alpha^\beta f(y)\,dy+x^2\int_{x/(1-\alpha)}^{x/(1-\beta)}\frac{1}{s^3}f(s)ds\\ &-(1-x)^2\int_{\alpha/(1-x)}^{\beta/(1-x)}f(t)\,dt.\end{aligned} \tag{37}$$

Here we made the substitution

$$s=\frac{x}{1-y}, \quad t=\frac{y}{1-x}. \tag{38}$$

Again, since f was supposed to be integrable, the right hand side depends continuously upon x, therefore so does the left hand side and f is continuous on $]0,1[$. But, then, the right hand side, and with it f on the left, is differentiable on $]0,1[$. Then again, the right hand side, and with it f on the left, is twice differentiable.

Let us differentiate (34), first with respect to x, and then the resulting equation

$$f'(x)-f\left(\frac{y}{1-x}\right)+\frac{y}{1-x}f'\left(\frac{y}{1-x}\right)=f'\left(\frac{x}{1-y}\right)$$

with respect to y in order to obtain

$$\frac{y}{(1-x)^2}f''\left(\frac{y}{1-x}\right)=\frac{x}{(1-y)^2}f''\left(\frac{x}{1-y}\right) \tag{39}$$

whenever

$$0<x<1, \quad 0<y<1, \quad x+y<1. \tag{40}$$

We again introduce new variables by (38). From (36) we see that $s\in]0,1[$,

$t\in]0,1[$. Conversely, if $s\in]0,1[$ and $t\in]0,1[$ are chosen arbitrarily, we can find x and y so that both (38) and (40) are satisfied. These x and y are, of course,

$$x=\frac{s-st}{1-st},\quad y=\frac{t-st}{1-st}.$$

So (39) goes over into

$$t(1-t)f''(t)=s(1-s)f''(s)=c\ (\text{constant})\quad \text{for all } s\in]0,1[,t\in]0,1[. \tag{41}$$

By integrating $t(1-t)f''(t)=c$ twice, we get

$$f(t)=c[t\log t+(1-t)\log(1-t)]+at+b.$$

If we substitute this into (33) and (34) we get $a=0$ and $b=0$, respectively, and we see that *the general integrable solution of* (33) *and* (34) *is given by*

$$f(t)=c[t\log t+(1-t)\log(1-t)]\quad \text{for all } t\in]0,1[. \tag{42}$$

Taking (29) and (32) into consideration, we get (30) by induction for all $n\geqslant 2$. The rest of Theorem 10 is obvious.

In many (but not all) respects the fundamental equation (34) of information behaves similarly to Cauchy's equation (2.2) (cf. Aczél–Daróczy 1975, pp. 79–102; Diderrich 1975, 1978, 1986; Daróczy–Maksa 1977; Maksa 1980).

Exercises and further results

1. (Robinson 1957.) Show that the only entire (everywhere analytic) functions $f:\mathbb{C}\to\mathbb{C}$ which satisfy

$$|f(z)|=|f(x)+f(iy)|\quad \text{for all } z=x+iy\in\mathbb{C}$$

are

$$f(z)=az,\quad f(z)=a\sin(bz)\quad \text{and}\quad f(z)=a\sinh(bz),$$

a being a complex constant and b a real one.

2. (Hille 1923.) Show that the only entire functions $f:\mathbb{C}\to\mathbb{C}$ satisfying

$$|f(z)|^2=|f(x)|^2+|f(iy)|^2\quad \text{for all } z=x+iy\in\mathbb{C}$$

are

$$f(z)=az,\quad f(z)=a\sin(bz)\quad \text{and}\quad f(z)=a\sinh(bz),$$

a being a complex constant and b a real one.

(For Exercises 3 and 4 see Haruki 1968*b*.)

3*. Show that the only entire functions $f:\mathbb{C}\to\mathbb{C}$ satisfying

$$|f(z_1+z_2)+f(z_1-z_2)|=|f(z_1+\bar{z}_2)+f(z_1-\bar{z}_2)|\,(z_1,z_2\in\mathbb{C})$$

are

$$f(z)=az+b,\quad f(z)=a\sin(cz)+b\cos(cz),\quad f(z)=a\sinh(cz)+b\cosh(cz),$$

a and b being complex constants and c a real one.

4*. Show that the only entire functions $f:\mathbb{C}\to\mathbb{C}$ satisfying

$$|f(z_1+z_2)|^2+|f(z_1-z_2)|^2\leqslant|f(z_1+\bar{z}_2)|^2+|f(z_1-\bar{z}_2)|^2\quad (z_1,z_2\in\mathbb{C})$$

are

$$f(z) = a\sinh(cz) + b\cosh(cz) \quad \text{or} \quad f(z) = az + b,$$

where $\operatorname{Re} c \geqslant 0$, $\operatorname{Im} c \geqslant 0$, but a, b, c are otherwise arbitrary complex constants.

5. Show that the image of $\mathbb{R}$ by a nowhere continuous automorphism of $\mathbb{C}$ is everywhere dense in $\mathbb{C}$.
6. In Theorem 7 a proof was given for the existence of a nontrivial *automorphism* of $\mathbb{C}$. Find a simpler proof for the existence of a nontrivial *endomorphism* of $\mathbb{C}$ (that is, it is not required to prove that the mapping is surjective (onto)).
7. Does there exist a nontrivial endomorphism of $\mathbb{C}$ into $\mathbb{C}$ which is continuous at a point P along a straight line going through P?
8. (E.g. Aczél–Daróoczy 1975, pp. 179–82.) Find all solutions $f:]0,1[\to \mathbb{R}$ of

$$f(x) + (1-x)^{\alpha} f\left(\frac{y}{1-x}\right) = f(y) + (1-y)^{\alpha} f\left(\frac{x}{1-y}\right),$$

where $\alpha \neq 1$. (Notice that no regularity is supposed.)

(For Exercises 9–11 see Maksa 1981*c*.)

9*. Find all continuous $f:]0,1[\to \mathbb{R}$ such that

$$f(1-x) + (1-x) f\left(\frac{y}{1-x}\right) = f(y) + (1-y) f\left(1 - \frac{x}{1-y}\right)$$

whenever $x, y, x+y \in]0,1[$.

10. Prove that all $f:]0,1[\to \mathbb{R}$ such that

$$f(1-x) + (1-x)^{\alpha} f\left(\frac{y}{1-x}\right) = f(y) + (1-y)^{\alpha} f\left(1 - \frac{x}{1-y}\right)$$

whenever $x, y, x+y \in]0,1[$ $(\alpha \neq 0, 1, 2)$ are of the form

$$f(x) = b(x^{\alpha} + (1-x)^{\alpha} - 1) \quad (x \in]0,1[),$$

where b is an arbitrary real constant.

11*. Find all $f:]0,1[\to \mathbb{R}$

$$f(1-x) + (1-x)^2 f\left(\frac{y}{1-x}\right) = f(y) + (1-y)^2 f\left(1 - \frac{x}{1-y}\right)$$

whenever $x, y, x+y \in]0,1[$.

12*. (Laczkovich 1981, 1984*a*.) Let $f: \mathbb{R} \to \mathbb{R}$ be such that

$$2f(x) \leqslant f(x+h) + f(x+2h) \quad \text{for all } x \in \mathbb{R}, \quad h \in \mathbb{R}_+.$$

Prove that f is nondecreasing.

13. (Dhombres 1979, pp. 1.18–1.2.) Prove or disprove the following: let $f: \mathbb{R} \to \mathbb{R}$ be a superadditive function ($f(x+y) \geqslant f(x) + f(y)$ for all $x, y \in \mathbb{R}$). Suppose that in some neighbourhood of $x_0 \in \mathbb{R}$ we have $f \geqslant g$ where $g(x_0) = -f(-x_0)$ and where g is differentiable at x_0. Then there exists a constant a in $\mathbb{R}$ such that $f(x) = ax$.
14. (Rădulescu 1980; Dhombres 1983.) Using Proposition 6, show that among all $f: \mathbb{R} \to \mathbb{R}$ both inequalities

$$f(x+y) \geqslant f(x) + f(y),$$
$$f(xy) \geqslant f(x) f(y),$$

hold for all $x, y \in \mathbb{R}$ if and only if $f(x) \equiv 0$ or $f(x) \equiv x$. Compare the results to Exercise 2.9.

15. Give counterexamples proving that the result in Exercise 14 does not hold if we change the direction in one (or both) of the inequalities.
16*. (Volkmann 1982; Dhombres 1983.) Let A be a ring and R be an ordered ring. We suppose that, for all z in R, we have $z^2 \geqslant 0$. Let $f: A \to R$ be such that, for all x, y in A,
$$f(x+y) \geqslant f(x) + f(y),$$
$$f(xy) \geqslant f(x)f(y).$$
Prove that f is a ring endomorphism.
17. (Dhombres 1971*b*.) Does the condition (d) at the beginning of Section 3 follow from (a), (b), and (c)? Same question for $\phi: \mathscr{A} \to \mathbb{C}$, where $\mathscr{A}$ is the algebra of Bohr almost periodic functions.
18. Let $\Phi: \mathbb{R} \to \mathbb{C}$ be a discontinuous character of $\mathbb{R}$, that is, a discontinuous solution of
$$\Phi(x+y) = \Phi(x)\Phi(y) \quad (x, y \in \mathbb{R}),$$
$$|\Phi(x)| = 1 \quad (x \in \mathbb{R}).$$
Prove or disprove that, for every $\varepsilon > 0$ and every n-tuple $(x_1, x_2, \ldots, x_n) \in \mathbb{R}^n$, there exists a continuous character χ such that
$$|\chi(x_j) - \Phi(x_j)| < \varepsilon \quad (j = 1, 2, \ldots, n).$$
19. Prove the following theorem of Kronecker: let $\mu_1, \ldots, \mu_n$ and $\Theta_1, \ldots, \Theta_n$ be real numbers. For the real numbers $\mu_1, \ldots, \mu_n, \Theta_1, \ldots, \Theta_n$, every equation
$$\sum_{k=1}^{n} d_k \mu_k = 0$$
(where $d_1, \ldots, d_n$ are integers), implies
$$\sum_{k=1}^{n} d_k \Theta_k = 0 \pmod{2\pi}$$
if and only if for every $\varepsilon > 0$ there exists a real number y such that the congruential inequalities
$$|\mu_k y - \Theta_k| \leqslant \varepsilon \pmod{2\pi}$$
hold for $k = 1, 2, \ldots, n$.
20. (Venkov 1970, pp. 34–5; Hardy–Wright 1979, pp. 377–8.) Prove the following result, also due to Kronecker. Let $\mu_1, \ldots, \mu_n$ be n real numbers. For every positive integer N, there exists a positive integer $q \leqslant N^n$ and integers $p_1, p_2, \ldots, p_n$, such that
$$|q\mu_k - p_k| \leqslant \frac{1}{N} \quad \text{for } k = 1, 2, \ldots, n.$$
Compare this result to that of Exercise 19, with
$$\Theta_k = 0 \quad (k = 1, 2, \ldots, n).$$
21. (Favard 1933, pp. 21–2.) Let $\lambda_1, \ldots, \lambda_n$ be n real numbers which are linearly independent over $\mathbb{Q}$. Consider the trigonometric polynomial
$$f(x) = \sum_{j=1}^{n} c_j \exp(i\lambda_j x).$$

Prove or disprove that

$$\sup_{x\in\mathbb{R}} |f(x)| = \sum_{j=1}^{n} |c_j|.$$

22. We keep the notation of the previous exercise but we now suppose that the real numbers $\lambda_1, \ldots, \lambda_n$ and π are linearly independent over $\mathbb{Q}$. Prove or disprove that

$$\sup_{u\in\mathbb{Z}} |f(u)| = \sum_{j=1}^{n} |c_j|.$$

23. Let $1, \mu_1, \ldots, \mu_k$ be linearly independent (if $\alpha_0 + \alpha_1\mu_1 + \cdots + \alpha_k\mu_k = 0$ and $\alpha_k \in \mathbb{Q}$, then $\alpha_0 = \alpha_1 = \cdots = \alpha_k = 0$). We define $P_n(n\mu_1, \ldots, n\mu_k) \in \mathbb{R}^k$. Prove or disprove that the sequence $\{P_n\}$ is dense in $\mathbb{R}^k/\mathbb{Z}^k$.
24. (Hardy–Wright 1979, pp. 378–82.) Put mirrors to the four sides of a square in $\mathbb{R}^2$. A ray of light emitted from a given pointform source inside the square in a given direction is reflected repeatedly by the mirrors. For completeness's sake suppose that the corners of the square reflect rays by turning them back on their paths. Does there exist a source and a direction of emission such that the path of the ray is dense in the square?
25. Let Δ^+ be the space of probability distribution functions vanishing at zero, that is,

$$\Delta^+ = \{F \mid F: \bar{\mathbb{R}}_+ = [0, \infty] \to [0, 1], F \text{ is left-continuous and nondecreasing}, \\ F(0) = 0, \quad F(\infty) = 1\}.$$

A binary operation τ on Δ^+ is said to be derivable from an operation on random variables if the following holds. There exists a Borel-measurable function $L: \mathbb{R}_+^2 \to \mathbb{R}_+$ such that, for any F, G in Δ^+, there exist two random variables X, Y defined in a common probability space $(\Omega, \mathscr{A}, P)$ and satisfying $\tau(F_X, F_Y) = F_{L(X,Y)}$. Here F_Z denotes the probability distribution corresponding to the random variable Z, that is, $F_Z(u) = P(\{w \mid Z(w) < u\})$. Show that neither the arithmetic mean $(F + G)/2$ nor the geometric mean $(FG)^{1/2}$ are derivable from operations of random variables. Show that this is also the case for any weighted quasilinear mean $f^{-1}((1 - \alpha)f(F) + \alpha f(G))$.

6

Conditional Cauchy equations. An application to geometry and a characterization of the Heaviside function

After having studied the domain of functional equations in previous chapters, the purpose of the present chapter is to motivate the study of functional equations on restricted domains, that is, when the validity of the equation is not assumed in full generality for all the possible values of the variables (conditional equations). Actually we deal with two very different examples. The first one is a fundamental result of projective geometry: it shows without any regularity assumption that bijective mappings of the real plane which map straight lines into straight lines (collineations) are, up to a translation, affine transforms. The second example uses conditional equations to give a characterization of the Heaviside function.

In our first investigation of vector addition, and of information contained in an event, we have encountered Cauchy's equation subjected to a condition concerning the variables x and y:

$$f(x+y)=f(x)+f(y) \quad \text{for all nonnegative } x \text{ and } y$$

(see (1.2) and (3.12)). The natural set of values for the variables x and y is the additive group $\mathbb{R}$, whereas the domain of the equation as determined by our problem is only $\bar{\mathbb{R}}_+$ ($\bar{\mathbb{R}}^2_+$, to be exact).

It is a frequent situation to get a functional equation with a restricted domain, restricted when compared to what could be considered the natural domain. Two names were coined to describe this phenomenon: '*functional equations on restricted domains*' (Kuczma 1978*b*) or '*conditional functional equations*' (Dhombres–Ger 1978). We shall use the second.

Let Z be a nonempty subset of the plane $\mathbb{R}^2$. We say that $f:\mathbb{R}\to\mathbb{R}$ satisfies a conditional Cauchy equation relative to Z if, for all pairs (x, y) in Z, we have

$$f(x+y)=f(x)+f(y).$$

For the sake of brevity, we shall sometimes say that f is ***Z*-additive** (cf. Section 7.1).

In the above case, $Z=\mathbb{R}_+^2$ and so Z was independent of the unknown function. It may be, however, that the subset Z is dependent upon the function f itself. Let us give an example. We start from a problem encountered in geometrical optics, which is also fundamental in affine geometry. This problem can now be treated directly, but the manner in which a conditional functional equation arises here is typical and that is the reason why we deal with the topic in this way.

We look for all bijective (that is, onto and one-to-one) mappings $T:\mathbb{R}^2\to\mathbb{R}^2$ which map straight lines into straight lines. The same classical question can be asked about affine spaces of dimension n on fields more general than $\mathbb{R}$ but, for the sake of notational simplicity, we shall restrict ourselves to the real two dimensional case. Any bijective linear mapping clearly maps straight lines into straight lines. Let

$$M_1=(0,0),\quad M_2=(1,0),\quad M_3=(0,1)\quad\text{and}\quad M_4=(1,1).$$

As M_1, M_2 and M_3 are not on the same straight line, the same is true for $T(M_1), T(M_2)$ and $T(M_3)$. Therefore, there exists a bijective linear mapping A such that $A(T(M_1))=M_1$, $A(T(M_2))=M_2$ and $A(T(M_3))=M_4$. Moreover, the composition of A and T is bijective and maps straight lines into straight lines. Since we wish to prove that T itself is a linear mapping, it follows that there is no loss of generality in supposing that $T(M_1)=M_1$, $T(M_2)=M_2$, and $T(M_3)=M_4$. We write $M=(x,y)$ and $T(x,y)=(F(x,y), G(x,y))$.

Recall that the collinearity of three points $M_1(x_1,y_1)$, $M_2(x_2,y_2)$ and $M_3(x_3,y_3)$, of the euclidean plane can be expressed by

$$\begin{vmatrix} x_1 & x_2 & x_3 \\ y_1 & y_2 & y_3 \\ 1 & 1 & 1 \end{vmatrix}=0.$$

As $(0,0)$, $(1,0)$ and $(x,0)$ are collinear and as we chose $G(0,0)=F(0,0)=0$, $F(1,0)=1, G(1,0)=0, F(0,1)=G(0,1)=1$, we get

$$0=\begin{vmatrix} G(0,0) & G(1,0) & G(x,0) \\ F(0,0) & F(1,0) & F(x,0) \\ 1 & 1 & 1 \end{vmatrix}=\begin{vmatrix} G(1,0) & G(x,0) \\ F(1,0) & F(x,0) \end{vmatrix}=-G(x,0),$$

that is, for all x,

$$G(x,0)=0. \tag{1}$$

In the same way, since $(0,0)$, $(0,1)$ and $(0,y)$ are collinear, we get

$$F(0,y)=G(0,y). \tag{2}$$

As T is injective, T maps parallel lines into parallel lines and so transforms parallelograms into parallelograms. Since the four points $(0,0), (x,0), (0,y)$ and (x,y) form a parallelogram, we get a system of two functional equations:

$$\left.\begin{aligned} F(x,y) &= F(x,0) + F(0,y) \\ G(x,y) &= G(x,0) + G(0,y) \end{aligned}\right\}. \tag{3}$$

Combining (1), (2) and (3), we deduce that

$$G(x,y) = F(0,y). \tag{4}$$

Another equation will be useful to determine F and G. As x runs through $\mathbb{R}$, (x,x) determines a line through $(0,0)$ and so does $(F(x,x), G(x,x))$. Hence there exist real constants α, β with $|\alpha| + |\beta| \neq 0$, such that

$$\alpha F(x,x) + \beta G(x,x) = 0.$$

In view of (3) and (4) this means

$$\alpha F(x,0) + (\alpha + \beta) F(0,x) = 0.$$

If we had $\alpha = 0$, then $|\alpha| + |\beta| = |\beta| \neq 0$ and so $F(0,x) = 0$. But, due to (4), this would give $G(x,y) \equiv 0$ which contradicts the bijectivity of T. If we had $\alpha + \beta = 0$, then $\alpha \neq 0$ and so $F(x,0) = 0$. However, due to (1), $G(x,0) = 0$. These two equations would again contradict the bijectivity of T. We may therefore write

$$F(0,x) = \gamma F(x,0) \quad (\gamma \neq 0). \tag{5}$$

We define a new function f by $f(x) = F(x,0)$ ($f(1) = 1$ because of the supposition $T(M_2) = M_2$), and we try to determine f with the aid of a single functional equation.

As $(x+y,0), (x,y)$ and $(0,x+y)$ are collinear, we get

$$0 = \begin{vmatrix} F(x+y,0) & F(x,y) & F(0,x+y) \\ G(x+y,0) & G(x,y) & G(0,x+y) \\ 1 & 1 & 1 \end{vmatrix},$$

that is, by (1), (3), (4) and (5),

$$0 = \begin{vmatrix} f(x+y) & f(x) + \gamma f(y) & \gamma f(x+y) \\ 0 & \gamma f(y) & \gamma f(x+y) \\ 1 & 1 & 1 \end{vmatrix},$$

$$f(x+y)[\gamma f(y) + \gamma f(x) + \gamma^2 f(y) - \gamma^2 f(y) - \gamma f(x+y)] = 0,$$

that is,

$$f(x+y)[f(x+y) - f(x) - f(y)] = 0.$$

This equation is a conditional Cauchy equation, the so-called *Mikusiński equation* (as J. Mikusiński was the first to propose this method for giving a quick proof of the fundamental theorem of affine geometry, see, e.g., Kuczma 1978*b*; Dhombres 1979, pp. 2.9–2.12):

$$f(x+y) = f(x) + f(y) \quad \text{if } f(x+y) \neq 0. \tag{6}$$

We now proceed to find the solution of (6). This equation is a conditional Cauchy equation relative to Z, which is related to the complement of the kernel of f:

$$\ker f = \{x \mid x \in \mathbb{R}, f(x) = 0\},$$

while

$$Z = \{(x, y) \mid x \in \mathbb{R}, y \in \mathbb{R}, f(x + y) \neq 0\}.$$

Here Z depends upon the unknown function f. Let us first prove that $\ker f$ is a subgroup of $\mathbb{R}$.

First, $0 \in \ker f$. Indeed, if there exists x with $f(x) \neq 0$, then $x + 0 \notin \ker f$ and thus $f(x + 0) = f(x) = f(x) + f(0)$, that is $f(0) = 0$.

Now let x and y be in $\ker f$. Due to (6), the hypothesis that $x + y \notin \ker f$ is a contradiction.

Let $x \in \ker f$ and suppose $-x \notin \ker f$. Equation (6) yields $f(-x) = f(-2x + x) = f(-2x) + f(x) = f(-2x)$. Therefore $f(-2x) \neq 0$. So, applying (6) again, we get $f(-x) = f(-2x) = f(-x - x) = 2f(-x)$. Therefore $f(-x) = 0$, which contradicts our supposition $-x \notin \ker f$. So $\ker f$ is a subgroup.

Let us now prove that, for all $x \in \mathbb{R}$,

$$f(2x) = 2f(x). \tag{7}$$

Equation (7) is true for all x in $\ker f$, since we have seen that $\ker f$ is a subgroup. Equation (7) is true if $x \notin \ker f$ and $2x \notin \ker f$, as it follows from (6) that

$$f(2x) = f(x) + f(x) = 2f(x).$$

Only the case remains where $f(2x) = 0$ and $f(x) \neq 0$. We prove that this case is impossible.

So we assume that there exists an x_0 with $f(2x_0) = 0$ and $f(x_0) = y_0 \neq 0$. Suppose first that there exists also an x such that $f(x) \neq 0$ and $f(x) \neq f(x_0)$. From (6) we get

$$f(x) = f(x - x_0 + x_0) = f(x - x_0) + f(x_0). \tag{8}$$

Thus $f(x - x_0) \neq 0$ and so

$$f(x - x_0) = f(x - 2x_0 + x_0) = f(x - 2x_0) + f(x_0). \tag{9}$$

But

$$f(x) = f(x - 2x_0 + 2x_0) = f(x - 2x_0) + f(2x_0)$$
$$= f(x - 2x_0).$$

From (9) we deduce that

$$f(x - x_0) = f(x) + f(x_0).$$

Using (8), the last equation yields $2f(x_0) = 2y_0 = 0$, which is a contradiction to $y_0 \neq 0$.

Therefore $f(x) = y_0$ whenever $x \notin \ker f$. We deduce that $f(x) = f(x/2) +$

$f(x/2)$. But $f(x/2)=0$ would contradict $f(x)=y_0\neq 0$ and $f(x/2)=y_0$ yields $y_0=2y_0$, which would also contradict $y_0\neq 0$. This concludes the proof of (7).

We proceed now to prove that $f(-x)=-f(x)$ for all x in $\mathbb{R}$. This is clearly true for all x in $\ker f$ as $\ker f$ is a group. If $x\notin\ker f$, then

$$f(x)=f(2x-x)=f(2x)+f(-x)=2f(x)+f(-x).$$

Thus

$$-f(x)=f(-x). \tag{10}$$

We are now ready to prove that every f satisfying (6) is, in fact, additive. Take two arbitrary real numbers x and y.

First suppose $y\notin\ker f$. Then, from (6) and (10),

$$f(y)=f(-x+y+x)=f(-x)+f(y+x)=-f(x)+f(y+x),$$

that is,

$$f(x+y)=f(x)+f(y).$$

By symmetry, the same is true for $x\notin\ker f$. Since $\ker f$ is a subgroup, if both x and y are in $\ker f$, then $f(x+y)=f(x)+f(y)$ is also true.

We have thus proved the following:

Theorem 1. *All functions $f:\mathbb{R}\to\mathbb{R}$ satisfying (6) are additive, that is, $f(x+y)=f(x)+f(y)$ for all real x, y.*

Going back to our problem from geometry, we shall now prove the following:

Proposition 2. *Let $T:\mathbb{R}^2\to\mathbb{R}^2$ be a bijection preserving collinearity and continuous at a point. Then T is linear.*

Proof. By a linear mapping we transform T as at the beginning of this chapter. We deduce, as there, that $f(x)=F(x,0)$ is additive, where $T(x,y)=(F(x,y),G(x,y))$. Continuity of T at one point of $\mathbb{R}^2$ leads to the continuity of f at some point. Therefore $f(x)=\alpha x$ for some constant α (Corollary 2.5). Using (1), (2), (3) and (4), we get that T itself is linear and so can be represented by a matrix which has to be nonsingular. Proposition 2 follows.

It is clear that we may weaken the continuity condition on T in Proposition 2 by using deeper regularity results from Corollary 2.5. However, no regularity is needed at all for the result to hold. Define a collineation on the real plane to be a bijection from $\mathbb{R}^2$ into $\mathbb{R}^2$ preserving collinearity. We can prove the following:

Theorem 3. *A collineation on the real plane is the composition of a translation and of a nonsingular linear transformation.*

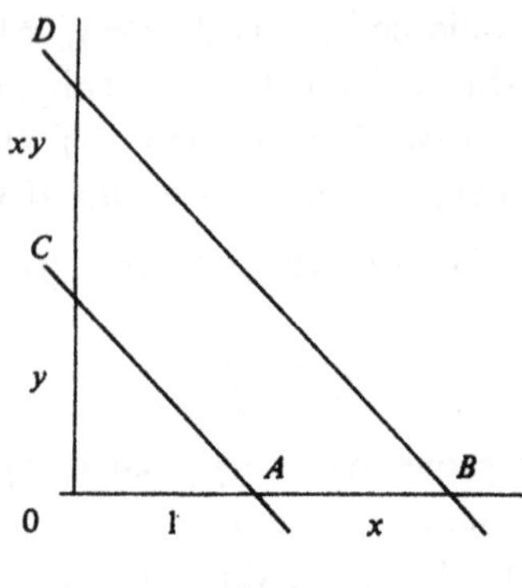

Fig. 8

Proof. We keep the same notation as in Proposition 2. The idea is to add one more functional equation for the additive function f, in order to avoid any regularity requirement. Let $A=(1,0), B=(x,0)$ and $C=(0,y)$. If BD is parallel to AC, then $D=(0,xy)$ (see Figure 8). We may apply the same calculation to $T(A)$, $T(B)$, $T(C)$ and $T(D)$. Using (1), (2) and (5), we get

$$f(x)f(y)=f(xy)f(1),$$

since T maps parallel lines onto parallel lines. We also had $f(1)=1$ (from $T(A)=A$) at the beginning of this chapter, so

$$f(xy)=f(x)f(y). \tag{11}$$

Moreover, we see from (6) and from Theorem 1 that f is additive. Therefore f is a ring automorphism of $\mathbb{R}$ (that is, satisfies (5.21) and (5.22)). Proposition 5.6 yields $f(x)=x$ or $f(x)=0$. The last case is impossible, since $f(1)=1$. Therefore $f(x)=x$. As in Proposition 2, this result yields that T is linear which concludes the proof of Theorem 3.

Theorem 3 has even been generalized to bijections which are supposed only to preserve collinearity for points lying on (at least) four straight lines (or on the union of three straight lines and a point). See also Aczél–Benz 1969; Aczél 1971*b* for generalizations.

The situation for the bicomplex plane $\mathbb{C}^2$ is completely different. In the proof of Theorem 5.7 we have constructed a nontrivial ring automorphism f of $\mathbb{C}$. This automorphism leads to a mapping $T:\mathbb{C}^2\to\mathbb{C}^2$ given by

$$T(x,y)=(f(x),f(y)).$$

This mapping T is a bijection. Moreover, it preserves the collinearity of points in the bicomplex plane because the algebraic conditions for the collinearity of three points is inherited by the transform, since f preserves addition and multiplication. This T is neither linear nor antilinear and presents a weird aspect: it is discontinuous at every point of the plane. It preserves, however, harmonic relations (and, more generally, rational valued cross-ratios) and conics.

In the next chapter, we shall systematically study conditional Cauchy equations. Let us conclude this section by giving the solution of a functional equation introduced by I. Fenyö and completely solved by W. Benz (1979*a*, 1981*b*). Our interest in the solution lies in the appearance of a conditional functional equation, for which we can apply a fixed point method, a common technique for functional equations in a single variable. On the other hand, the Heaviside function satisfies this equation.

We look for an $f:\mathbb{R}\to\mathbb{R}$ such that, for all x, y in $\mathbb{R}$,

$$f(x)f(y)=f(xy)f\left(\frac{x+y}{2}\right). \tag{12}$$

We introduce the set F_f of all x in $\mathbb{R}$ such that $f(x)\neq 0$. According to (12) x and y belong to F_f if and only if $(x+y)/2$ and xy belong to F_f. So it is convenient to introduce the following definition.

Define a subset F of $\mathbb{R}$ to be *regular* if the following holds: Both x and y belong to F if, and only if, both $(x+y)/2$ and xy belong to F.

The sets $F=\{0\}$, $F=\{1\}$ are the only singletons which are regular subsets. Any F_f, where f is a solution of (12), is a regular subset. We prove that a nonempty regular subset F, which is not a singleton, is not bounded from above.

Suppose first that there exists x such that $x\in F$ and $|x|>1$. As $x^2, x^4, \ldots, x^{2^n}$ are in F, it is clear that F is not bounded from above.

Suppose now that $|x|\leqslant 1$ for all $x\in F$. We have excluded the empty set and singletons as F and accordingly $F\nsubseteq\{-1,0,1\}$. Therefore, using that $t^2\in F$ if $t\in F$, we may also suppose that there exists an $x\in F$ such that $0<x<1$. Start from this x. There also exists an $y\in F$ such that $x^2>y$. For example, take $y=x^4$. Then use $x_1=x+(x^2-y)^{1/2}$. We see that $x_1>x$, $x_1^2>y$ and $x_1\in F$ (since $x_1'=x-(x^2-y)^{1/2}$ is such that $x_1x_1'=y$, $(x_1+x_1')/2=x$, and both x and y are in F). By induction, $\{x_n\}$, defined by $x_{n+1}=x_n+(x_n^2-y)^{1/2}$ for $n\geqslant 1$, is an increasing sequence of elements in F. If this sequence were bounded, the sequence $\{x_n\}$ would converge to a z satisfying $z=z+(z^2-y)^{1/2}$ which yields $z^2=y$. But $z^2\geqslant x_1^2>y$. This contradiction implies that $\{x_n\}$ is an increasing sequence of elements of F, which is not bounded.

We prove now the following:

Theorem 4. *A function $f:\mathbb{R}\to\mathbb{R}$ satisfies* (12) *if, and only if, there exists a regular subset F of $\mathbb{R}$, and two constants α, β such that*

(a) *if* $0 \notin F$, *or if* $0 \in F$ *and* $F \cap \mathbb{R}_- \neq \varnothing$, *then* $f(x) = \alpha\chi_F(x)$ *for all* x *in* $\mathbb{R}$,
(b) *if* $0 \in F$ *and* $F \cap \mathbb{R}_- = \varnothing$, *then* $f(x) = \alpha\chi_F(x)$ *for all* $x \neq 0$ *and* $f(0) = \beta$.

Here χ_F is the characteristic function of F ($\chi_F(x) = 1$ if $x \in F$, $\chi_F(x) = 0$ if $x \notin F$).

Proof. The case $F_f = \{0\}$ is contained in (b), as is the case $F_f = \{1\}$. The case $F_f = \varnothing$ is contained in (a). Therefore we may suppose for the proof that F is not bounded from above.

From (12), with $x = y$ in F_f, we deduce that

$$f(x) = f(x^2) \quad \text{on } F_f. \tag{13}$$

Therefore, for both x and y in F_f, using (13), we get

$$f\left(\frac{x+y}{2}\right) = \frac{f(x)f(y)}{f(xy)} = \frac{f(x^2)f(y^2)}{f(x^2y^2)} = f\left(\frac{x^2+y^2}{2}\right).$$

Thus f satisfies the conditional functional equation

$$f\left(\frac{x^2+y^2}{2}\right) = f\left(\frac{x+y}{2}\right) \quad \text{for all } x, y \text{ in } F_f. \tag{14}$$

Let us begin with the solution of (14) without any restriction on x and y. Then, with $x = y$, also (13) holds without restriction. The mapping $(x, y) \mapsto (u = (x+y)/2, v = xy)$ is from $\mathbb{R}^2$ onto the subset $\{(u, v) \in \mathbb{R}^2, v \leqslant u^2\}$. Equation (14) yields, for all u, v with $v \leqslant u^2$,

$$f(u) = f(2u^2 - v).$$

Using (13), we have, with $z = u^2 - v$,

$$f(u) = f(u^2) = f(u^2 + z) \quad (u \in \mathbb{R}, z \in \bar{\mathbb{R}}_+). \tag{15}$$

Let u, u' be any two numbers. As the intersection of the two intervals $[u^2, \infty[$ and $[u'^2, \infty[$ is never empty, we deduce from (15) that $f(u) = f(u')$. Therefore, the only solutions of (14), *with no restriction* on x and y, are the constant functions.

We now return to the *conditional* functional equation (14) and will prove that every solution is constant on F_f, except perhaps at 0. As previously, we get

$$f(t) = f(t^2 + z) \tag{16}$$

for all t in F_f but only for $z \geqslant 0$, subject to the condition that $t^2 - z$ is in F_f. At first it seems difficult to compare such conditions for two distinct t's directly. But a detour is possible. We start from $0 \leqslant x \leqslant y$, both x and y belonging to F_f. With $t = xy \in F_f$ and $z = (xy)^2 - x^4 \geqslant 0$, and as $t^2 - z = x^4 = (x^2)^2$ belongs to F_f, we deduce from (16)

$$f(xy) = f(2x^2y^2 - x^4). \tag{17}$$

But $2x^2y^2 - x^4 = x^2(2y^2 - x^2)$. Using the functional equation (12),

$$f(x^2)f(2y^2 - x^2) = f(y^2)f(2x^2y^2 - x^4). \tag{18}$$

With $u=y$ and $z=y^2-x^2$ ($z\geqslant 0$ and $u^2-z=x^2$ is in F_f), we may use (16) in order to get

$$f(2y^2-x^2)=f(y). \tag{19}$$

As $f(y)=f(y^2)\neq 0$ and $f(x)=f(x^2)\neq 0$ as well, we deduce from (18) and (19) that $f(x)=f(2x^2y^2-x^4)$. From (17) we get

$$f(x)=f(xy) \tag{20}$$

under the only restriction that x and y are both in F_f and $0\leqslant x\leqslant y$. Thus (20) is much more handy than (16).

Fix $x>0$ and take $y\geqslant x$ and $y\in F_f$. As F_f is not bounded from above, there exists a $z\in F_f$ such that $z\geqslant \sup(x/y, y/x)$. Then $xz\in F_f$, $xz\geqslant y$ and $yz\in F_f$, $yz\geqslant x$. Using (20) we obtain

$$f(y)=f(xz\cdot y)=f(x\cdot zy)=f(x).$$

This shows that f is constant on $F_f\cap\mathbb{R}_+$. We deduce, using (13), that f is constant on all F_f, except perhaps at 0.

In other words, there exists a nonzero constant α and $f(x)=\alpha\chi_{F_f}(x)$ for $x\neq 0$, where χ_{F_f} is the characteristic function of F_f. The value of f at 0 is rather arbitrary: clearly, if $0\notin F_f$, the value $f(0)$ is necessarily 0. But, if $0\in F_f$, two different cases may occur.

If F_f contains a negative element, say $-a$, then (12) yields

$$f(-a^{1/2})f(a^{1/2})=f(-a)f(0). \tag{21}$$

Both $f(-a)$ and $f(0)$ are different from 0, and so $-a^{1/2}$ and $a^{1/2}$ belong to F_f. Equation (21) yields $f(0)=\alpha$.

If F_f has no negative element, a direct computation shows that any nonzero value for $f(0)$ is possible for a solution of (12).

This concludes the proof of Theorem 4.

Finally, we give a characterization of the Heaviside function H where $H(x)=1$ for $x\geqslant 0$ and $H(x)=0$ for $x<0$:

Theorem 5. *The Heaviside function is the only right-continuous solution of* (12) *whose range contains* 0 *and* 1.

Proof. $\bar{\mathbb{R}}_+$ is a regular subset of $\mathbb{R}$ and so its characteristic function, the Heaviside function, satisfies (12), is right-continuous, and its range contains 0 and 1.

Conversely, let F be the regular subset, necessarily nonempty, associated with a right-continuous solution f of (12) taking somewhere the value 1 and somewhere else the value 0. As F cannot be a singleton, due to the right-continuity of f, we are sure that F is not bounded from above. Due to the right-continuity, for any x in F or in the complement $\mathscr{C}F$, there exists a

$\delta > 0$ (which may depend upon x) such that $[x, x+\delta[$ is a subset of F or of $\mathscr{C}F$ respectively. Now take $x_0 \in F$ and write $x_1 = \sup\{x \mid [x_0, x] \subset F, x \geqslant x_0\}$. In order to prove that $x_1 = \infty$, suppose for contradiction that x_1 is finite. Clearly, by the right continuity, $x_1 \in \mathscr{C}F$. Let $x_2 = \sup\{x \mid [x_1, x] \subset \mathscr{C}F, x \geqslant x_1\}$. As F is not bounded from above, x_2 is finite and we have already noticed that $x_2 > x_1$, and $x_2 \in F$. It is now easy to find a $y \in [x_0, x_1[$ such that $(y + x_2)/2$ belongs to $]x_1, x_2[$, that is, to $\mathscr{C}F$. But y and x_2 are in F, so that $(y + x_2)/2$ must be in F. The contradiction yields $x_1 = \infty$. In other words, we have proved that, if $x_0 \in F$, then also $[x_0, \infty[\subset F$. Now define $x_0 = \inf F$. Clearly $x_0 = -\infty$ is impossible, since f is somewhere equal to zero. We shall show that both $x_0 > 0$ and $x_0 < 0$ are also impossible.

Suppose $x_0 > 0$. Let $y \in F$. We may choose $\varepsilon > 0$ so that $x_0^2 + 2\varepsilon x_0 > y$. Therefore $x_0 + \varepsilon \in F$, and $(x_0 + \varepsilon)^2 > y$. But, with $x = x_0 + \varepsilon$, the number $x - (x^2 - y)^{1/2}$ is in F and is strictly smaller than x_0, which contradicts the fact that x_0 is a lower bound.

Suppose $x_0 < 0$. Let $y < 0$ be in F. We may choose $\varepsilon > 0$ so small that $x_0^2 + 2\varepsilon x_0 > y$. Therefore $x_0 + \varepsilon \in F$, $(x_0 + \varepsilon)^2 > y$. But with $x = x_0 + \varepsilon$, the number $x - (x^2 - y)^{1/2}$ is in F and is strictly smaller than x_0, which is a contradiction to the definition of x_0.

There only remains the possibility $x_0 = 0$ i.e. $f(x) = H(x)$ and this concludes the proof of Theorem 5.

Exercises and further results

(For Exercises 1–4, see Kuczma 1978*a*.)

1. Find the general solution of the following conditional Cauchy equation for $f: \mathbb{R} \to \mathbb{R}$:

$$f(x+y) = f(x) + f(y) \quad \text{if } f(x+y) + f(x) + f(y) \neq 0 \quad (x, y \in \mathbb{R})$$

(or, equivalently, $f(x+y)^2 = (f(x) + f(y))^2$ for $x, y \in \mathbb{R}$).

2. Let $(G, +)$ be a semigroup and $(F, +, \cdot)$ be a commutative ring without divisors of zero. Suppose that G is commutative and contains no elements of order 3. Does the following equation (ii) have solutions $f: G \to F$ which do not satisfy also equation (i)?

$$\text{(i)} \qquad f(x+y) = f(x) + f(y) \qquad (x, y \in G),$$
$$\text{(ii)} \qquad f(x+y)^2 = (f(x) + f(y))^2 \quad (x, y \in G).$$

3. Same question as in Exercise 2, but replace the condition that G is commutative by the assumption that it contains no elements of order 4 (and, as before, no element of order 3).

4. Find the general solution $f: G \to F$ of

$$f(x+y) = f(x) + f(y) \quad \text{if } f(x+y) + f(x) + f(y) \neq 0 \quad (x, y \in G)$$

where G and F are abelian groups.

5. Find the general solution of the following functional equation for $f:\mathbb{R}\to\mathbb{R}$:
$$f(x+y)^2=(f(x)f(y))^2 \quad (x,y\in\mathbb{R}).$$
(This is a conditional Cauchy equation too.)
6. Find the general solution $f:\mathbb{R}\to\mathbb{R}$ of
$$f(x+y)^n=(f(x)+f(y))^n \quad (x,y\in\mathbb{R}),$$
where n is a fixed positive integer.
7. Find the general solution $f:\mathbb{R}\to\mathbb{C}$ of the Lobachevskiĭ functional equation
$$f(x+y)f(x-y)=f(x)^2 \quad (x,y\in\mathbb{R}).$$
8*. Let F be a ring and $f:F\to F$ be additive. Suppose that, for each pair $(x,y)\in F$, we have *either*
$$f(xy)=f(x)f(y), \tag{22}$$
or
$$f(xy)=f(y)f(x). \tag{23}$$
Prove (Hua's theorem; cf. Hua 1949; Artin 1978, pp. 38–40 for a related result) that f is a ring endomorphism or a ring antiendomorphism (that is, that f is additive and either satisfies (22) *for all* x, $y\in F$ or satisfies (23) *for all* $x,y\in F$).
9*. (Reis–Shih 1977.) Let X^+ be the free group on an alphabet X, and $f:X^+\to X^+$ such that either
$$f(xy)=f(x)f(y) \quad \text{for all } x,y\in X^+,$$
or
$$f(xy)=f(y)f(x) \quad \text{for all } x,y\in X^+.$$
Show that f is an endomorphism or an antiendomorphism (defined as in Exercise 8).
10*. Let G be a group and F a field of characteristic different from 2. Show that the functional equation
$$f(xy)+f(yx)=f(x)f(y)+f(y)f(x)$$
has only homomorphisms or antihomomorphisms as solutions. From this deduce Hua's theorem (Exercise 8).
11. (Cf. Kannappan 1979*b*.) Let $f:G\to F$ where G is a group and F a field. Find all solutions of the functional equation
$$f(xy)+f(yx)=2f(x)f(y).$$
(Notice that this equation is obtained by adding (23) to (22) in Exercise 8.)
12. (Benz 1981*b*.) In Theorem 5, can the condition of right continuity everywhere be replaced by the right continuity at 0 and the existence of at least two points $x_1\neq x_2$ such that $f(x_1)f(x_2)\neq 0$?

7

Addundancy, extensions, quasi-extensions and extensions almost everywhere. Applications to harmonic analysis and to rational decision making

Motivated by the last chapter, we give the terminology to be used to classify conditional Cauchy equations, depending on whether the condition imposed upon the validity of the equation can be eliminated or not. This leads to notions such as addundancy, additive expandibility, quasi-extensions, and others. We develop them in order to state and prove the main theorems for conditional Cauchy equations. One application introduces the Fourier transform as a representation of the group algebra of all equivalence classes of Lebesgue integrable functions. A second application uses a conditional equation to solve a problem in rational group decision making.

7.1 Extensions and quasi-extensions

Keeping in mind the results proved in the previous chapters, here we introduce some terminology (see Daróczy–Losonczi 1967; and Dhombres–Ger 1978). Let Z be a nonempty subset of $\mathbb{R}^2$. We define

$$\Pi_1(Z) = \{x \mid \exists y:(x, y) \in Z\}: \text{ projection of } Z \text{ onto the } x\text{-axis},$$
$$\Pi_2(Z) = \{y \mid \exists x:(x, y) \in Z\}: \text{ projection of } Z \text{ onto the } y\text{-axis},$$
$$t(Z) = \{z \mid \exists x, \exists y:(x, y) \in Z,\ z = x + y\}: \text{ projection of } Z, \text{ parallel to the line } x + y = 0, \text{ onto the } x\text{-axis},$$
$$A(Z) = \Pi_1(Z) \cup \Pi_2(Z) \cup t(Z).$$

(In the present context all these sets are subsets of $\mathbb{R}$.) The conditional Cauchy equation relative to Z,

$$g(x + y) = g(x) + g(y) \quad \text{for all pairs } (x, y) \in Z, \tag{1}$$

could be stated for a g defined just on $A(Z)$ and not necessarily defined over the whole real axis. Indeed, equation (1) tells nothing about the values of g outside $A(Z)$ and therefore we cannot expect to derive from a solution of (1) any information outside $A(Z)$. However, it may be more convenient to let g be defined over all of $\mathbb{R}$. We introduce more terminology.

Let $\mathscr{Z}$ be a mapping which associates to any $g:\mathbb{R}\to\mathbb{R}$ a nonempty subset $\mathscr{Z}(g)$ of $\mathbb{R}^2$. We say that $\mathscr{Z}$ is additively *expandable* if, for any solution g of the *conditional Cauchy equation,*

$$g(x+y)=g(x)+g(y) \quad \text{for all pairs } (x,y)\in\mathscr{Z}(g),$$

there exists an additive $f:\mathbb{R}\to\mathbb{R}$ such that

$$f(x)=g(x) \quad \text{for all } x\in A(\mathscr{Z}(g)).$$

Such an f is called an *additive extension* (*extension*, for short) of g relative to $\mathscr{Z}$. If f is unique, we say that $\mathscr{Z}$ is *uniquely* additively expandable.

Let Z be a given nonempty subset of $\mathbb{R}^2$ and take for $\mathscr{Z}$ the constant mapping $\mathscr{Z}(g)=Z$ for all $g:\mathbb{R}\to\mathbb{R}$. If this $\mathscr{Z}$ is (uniquely) additively expandable, we just say that Z is (*uniquely*) *additively expandable.*

With this vocabulary, Theorem 2.1 can be restated in the following form:

$\mathbb{R}^2_+$ *is additively expandable.*

For a subset E of $\mathbb{R}$, we define $t^{-1}(E)$ as the set of all pairs (x,y) in $\mathbb{R}^2$ with $x+y\in E$. Define then $\mathscr{Z}_1$ as the mapping with $\mathscr{Z}_1(g)=t^{-1}(\mathscr{C}\ker g)$. (As before, $\mathscr{C}\ker g$ is the complement of $\ker g$.) Theorem 6.1 can be restated as follows:

Theorem 1. $\mathscr{Z}_1$ *is additively expandable.*

Of course, we often have to take into account regularity properties required from solutions of (1). Let R be a given nonempty class of functions $g:\mathbb{R}\to\mathbb{R}$. Let $\mathscr{Z}$ be a mapping as before, but now $\mathscr{Z}(g)$ is defined just for all g in R. We say $\mathscr{Z}$ is *additively R-expandable* if, for any solution g in R of

$$g(x+y)=g(x)+g(y) \quad \text{for all } (x,y)\in\mathscr{Z}(g),$$

there exists an additive f, also in the class R, such that

$$f(x)=g(x) \quad \text{for all } x\in A(\mathscr{Z}(g)),$$

(*uniquely* additively R-expandable if f is unique). In the same way, for a nonempty subset Z of $\mathbb{R}^2$ we get the definition of a (uniquely) *additively R-expandable set.* Two more definitions are useful.

A nonempty subset Z of the plane $\mathbb{R}^2$ is *addundant* (*or R-addundant*) if $A(Z)=\mathbb{R}$ and Z is additively expandable (or additively R-expandable).

Note. In previous papers (e.g. Dhombres–Ger 1978) the name 'redundant' was used for a somewhat more special concept. 'Addundant' is a contraction of 'additively redundant' (or of 'additively abundant').

In this section we shall investigate some examples of addundancy. First, we define the mapping $\mathscr{L}_2$ by $\mathscr{L}_2(g) = \mathbb{R} \times g(\mathbb{R})$ $(g(\mathbb{R}) := \{g(x) | x \in \mathbb{R}\})$. We get the following result.

Proposition 2. *$\mathscr{L}_2$ is not additively expandable.*

Proof. When $\mathscr{L}_2(g) = \mathbb{R} \times g(\mathbb{R})$, a $\mathscr{L}_2(g)$-additive function is exactly a solution of the following functional equation:

$$g(x + g(y)) = g(x) + g(g(y)) \quad \text{for all } (x, y) \text{ in } \mathbb{R}^2. \tag{2}$$

Let $g(x) = [x]$ be the integer part of x, that is, the greatest integer less than or equal to x. This g satisfies the functional equation (2) but is clearly not additive. (For the general solution of equation (2), see Exercise 7; for similar equations see Chapter 19.)

Theorem 3. *If C denotes the class of all real valued continuous functions over the real axis, then $\mathscr{L}_2$ is C-addundant.*

Proof. Our problem is to find all continuous $g: \mathbb{R} \to \mathbb{R}$ satisfying equation (2) and to prove that they are all of the form $g(x) = cx$ for some c in $\mathbb{R}$ (see Dhombres 1977c). First, by putting $x = 0$ into (2), we obtain $g(0) = 0$. Thus, if g is a constant, it has to be the zero constant. If g is not a constant, then, due to the continuity of g, its range has a nonempty interior F and $g(x + y) = g(x) + g(y)$ for every x in $\mathbb{R}$, every y in F. The set $G = \{y | g(x + y) = g(x) + g(y) \text{ for all } x \in \mathbb{R}\}$ is a subgroup of $\mathbb{R}$, and it contains the nonempty open set F. But the only subgroup of $\mathbb{R}$ which contains a nonempty open set is $\mathbb{R}$ itself. Therefore $G = \mathbb{R}$, so g is additive. Being

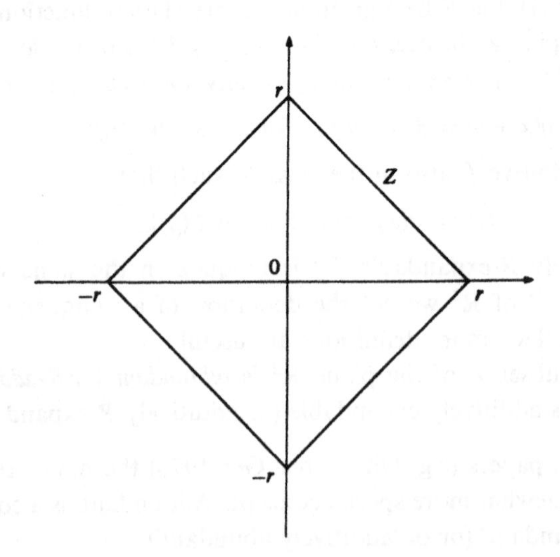

Fig. 9

continuous, g has to be of the form $g(x) = cx$ for some c in $\mathbb{R}$, which concludes the proof of Theorem 3.

The following is quite important (cf. Daróczy–Losonczi 1967 for a similar theorem; the proof below is due to Ng 1974*a*; Dhombres–Ger 1978; and J. Lawrence (unpublished)).

Theorem 4. *For any $r > 0$, the diamond $Z = \{(x, y) \mid |x| + |y| < r\}$ is uniquely additively expandable.*

This theorem means that, for every solution $g:]-r, r[\to \mathbb{R}$ of (see Figure 9)

$$g(x+y) = g(x) + g(y) \quad \text{for all } x, y \text{ satisfying } |x| + |y| < r, \tag{3}$$

there exists a unique solution $f: \mathbb{R} \to \mathbb{R}$ of the Cauchy equation such that

$$f(x) = g(x) \quad \text{for all } x \in]-r, r[. \tag{4}$$

Proof. Just as for the Cauchy equation on $\mathbb{R}^2$, we can prove that (3) implies

$$g(0) = 0$$

and

$$g(ns) = ng(s) \quad \text{whenever } n|s| < r. \tag{5}$$

Also (put $s = x/n$ into (5))

$$g\left(\frac{x}{n}\right) = \frac{1}{n} g(x) \quad \text{if } |x| < r. \tag{6}$$

Every real number t can be represented in the form

$$t = ny \quad \text{where } |y| < r \tag{7}$$

(and n is sufficiently large). Now, for an arbitrary $t \in \mathbb{R}$, represented in form (7), define

$$f(t) = ng(y) = ng\left(\frac{t}{n}\right) \quad (|y| < r). \tag{8}$$

Since every real number t can be represented in more than one way in form (7), we have to show that definition (8) is unambiguous. Indeed if, also,

$$t = mx \quad (|x| < r),$$

then $x/n = y/m$ and, by (6),

$$ng(y) = nmg\left(\frac{y}{m}\right) = mng\left(\frac{x}{n}\right) = mg(x)$$

and so f is well defined. If $|t| < r$, then (7) reduces to $t = t$ and (8) gives

$$f(t) = g(t) \quad \text{for } |t| < r,$$

which is (4).

We now prove that f satisfies Cauchy's equation. For arbitrary $u \in \mathbb{R}$, $v \in \mathbb{R}$ let n be so large that $(|u|+|v|)/n \in [0, r[$. Then, indeed,

$$f(u+v) = ng\left(\frac{u+v}{n}\right) = ng\left(\frac{u}{n}\right) + ng\left(\frac{v}{n}\right) = f(u) + f(v).$$

This proves the existence of f satisfying Cauchy's equation and (4). The uniqueness is easy. For every $t \in \mathbb{R}$ there exists a positive integer n, such that $|t|/n < r$. If $\tilde{f}$ also satisfies Cauchy's equation and (4), then, by (3) and (4),

$$f(t) = nf\left(\frac{t}{n}\right) = ng\left(\frac{t}{n}\right) = n\tilde{f}\left(\frac{t}{n}\right) = \tilde{f}(t) \quad \text{for all real } t.$$

This concludes the proof of Theorem 4. (The uniqueness could also have been proved easily in Theorem 2.1.)

A similar result holds for geometrical domains other than the open diamond used in Theorem 4, for example, for the closed diamond $\{(x, y) \mid |x| + |y| \leqslant r\}$, or for the closed triangle $\{(x, y) \mid x \geqslant 0,\ y \geqslant 0,\ x + y \leqslant r\}$, or for the hexagon/triangle $\{(x, y) \mid x, y, x + y \in I\}$ (I a proper interval with 0 in the interior or on the boundary), or for a circle of radius r and center 0, etc. (The domains should contain the origin or be connected and have 0 on the boundary.)

A more general treatment is provided by the following result, quoted here without proof. (Cf. Exercise 11. The fundamental idea of the proof is quite similar to that of Theorem 4. See Dhombres–Ger 1978 and Ng 1974*a*.) Let X be a nonempty subset of a divisible abelian group. We say that X is *full* if the following two conditions are satisfied:

(1) for all $x \in X$ and for all $\lambda \in Q \cap]0, 1[$ we have $\lambda x \in X$;
(2) for any pair $(x, y) \in X^2$ there exists an integer n, which may depend upon x and y, such that $(x + y)/n \in X$.

Theorem 5. *Let F and G be abelian divisible groups and let X be a full subset of G. Take the 'triangle' (hexagon)*

$$Z = \{(x, y) \mid x \in X, y \in X, x + y \in X\},$$

and suppose $f: G \to F$ is Z-additive. There exists an additive $g: G \to F$ such that $f = g$ on X. Moreover, g is unique if, and only if, the subgroup generated by X is G.

In other words, this Z is additively expandable, and it is uniquely additively expandable if, and only if, the subgroup generated by X is G.

We will see other domains which are not expandable in the above sense, but they are 'quasi-expandable', which we will define now (Daróczy–Losonczi 1967). (The lack of expansibility is mainly due to the role played by the origin for conditional Cauchy equations.)

Take a point $(a, b) \in Z$. A function f, *satisfying Cauchy's equation everywhere, is an (additive) quasi-extension of the solution g of a conditional Cauchy equation relative to Z if*

$$f(x) - f(a) = g(x) - g(a) \quad \text{for all } x \in \Pi_1(Z), \tag{9}$$

$$f(y) - f(b) = g(y) - g(b) \quad \text{for all } y \in \Pi_2(Z), \tag{10}$$

and

$$f(z) - f(a+b) = g(z) - g(a+b) \quad \text{for all } z \in t(Z). \tag{11}$$

If such an f exists, then Z is *quasi-expandable.*

This definition is *independent of the choice of the point* $(a, b) \in Z$. Indeed, let $(a', b') \in Z$ be another point. Then $a' \in \Pi_1(Z)$, $b' \in \Pi_2(Z)$, and $a' + b' \in t(Z)$. By (9), (10) and (11)

$$f(a') - f(a) = g(a') - g(a), \quad f(b') - f(b) = g(b') - g(b),$$

and

$$f(a' + b') - f(a+b) = g(a' + b') - g(a+b).$$

If we subtract these from (9), (10) and (11), we get

$$f(x) - f(a') = g(x) - g(a') \quad (x \in \Pi_1(Z)), \quad f(y) - f(b') = g(y) - g(b') \quad (y \in \Pi_2(Z)),$$
$$f(z) - f(a' + b') = g(z) - g(a' + b') \quad (z \in t(Z)),$$

that is, the same equations (9), (10) and (11) hold with (a', b') instead of (a, b).

Of course (9), (10) and (11) can also be written as

$$\left.\begin{aligned} &f(x) = g(x) + \alpha \quad (x \in \Pi_1(Z)), \quad f(y) = g(y) + \beta \quad (y \in \Pi_2(Z)), \\ &f(z) = g(z) + \alpha + \beta \quad (z \in t(Z)). \end{aligned}\right\} \tag{12}$$

If there exists an $f: \mathbb{R} \to \mathbb{R}$, which is additive and such that

$$f(x) = g(x) \quad \text{for all } x \text{ in } A(Z),$$

then we have called f an *(additive) extension* of our Z-additive function g. Every extension is a quasi-extension and, *if two of* $\Pi_1(Z) \cap \Pi_2(Z)$, $\Pi_1(Z) \cap t(Z)$, $\Pi_2(Z) \cap t(Z)$ *are nonempty, then all quasi-extensions relative to Z are extensions,* as can be seen from (12). In particular, *if* $(0, 0)$ *is in Z, or if* $]0, \varepsilon[^2 \subset Z$ *for some* $\varepsilon > 0$, *then all quasi-extensions are extensions.* This is what we meant by saying that the origin plays an important role here.

We need quasi-extensions because in many, even simple cases, no extensions exist, but in many of these there exist quasi-extensions.

For instance, the function g defined by

$$g(x) = \begin{cases} x + 1 & \text{for } x \in]2, 3[, \\ x + 2 & \text{for } x \in]4, 6[, \end{cases} \tag{13}$$

and taking arbitrary values elsewhere (on $\mathbb{R}$), is Z-additive, where $Z =]2, 3[^2$ as it satisfies

$$g(x + y) = g(x) + g(y) \quad \text{for all } (x, y) \in]2, 3[^2. \tag{14}$$

But g cannot be extended to an additive function for which

$$f(x) = g(x) \quad \text{whenever } x \in]2,3[\cup]4,6[= A(]2,3[^2). \tag{15}$$

Indeed, by (13) and (15), f would be continuous (and bounded) on an interval (say $]2,3[$) and thus, by Corollary 2.5, is of the form $f(x) = cx$ for all $x \in \mathbb{R}$. But no such function is an extension of (13). However, $f(x) = x$ ($x \in \mathbb{R}$) is a quasi-extension of the solution (13) of (14) relative to $]2,3[^2$ (take $a = b = \frac{5}{2}$ in (9) and (10) or $\alpha = \beta = -1$ in (12)).

Using quasi-extensions instead of extensions, there is no difficulty in defining a *(uniquely) quasi-expandable subset* Z of the plane $\mathbb{R}^2$ or in defining a *quasi-addundant* subset Z of the same plane.

For any $r > 0$, define the diamond $U_r(a,b) = \{(x,y) \mid |x-a| + |y-b| < r\}$ where (a,b) is a given point of $\mathbb{R}^2$. Notice that $\Pi_1(U_r(a,b)) =]a-r, a+r[$, $\Pi_2(U_r(a,b)) =]b-r, b+r[$, $t(U_r(a,b)) =]a+b-r, a+b+r[$. We prove the following:

Lemma 6. *$U_r(a,b)$ is uniquely quasi-expandable.*

We shall prove that there exists a *unique quasi-extension* of every solution of Cauchy's equation restricted to $U_r(a,b)$. By our definition,

$$g(x+y) = g(x) + g(y) \quad \text{whenever } (x,y) \in U_r(a,b). \tag{16}$$

But $(a,b) \in U_r(a,b)$. Thus, also,

$$g(a+b) = g(a) + g(b). \tag{17}$$

Choose s, t so that $|s| + |t| < r$, then

$$(s+a, t+b) \in U_r(a,b)$$

and we may substitute into (16)

$$x = s+a, \quad y = t+b \quad (s \in]-r,r[, \quad t \in]-r,r[, \quad |s| + |t| < r). \tag{18}$$

We get

$$g(s+t+a+b) = g(s+a) + g(t+b)$$

and, after subtraction of (17),

$$g(s+t+a+b) - g(a+b) = g(s+a) - g(a) + g(t+b) - g(b)$$
$$\text{whenever } |s| + |t| < r. \tag{19}$$

In particular, if $t = 0$, we get

$$g(s+a+b) - g(a+b) = g(s+a) - g(a) \quad \text{for all } s \in]-r,r[$$

and, if $s = 0$,

$$g(t+a+b) - g(a+b) = g(t+b) - g(b) \quad \text{for all } t \in]-r,r[.$$

In view of this, we may define

$$h(u) = g(u + a + b) - g(a + b) = g(u + a) - g(a)$$
$$= g(u + b) - g(b) \quad \text{for all } u \in]-r, r[\tag{20}$$

and (19) goes over into

$$h(s + t) = h(s) + h(t) \quad \text{whenever } |s| + |t| < r. \tag{21}$$

By Theorem 4, there exists a unique extension of every solution h to a solution $f: \mathbb{R} \to \mathbb{R}$ of (1) on $\mathbb{R}^2$. This is a quasi-extension of g. Indeed,

$$f(u) = h(u) \quad \text{if } u \in]-r, r[$$

and f satisfies Cauchy's equation; furthermore, we get, from (18) and (20),

$$f(x) - f(a) = f(s + a) - f(a) = f(s) = h(s) = g(s + a) - g(a)$$
$$= g(x) - g(a) \quad \text{for all } x \in]a - r, a + r[= \Pi_1(U_r(a, b)),$$
$$f(y) - f(b) = f(t + b) - f(b) = f(t) = h(t) = g(t + b) - g(b)$$
$$= g(y) - g(b) \quad \text{for all } y \in]b - r, b + r[= \Pi_2(U_r(a, b)),$$
$$f(z) - f(a + b) = f(u + a + b) - f(a + b) = f(u) = h(u) = g(u + a + b) - g(a + b)$$
$$= g(z) - g(a + b) \quad \text{for all } z \in]a + b - r, a + b + r[= t(U_r(a, b)),$$

so (9), (10) and (11) are satisfied. The uniqueness follows immediately from the uniqueness in Theorem 4 (since every quasi-extension of g must be an extension of h viewed as a solution of the conditional equation (21)).

This concludes the proof of the Lemma 6.

Notice that $U_r(a, b)$ is an (open) neighbourhood of (a, b). We can go over to other (connected and) open sets with aid of the following lemma:

Lemma 7. *Let $g: \mathbb{R} \to \mathbb{R}$ be a Z-additive function where Z is the union of two open subsets of $\mathbb{R}^2$ with nonempty intersection ($Z = Z_1 \cup Z_2, Z_1 \cap Z_2 \neq \varnothing$). If Z_1 and Z_2 are (uniquely) quasi-expandable, then so is the set Z and the three quasi-extensions are identical.*

Proof. Since Z_1 and Z_2 are open with nonempty intersection, $\Pi_1(Z_1) \cap \Pi_1(Z_2)$ contains a nonempty open interval I. Take an arbitrary $a \in I$. Let f_1 (resp. f_2) be a quasi-extension of $g: \mathbb{R} \to \mathbb{R}$ relative to Z_1 (resp. Z_2). By definition, f_1 and f_2 satisfy

$$f_1(x - a) = f_1(x) - f_1(a) = g(x) - g(a) \quad \text{for all } x \in \Pi_1(Z_1)$$

and

$$f_2(x - a) = f_2(x) - f_2(a) = g(x) - g(a) \quad \text{for all } x \in \Pi_1(Z_2).$$

If $x \in I$, then $x - a$ is in a neighbourhood J of the origin 0 and in this neighbourhood we have obtained $f_1 = f_2$. But $f_1 - f_2$ is additive and bounded above by 0 in this neighbourhood. Therefore, by Corollary 2.5, we have that $f_1(x)-f_2(x)=cx$ for some c in $\mathbb{R}$ and all x in $\mathbb{R}$. As $f_1(x)=f_2(x)$ for all x in J, we deduce that $c=0$. So f_1 and f_2 are indeed identical. This function is then a quasi-extension of g relative to $Z_1 \cup Z_2$.

As a consequence, we see that, for non-empty open Z, if there exists a quasi-extension relative to Z, then this quasi-extension is unique. This concludes the proof of Lemma 7.

Now we are ready to prove the following:

Theorem 8. *There exists a unique quasi-extension of every solution of Cauchy's equation relative to any connected open set $Z \subset \mathbb{R}^2$. Thus, every connected nonempty open subset of $\mathbb{R}^2$ is uniquely additively quasi-expandable.*

Proof. Since Z is open and connected, there exist open diamonds (squares) $U^{(1)}, U^{(2)}, \ldots, (\subseteq Z)$, such that $Z \subseteq \bigcup_{k=1}^{\infty} U^{(k)}$ and $(\bigcup_{k=1}^{n} U^{(k)}) \cap U^{(n+1)} \neq \emptyset$ $(n=1,2,\ldots)$. By Lemma 6 there exists exactly one quasi-extension f_k of g relative to $U^{(k)}$. By Lemma 7, $f_1=f_2=\cdots=f_n$ for all n. Thus $f=f_1$ is the unique quasi-extension of g relative to $\bigcup_{k=1}^{n} U^{(k)}$ for all n and so also the unique quasi-extension relative to Z. This concludes the proof of Theorem 8.

Note that the same argument shows that there is *at most* one quasi-extension relative to any nonempty *open* set in $\mathbb{R}^2$ even if connectedness is not supposed. All these are results of Daróczy and Losonczi (1967) with slightly simplified proofs. They also noticed that similar results are true if the values of g (and f) are in an arbitrary abelian topological group rather than $\mathbb{R}$, for instance in $\mathbb{R}^n$ (cf. Chapter 4).

Also, some boundary points may be attached to Z and Theorem 8 and the above note still remains true. At the end of Section 2 we will see an example where no quasi-extension exists (the set Z will not be connected).

7.2 Extensions almost everywhere and integral transforms

We shall now apply a conditional Cauchy equation in harmonic analysis. We need in this section some results from the Lebesgue integration theory.

Let $L^1(\mathbb{R})$ be the Banach space of all equivalence classes of Lebesgue integrable functions $f:\mathbb{R}\to\mathbb{C}$ equipped with the norm

$$\|f\| = \int_{\mathbb{R}} |f(x)|\,dx.$$

As is well known, the function $f*g$, the *convolution* of f and g in $L^1(\mathbb{R})$, defined by

$$f*g(x) = \int_{\mathbb{R}} f(t)g(x-t)\,dt,$$

is an element of $L^1(\mathbb{R})$. Moreover $L^1(\mathbb{R})$, with the operation $*$, is a complex Banach algebra without unit ($\|f*g\| \leqslant \|f\|\cdot\|g\|$ holds).

An important feature of this algebra, as was seen in the case of Bohr functions (cf. Section 5.3), is the set of all linear non-zero multiplicative functionals on L^1, that is, the *spectrum* of the Banach algebra $L^1(\mathbb{R})$. It is the set of all F such that $F \not\equiv 0$, with

$$F: L^1(\mathbb{R}) \to \mathbb{C},$$

$$F(\lambda f + \mu g) = \lambda F(f) + \mu F(g) \quad \text{for all } f, g \text{ in } L^1(\mathbb{R}), \text{ all } \lambda, \mu \text{ in } \mathbb{C}, \tag{22}$$

$$F(f*g) = F(f)F(g) \quad \text{for all } f, g \text{ in } L^1(\mathbb{R}) \tag{23}$$

and

$$|F(f)| \leqslant \|f\|_\infty \quad \text{for all } f \text{ in } L^1(\mathbb{R}). \tag{24}$$

(The inequality (24) can be written as $\|F\| \leqslant 1$.) Therefore, by a classical duality theorem in the Lebesgue theory, stating that $L^\infty(\mathbb{R})$ may be considered as the dual space of $L^1(\mathbb{R})$, there exists an $h: \mathbb{R} \to \mathbb{C}$, essentially bounded ($h \in L^\infty(\mathbb{R})$), measurable, and not 0 a.e. (almost everywhere), such that

$$F(f) = \int_{\mathbb{R}} f(x)h(x)dx.$$

(One can prove that (22) and (23) imply (24), even $\|F\| = 1$, see Exercise 13.)

Returning to the multiplicative condition (23) and using Fubini's theorem, we get the relation

$$\int_{\mathbb{R}} h(x)\left[\int_{\mathbb{R}} f(t)g(x-t)dt\right]dx = \left[\int_{\mathbb{R}} h(t)f(t)dt\right]\left[\int_{\mathbb{R}} h(u)g(u)du\right].$$

The left hand side can be written as

$$\int_{\mathbb{R}} f(t)\left[\int_{\mathbb{R}} h(x)g(x-t)dx\right]dt = \int_{\mathbb{R}} f(t)\left[\int_{\mathbb{R}} h(u+t)g(u)du\right]dt$$

and so, changing to double integrals, we get

$$\iint_{\mathbb{R}\times\mathbb{R}} f(t)g(u)[h(u+t) - h(u)h(t)]du\,dt = 0.$$

From the fact that the space of all linear combinations of functions of the form $f(t)g(u)$ is dense in $L^1(\mathbb{R}^2)$ (generalized Stone–Weierstrass theorem for $L^1(\mathbb{R}^2)$), we deduce that

$$h(u+t) = h(u)h(t) \text{ in the sense of } L^\infty(\mathbb{R}^2),$$

that is,

$$h(u+t) = h(u)h(t) \text{ almost everywhere in } \mathbb{R}^2. \tag{25}$$

Equation (25) is a conditional (exponential) Cauchy equation. Here $h: \mathbb{R} \to \mathbb{C}$. Define $\mathbb{C}^* = \mathbb{C}\setminus\{0\}$. If the solution h of (25) is equal to zero on a set of strictly positive Lebesgue measure, then $h=0$ almost everywhere (a.e.), which we exclude. If the solution h is 0 on a set of measure zero, then we modify h on this set so the range of h is $\mathbb{C}^*$ and h still satisfies (25) almost everywhere on $\mathbb{R}^2$.

Theorem 9. *If a function $h:\mathbb{R}\to\mathbb{C}$ satisfies (25) a.e. on $\mathbb{R}^2$, there exists a unique function $H:\mathbb{R}\to\mathbb{C}$ satisfying $H=h$ a.e. and the (exponential Cauchy) equation*

$$H(u+t)=H(u)H(t) \quad (u,t\in\mathbb{R}) \tag{26}$$

everywhere.

At first glance, Theorem 9 may look obvious. However its proof requires some attention. (For the proof below, cf. de Bruijn 1966 and Ger 1975. See also Jurkat 1965 and Exercise 12.)

Proof. Let $h:\mathbb{R}\to\mathbb{C}$ satisfy (25) almost everywhere on $\mathbb{R}^2$. This means that, for all pairs (x,y) *not belonging* to a subset Z' of Lebesgue measure zero in $\mathbb{R}^2$, we get

$$h(x+y)=h(x)h(y) \quad \text{for all } (x,y)\in Z=\mathbb{R}^2\backslash Z'. \tag{27}$$

As explained just before stating Theorem 9, once the possibility $h(x)=0$ a.e., $H(x)\equiv 0$ is recognized, we may suppose from now on that $h:\mathbb{R}\to\mathbb{C}^*$. Let X' be the set of all x in $\mathbb{R}$ such that the section

$$Z'_x=\{y\,|\,y\in\mathbb{R},(x,y)\in Z'\} \tag{28}$$

is not of Lebesgue measure zero. The set X' itself is of Lebesgue measure zero (using Fubini's theorem).

For any x in $\mathbb{R}$, we choose $i(x)$, a real number satisfying the two conditions

$$i(x)\notin X' \tag{29}$$

and

$$x-i(x)\notin X'. \tag{30}$$

Such a choice is always possible, as $X'\cup(x-X')$ is of Lebesgue measure zero, so certainly $X'\cup(x-X')\neq\mathbb{R}$. Let $Y_x=Z'_{i(x)}\cup[-i(x)+Z'_{x-i(x)}]$. (Remember that $t\pm U:=\{z\,|\,z=t\pm u,u\in U\}$.)

Our choice of $i(x)$ is such that Y_x is of Lebesgue measure zero. Now we may define $H:\mathbb{R}\to\mathbb{R}$ by

$$H(x)=\frac{h(x+y)}{h(y)} \quad (y\notin Y_x), \tag{31}$$

where y denotes any real number, not belonging to Y_x. We have proved that Y_x is of measure 0, so $Y_x\neq\mathbb{R}$.

In order that this definition of H make sense, we have to prove that the function H does not depend upon the particular choice of $y\notin Y_x$. Since

$y \notin Y_x$, we get $y \notin -i(x) + Z'_{x-i(x)}$. Therefore we compute, using (27) and $(x - i(x), i(x) + y) \in Z$, the complement of Z' in $\mathbb{R}^2$,

$$\frac{h(x+y)}{h(y)} = \frac{h(x - i(x) + i(x) + y)}{h(y)} = \frac{h(x - i(x))h(i(x) + y)}{h(y)}$$

and, as $y \notin Y_x$, we get $y \notin Z'_{i(x)}$ and thus $(i(x), y) \in Z$. Therefore we also get

$$\frac{h(x+y)}{h(y)} = \frac{h(x - i(x))h(i(x))h(y)}{h(y)} = h(x - i(x))h(i(x)).$$

This expression is independent of y.

Now we prove that H is a.e. equal to h. For any $x \notin X'$ let us choose y not in Y_x, and not in Z'_x, which we can do since $Y_x \cup Z'_x$ is of Lebesgue measure zero. Therefore, as $(x, y) \in Z$,

$$H(x) = \frac{h(x+y)}{h(y)} = \frac{h(x)h(y)}{h(y)} = h(x).$$

In other words, $H(x) = h(x)$ for all $x \notin X'$ or $H = h$ a.e.

To conclude our proof of Theorem 9, it must be proved that H satisfies (26). The trick is to use convenient translates. We start from real numbers u, v and a certain number of conditions will be imposed upon auxiliary real numbers x, y, z and t. We first require that

$$x \notin Y_{u+v}. \tag{32}$$

We deduce that

$$H(u+v) = \frac{h(u+v+x)}{h(x)}.$$

Then we require

$$y \notin Y_v \tag{33}$$

and

$$z \notin Y_u, \tag{34}$$

so that $H(v) = h(v+y)/h(y)$ and $H(u) = h(u+z)/h(z)$. Let $s = y - x$ and require that $s \notin Z'_x$ or

$$y \notin x + Z'_x. \tag{35}$$

We choose $t = z - v - x - s$, so that $z - t = v + y$ and require $t \notin Z'_{v+y}$, which amounts to

$$z \notin v + y + Z'_{v+y}. \tag{36}$$

We write

$$\frac{H(u+v)}{H(u)H(v)} = \frac{h(u+v+x)}{h(x)} \cdot \frac{h(y)}{h(v+y)} \cdot \frac{h(z)}{h(u+z)}$$

and obtain that

$$\frac{H(u+v)}{H(u)H(v)} = \frac{h(u+v+x)}{h(x)} \cdot \frac{h(x)h(s)}{h(z)} \cdot h(t) \cdot \frac{h(z)}{h(u+z)}$$
$$= \frac{h(u+v+x)h(s)h(t)}{h(u+z)}.$$

Suppose $(s,t) \in Z$ which amounts to $t \notin Z'_{-x+y}$ and can be written as

$$z \notin v + y + Z'_{-x+y} \tag{37}$$

and $(u+v+x, s+t) \in Z$ which means $t \notin -y+x+Z'_{u+v+x}$ and

$$z \notin v + x + Z'_{u+v+x}. \tag{38}$$

Thus

$$\frac{H(u+v)}{H(u)H(v)} = \frac{h(u+v+x)h(s+t)}{h(u+z)} = \frac{h(u+v+x+s+t)}{h(u+z)} = \frac{h(u+z)}{h(u+z)} = 1.$$

Equation (26) is thus satisfied as soon as we prove that we can find x, y, z in $\mathbb{R}$ satisfying (32), (33), (34), (35), (36), (37) and (38). First we require that all subsets occurring in these seven conditions be of Lebesgue measure zero. This is already true for (32), (33) and (34). For (35), this requires

$$x \notin X'. \tag{39}$$

For (36),

$$y \notin -v + X'. \tag{40}$$

For (37),

$$y \notin x + X' \tag{41}$$

and, for (38),

$$x \notin -u - v + X'. \tag{42}$$

We can now choose x as described above, since (32), (39) and (42) only require that x should not belong to some subset of Lebesgue measure zero. The second specification is for y, which is possible for the same reason, watching out for (33), (35), (40) and (41). The third and last specification is for z which is also possible if we satisfy (34), (36), (37) and (38). This concludes the proof of Theorem 9.

Of course, equation (26) for $H: \mathbb{R} \to \mathbb{C}$ has already been solved in Section 5.1 (Theorem 5.3). Moreover, here, as h, also H has to belong to $L^\infty(\mathbb{R})$, that is, H is essentially bounded and Lebesgue measurable. Theorem 5.4 yields

$$H(t) = e^{\lambda t} \quad \text{for some } \lambda \in \mathbb{C},$$

or

$$H(t) = 0.$$

Since h was assumed not to be equal to zero almost everywhere, we exclude the last case. Since H is essentially bounded on $\mathbb{R}$, therefore λ is necessarily of the form $\lambda = i\beta$ where $\beta \in \mathbb{R}$. We write, for notational purposes,

$$\lambda = 2i\pi a \quad (a \in \mathbb{R}).$$

Going back to our functional F, acting on the convolution algebra $L^1(\mathbb{R})$, we have obtained

$$F(f) = \int_{-\infty}^{\infty} f(x) \exp(2i\pi a x)\, dx.$$

This integral is usually denoted by $\hat{f}(a)$. It is called the *Fourier integral* of f computed in a.

We have proved that every element F of the spectrum of the convolution algebra $L^1(\mathbb{R})$ has to be of the form

$$F(f) = \hat{f}(a)$$

for some a in $\mathbb{R}$.

Conversely, let $a \in \mathbb{R}$ be given. The mapping $f \mapsto \hat{f}(a)$ from $L^1(\mathbb{R})$ into $\mathbb{C}$ is well defined. We even notice

$$|\hat{f}(a)| \leqslant \int_{-\infty}^{\infty} |f(x)|\, dx.$$

So the mapping $f \mapsto \hat{f}$ is linear. It is a multiplicative mapping

$$(f * g)(a) = \hat{f}(a)\hat{g}(a) \quad \text{for } f, g \text{ in } L^1(\mathbb{R}),$$

as can easily be seen by using double integrals.

In order to prove that this mapping defines an element of the spectrum of $L^1(\mathbb{R})$, it remains to be proved that it is not the zero mapping. A classical computation solves this as, for $f(x) = e^{-x^2}$, we obtain

$$\hat{f}(a) = e^{-a^2}.$$

For two distinct a's, a_1 and a_2, the previous function f yields $\hat{f}(a_1) \neq \hat{f}(a_2)$.

In other words, we have established a bijection between the spectrum of $L^1(\mathbb{R})$ and the set of all real numbers $\mathbb{R}$. We state:

Theorem 10. *The spectrum of the convolution algebra $L^1(\mathbb{R})$ can be identified with the set $\mathbb{R}$ by identifying $a \in \mathbb{R}$ with the mapping*

$$f \mapsto \hat{f}(a).$$

In the proof of Theorem 10, we have used Theorem 9. Conversely, under some restrictions, Theorem 9 follows from Theorem 10, see Exercise 14.

If some natural topology (weak star topology) is introduced in the spectrum, it is convenient to compare it to the usual topology of $\mathbb{R}$. We find that they coincide and $a \to \hat{f}(a)$ is continuous. Moreover, the *Fourier transform*

$\hat{f}$ of f is an element of the algebra $C_0(\mathbb{R})$ of all complex valued continuous functions of a real variable x, which tend to zero when x goes to ∞ or to $-\infty$. Theorem 10 shows that the mapping $f \mapsto \hat{f}$ from the convolution algebra $L^1(\mathbb{R})$ into the algebra $C_0(\mathbb{R})$ is a linear homomorphism. A nice result on the Fourier integral is that the mapping $\hat{f} \mapsto f$ is also an injection from $L^1(\mathbb{R})$ into $C_0(\mathbb{R})$. However, it is not a surjection (Rudin 1987, p. 185).

We mention here that Theorem 10 is basic to harmonic analysis and can be generalized to any locally compact abelian group. If we start from the group $\mathbb{R}$, equipped now with the discrete topology $\mathbb{R}_d$, we define $L^1(\mathbb{R}_d)$ and its spectrum. The identification of the spectrum, similar to that in Theorem 10, would lead us back to the Bohr group of $\mathbb{R}$ which we obtained in a different way in Section 5.3. (See also Exercise 10.22.)

A further comment on Theorem 9 might be useful. An irritating aspect in the statement made there is that the subset of $\mathbb{R}$, on which $H = h$, is not clearly related to the subset Z of $\mathbb{R}^2$ on which h is Z-additive. It certainly does not have to be $A(Z)$. In other words, H is not an extension or quasi-extension of h relative Z. In fact, in this case there may not exist any extension (or even quasi-extension) at all. We show this, equivalently, for solutions $f: \mathbb{R} \to \mathbb{R}$ of

$$f(x+y) = f(x) + f(y) \quad \text{a.e.,} \tag{43}$$

instead of (25). So now (43) *is true for all* (x, y) *in* $Z \subset \mathbb{R}^2$ *where* $\mathbb{R}^2 \backslash Z$ *is of Lebesgue measure zero.* By a similar proof as for Theorem 9, *there exists an* $F: \mathbb{R} \to \mathbb{R}$, *satisfying the Cauchy equation*

$$F(x+y) = F(x) + F(y) \quad \textit{for all } x, y \textit{ in } \mathbb{R}^2 \tag{44}$$

and such that

$$F(x) = f(x) \quad \text{a.e.,} \tag{45}$$

that is, for all $x \in Z_1$ where $\mathbb{R} \backslash Z_1$ is of Lebesgue measure zero.

However, Z_1 does not necessarily contain $A(Z)$. We give an example where $A(Z) = \mathbb{R}$ and $Z_1 \neq \mathbb{R}$. Take (Figure 10)

$$\Delta = \Delta_1 \cup \Delta_2 \cup \Delta_1' \cup \Delta_2' \cup \Delta_1'' \cup \Delta_2''$$

where, for $j = 1$ or 2,

$$\Delta_j = \{(x, y) \mid x = j\},$$
$$\Delta_j' = \{(x, y) \mid y = j\},$$
$$\Delta_j'' = \{(x, y) \mid x + y = j\},$$

and define Z in $\mathbb{R}^2$ so that

$$Z = (\mathbb{R}^2 \backslash \Delta) \cup \{(1, 1)\}.$$

Clearly $\mathbb{R}^2 \backslash Z$ is of Lebesgue measure zero. Now define $f: \mathbb{R} \to \mathbb{R}$ to be equal

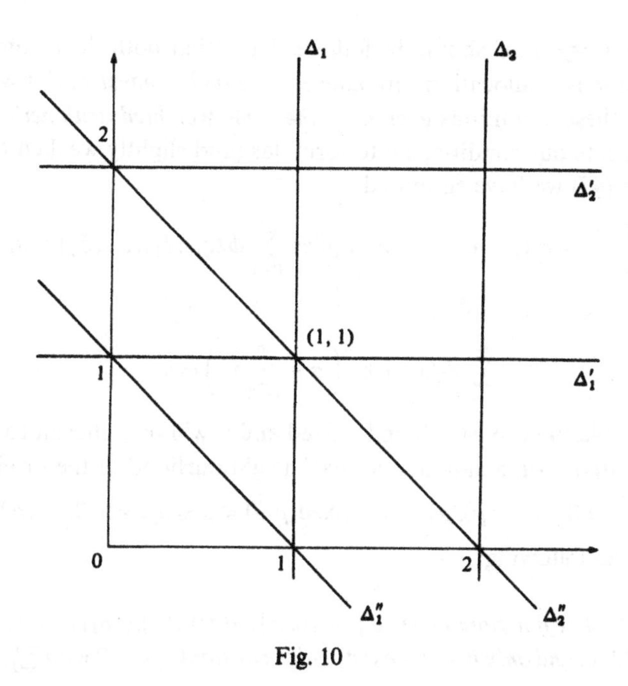

Fig. 10

to zero for all x in $\mathbb{R}$ except for $x=1$ where $f(1)=1$ and $x=2$ where $f(2)=2$. This function is Z-additive (we have $f(1+1)=f(2)=2=1+1=f(1)+f(1)$ and elsewhere for $(x,y)\in Z$ we get $f(x)=f(y)=f(x+y)=0$). But $A(Z)=\mathbb{R}$ and clearly f is not additive on all of $\mathbb{R}$ (example of V. Zinde–Walsh, modified; see Aczél 1980*c*).

7.3 Consensus allocations

We end this chapter with an application of a conditional Cauchy equation to rational group decision making.

The problem is the following (see Aczél–Wagner 1981; Aczél–Kannappan–Ng–Wagner 1981): A fixed amount σ of money or other quantifiable resource is to be allocated to a fixed number $m>2$ of competing projects. Each member of a group of n decision makers makes recommendations, the jth decision maker allocating, say, ξ_{ij} to the ith project in order to establish the 'consensus' allocation

$$\phi_i(x_i)=\phi_i(\xi_{i1},\ldots,\xi_{in})\quad(i=1,2,\ldots,m),\tag{46}$$

where $x_i=(\xi_{i1},\ldots,\xi_{in})$. (For easy recognition, we denote here all (real) scalars by Greek letters, vectors by Latin letters, except that, when a Greek letter stands where a vector should be, $\sigma=(\sigma,\sigma,\ldots,\sigma)$, etc., is meant.) We suppose that

$$\phi_i(0)=\phi_i(0,\ldots,0)=0\quad(i=1,2,\ldots,)\tag{47}$$

(‘*unanimity in rejection*’ should be followed) and that both the recommended and the consensus allocations are *nonnegative and exhaust* σ. We will prove that under these conditions *each* ϕ_i *is the same weighted arithmetic mean.*

We translate our conditions into formulas (and slightly weaken them). In addition to (47) we have supposed

$$\sum_{i=1}^{m} \xi_{ij} = \sigma \ (j = 1, 2, \ldots, n) \text{ implies } \sum_{i=1}^{m} \phi_i(\xi_{i1}, \xi_{i2}, \ldots, \xi_{in}) = \sigma, \tag{48}$$

or, with $x_i = (\xi_{i1}, \xi_{i2}, \ldots, \xi_{in})$,

$$\sum_{i=2}^{m} \phi_i(x_i) + \phi_1\left(\sigma - \sum_{i=2}^{m} x_i\right) = \sigma. \tag{49}$$

We will be able to keep $m(>2)$ and σ fixed and it will be sufficient to suppose the nonnegativity on a subinterval (half-neighbourhood of the origin):

$$\phi_i(x) \geqslant 0 \quad \text{if } x \in [0, \rho]^n \text{ for a fixed } \rho \ (0 < \rho \leqslant \sigma; i = 1, 2, \ldots, m). \tag{50}$$

We prove the following:

Theorem 11. *The functions* $\phi_i: [0, \sigma]^n \to [0, \sigma]$ *satisfy* (47), (48) [*or, equivalently,* (49)] *and* (50) *if, and only if, there exist scalar constants* $\omega_j \geqslant 0$ *with* $\sum_{j=1}^{n} \omega_j = 1$ *such that*

$$\phi_i(x) = \phi_i(\xi_1, \xi_2, \ldots, \xi_n) = \sum_{j=1}^{n} \omega_j \xi_j \quad (i = 1, \ldots, m; x \in [0, \sigma]^n). \tag{51}$$

Proof. Put $x_2 = x_3 = \cdots = x_n = 0$ into (49) and take (47) into consideration in order to get $\phi_1(\sigma) = \sigma$ and, similarly,

$$\phi_i(\sigma) = \sigma \quad (i = 1, 2, \ldots, m) \tag{52}$$

(‘unanimity on overwhelming merit’). Also, substituting $x_3 = \cdots = x_m = 0$ into (49), we get, with (47),

$$\phi_1(\sigma - x) = \sigma - \phi_2(x)$$

and, with this and (47), putting $x_4 = \cdots = x_m = 0$, $x_2 = x$, $x_3 = y$ into (49), the Pexider equation

$$\phi_2(x + y) = \phi_2(x) + \phi_3(y) \quad (x, y, x + y \in [0, \sigma]^n), \tag{53}$$

cf. (4.30). As in Section 4.3, we put here $x = 0$ and get, using (47) again, $\phi_2(y) = \phi_3(y)$ and, in general,

$$\phi_1 = \phi_2 = \cdots = \phi_m =: \phi, \tag{54}$$

which reduces (53) to

$$\phi(x + y) = \phi(x) + \phi(y) \quad \text{for } x, y, x + y \in [0, \sigma]^n.$$

This is (4.10) ‘on a restricted domain’ (a conditional multiplace Cauchy equation). By Theorem 5 (and in analogy to Theorem 4), ϕ has an additive extension. Moreover, by the sequel of Corollary 4.2, this extension, and thus

ϕ itself, is of the form

$$\phi(x) = (w, x) = \sum_{j=1}^{n} \omega_j \xi_j.$$

In view of (50) and (52), we have $\omega_j \geqslant 0$ and $\sum_{j=1}^{n} \omega_j = 1$. This and (54) gives the result since (51) clearly satisfies all our conditions.

There exist other solutions if $m \leqslant 2$, for instance,

$$\phi_1(x) = \phi_2(x) = \left[\frac{1}{n}\sum_{j=1}^{n}\left(\xi_j - \frac{\sigma}{2}\right)^3\right]^{1/3} + \frac{\sigma}{2}$$

for $m = 2$, where

$$\sigma = (\xi_1 + \xi_2 + \cdots + \xi_n)/n.$$

Exercises and further results

1. Give an example of a subset Z of $\mathbb{R}^2$ which is not additively quasi-expandable.
2. Show by an example that, when $(0, 0)$ belongs to the boundary of Z, there may exist a quasi-extension which is not an extension.
3. Is the set $Z = [1, 2]^2$ uniquely quasi-expandable? Is it (uniquely) expandable?
4. Show that there exists a subset of $\mathbb{R}^2$ which is (quasi-) expandable but not uniquely (quasi-) expandable.
5. Give a simpler example than that at the end of Section 2 of an open subset of $\mathbb{R}^2$ which is not quasi-expandable.
6. (Chernoff 1975.) Find all solutions $f:\mathbb{R} \to \mathbb{R}$ of the functional equation

$$f(x + y^n) = f(x) + (f(y))^n \quad (x \in \mathbb{R};\, y \in \bar{\mathbb{R}}_+),$$

where n is a positive integer.

(For Exs. 7–10, see Dhombres 1979, pp. 3.13–3.22.)

7. Let G be an abelian group and $f:G \to G$. Find all solutions of

$$f(x + f(y)) = f(x) + f(f(y)) \quad (x, y \in G).$$

(Cf. Proposition 2.)

8. Prove or disprove that all continuous solutions $f:\mathbb{R} \to \mathbb{R}$ of the functional equation

$$f(x + y - f(y)) = f(x) + f(y - f(y)) \quad (x, y \in \mathbb{R})$$

are additive.

9. For any mapping $g:G \to G$, where G is an abelian group, we define a function $\mathscr{X}$ by $\mathscr{X}(g) = h(G)$ where $h(x) = x - g(x)$. Prove or disprove that $\mathscr{X}$ is not additively expandable.
10. Let H be a closed subgroup of an abelian topological group G. Suppose that $g:H \to H$ is a continuous bijection. Prove or disprove the equivalence of the following two assertions:

(a) There exists a continuous $f:G \to H$ such that $f(x + z) - f(x) = g(z)$ for all x in G, z in H.

(b) g is additive and there exists a continuous lifting relative to H.

11. (Dhombres–Ger 1978.) Prove Theorem 5.
12. Modify the proof of Theorem 9 in order to obtain the following result:

    Let G be a locally compact group and F a group. Let $h:G \to F$ be such that $h(xy) = h(x)h(y)$ for $(x, y) \in Z$ where $(G \times G) \setminus Z$ is of Haar measure zero. There exists a homomorphism $H:G \to F$ such that $H = h$ except possibly on a subset of Haar measure zero.
13. Show that (22) and (23) imply (24), even $\|F\| = 1$. In other words, show that every complex not identically zero linear multiplicative functional on $L^1(\mathbb{R})$ is of norm 1.
14. Give a direct proof of Theorem 9 from Theorem 10, assuming that h is essentially bounded and Lebesgue measurable on $\mathbb{R}$. Deduce that H is continuous.

15*. (Weil 1940, pp. 48–52; Dhombres 1979, pp. 1.6–1.9.) Let E and F be two subsets of positive Lebesgue measure in $\mathbb{R}$. Let f (resp. g) be the characteristic function of E (resp. F), that is, $f(x) = 1$ if $x \in E$ and $f(x) = 0$ if $x \notin E$. Show that the convolution h, where

$$h(x) = \int_{-\infty}^{\infty} f(x - y)g(y)dy,$$

is a continuous function, not identically zero (use the bijective property of the mapping $f \mapsto \hat{f}$) whose support is a subset of $E + F$. Deduce that $E + F$ contains a nonempty open interval. Apply this result to give another proof of Theorem 2.6.

8

D'Alembert's functional equation. An application to noneuclidean mechanics

In the first chapter on the parallelogram law for addition of vectors, we introduced a functional equation, the d'Alembert equation, which has the cosine functions as bounded regular solutions. We now find all continuous solutions of this equation. Going back to applications, we study composition of forces in noneuclidean geometry and mechanics. At the end of the chapter, with the help of some techniques from harmonic analysis, we characterize the functions generalizing the cosine functions to more general topological groups.

As promised in Chapter 1, we now determine all continuous solutions $f:\bar{\mathbb{R}}_+ \to \mathbb{R}$ of d'Alembert's equation (1.6), that is, of

$$f(\phi+\psi)+f(\phi-\psi)=2f(\phi)f(\psi) \tag{1}$$

(d'Alembert 1750, 1769; Cauchy 1821) on the domain $\{(\phi,\psi)|0\leqslant\psi\leqslant\phi\}$. This restriction is necessary because f is only defined on $\bar{\mathbb{R}}_+$. Of course,

$$f(\phi)=0 \quad \text{for all } \phi\geqslant 0 \tag{2}$$

gives a solution (it is the first formula in (1.7)). From now on – till the statement of Theorem 1 – we disregard this solution. Then, putting $\psi=0$ into (1), we get

$$f(0)=1. \tag{3}$$

Since f was supposed to be continuous on $\bar{\mathbb{R}}_+$, there exists a positive C such that

$$f(\phi)>0 \quad \text{for every } \phi\in[0,C]. \tag{4}$$

We distinguish two cases:

$$f(C)>1 \tag{5}$$

and

$$f(C)\leqslant 1. \tag{6}$$

The two cases can be treated quite similarly ((6) is slightly more difficult). Condition (5) leads to the continuous solution

$$f(\phi) = \cosh(c\phi) \quad (\phi \in \bar{\mathbb{R}}_+) \tag{7}$$

(the second formula in (1.7)) with $c \neq 0$, while (6) leads to

$$f(\phi) = \cos(c\phi) \quad (\phi \in \bar{\mathbb{R}}_+) \tag{8}$$

as continuous solution (the third formula in (1.7)). We will give the details for case (6).

Because of (4) and (6), there exists an angle $\alpha \in [0, \pi/2[$ such that

$$\cos\alpha = f(C). \tag{9}$$

With $\psi = \phi = x/2$, and taking (3) into consideration, from (1) we obtain

$$f(x) + 1 = 2f\left(\frac{x}{2}\right)^2. \tag{10}$$

Putting $x = C$ here, we get, with (9),

$$f\left(\frac{C}{2}\right) = \left(\frac{1 + f(C)}{2}\right)^{1/2} = \left(\frac{1 + \cos\alpha}{2}\right)^{1/2} = \cos\frac{\alpha}{2}. \tag{11}$$

Here only positive square roots come into consideration, because both extremities of (11) are positive in consequence of (4) and of $\alpha \in [0, \pi/2[$. Similarly, it follows from (4), (9), (10) and from (11), by induction, that

$$f\left(\frac{C}{2^n}\right) = \cos\frac{\alpha}{2^n} \quad \text{for all } n = 0, 1, 2, \ldots. \tag{12}$$

Now substitute

$$\phi = m\frac{C}{2^n}, \quad \psi = \frac{C}{2^n}$$

into (1), where m is a positive integer, in order to get

$$f\left(\frac{m+1}{2^n}C\right) + f\left(\frac{m-1}{2^n}C\right) = 2f\left(\frac{m}{2^n}C\right)f\left(\frac{1}{2^n}C\right). \tag{13}$$

We prove, by induction with respect to k, that

$$f\left(\frac{k}{2^n}C\right) = \cos\frac{k}{2^n}\alpha \quad \text{for all } k = 0, 1, 2, \ldots \quad \text{and} \quad n = 0, 1, 2, \ldots. \tag{14}$$

Indeed, (14) is true for $k = 0$ and for $k = 1$ by (3) and (12). Suppose now that it holds for $k = m - 1$ and $k = m$. Then, by (13),

$$f\left(\frac{m+1}{2^n}C\right) = 2f\left(\frac{m}{2^n}C\right)f\left(\frac{1}{2^n}C\right) - f\left(\frac{m-1}{2^n}C\right)$$

$$= 2\cos\left(\frac{m}{2^n}\alpha\right)\cos\frac{\alpha}{2^n} - \cos\left(\frac{m-1}{2^n}\alpha\right) = \cos\left(\frac{m+1}{2^n}\alpha\right)$$

and (14) is proved. In other words, *for all nonnegative dyadic fractions* δ (that is, $\delta = m/2^n$ where m and n are nonnegative integers)

$$f(\delta C) = \cos \delta\alpha. \tag{15}$$

Since both f and cos are continuous on $\bar{\mathbb{R}}_+$, by putting into (15) a sequence $\{\delta_m\}$ of dyadic fractions tending to the (arbitrary) positive number x, we get $f(Cx) = \cos \alpha x$ or, with $\alpha/C = c$, $Cx = \phi$, $f(\phi) = \cos c\phi$ for all $\phi \geqslant 0$, that is, (8) holds. Similarly, in the case (5) we get (7). Conversely, (7) and (8) satisfy (1). So we have proved the following:

Theorem 1. *The general continuous solution* $f:\bar{\mathbb{R}}_+ \to \mathbb{R}$ *of d'Alembert's equation* (1) *is given by* (2), (7), *and* (8), *where c is an arbitrary real constant.*

Of course, the same arguments work if (cf. Chapter 1) the variables are restricted also by $\phi \leqslant \pi/4$ in the conditional equation (1) and by $\phi \leqslant \pi/2$ in (2), (7) and (8).

For Cauchy's equations, for instance (1.2), all continuous solutions could also have been determined by a similar (even simpler) method. On the other hand, the above method gives the same general continuous solution for d'Alembert's equation (1) without the restriction $0 \leqslant \psi \leqslant \phi$. The continuous solutions would again be given by (2), (7), and (8), for all $\phi \in \mathbb{R}$. In Chapter 13 we will determine the general solutions of d'Alembert's equation under weaker conditions (in wider classes of admissible functions).

With Theorem 1 above, we have filled the last gap in the axiomatic characterization of the vector addition, given in Chapter 1. This, in turn, leads to an application in noneuclidean geometry and mechanics.

In noneuclidean geometry, the vectors (forces) cannot, of course, undergo parallel displacements anymore. They can, however, be shifted along lines. Therefore it is possible to apply, unchanged, the results and proofs of Chapter 1 concerning vectors of equal lengths, but no longer those concerning perpendicular vectors. Nevertheless, also in noneuclidean geometry, the result still applies that the component of a vector $\mathbf{p}$, in a direction forming an angle $\alpha(<\pi/2)$ with the vector, has the length

$$|\mathbf{p}| \cos \alpha \tag{16}$$

($|\mathbf{p}|$ being the length of $\mathbf{p}$) which, in euclidean geometry, is essentially equivalent to the parallelogram rule for orthogonal vectors. For what follows, cf. Picard 1922*a*, 1928, pp. 11–17; Aczél 1966*c*, pp. 125–8.

Now let $\mathbf{r}$ be the resultant of two unit vectors $\mathbf{p}$ and $\mathbf{q}$ perpendicular to the line $\bar{A}A$ at the points A and $\bar{A}$. The resultant is obviously ('by symmetry', cf. (i) in Chapter 1) perpendicular to $\bar{A}A$ at the midpoint C between A and $\bar{A}$ (see Figure 11).

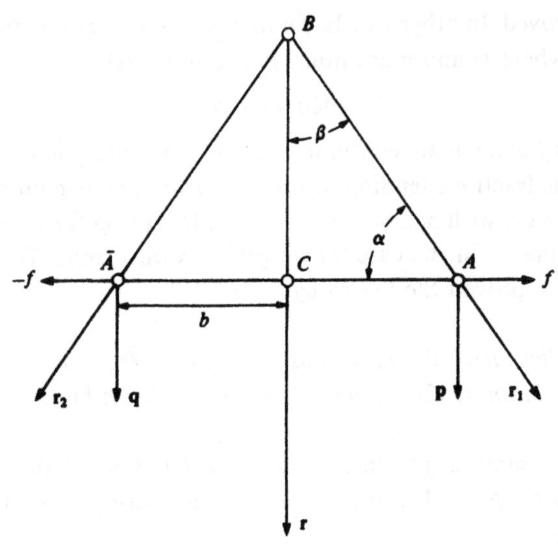

Fig. 11

This system of forces is not changed by adding at A and $\bar{A}$ the two forces $\mathbf{f}$ and $-\mathbf{f}$ (of the same length and opposite direction, inverses to each other) lying in the line $\bar{A}A$.

Let $\mathbf{r}_1$ (the resultant of $\mathbf{f}$ and $\mathbf{p}$) and $\mathbf{r}_2$ (the resultant of $-\mathbf{f}$ and $\mathbf{q}$) form angles $-\alpha$ and $-(\pi-\alpha)$ with $\bar{A}A$ respectively. We denote by β half the angle between $\mathbf{r}_1$ and $\mathbf{r}_2$, and by $\mathbf{r}$ the resultant of $\mathbf{r}_1$ and $\mathbf{r}_2$. Thus (writing + for vector addition here)

$$\mathbf{p}+\mathbf{f}=\mathbf{r}_1,\quad -\mathbf{f}+\mathbf{q}=\mathbf{r}_2,\quad \mathbf{r}_1+\mathbf{r}_2=\mathbf{r},\quad |\mathbf{p}|=|\mathbf{q}|=1,\quad |\mathbf{r}_1|=|\mathbf{r}_2|,$$
$$(\mathbf{r}_1,\mathbf{f})\sphericalangle=\alpha=(-\mathbf{f},\mathbf{r}_2)\sphericalangle,\quad (\mathbf{r}_1,\mathbf{r}_2)\sphericalangle=2\beta.$$

From this, from (16), and from the result in Chapter 1 about the resultant of two vectors of equal length, we get

$$1=|\mathbf{p}|=|\mathbf{q}|=|\mathbf{r}_1|\sin\alpha=|\mathbf{r}_2|\sin\alpha,$$
$$|\mathbf{p}+\mathbf{q}|=|(\mathbf{p}+\mathbf{f})+(-\mathbf{f}+\mathbf{q})|=|\mathbf{r}_1+\mathbf{r}_2|=|\mathbf{r}|=2|\mathbf{r}_1|\cos\beta=2\frac{\cos\beta}{\sin\alpha}. \tag{17}$$

Denote by $|\bar{A}A|$ the length of the line segment $\bar{A}A$, etc. Since $\mathbf{r}$ is completely determined by $|\bar{A}A|=2|CA|=2b$, the ratio $\cos\beta/\sin\alpha$ depends only upon b. This dependence is also continuous. We write

$$f(b):=\frac{\cos\beta}{\sin\alpha}. \tag{18}$$

Thus, from (17), for two forces $\mathbf{p}$ and $\mathbf{q}$, of equal length, acting at A and $\bar{A}$, the length of the resultant $\mathbf{r}$ is given by

$$|\mathbf{r}| = |\mathbf{p}+\mathbf{q}| = 2f(b). \tag{19}$$

Now, at points $A', \bar{A}'$ of the line $\bar{A}A$, with $|\bar{A}'C| = |CA'| = b' > b$, draw two additional unit vectors $\mathbf{p}'$ and $\mathbf{q}'$ perpendicular to $\bar{A}A$ (Figure 12) and let $\mathbf{r}', \mathbf{p}'', \mathbf{q}'', \mathbf{s}$ be the resultants of $\mathbf{p}, \mathbf{q}'$, of $\mathbf{p}, \mathbf{p}'$, of $\mathbf{q}, \mathbf{q}'$, and of $\mathbf{p}'', \mathbf{q}''$, acting at $C, A'', \bar{A}''$ and C respectively. All this gives, with (19),

$$\mathbf{p}+\mathbf{q}=\mathbf{r},\quad \mathbf{p}'+\mathbf{q}'=\mathbf{r}',\quad |\mathbf{r}|=2f(b)\quad |\mathbf{r}'|=2f(b'),\quad |\mathbf{r}+\mathbf{r}'|=2f(b)+2f(b'),$$
$$\mathbf{p}+\mathbf{p}'=\mathbf{p}'',\quad \mathbf{q}+\mathbf{q}'=\mathbf{q}'',\quad |\mathbf{p}''|=|\mathbf{q}''|=2f(y),\quad |\mathbf{s}|=|\mathbf{p}''+\mathbf{q}''|=4f(y)f(x),$$

where

$$y=|AA''|=|A''A'|=|\bar{A}'\bar{A}''|=|\bar{A}''\bar{A}|=\frac{b'-b}{2},\quad x=|CA''|=|\bar{A}''C|=\frac{b'+b}{2},$$

that is,

$$x+y=b',\quad x-y=b$$

and that is why

$$|\mathbf{s}|=|(\mathbf{p}+\mathbf{p}')+(\mathbf{q}+\mathbf{q}')|=|\mathbf{p}''+\mathbf{q}''|=4f(x)f(y).$$

On the other hand,

$$|\mathbf{s}|=|(\mathbf{p}+\mathbf{q})+(\mathbf{p}'+\mathbf{q}')|=|\mathbf{r}+\mathbf{r}'|=2f(x-y)+2f(x+y).$$

(Note that we again used the mediality $(\mathbf{p}+\mathbf{q})+(\mathbf{p}'+\mathbf{q}')=(\mathbf{p}+\mathbf{p}')+(\mathbf{q}+\mathbf{q}')$.)

So we have d'Alembert's equation:

$$f(x+y)+f(x-y)=2f(x)f(y)\quad (0 \leqslant y \leqslant x).$$

The function f is continuous and, since $f(x)\equiv 0$ contradicts (18), we get from

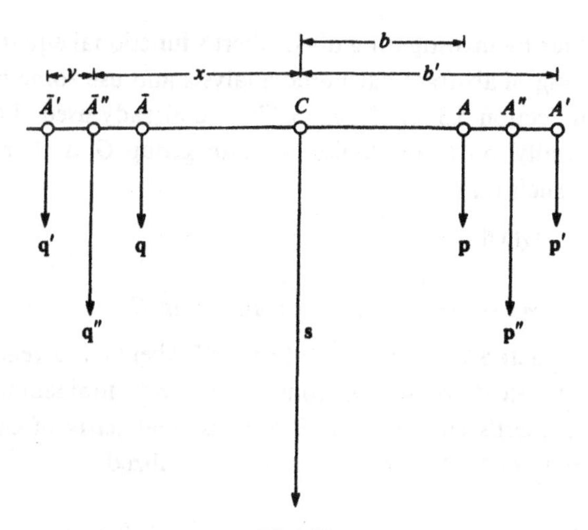

Fig. 12

Theorem 1

$$f(x)\begin{cases} = \cosh(x/d) \\ \text{or} \quad = 1 \\ \text{or} \quad = \cos(x/d) \end{cases}$$

($d \neq 0$), that is, cf. (18) and Figure 11,

$$\frac{\cos\beta}{\sin\alpha}\begin{cases} = \cosh(b/d) \\ \text{or} \quad = 1 \\ \text{or} \quad = \cos(b/d). \end{cases} \tag{20}$$

These are indeed the basic formulae of trigonometry in hyperbolic, euclidean, and elliptic geometry respectively.

In particular, in hyperbolic geometry, when we move the point B of Figure 11 to infinity, then $\beta \to 0$ and $\alpha = \Pi(b)$ becomes the *'angle of parallelism'* belonging to b. In this case, (20) gives

$$\frac{1}{\sin \Pi(b)} = \cosh\frac{b}{d},$$

or

$$\tan\tfrac{1}{2}\Pi(b) = e^{b/d},$$

which are further fundamental formulae of hyperbolic geometry.

Also, from (20), in the hyperbolic case (when $b \neq 0$),

$$\sin\left(\frac{\pi}{2} - \beta\right) = \cos\beta > \sin\alpha, \quad \frac{\pi}{2} - \beta > \alpha, \quad \alpha + \beta < \frac{\pi}{2},$$

that is, the sum of angles in a right triangle is smaller than π. By dividing an arbitrary triangle into two right ones, we see that the same holds in every triangle, again a characteristic of hyperbolic geometry,

We end this chapter by investigating d'Alembert's functional equation in the more general setting of abstract harmonic analysis and use some basic facts of that theory. In Section 5.3 (cf. (5.26), (5.27)), we already used characters χ on $\mathbb{R}$. More generally, on a topological abelian group G, a character is a function $\chi: G \to \mathbb{C}$ such that

$$|\chi(x)| = 1 \tag{21}$$

and

$$\chi(x+y) = \chi(x)\chi(y) \quad \text{for all } x, y \text{ in } G. \tag{22}$$

All continuous characters of $\mathbb{R}$ are of the form e^{icx} where c is a real constant (use Theorem 5.4). Real, bounded, continuous and nonidentically zero solutions of d'Alembert's equation on $\mathbb{R}$ are just real parts of continuous characters (cf. Theorem 1). This result can be generalized.

Theorem 2. *Let G be a connected and locally compact abelian group. A function*

$f: G \to R$ is the real part of a continuous character of G if and only if f is continuous, not identically 0, bounded, and satisfies d'Alembert's functional equation:

$$f(x+y)+f(x-y)=2f(x)f(y) \quad \textit{for all } x, y \textit{ in } G. \tag{23}$$

(A technique using Haar measures could be applied here to reduce the regularity assumption (continuity) for f and still get the same result, cf. Chapter 14.)

Proof. Let χ be a continuous character on G. We clearly have $\chi(0)=1$ (see (22) and use (21) to eliminate $\chi \equiv 0$). By (21), χ is bounded and, by (22), $\chi(-y)=\bar{\chi}(y)$. Therefore, if we set $f(x)=\operatorname{Re}\chi(x)$, then $f(x)\not\equiv 0$ and we get

$$f(x+y)=f(x)f(y)-(\operatorname{Im}\chi(x))(\operatorname{Im}\chi(y))$$

and

$$f(x-y)=f(x)f(-y)-(\operatorname{Im}\chi(x)(\operatorname{Im}\bar{\chi}(y))=f(x)f(y)+(\operatorname{Im}\chi(x)(\operatorname{Im}\chi(y)).$$

So f satisfies (23).

Conversely, let $f: G \to R$ satisfy (23) and be both bounded and continuous. With $y=0$ we get $f(0)=1$. If $f\equiv 1$, then the theorem is clearly true. So we may suppose $f(x)\not\equiv 1$. Following a method due to O'Connor (1977*b*) we shall first show that the absolute value of f is bounded by one, and that f has at least one zero and is positive definite on G.

Suppose for contradiction that $|f(x_0)|=1+\alpha$, $\alpha>0$, for some x_0 in G. Let us prove by induction that $|f(2^n x_0)|>1+2^n\alpha$ $(n=1,2,\ldots)$. By (23), $f(2x_0)=2f(x_0)^2-1$ and so $f(2x_0)=2(1+\alpha)^2-1>1+2\alpha$, so the statement holds for $n=1$. Assuming now that the result would be true for $n-1$, we compute that

$$f(2^n x_0)=2|f(2^{n-1}x_0)|^2-1>2(1+2^{n-1}\alpha)^2-1>1+2^n\alpha$$

which concludes the induction. But this result contradicts the fact that f is bounded. Thus $|f(x)|\leqslant 1$ for all x in G.

As f is not identically equal to 1, let y_0 be such that $f(y_0)=\beta<1$. We may always pick a natural number n so that $\beta<\cos(\pi/2^n)<1$. As $f(0)=1$, f is continuous and, since G is connected, there exists a z_0 in G with $f(z_0)=\cos(\pi/2^n)$. From (23), as in (10), we deduce that

$$f(2z_0)=2f(z_0)^2-1=\cos\frac{\pi}{2^{n-1}}.$$

In the same way,

$$f(2^2 z_0)=\cos\frac{\pi}{2^{n-2}}$$

and so on; finally,

$$f(2^n z_0) = \cos \pi = -1.$$

As $f(0)=1$ and f is continuous, there exists a t_0 in G such that $f(t_C)=0$. Now, equation (23) with $g(x)=f(t_0-x)$ gives

$$2g(x)g(y)=f(2t_0-x-y)+f(y-x).$$

But, replacing x by t_0-x and y by t_0 in (23) gives $f(2t_0-x)=-f(x)$ for all x. Also, putting $x=0$ into (23), we deduce from $f(0)=1$ the evenness of f:

$$f(y)+f(-y)=2f(y) \quad \text{and so} \quad f(-y)=f(y).$$

Thus

$$2g(x)g(y)=-f(x+y)+f(y-x),$$

and

$$2g(x)g(y)+2f(x)f(y)=f(y-x)+f(x-y)=2f(x-y),$$

so that we have a formula for $f(x-y)$. This will be convenient for showing that f is positive definite. Let $z_1, z_2, .., z_n$ be n complex numbers, $x_1, \ldots, x_n$ be n elements of G,

$$\sum_{k=1}^{n}\sum_{i=1}^{n} z_i\bar{z}_k f(x_i-x_k)=\left(\sum_{i=1}^{n} g(x_i)z_i\right)\left(\sum_{k=1}^{n} g(x_k)\bar{z}_k\right)+\left(\sum_{i=1}^{n} f(x_i)z_i\right)\left(\sum_{k=1}^{n} f(x_k)\bar{z}_k\right)$$

$$=\left|\sum_{i=1}^{n} f(x_i)z_i\right|^2+\left|\sum_{i=1}^{n} g(x_i)z_i\right|^2.$$

The right hand side is positive or zero. Therefore so is the left hand side, that is, f is *positive definite*. A famous theorem of S. Bochner (cf. Exercise 10.18) asserts that, for every continuous positive definite function over a locally compact abelian group, there exists a unique nonnegative bounded Radon measure μ on the dual topological group $G^\wedge$ of G such that, for all x in G,

$$f(x)=\int_{G^\wedge} \chi(x)d\mu(\chi), \tag{24}$$

that is, f is the Fourier transform of the Radon measure μ.

For the reader unfamiliar with harmonic analysis, we interpret (24) in the case $G=\mathbb{R}/(2\pi\mathbb{Z})$. There $G^\wedge=\mathbb{Z}$ and $\chi(x)=e^{inx}$ for some integer n. So (24) goes over into

$$f(x)=\sum_{n=-\infty}^{\infty} \mu(n)e^{inx},$$

where $\mu(n)\geqslant 0$ and $\sum_{n=-\infty}^{\infty}\mu(n)$ is finite (equal to 1, since $f(0)=1$).

We now express the functional equation (23) in terms of μ. A direct use of (24), with equality in the Cauchy–Schwarz inequality, gives Theorem 2 in the case where G is separable, as proved by O'Connor (1977*b*). We can do without the separability assumption if we reason as follows: we can write (23) as

$$2\left(\int_{G^\wedge} \chi(x)d\mu(\chi)\right)\left(\int_{G^\wedge} \chi'(y)d\mu(\chi')\right) = \int_{G^\wedge} \chi(x)\chi(y)d\mu(\chi) + \int_{G^\wedge} \chi(x)\overline{\chi(y)}d\mu(\chi)$$

and so

$$\int_{G^\wedge} \chi(x)\left[\int_{G^\wedge} \chi'(y)d\mu(\chi') - \frac{\chi(y)+\bar{\chi}(y)}{2}\right]d\mu(\chi) = 0. \tag{25}$$

Let $F(\chi) = \int_{G^\wedge} \chi'(y)d\mu(\chi') - (\chi(y) + \bar{\chi}(y))/2$, where y is considered to be a parameter. Then F is continuous on $G^\wedge$ and (25) states that the Fourier transform (on $G^\wedge$) of the product of F and of the Radon measure μ is identically 0. But the Fourier transform is a one-to-one mapping (bijection) for bounded Radon measures. So

$$f(y) = \int_{G^\wedge} \chi'(y)d\mu(\chi') = \frac{\chi(y)+\bar{\chi}(y)}{2} \quad (\mu\text{-almost-everywhere with respect to } \chi).$$

Notice that the left hand side is independent of χ. Therefore, for each given $y \in G$, $\operatorname{Re}\chi(y) = (\chi(y) + \bar{\chi}(y))/2$ is a constant on the support of μ. Let $\chi \neq \chi'$, and $\chi \neq \bar{\chi}'$; there exists y in G such that $\operatorname{Re}\chi(y) \neq \operatorname{Re}\chi'(y)$. This implies that the support of μ is reduced to at most two points, a character χ_0 and its conjugate. In other words, $\mu = a\chi_0 + b\bar{\chi}_0$ $(a, b \in \mathbb{C})$. But μ is positive, therefore $a \geqslant 0$, $b \geqslant 0$. Thus

$$f(x) = (a+b)\operatorname{Re}\chi_0(x) + i(a-b)\operatorname{Im}\chi_0(x).$$

However, f is real valued by supposition, therefore $a = b$. From $f(0) = 1$ we have now $a + b = 1$. Thus

$$f(x) = \operatorname{Re}\chi_0(x).$$

This concludes the proof of Theorem 2.

We will come back in Chapter 13 to the real and complex valued solutions of (23) on a group, without any regularity assumptions.

Exercises and further results

1. Determine all continuous solutions $f:\mathbb{R} \to \mathbb{R}$ of

$$f(x+y) + f(x-y) = 2f(x) \quad (x, y \in \mathbb{R})$$

by the method used in the proof of Theorem 1.

2. Same question for Lobachevskiĭ's equation

$$f(x+y)f(x-y) = f(x)^2 \quad (x, y \in \mathbb{R}).$$

3. Determine all continuous solutions $f:\mathbb{R} \to \mathbb{R}$ of

$$f(x+y) - f(x-y) = 2f(y) \quad (x, y \in \mathbb{R}).$$

4. Determine all continuous solutions $f:\mathbb{R}\to\mathbb{R}_+$ of the functional equation
$$f(x+y)-f(x-y)=4(f(x)f(y))^{1/2}\quad (x,y\in\mathbb{R}).$$
5. Find all continuous solutions $f:\bar{\mathbb{R}}_+\to\mathbb{R}$ of the functional equation
$$f(x)^2+f(y)^2=f(x+y)f(x-y)+1\quad (0\leqslant y\leqslant x).$$
Specify all continuous solutions $f:\bar{\mathbb{R}}_+\to\bar{\mathbb{R}}_+$.

(For Exercise 6–8 see Corovei 1976.)

6. Let G be an abelian group and let F be a (skew)field. Suppose that $f:G\to F$ satisfies
$$f(xy)+f(xy^{-1})=2f(x)f(y) \tag{26}$$
for all $x\in G$, $y\in S$, where S is a subsemigroup of G such that $G=SS^{-1}$. Show that f satisfies (26) for all $x\in G$, $y\in G$.
7. Same question as in the previous exercise except that this time equation (26) is supposed to hold for all $x\in S$, $y\in G$.
8*. Let now G be a group (not necessarily abelian) and S a subsemigroup such that $G=SS^{-1}S$. Show that every function $f:G\to F$, which satisfies (26) for $x\in S$, $y\in SS^{-1}$ and $f(xyz)=f(xzy)$ for all $x,y,z\in G$, satisfies (26) on all of G.
9. (Ljubenova 1981.) Prove Theorem 2 (without the assumption that G is connected and locally compact; let G just be a topological abelian group) by a direct computation using for some $x_0\in G$ with $|f(x_0)|<1$
$$g(x)=\tfrac{1}{2}(f(x-x_0)-f(x+x_0)]/(1-f(x_0)^2)^{1/2}$$
and $\chi(x)=f(x)+ig(x)$.
10. Prove Theorem 2 by applying the Cauchy–Schwarz inequality directly to (24) in the case where $G=\mathbb{R}^2$.
11. Let G be an abelian topological group G. Prove that every bounded solution $\chi:G\to\mathbb{C}$ of (22), which is not identically 0, is a character of G.
12. Find all continuous characters of $\mathbb{R}^n$.
13. Find all continuous solutions $g:\mathbb{R}^n\to\mathbb{R}$ of the d'Alembert functional equation
$$f(x+y)+f(x-y)=2f(x)f(y).$$
14. Find all continuous characters of the topological group $\mathbb{R}/\mathbb{Z}$.
15. Find all characters of $\mathbb{Z}$. Compare the result to Exercise 14.
16. Let n be an integer greater than 1 and consider the finite cyclic group $\mathbb{Z}_n=\mathbb{Z}/n\mathbb{Z}$. We define (cf. Sections 5.3, 10.5) a binary operation $*$ on the set $\mathbb{Z}_n^\wedge$ of all characters of $\mathbb{Z}_n$ by
$$\chi_1*\chi_2(x)=\chi_1(x)\chi_2(x)\quad (x\in\mathbb{Z}_n).$$
Prove or disprove that $(\mathbb{Z}_n^\wedge,*)$ is isomorphic to $\mathbb{Z}_n$.
17*. Let $(G,+)$ be a finite abelian group and let $(G^\wedge,*)$ be its dual group, that is $G^\wedge$ is the set of all characters of G and $G^\wedge$ is equipped with the binary operation $*$ defined by
$$\chi_1*\chi_2(x)=\chi_1(x)\chi_2(x)\quad (x\in G).$$
Prove or disprove that $(G^\wedge,*)$ is isomorphic to $(G,+)$.
18. Let G be a topological group. Show that every continuous character on a dense subgroup can be extended in a unique way to a continuous character on all of G.

19*. (Dixmier 1957.) Let G be a locally compact abelian group. As usual, $G^\wedge$ denotes the set of all continuous characters of G. We define (cf. Exercises 16, 17 and Sections 5.3, 10.5) a binary operation on $G^\wedge$ by

$$\chi_1 * \chi_2(x) = \chi_1(x)\chi_2(x).$$

By C we denote the set of all continuous solutions $f: G \to \mathbb{R}$ of Cauchy's equation

$$f(x+y) = f(x) + f(y).$$

The relation $\chi(x) = \exp(if(x))$ defines a mapping from C into $G^\wedge$. Show that this mapping is surjective if and only if $G^\wedge$ is the union of one parameter subgroups (that is, of ranges of continuous homomorphic maps of $\mathbb{R}$).

20. Find all continuous characters of the topological group $\mathbb{Q}$ (the set of rational numbers with the usual topology induced by the topology of $\mathbb{R}$ and with addition as operation).

(For Exercises 21–22, see, e.g., Hewitt–Ross 1963, pp. 401–5, 414–15.)

21*. Find all characters of the group $\mathbb{Q}$.

22*. Let $x_1, \ldots, x_n$ be n rational numbers. Let Φ be a character of $\mathbb{Q}$. Show that there exists a continuous character χ of $\mathbb{Q}$ (as in Exercise 20) such that

$$\chi(x_i) = \Phi(x_i) \quad (i = 1, 2, \ldots, n).$$

23. Let G be an abelian group and S a subsemigroup of G such that $G = S \cdot S^{-1}$. Let $f: G \to F$ be not identically zero, where F is a field with characteristic different from 2. If there exists $g: S \to F$ such that

$$f(xy) + f(xy^{-1}) = 2f(x)g(y) \quad (x \in G,\ y \in S),$$

prove that there exists an extension $h: G \to F$ of g such that

$$f(xy) + f(xy^{-1}) = 2f(x)h(y) \quad (x \in G,\ y \in G).$$

9

Images of sets and functional equations. Applications to relativity theory and to additive functions bounded on particular sets

This chapter is divided into two parts. The first part solves a chronogeometrical problem originating in relativity theory and deals with a functional equation, formally similar to the Cauchy equation, but where subsets interplay with the variable. The second part also deals with subsets and provides a converse theorem to the important regularity result which was proved in Chapter 2, namely that an additive real valued function, bounded above on a set $E \subset \mathbb{R}$ of positive Lebesgue measure, is continuous.

9.1 Equations containing images of sets and chronogeometry

Up to now, we have been working with functions defined for real and complex variables or, more generally, on groups or even on monoids and groupoids (see also Chapters 15, 18 and 19). A large class of interesting functions act on a family F of subsets of a given set E. For example, we may look for $f: F \to \mathbb{R}$ such that, for all X, Y in F such that $X \cap Y = \varnothing$, we have

$$f(X \cup Y) = f(X) + f(Y).$$

Here too, some regularity has to be assumed to avoid unpleasant solutions. In this generality it is the object of *measure theory* to describe all regular f satisfying the above equation, which could also be considered as a conditional functional equation.

Some other functional equations, which we shall now study, use the interplay between subsets of $\mathbb{R}$ (or of $\mathbb{R}^n$) and real numbers.

We recall some notation. Let E be a subset of $\mathbb{R}^n$, and $x \in \mathbb{R}^n$. We denote by $E + x$ the subset of all $y + x$, where y goes through E, that is, the translate of E. By $x - E$ we mean the subset of all $x - y$, where y runs through E. As before, if $S \subset \mathbb{R}^n$ and $f: \mathbb{R}^n \to \mathbb{R}$, the set $f(S)$ is the image of S under f. So we have now real valued functions on $\mathbb{R}^n$, but consider equations involving images of sets under them.

Our aim is to study functions f which, for a given subset of $\mathbb{R}^n$ and for all x in some prescribed subset of $\mathbb{R}^n$, satisfy

$$f(E+x)=f(E)+f(x). \tag{1}$$

Equation (1) is an equality for sets. For ordinary equations, a natural technique is to derive results from inequalities in both directions as an equivalent of equality. Similarly, here we shall use the inclusions

$$f(E+x)\subseteq f(E)+f(x) \tag{2}$$

and

$$f(E+x)\supseteq f(E)+f(x). \tag{3}$$

Right now the subset E will not be arbitrary but a nonempty open convex cone of $\mathbb{R}^n$ ($n\geqslant 1$), with vertex at the origin (but $0\notin E$). A *cone* (with vertex at the origin) is a subset C with the property that $x\in C$ implies $\lambda x\in C$ for all $\lambda>0$. (A set C is *convex* if, for all x, y in C, $[x,y]=\{\lambda x+(1-\lambda)y \mid 0\leqslant\lambda\leqslant 1\}\subset C$. A *ray* for $x\in C$ ($x\neq 0$), is the set of all $\lambda x\in C$ ($\lambda\in\mathbb{R}_+$).

We will also restrict f to be defined only on E ($f:E\to\mathbb{R}$) and suppose that (1) is valid for all x in E.

In order to easily state some results, let us recall that a convex cone E (with $0\notin E$) generates a partial (and strict) order $\succ$ on $\mathbb{R}^n$ by

$$x\succ y \quad \text{if and only if } x-y\in E.$$

We will call a map $f:E\to\mathbb{R}$ *order preserving* if

$$x\succ y\Rightarrow f(x)>f(y)$$

and *order reversing* if

$$x\succ y\Rightarrow f(x)<f(y).$$

Our first result reduces the case $n>1$ to the case $n=1$.

Proposition 1. *Let $f:E\to\mathbb{R}$ be a continuous and not identically zero function defined on a given open and nonempty convex cone E in $\mathbb{R}^n$ with vertex at 0 ($0\notin E$). Suppose that, at least for some x in E, $f(E+x)$ differs from $\mathbb{R}$. Then every solution f of* (1) *for all x in E is (a) order preserving or order reversing and (b) for every ray $\mathscr{E}$ of E,*

$$f(\mathscr{E}+x)=f(\mathscr{E})+f(x) \quad \textit{for all } x\in E. \tag{4}$$

Proof. Let $F=f(E)$. We first notice that F is a connected subset of $\mathbb{R}$ because f is continuous and E is a convex and therefore connected subset of $\mathbb{R}^n$. We also notice that F is a semigroup in $\mathbb{R}$. Indeed, for $x\in E$, $x'\in E$, equation (1), in the form (3), yields $f(x)+f(x')\in f(E+x)$. But $E+x\subseteq E$ as E is a convex cone and $x\in E$. (If $x\in E$ and $y\in E$ then $(x+y)/2$ is in E and so is $2(x+y)/2=x+y$.) Therefore $f(E+x)\subseteq f(E)=F$, that is, $F+F\subset F$. As a consequence, $ny\in F$ for all positive integers n and all y in F. But $F=\{0\}$ is impossible, as f is not identically zero. Therefore F is not bounded.

Let us show that F has to be bounded either from below or from above. Otherwise, since F is connected, we would have $F = \mathbb{R}$. Then it would follow from equation (1) that

$$f(E + x) = F + f(x) = \mathbb{R} \quad \text{for all } x \in E, \tag{5}$$

contrary to the supposition.

The only remaining case is where F is bounded from one side and not bounded from the other. Equation (1) remains unchanged when f is replaced by $-f$, so, without loss of generality, we may suppose F bounded from below and not bounded from above.

In this situation, $a = \inf F$ is a finite real number. It is first clear that $a \geqslant 0$ because, if there existed an $x_0 \in E$ with $f(x_0) < 0$, then $nf(x_0)$ would also belong to F for every positive integer n, contradicting the fact that $a = \inf F$ is finite.

In order to prove $a = 0$, suppose for contradiction $a > 0$. Then there would exist an $x_0 \in E$ for which $f(x_0) < 2a$. For any x_1 with $x_1 \prec x_0$, we get $f(x_0) \in f(x_1 + E)$. Therefore

$$\inf f(x_1 + E) < 2a. \tag{6}$$

But using (1), $\inf f(x_1 + E) = f(x_1) + \inf F = f(x_1) + a \geqslant 2a$, which contradicts (6). Therefore $a = 0$.

Suppose that x and x' are in E, with $x \succ x'$. By definition, $x \in x' + E$ and $f(x) \in f(x' + E) \subset f(x') + F$ due to (2). In other words, $f(x) - f(x') \in F$. To prove that f is order preserving, it is sufficient to establish $F \subset \mathbb{R}_+$.

But F is connected and not bounded from above and $a = 0$. So we have either $F = \mathbb{R}_+$ or $F = \bar{\mathbb{R}}_+$.

We shall now prove $F = \mathbb{R}_+$, by showing that the set of all zeros of f is empty, that is, $\ker f = \varnothing$. Let $x \in \ker f$. From (1) we get $f(E + x) = F$, so there exists a $y \in E + x \subseteq E$ such that $f(y) = 0$. The subset $E \cap (y - E)$ contains x and is an open subset of E. But for any z in $E \cap (y - E)$ we have $f(z) \leqslant f(y) = 0$. As $f(z) \in \bar{\mathbb{R}}_+$, we have $f(z) = 0$. The subset $E \cap (y - E)$ is a neighbourhood of x and is contained in $\ker f$. The point x being arbitrary in $\ker f$, this set is open. Since f is continuous, $\ker f$ is closed. As E is connected, this implies either $\ker f = E$, which case is excluded, since f is, by supposition, not identically zero or $\ker f = \varnothing$, as asserted.

We have proved that $F = \mathbb{R}_+$ and that f is order preserving and get

$$\lim_{\substack{x \to 0 \\ x \in E}} f(x) = 0.$$

Let $\mathscr{E}$ be any ray of E, generated by some x in E. Let us prove by contradiction that $f(\mathscr{E})$ is not bounded from above. Suppose $b \geqslant f(\lambda x)$ for all $\lambda > 0$. For any y in E, $(E + y) \cap \mathscr{E}$ is nonempty as can be seen in the plane generated by the ray $\mathscr{E}$ and by y. Let z be an element of $(E + y) \cap \mathscr{E}$. Since f

is order preserving, we get

$$f(y) < f(z) \leqslant b.$$

But y was an arbitrary element of E, so this shows that $\sup F \leqslant b$, contradicting $\sup F = \infty$, which we have already established.

Now let λ_0 and λ_1 be two positive numbers such that $\lambda_0 < \lambda_1$. For the order $\prec$, we deduce that $\lambda_0 x \prec \lambda_1 x$. Therefore $f(\lambda_0 x) < f(\lambda_1 x)$. The ray $\mathscr{E}$ itself is connected and so is $f(\mathscr{E} + \lambda x)$, for each $\lambda > 0$, by the continuity of f. As $f(\mathscr{E})$ is not bounded and f is order preserving, we deduce that

$$f(\mathscr{E} + \lambda x) =]f(\lambda x), \infty[.$$

In the same way, and because $\lim\limits_{\substack{x \in E \\ x \to 0}} f(x) = 0$, we get

$$f(\mathscr{E}) =]0, \infty[.$$

Thus

$$f(\mathscr{E} + \lambda x) = f(\mathscr{E}) + f(\lambda x).$$

As λx is an arbitrary element of $\mathscr{E}$, equation (4) is valid for any x in $\mathscr{E}$. This, and the possibility of replacing f by $-f$, mentioned above, complete the proof of Proposition 1.

Equation (4) is nothing but equation (1) under the assumption that $n = 1$ and $E =]0, \infty[= \mathbb{R}_+$. The order generated by this E is the natural order of $\mathbb{R}$ and so 'order preserving' means just 'strictly increasing'. We now proceed to the solution in this case.

Theorem 2. *A continuous function $f: E =]0, \infty[\to \mathbb{R}$ is a solution of*

$$f(]x, \infty[) = f(x) + f(]0, \infty[) \quad \text{for all } x \in]0, \infty[\tag{7}$$

if, and only if, f has one of the following four properties:

(A) $f \equiv 0$,
(B) *f is a strictly increasing function from $]0, \infty[$ onto $]0, \infty[$,*
(C) *f is a strictly decreasing function from $]0, \infty[$ onto $]-\infty, 0[$,*
(D) *$f(]x, \infty[) = \mathbb{R}$ for all x in $]0, \infty[$.*

Clearly no function can be in more than one of these categories.

Proof. We first prove that all functions of the forms (A), (B), (C) and (D) are solutions of (7). This is obvious for (A). For (B), we notice $f(]x, \infty[) =]f(x), \infty[$ and $f(]0, \infty[) =]0, \infty[$ which gives (7). For (C), in the same way, $f(]x, \infty[) =]-\infty, f(x)[$ and $f(]0, \infty[) =]-\infty, 0[$. In the last case (D), $f(]0, \infty[) \supset f(]x, \infty[)$ for any $x > 0$ and so $f(]0, \infty[) = \mathbb{R}$ as well, yielding (7).

Conversely, let f be a continuous solution of (7). As in Proposition 1, we define $F = f(]0, \infty[)$.

If $\inf F = a$ is finite, we deduce, as in Proposition 1, that $a = 0$. If (A) is excluded, we also get $0 \notin F$. As f is strictly increasing on $]0, \infty[$, and $\sup F = \infty$, by continuity we conclude that f is of the form (B).

If $\inf F = -\infty$ and $\sup F = b < \infty$, changing f into $-f$, we get (C).

If $\inf F = -\infty$ and $\sup F = \infty$, by continuity we get $F = \mathbb{R}$ and, from (7), $f(]x, \infty[) = \mathbb{R}$ for all x in $]0, \infty[$, which is case (D).

This concludes the proof of Theorem 2.

As a consequence, we may state the following:

Theorem 3. *In Proposition* 1, *the necessary conditions* (*a*) *and* (*b*) *are also sufficient.*

Proof. We have already proved that the conditions are necessary (Proposition 1). Conversely, under the hypotheses of Proposition 1, let $f: E \to \mathbb{R}$ be an order preserving or order reversing function satisfying (4) on every ray $\mathscr{E}$ of E. Using Theorem 2, we can see that, for each ray $\mathscr{E}$, one of the following three cases holds: $f(\mathscr{E}) = 0$ or $f(\mathscr{E}) =]0, \infty[$ or $f(\mathscr{E}) =]-\infty, 0[$. The reason is that in the last case, (D), that is, $f(\mathscr{E}) = \mathbb{R}$, we have $f(\mathscr{E} + x) = \mathbb{R}$ for any x in $\mathscr{E}$ by (7), and, so much the more, $f(E + x) = \mathbb{R}$, which is excluded by the hypothesis we made in Proposition 1. There is a natural bijection between all the rays $\mathscr{E}$ of E and the set $E_1 = B_1 \cap E$ where $B_1 = \{x \mid x \in \mathbb{R}^n, \|x\| = 1\}$ ($\|\cdot\|$ being, for example, the euclidean norm). Let A be the set of all x in E_1 such that the ray $\mathscr{E}_x$, containing x, satisfies $f(\mathscr{E}_x) = 0$. Let B (resp. C) be the set of all x in E_1 such that the ray $\mathscr{E}_x$ satisfies $f(\mathscr{E}_x) =]0, \infty[$ (resp. $]-\infty, 0[$). The union of the disjoint A, B and C is all of E_1. Notice that the order induced by $\prec$ onto any ray is the usual order on the ray. Therefore, on a ray $\mathscr{E}_x$ for which $x \in A$, the function f can be neither order preserving nor order reversing, and so $A = \varnothing$. Let us suppose $B \neq \varnothing$. Let $x \in B$. This implies that f is order preserving. If we had supposed $C \neq \varnothing$, then it would be order reversing. Thus either $B = E_1$ or $C = E_1$.

If $B = E_1$, then, for any x in E and any y in $E + x$, $f(y) > f(x)$. Let $\mathscr{E}$ be the ray in E containing x. We get $f(\mathscr{E} + x) =]f(x), \infty[$. Therefore $f(E + x) =]f(x), \infty[$. Now $f(E) =]0, \infty[$ since, for any ray $\mathscr{E}$, $f(\mathscr{E}) =]0, \infty[$ and E is the union of all such rays. Thus (1) is satisfied.

The same result holds in the case $C = E_1$. This concludes the proof of Theorem 3.

It might be useful to reformulate Theorem 3.

Theorem 4. *A continuous* $f: E \to \mathbb{R}$, *where* E *is an open convex nonempty cone of* $\mathbb{R}^n$ *with vertex at* 0 ($0 \notin E$) *satisfies* (1) *for all* x *in* E *if, and only if,* f *is of one* *of the following four mutually exclusive forms:*

(A) $f \equiv 0$,

(B) $f(E) =]0, \infty[$, *f is order preserving and unbounded from above on any ray of E,*

(C) $f(E) =]-\infty, 0[$, *f is order reversing and unbounded from below on any ray of E,*

(D) $f(E + x) = \mathbb{R}$ *for all x in E.*

This result is easy to deduce from Proposition 1 and Theorem 3. Just notice that, for any f in the cases (A), (B) or (C), $\lim_{x\in E, x\to 0} f(x) = 0$. In order to prove this assertion in case (B), take $\varepsilon > 0$ and $y \in E$ such that $f(y) \leqslant \varepsilon$. In the punctured neighborhood $E \cap (y - E)$ of 0, f is between 0 and ε, which yields the proof.

Notice that, in case (D), we do not necessarily have $f(\mathscr{E} + x) = \mathbb{R}$ for any ray $\mathscr{E}$ in E. Take, for example, $E = \mathbb{R}_+^2$. The function given by $f(x, y) = \log(y/x)$ for $(x, y) \in E$ is a continuous solution of (1), of the form (D). However, for any ray $\mathscr{E}$ enclosing an acute angle θ $(0 < \theta < (\pi/2))$ with the x-axis we get $f(\mathscr{E}) = \log(\tan\theta)$. In other words, Proposition 1 is no longer valid when $f(E + x) = \mathbb{R}$ for all x in E.

If $|a, b\|$ is an interval $(-\infty < a < b < \infty)$, where $|$ and $\|$ are independently] or [, Theorem 4 leads to a characterization of monotonic functions (in the wider sense). Let $f: |a, b\| \to \mathbb{R}$ be a continuous function whose range is a proper subset of $\mathbb{R}$. Then it satisfies the functional equation

$$f(|a, b\|) + f(x) = f(|x, b\|) \quad \text{for all } x \text{ in } |a, b\| \tag{8}$$

if and only if the following is true:

(i) If $| = [$ and $\| =]$, then $f \equiv 0$.

(ii) If $| = [$ and $\| = [$, then either $f \equiv 0$, or f is strictly increasing with $f(a) = 0$, $\lim_{x\to b} f(x) = \infty$, or f is strictly decreasing with $f(a) = 0$, $\lim_{x\to b} f(x) = -\infty$.

(iii) If $| =]$ and $\| = [$, then either $f \equiv 0$, or f is strictly increasing with $\lim_{x\to a} f(x) = 0$, $\lim_{x\to b} f(x) = \infty$, or f is strictly decreasing with $\lim_{x\to a} f(x) = 0$, $\lim_{x\to b} f(x) = -\infty$.

(iv) If $| =]$ and $\| =]$, then $f \equiv 0$.

The proof of (iii) is easy, applying Theorem 2 after a homomorphical change of variable mapping $]a, b[$ onto $]0, \infty[$. A theorem similar to but easier than Theorem 2 would lead to the case (ii). Let us prove (i) directly. From (8) we deduce that, for y_1, y_2 in $f([a, b])$, $y_1 + y_2$ is also in $f([a, b])$. Therefore, $ny \in f([a, b])$ for any positive integer n and any y in $f([a, b])$. But f is continuous, and so $f([a, b])$ is bounded both from below and from above. In other words, $f([a, b]) = 0$ (because, for $y \neq 0$, $\{ny \mid n = 1, 2, \dots\}$ is not bounded).

We also prove (iv) directly. As f is continuous, and $f(]x,b]) \subset f([x,b])$, the right hand side of (8) is a bounded set for every x in $]a,b]$. But $f(]a,b])$ cannot be bounded if it is not reduced to $\{0\}$, because it is a semigroup of $\mathbb{R}$.

The study of (1) when E is a closed nonempty convex subset of $\mathbb{R}^n$ is more intricate. The general case, without convexity, appears to be rather difficult (Dhombres 1984*b*).

The functional equation (1) arose in connection with a problem in relativity theory (see Alexandrov 1967 and 1970). He studied a 'geometry' in $\mathbb{R}^n$ (or more exactly in the associated affine n-space) based on a relation of precedence of points, a '*chronogeometry*'. When $n = 4$, the points of $\mathbb{R}^4$ are called events. Every event is the vertex of a 'cone' C_x of events preceded by x, for example the cone determined by the propagation of light. This structure of precedence determines the spatial–temporal structure as well as the metric relations.

Precedence of x to y is formally defined by $y \in C_x$. Relevant to relativity theory is the hypothesis that C_x is just the translate of a given cone C with vertex at $0: C_x = C + x$ ('cone of light'). Then it can be shown that for the precedence relation to be transitive, it is enough to suppose that the cone C is convex.

In this context, the interesting mappings $f: \mathbb{R}^n \to \mathbb{R}^n$ are those which keep the cones invariant. More exactly, we look for those *bijections* f for which $C_{f(x)} = f(C_x)$. A functional equation characterizes this property:

$$C + f(x) = f(C + x). \tag{9}$$

Alexandrov (1967) has proved, in the case $n > 2$, by geometric arguments that (9) implies (for bijections, but with no other regularity assumption on f), that f is affine ($f(x) = Ax + b$, $\det A \neq 0$) if the cone C has the following properties (cf. Exercise 9):

$$C \text{ does not lie in a plane,} \tag{10}$$

$$\text{the closed convex hull of } C \text{ does not contain straight lines,} \tag{11}$$

$$\text{the closed convex hull of } C \text{ is not the cartesian product of a ray and an } (n-1) \text{ dimensional cone.} \tag{12}$$

As a consequence, one obtains a characterization of Lorentz transformations as the mappings preserving the system of elliptic cones.

If we put some further restrictions upon C, which we shall not describe here, a little more can be achieved, namely it can be proved that any bijection for $f: \mathbb{R}^n \to \mathbb{R}$ satisfying

$$f(C + x) = g(x) + C \tag{13}$$

for some (unknown) function g has to be affine ($f(x) = Ax + b$, $\det A \neq 0$). Equation (13) is in a way, a Pexider generalization of (9), cf. Section 4.3.

In order to obtain a somewhat more general solution f, Alexandrov proposed to study the functional equation

$$f(C+x)=f(C)+f(x), \tag{14}$$

instead of (9). We already gave results when $f:\mathbb{R}^n \to \mathbb{R}$ was supposed to be continuous (Theorem 4) on an open and nonempty convex cone C. A little more can be achieved when $n=1$ under the assumption that f is bounded from above. However, for $f:\mathbb{R}^n \to \mathbb{R}^n$ with $n \neq 1$, equations of the form (14) have not been completely solved, not even under regularity conditions like continuity (see Täuber 1973; Forti 1977, 1978; Täuber–Neuhaus 1978*a*, *b*; Borelli Forti–Forti 1979; Dhombres 1982*a*, 1984*b*).

9.2 Sets on which bounded additive functions are continuous

In the case of the Cauchy equation, there is another way to consider the interplay between subsets of $\mathbb{R}$ and real numbers. We have already proved that a solution of the Cauchy equation, bounded from above on a set E of positive Lebesgue measure is necessarily continuous (Theorem 2.8). Can, in this statement, sets of positive measure be replaced by other classes of sets? In other words, we look for a converse of Theorem 2.8 in terms of the subset where f is bounded.

Let $f:\mathbb{R} \to \mathbb{R}$ be an additive function, not identically zero, and consider, for an $M \in \mathbb{R}$, the subset

$$E=\{x \mid x \in \mathbb{R}; f(x) \leqslant M\}. \tag{15}$$

E possesses some interesting properties.

If x, y are in E and $\alpha \in \mathbb{Q} \cap [0,1]$, then $\alpha x+(1-\alpha)y$ is also in E. A subset of $\mathbb{R}$ which has this property is called $\mathbb{Q}$-*convex*.

Take $x_0 \in E$ such that $f(x_0)<M$ and let x be a point in $\mathbb{R}$. There is an $\alpha_x>0$ in $\mathbb{Q}$ such that $f(x_0)+\alpha_x|f(x)| \leqslant M$. Therefore, for any α in $\mathbb{Q}$, $0 \leqslant \alpha \leqslant \alpha_x$, $x_0+\alpha x \in E$. A subset E of $\mathbb{R}$ with this property at a point x_0 is called $\mathbb{Q}$-*radial* at x_0.

Clearly, E is not bounded (we use $f(\alpha x)=\alpha f(x)$ for all $\alpha \in \mathbb{Q}$ and the existence of an $x \in \mathbb{R}^*$ such that $f(x) \leqslant 0$). These three properties almost characterize those subsets of $\mathbb{R}$ which are of the form (15) for some additive $f:\mathbb{R} \to \mathbb{R}$. The exact answer is the following (see Kuczma 1970):

Proposition 5. *Let E be $\mathbb{Q}$-radial at some point x_0 in E and let us suppose E to be $\mathbb{Q}$-convex and nonempty. Then there exists a not identically zero additive $f:\mathbb{R} \to \mathbb{R}$ and constants M, N $(M>N)$ such that either f is continuous and $E \supseteq \{x \mid x \in \mathbb{R}; N<f(x)<M\}$ or f is discontinuous and $E \subseteq \{x \mid x \in \mathbb{R}; f(x) \leqslant M\}$.*

It should be noted that, in Proposition 5, the two possible cases are mutually exclusive. The set $\{x \mid x \in \mathbb{R}; f(x) \leqslant M\}$ for a discontinuous and additive $f:\mathbb{R}\to\mathbb{R}$ cannot contain an open and nonempty set of the form $\{x|x\in\mathbb{R}; N<g(x)<M\}$ for a continuous not identically zero and additive g with $N<M$ (cf. Corollary 2.5).

The essential part of the proof of Proposition 5 is the case where $x_0=0$. In fact, when $x_0\neq 0$, the set $E-x_0$ is $\mathbb{Q}$-radial at 0, $\mathbb{Q}$-convex and nonempty. Therefore, if we already have Proposition 5 in the case $x_0=0$, we get either

$$E\supseteq x_0+\{x|x\in\mathbb{R}; N<f(x)<M\}=\{x|x\in\mathbb{R}; N'<f(x)<M'\}$$

(where we have written $N'=N+f(x_0)$; $M'=M+f(x_0)$) and f is continuous or

$$E\subseteq x_0+\{x|x\in\mathbb{R}; f(x)\leqslant M\}\subset\{x|x\in\mathbb{R}, f(x)\leqslant M'\},$$

with $M'=M+f(x_0)$, and f is discontinuous, as asserted.

So we may now suppose that 0 is the point x_0 where E is $\mathbb{Q}$-radial. Note that this implies that $0\in E$ and the $\mathbb{Q}$-convexity of E yields $\alpha x\in E$ for all α with $0\leqslant\alpha\leqslant 1$; $\alpha\in\mathbb{Q}$ and $x\in E$.

First we suppose that E is a convex subset of $\mathbb{R}$ (that is, $ax+(1-a)y\in E$ for all x, y in E, and for all a in $\mathbb{R}$ such $0\leqslant a\leqslant 1$). Convex subsets of $\mathbb{R}$ are closed, open or half-open (finite or infinite) intervals (possibly $\mathbb{R}$ itself). Therefore in all cases $E\supseteq\{x|x\in\mathbb{R}, N<x<M\}$ for some constants M, N with $M>N$. With $f(x)=x$, we get $E\supseteq\{x|x\in\mathbb{R}; N<f(x)<M\}$. Thus we have the first alternative in Proposition 5.

We now suppose that the subset E of $\mathbb{R}$ is not convex. As $0\in E$, we either have $0<x_1<x_2$, $x_1\notin E$, $x_2\in E$ or $0>x_1>x_2$, $x_1\notin E$, $x_2\in E$. There is no loss of generality (change E to $-E$) in dealing only with the first case. We will now obtain the second alternative in Proposition 5.

Clearly, x_1 and x_2 have to be independent over $\mathbb{Z}$. Indeed, $n_1x_1+n_2x_2=0$ with $0<x_1<x_2$ leads to $x_1=(-n_2/n_1)x_2$ and $0\leqslant -n_2/n_1\leqslant 1$. By $\mathbb{Q}$-convexity, $(-n_2/n_1)x_2$ has to belong to E, which is a contradiction.

Let $\mathscr{E}(x_1,x_2)$ be the subset of $\mathbb{R}$ consisting of all x such that $x=\alpha_1x_1+\alpha_2x_2$ where α_1,α_2 are in $\mathbb{Q}$. Our first aim is to construct an additive function g over $\mathscr{E}(x_1,x_2)$ which is bounded from above over $E\cap\mathscr{E}(x_1,x_2)$. Lemma 2.2 shows that the additive g is determined by choosing $g(x_1)=r_1$ and $g(x_2)=r_2$. This choice should be such that g remains bounded from above on $E\cap\mathscr{E}(x_1,x_2)$, that is, $\sup(\alpha_1r_1+\alpha_2r_2)$ should be finite for all α_1,α_2 in $\mathbb{Q}$ whenever $\alpha_1x_1+\alpha_2x_2$ lies in E.

To show that this is possible, let us first try to write x_1, which is not in E, as a $\mathbb{Q}$-convex combination of $\alpha_1x_1+\alpha_2x_2$ and some rational multiple of x_2. We easily get

$$x_1=\frac{\alpha_1x_1+\alpha_2x_2}{\alpha_1}+\left(1-\frac{1}{\alpha_1}\right)\frac{\alpha_2}{\alpha_1-1}(-x_2).$$

Suppose that $\alpha_1>1$ and $\alpha_1x_1+\alpha_2x_2\in E$. We can not have $\alpha_2(-x_2)/(\alpha_1-1)\in E$, because in this case x_1 is a $\mathbb{Q}$-convex combination of elements in E and so belongs to E, but $x_1 \notin E$. However, when $|\alpha_2/(\alpha_1 - 1)|$ is small enough, we obtain that $\alpha_2(-x_2)/(\alpha_1 - 1) \in E$ due to the $\mathbb{Q}$-radiality of E at the point 0 (apply $\mathbb{Q}$-radiality to $-x_2$ if $\alpha_2 > 0$, and to x_2 if $\alpha_2 < 0$). We conclude by showing that there exists a positive

$$\beta = \inf\left\{\frac{\alpha_2}{\alpha_1 - 1}\,\middle|\, \alpha_1 x_1 + \alpha_2 x_2 \in E, \quad 1 < \alpha_1 \in \mathbb{Q}, 0 < \alpha_2 \in \mathbb{Q}\right\}.$$

We immediately get, for $\alpha_1 > 1$ and $\alpha_2 > 0$, α_1, α_2 in $\mathbb{Q}$, and $\alpha_1 x_1 + \alpha_2 x_2 \in E$,

$$\alpha_1 - \frac{\alpha_2}{\beta} \leqslant 1. \tag{16}$$

If we take $g(x_1) = r_1 = 1$ and $g(x_2) = r_2 = -1/\beta$, we get for g the bounds we were looking for. It only remains to verify that the inequality (16) still holds for other rational values of α_1 and α_2 such that $\alpha_1 x_1 + \alpha_2 x_2 \in E$, but either $\alpha_2 \leqslant 0$ or $\alpha_1 \leqslant 1$.

For $\alpha_2 = 0$, (16) is obviously true with $\alpha_1 \leqslant 1$ and $\alpha_1 \geqslant 1$ is impossible since $\alpha_1 x_1 \in E$ implies by $\mathbb{Q}$-convexity, $x_1 \in E$ which is contrary to our hypothesis.

For $\alpha_1 \leqslant 1$ and $\alpha_2 \geqslant 0$, the inequality (16) is obvious.

For $\alpha_1 \leqslant 1$ and $\alpha_2 < 0$, we shall prove the inequality (16) by contradition. So we suppose there exist α_1, α_2 in $\mathbb{Q}$ such that $\alpha_1 \leqslant 1$, $\alpha_2 < 0$, $\alpha_1 x_1 + \alpha_2 x_2 \in E$ but $\alpha_1 - (\alpha_2/\beta) > 1$. We try, in this new case, to express x_1 in the form $\gamma y_1 + (1-\gamma) y_2$ with $\gamma \in Q \cap [0,1]$ and $y_1, y_2 \in E$. We naturally wish to keep y_1 as $\alpha_1 x_1 + \alpha_2 x_2$ and look for some y_2 of the form $\alpha_1' x_1 + \alpha_2' x_2$, where α_1', α_2' shall have to be determined in $\mathbb{Q}$ so that $\alpha_1' x_1 + \alpha_2' x_2 \in E$. To get more freedom, we work with $y_1 = (1/\delta)(\alpha_1 x_1 + \alpha_2 x_2)$ and $y_2 = (1/\delta)(\alpha_1' x_1 + \alpha_2' x_2)$ for some $\delta \in \mathbb{Q}$, $\delta \geqslant 1$. Therefore, we look for some $\gamma \in \mathbb{Q} \cap [0,1]$ such that

$$x_1 = \gamma \frac{\alpha_1 x_1 + \alpha_2 x_2}{\delta} + (1-\gamma)\frac{\alpha_1' x_1 + \alpha_2' x_2}{\delta}.$$

We get as appropriate values for γ and δ

$$\gamma = \frac{\alpha_2'}{\alpha_2' - \alpha_2} \quad \text{and} \quad \delta = \frac{\alpha_2'\alpha_1 - \alpha_2\alpha_1'}{\alpha_2' - \alpha_2}.$$

Because $\alpha_2 < 0$, it is enough to require $\alpha_2' \geqslant 0$ in order to have $0 \leqslant \gamma \leqslant 1$. The condition $\delta \geqslant 1$ can now be transformed into $\alpha_2'(\alpha_1 - 1) \geqslant \alpha_2(\alpha_1' - 1)$ which, if we suppose $\alpha_1' > 1$, amounts to

$$\frac{\alpha_2'}{\alpha_1' - 1} \leqslant \frac{\alpha_2}{\alpha_1 - 1}. \tag{17}$$

(under the convention that $\alpha_2/(\alpha_1 - 1) = \infty$ if $\alpha_1 = 1$). Our hypothesis was $\alpha_2/(\alpha_1 - 1) > \beta$. Therefore, due to the definition of β as a g.l.b., we may find α_1', α_2' in $\mathbb{Q}$, $\alpha_1' > 1$, $\alpha_2' > 0$, satisfying $\alpha_1' x_1 + \alpha_2' x_2 \in E$, for which (17) is satisfied.

This is a contradiction, since x_1 appears as a $\mathbb{Q}$-convex combination of elements of E. Hence (16) is true in this case as well.

For $\alpha_1 > 1$ and $\alpha_2 < 0$, clearly the inequality (16) is not satisfied and $\alpha_1 - (\alpha_2/\beta) > 1$. We proceed in the same way as in the case $\alpha_1 < 1$ and $\alpha_2 < 0$. But this time our hypothesis is $\beta > \alpha_2/(\alpha_1 - 1)$ and we look for $\alpha_1' > 1$, $\alpha_2' > 0$, $\alpha_1' \in \mathbb{Q}$, $\alpha_2' \in \mathbb{Q}$, $\alpha_1' x_1 + \alpha_2' x_2 \in E$ for which we have the inequality opposite to (17)

$$\frac{\alpha_2'}{\alpha_1' - 1} \geqslant \frac{\alpha_2}{\alpha_1 - 1}. \tag{18}$$

As $\alpha_2'/(\alpha_1' - 1) \geqslant \beta > \alpha_2(\alpha_1 - 1)$, we may find α_1', as required, and get a contradiction, so that the case $\alpha_1 > 1$, $\alpha_2 < 0$ is not possible.

Now we have completed our construction of a $g:\mathscr{E}(x_1, x_2) \to \mathbb{R}$, which is additive on $\mathscr{E}(x_1, x_2)$ and bounded above by 1 on $\mathscr{E}(x_1, x_2) \cap E$. By Theorem 7.5, there exists an additive extension f of g to all of $\mathbb{R}$, which is still bounded above by 1 on E.

In other words,

$$E \subseteq \{x \mid x \in \mathbb{R}; f(x) \leqslant 1\}.$$

For $0 < x_1 < x_2$, we get $f(x_1) = 1$ and $f(x_2) = -1/\beta < 0$. Therefore f cannot be of the form $f(x) = ax$ for some a on $\mathbb{R}$ and so cannot be continuous (Corollary 2.5).

This concludes the proof of Proposition 5.

In order to state our first converse theorem, we define the $\mathbb{Q}$-convex hull $\mathbb{Q}(E)$ of a set E to be the smallest $\mathbb{Q}$-convex set containing E.

Theorem 6. *Let E be a nonempty subset of $\mathbb{R}$. The following two properties for E are equivalent:*

(i) *Every $f:\mathbb{R} \to \mathbb{R}$, additive and bounded above on E, is continuous.*
(ii) *For every subset F of $\mathbb{R}$ containing a Hamel basis, the set $\mathbb{Q}((E + F) \cup (E - F))$ contains a subset of positive Lebesgue measure.*

Proof. To prove that (i) implies (ii) we will show that, if (ii) is not satisfied, then (i) is not satisfied. We thus suppose that there exists a subset F of $\mathbb{R}$, containing a Hamel basis and such that $\mathbb{Q}((E + F) \cup (E - F))$ contains no subset of positive Lebesgue measure.

We first notice that $\mathbb{Q}((E + F) \cup (E - F))$ is $\mathbb{Q}$-radial at any point t of E (and $E \subset \mathbb{Q}((E + F) \cup (E - F))$. For that, we start from an x in $\mathbb{R}^* = \mathbb{R} \setminus \{0\}$, and can write

$$x = \alpha_1 x_1 + \cdots + \alpha_n x_n,$$

where $\alpha_i > 0$ and $\alpha_i \in \mathbb{Q}$; $x_1, x_2, \ldots, x_n \in F \cup (-F)$. Such a representation is always possible, as F contains a Hamel basis for $\mathbb{R}$. Therefore, for

$\sum_{i=1}^{n}\alpha_i = \alpha_0 > 0$ and for every t in E, we get

$$\frac{x}{\alpha_0} + t = \frac{\alpha_1}{\alpha_0}(x_1 + t) + \cdots + \frac{\alpha_n}{\alpha_0}(x_n + t),$$

proving that $x/\alpha_0 + t \in \mathbb{Q}((E+F)\cup(E-F))$. But $t \in E \subset \mathbb{Q}((E+F)\cup(E-F))$ and from the $\mathbb{Q}$-convexity of $\mathbb{Q}((E+F)\cup(E-F))$ we deduce that

$$\alpha\left(\frac{x}{\alpha_0} + t\right) + (1-\alpha)t = \alpha\frac{x}{\alpha_0} + t \in \mathbb{Q}((E+F)\cup(E-F)) \tag{19}$$

for all $x \in R^* = R\setminus\{0\}$, $t \in E$ and for all $\alpha \in \mathbb{Q}$ with $0 \leqslant \alpha \leqslant 1$. The last relation (19) remains true when $x = 0$. So we have the $\mathbb{Q}$-radiality of $\mathbb{Q}((E+F)\cup(E-F))$ at any t (of E). Therefore $\mathbb{Q}((E+F)\cup(E-F))$ is a nonempty, $\mathbb{Q}$-convex and $\mathbb{Q}$-radial subset of $\mathbb{R}$. Proposition 5 yields the form of such a set. As it cannot contain, by our hypothesis, a subset of positive Lebesgue measure, it cannot include a subset of the form $\{x \mid x \in \mathbb{R};\ N < f(x) < M\}$ for a continuous nonidentically zero and additive $f:\mathbb{R}\to\mathbb{R}$. Therefore there exists a discontinuous additive $f:\mathbb{R}\to\mathbb{R}$, and an M such that

$$E \subseteq \mathbb{Q}((E+F)\cup(E-F)) \subseteq \{x \mid x \in \mathbb{R};\ f(x) \leqslant M\}.$$

Thus the property (i) is not satisfied.

We now prove in a direct way that Property (ii) implies Property (i). Let $f:\mathbb{R}\to\mathbb{R}$ be an additive function, bounded above on E by some M. Let $F = \{x \mid x \in \mathbb{R};\ |f(x)| \leqslant 1\}$. This set F contains a Hamel basis for $\mathbb{R}$. To see this, let h_0 be an element of a Hamel basis H. For some $\alpha \neq 0$ in $\mathbb{Q}$, $|f(\alpha h_0)| \leqslant 1$. But we have already noted that $(H\setminus\{h_0\})\cup\{\alpha h_0\}$ is another Hamel basis. As in the proof of Proposition 2.12 and Corollary 2.13, we deduce the existence of a Hamel basis H with $H \subset F$.

Moreover, f is bounded above by $M+1$ on $(E+F)\cup(E-F)$, as is easily seen. Thus f is bounded above by $M+1$ on $\mathbb{Q}((E+F)\cup(E-F))$, which means that f is bounded above on a set of positive Lebesgue measure. Theorem 2.8 yields the continuity of f and concludes the proof of the Theorem 6.

Note 1. We could easily replace Property (ii) in Theorem 6 by the following analogous property:

(iii) *For every subset F of $\mathbb{R}$, $\mathbb{Q}$-radial at some point, the $\mathbb{Q}$-convex hull of $E-F$ contains nonempty open convex subset.*

A simpler result would be expected if we supposed a bilateral boundedness condition. In fact, then we do not need all subsets F containing a Hamel basis for $\mathbb{R}$.

Theorem 7. *A nonempty subset E of R has the property that any additive*

$f:\mathbb{R}\to\mathbb{R}$, bounded in absolute value on E, is continuous on $\mathbb{R}$ if, and only if, the $\mathbb{Q}$-convex hull of $E-E$ contains a subset of positive Lebesgue measure.

Proof. To prove the sufficiency, let us start with an $f:\mathbb{R}\to\mathbb{R}$, additive and such that $|f(x)|\leqslant M$ for all x in E. Clearly, $|f(x)|\leqslant 2M$ on $E-E$ and, by $f(\alpha x)=\alpha f(x)$ for $\alpha\in\mathbb{Q}$, the same bound occurs for $\mathbb{Q}(E-E)$ and so for f on a set of positive Lebesgue measure. Theorem 2.8 yields the continuity of f on $\mathbb{R}$. We shall prove the necessity by contradiction. We thus suppose that $\mathbb{Q}(E-E)$ contains no subset of positive Lebesgue measure but that any additive $f:\mathbb{R}\to\mathbb{R}$, bounded in absolute value on $E(\neq\varnothing)$, is continuous. Clearly, E does not reduce to a singleton. Moreover, an additive $f:\mathbb{R}\to\mathbb{R}$, equal to zero on $E-E$, is constant on E and so continuous everywhere, which implies that f is zero everywhere, as E contains more than one element. Corollary 2.13 yields that $E-E$ contains a Hamel basis H and so, by symmetry, it contains also $-H$. The same argument as in Theorem 6 proves that $\mathbb{Q}(E-E)$ is $\mathbb{Q}$-radial at 0. But $\mathbb{Q}(E-E)$, which is $\mathbb{Q}$-convex, contains no subset of positive Lebesgue measure. Proposition 5 yields a discontinuous additive $f:\mathbb{R}\to\mathbb{R}$, bounded from above on $\mathbb{Q}(E-E)$ by some constant M, and M can be chosen to be positive. But $\mathbb{Q}(E-E)=-\mathbb{Q}(E-E)$ and so, as $f(-x)=-f(x)$, we deduce that $|f(x)|\leqslant M$ on $\mathbb{Q}(E-E)$. Therefore $|f(x)|\leqslant M+|f(x_0)|(x\in E)$, where x_0 is any given point of E. This is a contradiction to our hypothesis and ends the proof.

Note 2. A subset E, which has the property (i) of Theorem 6, is such that any additive $f:\mathbb{R}\to\mathbb{R}$, bounded in absolute value on E, is continuous. The converse is not true. Let us consider the following example. Take $E=\{x\,|\,x\in\mathbb{R}; f(x)\leqslant M\}$ for some constant M and some discontinuous additive $f:\mathbb{R}\to\mathbb{R}$. The subset E is not empty. By definition, E does not have the property (i) of Theorem 6. However, let $g:\mathbb{R}\to\mathbb{R}$ be additive and bounded in absolute value on E. Since $g(\alpha x)=\alpha g(x)$ and $\alpha x\in E$ for all α in some unbounded subset of $\mathbb{Q}$, g is therefore necessarily zero on E and so is continuous on $\mathbb{R}$. In other words, the fact that $\mathbb{Q}(E-E)$ contains a subset of positive Lebesgue measure does not imply the same thing for $\mathbb{Q}((E+F)\cup(E-F))$ where F is a subset of $\mathbb{R}$ containing a Hamel basis.

Note 3. For a given Hamel basis H and an element h of H, the projection along $h\in H$ is the application $x\mapsto\alpha_h$ where α_h is the rational coefficient (possibly 0) of h in the unique decomposition of x as a finite rational combination of elements of H. When E is of positive Lebesgue measure, for any Hamel basis H, and any given element h of H, the projection along this element cannot remain bounded in absolute value (or even bounded above) on E. This is an easy consequence of Theorem 7.

Note 4. Theorem 6 and 7 can be extended in many ways: for instance to more general domains and ranges (see Ger–Kuczma 1970; Kuczma–Smital 1976; and Dhombres 1979, pp. 4.46, 4.47) or to inequalities in place of equality in the Cauchy equation, for example to the so-called Jensen convex functions (see, e.g., Kuczma 1973*a*, 1985; Smital 1976*b*; Ger 1978*c*). A function $f:I\to\mathbb{R}$ is *Jensen convex* if

$$f\left(\frac{x+y}{2}\right)\leqslant\frac{f(x)+f(y)}{2}\quad\text{for all } x,y\in I \tag{20}$$

(I may, for instance, be a real interval or its intersection with $\mathbb{Q}$ or a convex set in $\mathbb{R}^n$, etc.).

Exercises and further results

(For Exercises 1 and 2, see Dhombres 1982*a*, 1984*b*.)

1. Let $E=]0,\infty[$. Find all those solutions $f:E\to\mathbb{R}$ of the functional equation
$$f(x+E)=f(x)+f(E)\quad(x\in E),$$
for which $f(E)$ is a connected proper subspace of $\mathbb{R}$.
2. What happens if we replace $E=]0,\infty[$ in Theorem 2 by $E=[0,\infty[$? That is, find all continuous solutions $f:\bar{\mathbb{R}}_+\to\mathbb{R}$ of
$$f(x+\bar{\mathbb{R}}_+)=f(x)+f(\bar{\mathbb{R}}_+)\quad(x\in\bar{\mathbb{R}}_+).$$

(For Exercises 3, 4 and 5, see Täuber-Neuhaus 1978*a*.)

3. Let $f:\mathbb{R}_+\to\mathbb{R}_+$ be a surjection. Show that f is strictly increasing if, and only if, f satisfies the functional equation
$$f(x+\mathbb{R}_+)=f(x)+f(\mathbb{R}_+)\quad(x\in\mathbb{R}_+).$$
4. Find all strictly increasing solutions $f:\mathbb{R}_+\to\mathbb{R}$ such that
$$f\left(\frac{x+\mathbb{R}_+}{2}\right)=\frac{f(x)+f(\mathbb{R}_+)}{2}\quad(x\in\mathbb{R}_+).$$
5. Same question as in Exercise 4 with $\mathbb{R}_+$ replaced by $\bar{\mathbb{R}}_+$.

6*. (Borelli Forti 1980.) Let E be a proper open convex cone in $\mathbb{R}^2$ with vertex at 0 and $f:E\to\mathbb{R}^2$. Write
$$f(x_1,x_2)=(f_1(x_1,x_2),f_2(x_1,x_2)).$$
Suppose $f_i:E\to\mathbb{R}_+$ are solutions ($i=1,2$) of
$$f_i(x+E)=f_i(x)+f_i(E)\quad(x\in E).$$
Prove or disprove that f satisfies
$$f(x+E)=f(x)+f(E)\quad(x\in E)$$
if and only if f_1 depends only upon x_1 and f_2 depends only upon x_2 *or* f_1 depends only upon x_2 and f_2 depends only upon x_1.

7*. (Alexandrov 1970.) Let $E=]0,\infty[$. Find all solutions $f:E\to\mathbb{R}$, bounded from below, of the functional equation
$$f(x+E)=f(x)+f(E)\quad(x\in E).$$

8. Find a nonconstant periodic Jensen convex function $f:\mathbb{R}\to\bar{\mathbb{R}}_+$ which is not additive.
9. (Alexandrov 1967.) Prove that, for a cone having properties (10), (11) and (12), all bijections $f:\mathbb{R}^n\to\mathbb{R}^n$ $(n>2)$, which satisfy (9), are nonsingular affine transforms $(f(x)=Ax+b, \det A\neq 0)$.
10. Let E be a subset of $\mathbb{R}$ satisfying condition (ii) of Theorem 6. Prove or disprove that every additive $f:\mathbb{R}\to\mathbb{R}$, which is bounded from below on E, is continuous.
11. (Sierpinski 1920*b*.) Let $f:\mathbb{R}^n\to\mathbb{R}$ be a Jensen convex function. Suppose there exists a measurable function $g:\mathbb{R}^n\to\mathbb{R}$ majorizing f, that is,
$$f(x)\leqslant g(x)\quad (x\in\mathbb{R}^n).$$
Show that f is continuous and convex (that is,
$$f(\lambda x+(1-\lambda)y)\leqslant\lambda f(x)+(1-\lambda)f(y),$$
where $\lambda\in]0,1[$; $x, y\in\mathbb{R}^n$).
12. (Ger–Kuczma 1970.) Let E be a nonempty subset of $\mathbb{R}$. Prove that every Jensen convex function $f:\mathbb{R}\to\mathbb{R}$, which is bounded above on E, is continuous if, and only if, E has property (i) (or (ii)) of Theorem 6.
13*. (Kuczma 1973*a*.) Let E be a continuous cartesian curve in $\mathbb{R}^2$ which is different from a straight segment or a straight line (that is E is the graph of $\phi: I\to\mathbb{R}$, where I is a nonempty open interval of $\mathbb{R}$ and ϕ is not a linear function). Prove or disprove that every additive $f:\mathbb{R}^2\to\mathbb{R}$, bounded from above on E, is continuous.
14. (Smital 1976*b*.) Let E be a measurable subset of $\mathbb{R}$ with positive Lebesgue measure. Show that every Jensen convex function $f:\mathbb{R}\to\mathbb{R}$, bounded below on E, is locally bounded below on $\mathbb{R}$.
15*. (Smital 1968.) Show that there exists a discontinuous additive $f:\mathbb{R}\to\mathbb{R}$ such that its graph intersects every Borel subset of $\mathbb{R}^2$ whose projection on the x-axis is not countable. Show that for all such functions $f(\mathbb{R})=\mathbb{R}$ and that they have the Darboux property (that is S is a connected subset of $\mathbb{R}$, then $f(S)$ is connected).
16. (Gajda 1984*a*.) Let E, F be linear real topological spaces and $f:E\to F$ an additive function. Suppose that there exist a non-identically zero linear functional $\phi:F\to\mathbb{R}$ and a positive number α such that
$$\phi(f(x))\leqslant\alpha$$
for all x is a given nonempty open subset U of E. Show that there exists a continuous and additive $g:E\to F$ such that
$$\phi(f(x)-g(x))=0\quad (x\in E).$$

10

Some applications of functional equations in functional analysis, in the geometry of Banach spaces and in valuation theory

In functional analysis, in particular in the geometry of Banach spaces, functional equations appear quite often. In the first five sections of the present chapter, we shall investigate five different types of such occurrences. Of course we will use fundamental concepts and results of functional analysis (in particular in Sections 1, 4, 5).

The first two applications deal with the characterization of extreme points and rays for some convex subsets of function spaces. Quite often such extreme points can be computed with the aid of functional equations. The third and fourth application will lead us to a characterization of strictly convex normed spaces. The more special inner product spaces will be characterized in Chapter 11. The fifth application concerns harmonic analysis again. The norms, which play a quite important role in all this, are, of course, generalizations of the absolute value from $\mathbb{R}$ or $\mathbb{C}$ to general linear spaces.

Another generalization, this time to fields, is the valuation, with which we deal for the fields $\mathbb{Q}$ and $\mathbb{R}$ in Section 6.

10.1 Functional equations and extreme points

Let E be a real or complex linear space and C be a convex subset of E (that is, as defined in Section 9.1, for all x, y in C, the segment $[x, y]$ lies in C, where $[x, y] = \{z \mid z = \lambda x + (1-\lambda)y \in E,\ 0 \leqslant \lambda \leqslant 1\}$).

An *extreme point* of C is an element of C which can never be written as a proper convex combination of two distinct points of C. It is not difficult to see that this definition can be written in the following convenient form. The point $x \in C$ is an extreme point of C if, and only if, $x = (x_1 + x_2)/2$ ($x_1 \in C$, $x_2 \in C$), implies $x = x_1 = x_2$. For instance, if C is a triangle, a convex subset of $E = \mathbb{R}^2$, the extreme points of C obviously are its three vertices. On the other hand, the open unit disc in $\mathbb{R}^2$ has no extreme points.

An important theorem, which we shall not use in the sequel, states that some convex subsets of (possibly infinite dimensional) real linear spaces are generated by their extreme points. In fact, a compact convex set in a locally convex real Hausdorff space is the closure of the convex hull of its extreme points (Krein–Milman theorem; for a proof, we refer to Dunford–Schwartz 1958, p. 440). This explains the advantage of finding the extreme points of convex sets. We shall investigate, however, two cases of convex subsets which are not compact with respect to the natural topology which they inherit from their embedding in some topological linear space. We will see that, in these cases too, it is important to characterize the extreme points.

Let X and Y be two compact Hausdorff spaces. Let $C(X)$ be the Banach space of all complex valued continuous functions over X, equipped with the topology of the uniform norm

$$\|f\| = \|f\|_\infty = \sup_{x \in X} |f(x)|.$$

(We will use $\|f\|$ for $\|f\|_\infty$ when there is no danger of misunderstanding.) By $C_R(X)$ we mean the subset of all real valued functions in $C(X)$. We denote by $B_1(X, Y)$ the unit ball in the space of all bounded linear operators from $C(X)$ into $C(Y)$ ($P \in B_1(X, Y)$ if $P: C(X) \to C(Y)$ is a linear operator such that $\|P\| \leqslant 1$). Let $B_1^+(X, Y)$ be the subset of all bounded linear operators P from $C(X)$ into $C(Y)$ such that $\|P\| = 1$ and $P(\mathbf{1}) = \mathbf{1}$, where $\mathbf{1}$ denotes the constant function which equals 1 everywhere where it is defined.

The subset $B_1^+(X, Y)$ (B_1^+ for short) is a convex subset of the unit ball $B_1(X, Y)$. A special case of the following theorem was first proved in the real case by A. and C. Ionescu–Tulcea (1962), then extended by Phelps (1963) to the complex case.

Theorem 1. *An operator P in $B_1^+(X, Y)$ is an extreme point of $B_1^+(X, Y)$ if, and only if, P is a multiplicative operator, that is, satisfies the functional equation*

$$P(fg) = Pf \cdot Pg \tag{1}$$

for all f, g in $C(X)$.

Proof. We first prove that every operator P in B_1^+ is nonnegative, that is, it transforms a nonnegative f in $C(X)$ into a nonnegative Pf (in $C(Y)$). We say that f is nonnegative and write $f \geqslant 0$, if $f(x) \geqslant 0$ for all x in X. We then denote a nonnegative operator by $P \geqslant 0$. We will need the following:

Lemma 2. *If $P \in B_1^+$, then $P \geqslant 0$.*

Suppose for contradiction that, for some nonnegative f in $C(X)$, there exists a $y_0 \in Y$ with $Pf(y_0) \notin \bar{\mathbb{R}}_+$. Since f belongs to $C(X)$, the values $f(x)$ belong to a bounded subset of $\bar{\mathbb{R}}_+$. We can find an $\alpha\in\mathbb{C}$ and $r\in\mathbb{R}_+$ such that

$$|f(x)-\alpha|\leqslant r,\quad (x\in X) \tag{2}$$

and

$$|Pf(y_0)-\alpha|>r.$$

Then $\|f-\alpha\mathbf{1}\|=\|P\|\ \|f-\alpha\mathbf{1}\|\geqslant\|P(f-\alpha\mathbf{1})\|$ and $\|P(f-\alpha\mathbf{1})\|\geqslant|Pf(y_0)-\alpha|>r$. Thus $\|f-\alpha\mathbf{1}\|>r$, which contradicts (2) and thus proves Lemma 2.

Lemma 2 yields the 'only if' part of the following convenient characterization of $B_1^+(X,Y)$:

Lemma 3. *A linear operator $P:C(X)\to C(Y)$ belongs to $B_1^+(X,Y)$ if, and only if, $P\geqslant 0$ and $P(\mathbf{1})=\mathbf{1}$.*

To prove the 'if' part, suppose $P\geqslant 0$ and $P(\mathbf{1})=\mathbf{1}$. The classical Cauchy–Schwarz argument using the nonnegativity of P proves that, for f_1,f_2 in $C(X)$, we get

$$|P(f_1\bar{f}_2)|^2\leqslant P(|f_1|^2)P(|f_2|^2). \tag{3}$$

Every (complex valued) f in $C(X)$ can be written in the form $f=g\bar{h}$, where g and h are functions in $C(X)$ and $|g|=|h|$. Thus, with $f_1=g$, $f_2=h$ and $P\geqslant 0$, (3) yields

$$|Pf|\leqslant P(|f|). \tag{4}$$

As $|f|\leqslant\|f\|\mathbf{1}$, we deduce from (4) and from $P\geqslant 0$ that $|Pf|\leqslant P(|f|)\leqslant\|f\|P(\mathbf{1})=\|f\|\mathbf{1}$. Thus $\|\mathbf{P}\|\leqslant 1$. As $P(\mathbf{1})=\mathbf{1}$, $\|P\|=1$. In other words, $P\in B_1^+(X,Y)$, as asserted.

We may now proceed to the *proof of Theorem* 1. Let first P be an extreme point of $B_1^+(X,Y)$. Fix some g in $C(X)$, where $0\leqslant g\leqslant\mathbf{1}$ and consider the linear operator $Q:C(X)\to C(Y)$, defined by

$$Qf=P(fg)-Pf\cdot Pg.$$

We also define $P_1=P+Q$ and $P_2=P-Q$. Both P_1 and P_2 are linear operators. Because $Q(\mathbf{1})=0$, we deduce that $P_1(\mathbf{1})=P_2(\mathbf{1})=\mathbf{1}$.

Take $f\geqslant 0$ in $C(X)$. We compute P_1f:

$$P_1f=(\mathbf{1}-Pg)Pf+P(fg).$$

As $0\leqslant Pg\leqslant\mathbf{1}$, $0\leqslant Pf$ and $0\leqslant P(fg)$, we conclude that $P_1f\geqslant 0$, that is, $P_1\geqslant 0$. In the same way we compute P_2f, using the linearity of P:

$$P_2f=P(f(\mathbf{1}-g))+Pf\cdot Pg.$$

As $0\leqslant f(\mathbf{1}-g)$, we have $0\leqslant P(f(\mathbf{1}-g))$ and $Pf\cdot Pg\geqslant 0$. Thus $P_2f\geqslant 0$, that is, $P_2\geqslant 0$.

Lemma 3 proves that P_1 and P_2 belong to $B_1^+(X, Y)$. But $P=(P_1+P_2)/2$. As P is an extreme point of $B_1^+(X, Y)$, we deduce that $P=P_1=P_2$. This result implies $Q=0$ and so, for all f in $C(X)$ and for the fixed g,

$$P(fg)=Pf\cdot Pg. \tag{5}$$

Using the linearity of the operator P, we first extend (5) to all real valued g in $C(X)$ and then extend it to any f in $C(X)$ and any g in $C(X)$. This proves the 'only if' part of Theorem 1.

Conversely, let P be an operator in $B_1^+(X, Y)$ satisfying (1). Suppose $P=(P_1+P_2)/2$, where P_1 and P_2 are linear operators in $B_1^+(X, Y)$. First let f be a real function in $C(X)$. We compute

$$4(Pf)^2=(P_1f+P_2f)^2=(P_1f)^2+(P_2f)^2+2P_1f\cdot P_2f.$$

From (3), using $P_1(\mathbf{1})=P_2(\mathbf{1})=\mathbf{1}$, we notice that

$$(P_1f)^2\leqslant P_1(f^2) \quad\text{and}\quad (P_2f)^2\leqslant P_2(f^2).$$

Thus

$$4(Pf)^2\leqslant P_1(f^2)+P_2(f^2)+2P_1f\cdot P_2f. \tag{6}$$

On the other hand,

$$4P(f^2)=2P_1(f^2)+2P_2(f^2). \tag{7}$$

But $(Pf)^2=P(f^2)$ because P satisfies (1) and so (6) and (7) yield

$$(P_1f-P_2f)^2\leqslant 0.$$

Thus $P_1f=P_2f$. Because f is an arbitrary real function in $C(X)$. and P_1, P_2 are nonnegative linear operators, we see that $P_1f=P_2f$ for all $f\in C(X)$, that is, $P_1=P_2$. Therefore P is an extreme point of $B_1^+(X, Y)$, which concludes the proof of Theorem 1.

The characterization of the set of all extreme points of $B_1(X, Y)$, instead of those of $B_1^+(X, Y)$, appears to be a much more difficult problem and is still unsolved in this general setting. However, many special results exist for some classes of compact spaces X and Y.

In particular, when Y is reduced to a singleton $\{y\}$, the algebra $C(Y)$ is isomorphic to $\mathbb{C}$. In this context an element of $B_1(X, Y)$ is a complex valued Radon measure of norm at most 1. We write $M_1(X)$ instead of $B_1(X, Y)$ and $M_1^+(X)$ for the subset of all nonnegative Radon measures with total measure 1 (cf. Lemma 3).

Proposition 4. *An element μ of $M_1(X)$ is an extreme point of $M_1(X)$ if, and only if, $|\mu(\mathbf{1})|=1$ and, for all f, g in $C(X)$, the functional equation*

$$\mu(\mathbf{1})\mu(fg)=\mu(f)\mu(g) \tag{8}$$

holds.

Proof. Let μ be an extreme point of $M_1(X)$. By contradiction we show that $\|\mu\| = 1$. In fact, if $0 < \|\mu\| < 1$, we may obviously write μ as a proper convex combination of two distinct measures of $M_1(X)$:

$$\mu = \|\mu\| \mu_1 + (1 - \|\mu\|)\mu_2.$$

The same is true when $\|\mu\| = 0$. Now we introduce the measure $|\mu|$ defined by $|\mu|(f) = \sup_{|g| \leqslant |f|} |\mu(g)|$. (Note that $|\mu|$ is not the same as $\|\mu\|$ which is a number and that $|\mu|(f)$ is not the same as $|\mu(f)|$.) Next we prove that $|\mu|$ is an extreme point of $M_1(X)$. Suppose that $|\mu| = (\mu_1 + \mu_2)/2$, where the μ_i $(i = 1, 2)$ are in $M_1(X)$. The Radon–Nikodym theorem proves the existence of a measurable $h: X \to \mathbb{C}$ such that $\mu = h|\mu|$ (that is, $\mu(f) = |\mu|(hf)$) and for all x in X, $|h(x)| = 1$. Thus $\mu = (h\mu_1 + h\mu_2)/2$. But $h\mu_1$ and $h\mu_2$ belong to $M_1(X)$, as do μ_1 and μ_2. The extremality of μ yields $h\mu_1 = h\mu_2$, that is, $\mu_1 = \mu_2$. Therefore $|\mu|$ is an extreme point of $M_1(X)$.

Since $|\mu|$ is a positive Radon measure and since $\|\mu\| = 1$, therefore $|\mu|(1) = 1$. So $|\mu|$ is also an extreme point of $M_1^+(X)$. Theorem 1 then shows that, for all f, g in $C(X)$,

$$|\mu|(fg) = |\mu|(f)\,|\mu|(g).$$

From $\mu(f) = |\mu|(hf)$ and from $h\bar{h} = 1$ we see that $|\mu|(f) = \mu(f\bar{h})$. Thus we get

$$\mu(fg\bar{h}) = \mu(f\bar{h})\mu(g\bar{h}). \tag{9}$$

If h were a continuous function, we would immediately deduce, with $g = h$, that $\mu(f) = \mu(f\bar{h})\mu(1)$ for all $f \in C(X)$. In general, h is only supposed to be measurable but can be approximated by a sequence of continuous functions $\{g_n\}$. We obtain, letting n go to infinity, the same result,

$$\mu(f) = \mu(f\bar{h})\mu(1).$$

Multiplying (9) by $\mu(1)^2$ gives (8).

Now we have not only $|\mu|(1) = 1$ (that is, $\int 1 d|\mu| = 1$) but also $|\mu(1)| = 1$ (that is, $|\int 1 d\mu| = 1$). Indeed $\mu(f\bar{h}) = |\mu|(f)$ and, as we have seen, $\mu(f) = \mu(f\bar{h})\mu(1)$. Thus $\mu(f) = \mu(1)|\mu|(f)$, that is, $\mu = \mu(1)|\mu|$. Since μ and $|\mu|$ have the same norm, we get $|\mu(1)| = 1$.

Conversely, let $\mu \in M_1(X)$ satisfy the functional equation (8) and $|\mu(1)| = 1$. Suppose that $\mu = (\mu_1 + \mu_2)/2$, where μ_1, μ_2 are measures in $M_1(X)$. We write

$$\mu(1) = \frac{\mu_1(1) + \mu_2(1)}{2}.$$

But $|\mu(1)| = 1$, $|\mu_1(1)| \leqslant 1$ and $|\mu_2(1)| \leqslant 1$. The extreme points of the unit disk of $\mathbb{C}$ are the complex numbers of modulus 1. Therefore $\mu(1) = \mu_1(1) = \mu_2(1)$. We may use $\nu = \mu/(\mu(1))$, $\nu_1 = \mu_1/(\mu(1))$, $\nu_2 = \mu_2/(\mu(1))$. We have $\nu = (\nu_1 + \nu_2)/2$ and the three measures ν, ν_1, ν_2 belong to $M_1^+(X)$. Equation (8) shows that ν satisfies $\nu(fg) = \nu(f)\nu(g)$. We use Theorem 1 to show that ν is extreme in

$M_1^+(X)$ and so $\nu_1 = \nu_2 = \nu$. This proves $\mu = \mu_1 = \mu_2$ and so μ itself is an extreme point of $M_1(X)$, as asserted.

We shall now determine all solutions of the functional equation (8).

We will need the *Dirac measure* δ_{x_0} at x_0 in X, which is defined by $\delta_{x_0}(f) = f(x_0)$ for all f in $C(X)$.

Lemma 5. *A complex valued bounded Radon measure μ on X satisfies* (8) *for all f, g in $C(X)$ if, and only if, $\mu = \alpha\delta_{x_0}$ where δ_{x_0} is the Dirac measure at some x_0 in X and α is a complex number.*

Proof. If $\mu = \alpha\delta_{x_0}$, then $\mu(\mathbf{1}) = \alpha$ and $\mu(\mathbf{1})\mu(fg) = \alpha^2(fg)(x_0)$. But $\mu(f)\mu(g) = \alpha^2 f(x_0)g(x_0)$, so that (8) is satisfied.

Conversely, let μ be a solution of (8). If $\mu(\mathbf{1}) = 0$, with $f = g$ in (8), we see that $\mu(f) = 0$ for all f in $C(X)$. Thus $\mu = 0$ and we may just take $\alpha = 0$ and any x_0 in X. If $\mu(\mathbf{1}) \neq 0$, then $\nu = \mu/\mu(\mathbf{1})$ satisfies the functional equation (8) and $\nu(\mathbf{1}) = 1$. We prove that the support of ν is a singleton. For contradiction, suppose that x_1, x_2 are two distinct points in the support of ν. The topological space X being compact and Hausdorff, there exist two closed neighbourhoods U_1, U_2 of x_1 and x_2, respectively, such that $U_1 \cap U_2 = \varnothing$. For $i = 1, 2$, as x_i is in the support of ν, there exist $f_i \in C(X)$, the support of which is in U_i, so that $\nu(f_i) \neq 0$. But $f_1 f_2 = 0$ and so (9) is contradicted:

$$0 = \nu(f_1 f_2) = \nu(f_1)\nu(f_2).$$

Let x_0 be the support of ν. Then $\nu = \alpha\delta_{x_0}$. But, since $\nu(\mathbf{1}) = 1$, we have $\nu = \delta_{x_0}$. With $\mu(\mathbf{1}) = \alpha \in \mathbb{C}$, we finally obtain $\mu = \alpha\delta_{x_0}$, which concludes the proof of Lemma 5. Notice that x_0 is uniquely defined in X if $\alpha \neq 0$.

Proposition 6. *Let $P: C(X) \to C(Y)$ be a bounded linear operator. We suppose that there exists no y in Y such that $Pf(y) = 0$ for all f in $C(X)$. Then P satisfies, for all f, g in $C(X)$,*

$$P(\mathbf{1})P(fg) = Pf \cdot Pg \tag{10}$$

if, and only if, $Pf(y) = \alpha(y)f(\phi(y))$ for all $y \in Y$, where α is a nonvanishing function in $C(Y)$ and $\phi: Y \to X$ a continuous function.

Proof. Suppose that P, defined on $C(X)$, has the following form for a continuous $\phi: Y \to X$:

$$Pf(y) = \alpha(y)f(\phi(y)), \quad \alpha \in C(Y). \tag{11}$$

Then $Pf \in C(Y)$ and (10) is satisfied. The fact that there exists no y in Y such that $Pf(y) = 0$ for all f in $C(X)$ is equivalent to the fact that α does not vanish. Notice that $\alpha(y) = P(\mathbf{1})(y)$ when P is of the form (11).

Now let the solution $P: C(X) \to C(Y)$ of (10) be a bounded linear operator with the property as stated in Proposition 6. We fix y in Y. The mapping $f \mapsto Pf(y)$ is a bounded linear functional on $C(X)$.

By the Riesz representation theorem, there exists a bounded Radon measure μ, acting on $C(X)$, such that $\mu(f) = Pf(y)$. As P satisfies (10), the measure μ satisfies (8). Lemma 5 yields $\mu = \alpha(y)\delta_{\phi(y)}$ where $\phi(y) \in X$ and is well defined if $\alpha(y) \neq 0$. But, as $Pf(y) \neq 0$ for at least some f in $C(X)$, we know that $\alpha(y) \neq 0$. Thus P has the form (11). It remains to be shown that α and ϕ have the required properties. As $\alpha(y) = P(\mathbf{1})(y)$, we have $\alpha \in C(Y)$ (and α is nonvanishing). We prove that $\phi: Y \to X$ is continuous. Let y_γ be a net converging towards y in Y. The net $x_\gamma = \phi(y_\gamma)$ belongs to X. As X is compact, there exists a subnet y_β such that x_β converges to x in X. For any such subnet $\alpha(y_\beta)f(\phi(y_\beta))$ converges to $\alpha(y)f(x)$. But $\alpha(y_\beta)f(\phi(y_\beta)) = Pf(y_\beta)$ converges towards $Pf(y) = \alpha(y)f(\phi(y))$. As $\alpha(y) \neq 0$, we get $f(\phi(y)) = f(x)$ for all f in $C(X)$. As $C(X)$ separates points of X, we deduce that $x = \phi(y)$. In other words, ϕ is continuous, which concludes the proof of Proposition 6.

If $|\alpha(y)| = 1$ for all y in Y, $\alpha \in C(Y)$ and $\phi: Y \to X$ is continuous, then an operator of the form (11) is called a *nice operator*. A combination of Theorem 1 and Proposition 6 yields that an operator in $B_1^+(X, Y)$ is extreme if, and only if, it is nice; in this case $P(\mathbf{1})(y) = \alpha(y) = 1$ for all $y \in Y$. The same statement is true when we replace $B_1^+(X, Y)$ by $B_1(X, z)$ where z is a singleton (Proposition 4). It has been shown by a counterexample that this statement does not hold in general for $B_1(X, Y)$ (Sharir 1976). A reason for the different behaviour of B_1 and B_1^+ is that there is a simple bijection of the set of extreme points of $M_1^+(X)$ onto X, but this is not the case for $M_1(X)$. However, the equivalence is true for $B_1(X, Y)$ in some rather special cases, for example if X is dispersed or if Y is extremely disconnected (Sharir 1972). In the real case, that is, for $P: C_{\mathbb{R}}(X) \to C_{\mathbb{R}}(Y)$, the statement is true if X is metrizable (Blumenthal–Lindenstrauss–Phelps 1965) but does not hold in general (Sharir 1977). The general study of the functional equation (10) with respect to extremality may still bring new results. (For a survey of nice operators and extreme operators, see Brillouët 1979.)

10.2 Totally monotonic functions and extreme rays

Our second example for functional equations and extreme elements (rays) will deal with totally monotonic functions. Take $f: I \to \mathbb{R}$, where $I =]-\infty, 0[$. We say that f is *totally monotonic* if, for all $x \in I$, for all nonnegative integers n and for all $\alpha > 0$ such that

$$x, x + \alpha, x + 2\alpha, \ldots, x + n\alpha \in I,$$

we have

$$\Delta_\alpha^n f(x) := \sum_{k=0}^{n} (-1)^{n-k} \binom{n}{k} f(x+k\alpha) \geqslant 0, \tag{12}$$

where of course,

$$\binom{n}{k} = \frac{n!}{k!(n-k)!}.$$

From (12) with $n=0$, $f(x)\geqslant 0$ for all x in I, so f is *nonnegative* on I, and, with $n=1$, $f(x+\alpha)\geqslant f(x)$ for all x in I, with $x+\alpha \in I$. In other words, f is an increasing function over I.

The inequality with $n=2$ gives $f(x)-2f(x+\alpha)+f(x+2\alpha)\geqslant 0$. In other words,

$$f\left(\frac{y_1+y_2}{2}\right) \leqslant \frac{f(y_1)+f(y_2)}{2} \quad \text{for all } y_1, y_2 \text{ in } I. \tag{13}$$

A function $f: I \to \mathbb{R}$, satisfying (13), is called *Jensen convex* (cf. (9.20) and Exercise 9.8). Since f is increasing and nonnegative, it is bounded on any closed subinterval of $]-\infty, 0[$. From this one can show that f is continuous. Therefore, f is *convex* in the usual sense, that is,

$$f(\lambda y_1 + (1-\lambda)y_2) \leqslant \lambda f(y_1) + (1-\lambda) f(y_2) \quad \text{for all } y_1, y_2 \text{ in } I \text{ and for all } \lambda \in [0,1].$$

We will show that every function $f: I \to \mathbb{R}$ with nonnegative derivatives of all orders is totally monotonic. For instance $f(x) = -1/x$ $(x<0)$, or $f(x) = e^{\lambda x}$ $(x \in \mathbb{R})$ with $\lambda \geqslant 0$, are such functions. Since

$$\Delta_\alpha^1 f(x) = f(x+\alpha) - f(x),$$

we deduce

$$\Delta_\alpha^1 f(x) = \alpha f'(x+\theta\alpha) \geqslant 0 \quad \text{where } 0<\theta<1,$$

because $f'(x)\geqslant 0$ for all x in I. So (12) holds for $n=1$. In the same way, for $n=2$,

$$\Delta_\alpha^2 f(x) = \Delta_\alpha^1(\Delta_\alpha^1 f(x))$$

and, using the same formula, we deduce that $\Delta_\alpha^2 f(x) \geqslant 0$. More generally, $\Delta_\alpha^n f(x) = \Delta_\alpha^1(\Delta_\alpha^{n-1} f(x)) \geqslant 0$. In other words, we have proved that any function $f: I \to \mathbb{R}$ with positive derivatives of all orders is a totally monotonic function. Let C be the set of all totally monotonic functions over I. In the same way we prove that C is stable under Δ_α^1, that is, $\Delta_\alpha^1 f \in C$ if $f \in C$ and, more generally, stable under Δ_α^n.

We wish to obtain a representation theorem for all totally monotonic functions. A nice technique consists of studying the extreme elements of the set C of all totally monotonic functions over I. Let us be more specific.

The set C is a proper cone with vertex at 0, which means that $f \in C$ implies $\lambda f \in C$ for all $\lambda \geqslant 0$ and f and $-f$ are in C if and only if $f \equiv 0$. This is obvious for C, since all f in C must be nonnegative. That C is convex, is clear.

For a convex cone with vertex at 0, the notion of extreme point is not suitable because the only extreme point is the vertex 0. The corresponding notion is that of *extreme ray*. In fact, a proper convex cone C induces a partial order on C according to

$$f \succ g \quad \text{if } f-g \in C$$

for f,g in C (see Section 9.1).

We will say that f belongs to an *extreme ray* if f is not identically zero and, for any g in C, the relation $f \succ g$ implies $g=\lambda f$ for some $\lambda \geqslant 0$. It is clear that this is a property of the ray containing f. The reader will easily show that the intersection of extreme rays with hyperplanes not going through 0 provides extreme points of the intersection. (See Exercise 2).

Our purpose is to characterize the extreme rays of the cone C of all totally monotonic functions on I.

Theorem 7. *The extreme rays of C are exactly those containing a function of the form $e^{\lambda x}$ where $\lambda \geqslant 0$.*

Let f be in C, not identically 0, and belonging to some extreme ray of C. Let $\beta \geqslant 0$. We consider $f_\beta(x) = f(x-\beta)$. Clearly, $f_\beta \in C$. Define F by

$$F = f - f_\beta.$$

This F belongs to C. In fact, we get

$$\begin{aligned}\Delta^1_\alpha F(x) &= f(x+\alpha) - f(x) - (f(x+\alpha-\beta) - f(x-\beta))\\ &= \Delta^1_\alpha f(x) - \Delta^1_\alpha f(x-\beta).\end{aligned}$$

As $f_\beta \in C$, $\Delta^1_\alpha f(x) - \Delta^1_\alpha f(x-\beta) = \Delta^1_\beta(\Delta^1_\alpha f_\beta)(x) \geqslant 0$ because $\Delta^1_\alpha f_\beta \in C$. In the same way, $\Delta^n_\alpha F(x) \geqslant 0$ for all integers n, and so $F \in C$. For the order defined by C, we get

$$f \succ f_\beta \quad \text{for all } \beta \geqslant 0.$$

But, as f is on an extreme ray of C, we deduce the existence of a nonnegative λ, depending only upon β, such that

$$f_\beta = \lambda f$$

or

$$f(x-\beta) = \lambda(\beta) f(x) \quad (x<0), \tag{14}$$

which is a functional equation. The left hand side has a (finite) limit when x goes to 0 ($x<0$). So the right hand side has one too. If $\lambda(\beta) \neq 0$, this yields the existence of

$$\lim_{x \to 0-} f(x) = f(0-),$$

nonnegative since $f \geqslant 0$. If $\lambda(\beta) = 0$ for all $\beta > 0$, we deduce from (14), by the continuity of f, that f is identically zero, which was excluded. Therefore $f(0-)$ always exists and is positive (if we exclude $f \equiv 0$).

As we wish to determine an extreme ray, rather than a point on that ray, we may suppose, without loss of generality, that $f(0-)=1$. Using this limit in (14), when x tends to 0 increasingly, we obtain $f(-\beta)=\lambda(\beta)$, which in turn yields $f(x-\beta)=f(-\beta)f(x)$. Finally for f we get a Cauchy's exponential equation restricted to the domain $]-\infty,0]^2$. An easy adaptation of Theorem 3.5 ($\bar{\mathbb{R}}_+$ being replaced by $\bar{\mathbb{R}}_- =]-\infty,0]$) yields for some $\lambda \in \mathbb{R}$

$$f(x)=e^{\lambda x}, \quad x \in]-\infty,0].$$

As an element of C, f is increasing, which in turn yields $\lambda \geqslant 0$ and concludes the proof of Theorem 7.

We cannot apply the Krein–Milman theorem in the present setting, since the convex cone C is not compact. However, a skillful technique due to Choquet gives the possibility to use the extreme rays of C in order to recover all of C. Here we shall not give this technique, which goes beyond the scope of this book (see Choquet 1969), but just state the Bernstein representation theorem for totally monotonic functions.

Theorem 8. *A function $f: I =]-\infty,0[\to \mathbb{R}$ is totally monotonic if, and only if, there exists a nonnegative Radon measure μ such that*

$$f(x)=\int_0^\infty e^{\lambda x}\, d\mu(\lambda), \quad x<0.$$

One generally prefers to define *completely monotonic* functions on $\mathbb{R}_+$ instead of totally monotonic functions on $]-\infty,0[$. The link is that g is completely monotonic on $\mathbb{R}_+$ if and only if f, where $f(x)=g(-x)$, is totally monotonic on $]-\infty,0[$. Theorem 8 easily yields the representation theorem for completely monotonic functions g. In particular, a C^∞ completely monotonic function g on $\mathbb{R}_+$ is such that $(-1)^k g^{(k)}(x) \geqslant 0$ $(x>0; k=0,1,2,\ldots)$.

10.3 A characterization of strictly convex normed spaces

Let $(E, \|\cdot\|)$ be a real normed space. We call it *strictly convex* if every point of its unit sphere is an extreme point of the unit ball. In other words, if $\|x\| \leqslant 1$ and $\|y\| \leqslant 1$, then $\|(x+y)/2\| = 1$ implies $x=y$. The technique in the following result is due to Fischer and Muszély 1967 (cf. Kuczma 1978*a*).

Theorem 9. *A real normed space $(E, \|\cdot\|)$ is strictly convex if, and only if, the only continuous functions $f: \mathbb{R} \to E$ satisfying*

$$\|f(x+y)\| = \|f(x)+f(y)\| \quad \textit{for all } x,y \textit{ in } \mathbb{R} \tag{15}$$

are the additive ones, that is, the continuous solutions of the Cauchy equation

$$f(x+y)=f(x)+f(y) \quad (x,y\in\mathbb{R}). \tag{16}$$

Proof. Suppose first that E is strictly convex and let $f:\mathbb{R}\to E$ be a continuous function satisfying (15).

We start from the obvious $\|f(2x)\|=2\|f(x)\|$ and proceed to estimate $\|f(3x)\|$:

$$\|f(3x)\|=\|f(2x+x)\|=\|f(2x)+f(x)\|\leqslant\|f(2x)\|+\|f(x)\|=3\|f(x)\|. \tag{17}$$

The next computation is done by estimating $f(4x)$ in two different ways:

$$\|f(4x)\|=\|f(3x)+f(x)\|\leqslant\|f(3x)\|+\|f(x)\|\leqslant 4\|f(x)\| \tag{18}$$

and

$$\|f(4x)\|=\|f(2x)+f(2x)\|=4\|f(x)\|. \tag{19}$$

Therefore, using (19), the inequalities in (18) are in fact equalities, so that

$$\|f(3x)\|=3\|f(x)\|.$$

For the same reason, the inequality in (17) is also an equality,

$$\|f(2x)+f(x)\|=\|f(2x)\|+\|f(x)\|,$$

or

$$\|f(2x)+f(x)\|=\left\|\frac{f(2x)}{2}\right\|+\left\|\frac{f(2x)}{2}\right\|+\|f(x)\|.$$

But, by the triangle inequality,

$$\begin{aligned}\|f(2x)+f(x)\|&\leqslant\left\|\frac{f(2x)}{2}\right\|+\left\|\frac{f(2x)}{2}+f(x)\right\|\leqslant\left\|\frac{f(2x)}{2}\right\|+\left\|\frac{f(2x)}{2}\right\|+\|f(x)\|\\&=\|f(2x)+f(x)\|.\end{aligned}$$

Therefore

$$\left\|\frac{f(2x)}{2}+f(x)\right\|=\left\|\frac{f(2x)}{2}\right\|+\|f(x)\| \tag{20}$$

and, since $\|f(2x)/2\|=\|f(x)\|$, the strict convexity of E yields

$$f(2x)=2f(x).$$

Suppose now that $f(nx)=nf(x)$ for some integer $n\geqslant 2$. We compute

$$\begin{aligned}\|f((n+1)x)\|&=\|f(nx)+f(x)\|\\&=\|nf(x)+f(x)\|\\&=(n+1)\|f(x)\|\end{aligned}$$

and

$$\begin{aligned}\|f((n+2)x)\|&=\|f(nx)+f(2x)\|\\&=\|nf(x)+2f(x)\|=(n+2)\|f(x)\|.\end{aligned}$$

Again, by the triangle inequality,

$$\begin{aligned}\|f((n+2)x)\| &= \|f((n+1)x)+f(x)\| \\ &\leqslant \|f((n+1)x)\| + \|f(x)\| \\ &= (n+2)\|f(x)\|.\end{aligned}$$

As the two extreme terms are equal, we get an equality sign everywhere,

$$\|f((n+1)x)+f(x)\| = \|f((n+1)x)\| + \|f(x)\|. \tag{21}$$

Now

$$\begin{aligned}\|f((n+1)x)+f(x)\| &= (n+1)\left\|\frac{f((n+1)x)}{n+1}\right\| + \|f(x)\| \\ &\geqslant \frac{n}{n+1}\|f((n+1)x)\| + \left\|\frac{f((n+1)x)}{n+1} + f(x)\right\| \\ &\geqslant \|f((n+1)x)+f(x)\|.\end{aligned}$$

The intermediate inequality becomes

$$\left\|\frac{f((n+1)x)}{n+1} + f(x)\right\| = \left\|\frac{f((n+1)x)}{n+1}\right\| + \|f(x)\|.$$

Strict convexity of E yields

$$f((n+1)x) = (n+1)f(x)$$

and concludes the induction. As usual, at once we get

$$f\left(\frac{x}{m}\right) = \frac{1}{m}f(x).$$

Therefore, for every positive rational r,

$$f(rx) = rf(x). \tag{22}$$

Also, from (15), we have $\|f(0)\| = 0$, or $f(0) = 0$.

Using the continuity of f, we deduce that

$$f(x) = xf(1) \quad \text{for all } x \geqslant 0.$$

Moreover,

$$0 = \|f(0)\| = \|f(x-x)\| = \|f(x)+f(-x)\|,$$

so that f is odd. In other words, for all x in $\mathbb{R}$,

$$f(x) = xf(1)$$

and so f is additive

$$f(x+y) = f(x) + f(y) \quad (x, y \in \mathbb{R}).$$

Conversely, suppose that E is not a strictly convex space. We shall construct a continuous $f: \mathbb{R} \to E$, satisfying (15), which is not additive. This will conclude the proof of Theorem 9.

The supposition that E is not strictly convex yields two elements a, b in

E, with $\|a\| = \|b\| = 1$, $a \neq b$ and

$$\|a+b\| = \|a\| + \|b\| = 2.$$

It is clear that a and b are linearly independent vectors of E (if $a = \lambda b$, then $|\lambda| = 1$ and $|\lambda + 1| = 2$, so that $\lambda = 1$ but a and b are different).

Moreover, when λ and μ are of the same sign (both positive or both negative), we get

$$\|\lambda a + \mu b\| = |\lambda + \mu|. \tag{23}$$

Indeed, if, for example, $\mu > \lambda \geqslant 0$, we can write

$$\begin{aligned}\|\lambda a + \mu b\| &= \|\mu(a+b) - (\mu - \lambda)a\| \\ &\geqslant \mu\|a+b\| - (\mu-\lambda)\|a\|\end{aligned}$$

and thus

$$\|\lambda a + \mu b\| \geqslant \mu 2 - (\mu - \lambda)1 = \lambda + \mu.$$

But the converse inequality is obvious, so that (23) is true.

Now take for $g: \mathbb{R} \to \mathbb{R}$ any non identically 0, continuous and additive function (that is, $g(x) = cx$, for some $c \in \mathbb{R}$, $c \neq 0$). We define $f: \mathbb{R} \to E$ by

$$f(x) = \begin{cases} g(x)a & \text{if } |g(x)| \leqslant 1 \\ s(g(x))a + [g(x) - s(g(x))]b & \text{if } |g(x)| > 1, \end{cases}$$

where, by convention,

$$\begin{aligned} s(g(x)) &= 1 \quad && \text{if } g(x) \geqslant 0, \\ s(g(x)) &= -1 && \text{if } g(x) < 0. \end{aligned}$$

As g is continuous, so is f (if $x_n \to x$ and $|g(x)| = 1$, $(g(x_n) - s(g(x_n))b$ converges to 0). We shall make use of the fact that the sum of the coefficients of a and b is equal to $g(x)$, in order to show that (15) is valid, but additivity is not. Let us prove that f satisfies (15). We notice that

$$\|f(x)\| = |g(x)|. \tag{24}$$

This is obvious if $|g(x)| \leqslant 1$. If $|g(x)| > 1$, using (23), we get

$$\|s(g(x))a + [g(x) - s(g(x))]b\| = |s(g(x)) + g(x) - s(g(x))| = |g(x)|.$$

We now compute $\|f(x) + f(y)\|$. We have to make a distinction among four cases according to the relationship of $|g(x)|$ and $|g(y)|$ to 1.

If $|g(x)| \leqslant 1$ and $|g(y)| > 1$, then $f(x) + f(y) = (g(x) + s(g(y)))a + [g(y) - s(g(y))]b$. Both coefficients of a and b have the same sign. Therefore we get, using (23),

$$\|f(x) + f(y)\| = |g(x) + s(g(y)) + g(y) - s(g(y))| = |g(x) + g(y)| = |g(x+y)|.$$

If $|g(x)| > 1$ and $|g(y)| \leqslant 1$, the same result holds by symmetry in x and y.

If $|g(x)| > 1$ and $|g(y)| > 1$, then

$$f(x) + f(y) = (s[g(x)] + s[g(y)])a + (g(x+y) - s[g(x)] - s[g(y)])b.$$

Either $s[g(x)] = -s[g(y)]$, in which case $\|f(x)+f(y)\| = |g(x+y)|$ or $s[g(x)] = s[g(y)]$ which also yields $\|f(x)+f(y)\| = |g(x+y)|$.

If $|g(x)| \leqslant 1$ and $|g(y)| \leqslant 1$, then

$$\|f(x)+f(y)\| = \|g(x)a + g(y)a\| = |g(x)+g(y)| = |g(x+y)|.$$

We have now proved, in all cases,

$$\|f(x)+f(y)\| = |g(x+y)|.$$

But $\|f(x+y)\| = |g(x+y)|$, by (24). Thus (15) is satisfied for all x, y in $\mathbb{R}$. The last step is to prove that f is not an additive function. Changing g into $-g$ if necessary, there exists an x_0 in $\mathbb{R}$ such that $g(x_0) = \alpha > 0$. For n large enough, $g(nx_0) > 1$. So there exist x, y with

$$g(x) > 1, \quad g(y) > 1 \quad \text{and} \quad g(x+y) > 1.$$

In this case,

$$f(x) + f(y) = 2a + (g(x+y) - 2)b.$$

But

$$f(x+y) = a + (g(x+y) - 1)b.$$

Thus

$$\|f(x) + f(y) - f(x+y)\| = \|a - b\| \neq 0$$

and f is not additive, which concludes the proof of Theorem 9.

The continuity assumption for f in Theorem 9 can be considerably weakened (cf. Dhombres 1979, pp. 2.23–2.25). But it remains an open problem whether the statement of Theorem 9 would be valid without any regularity conditions on f.

10.4 Isometries in real normed spaces

From the beginnings of linear analysis, it has been known (Mazur–Ulam 1932) that an isometry f of a real normed linear space onto another necessarily satisfies the so-called Jensen equation:

$$f\left(\frac{x+y}{2}\right) = \frac{f(x)+f(y)}{2} \tag{25}$$

(the equality case of (13) or of (9.20); cf. (15.19)).

When $(E, \|\cdot\|)$ and $(F, \|\cdot\|)$ are two real normed spaces, an isometry $f: E \to F$ is defined by the functional equation

$$\|f(x) - f(y)\| = \|x - y\|. \tag{26}$$

Proposition 10. *If f is onto (surjective), then* (26) *implies* (25).

Proof. We prove (25) by showing that the relation 'midpoint' is invariant under f. We therefore look for a purely metric characterization of a midpoint in a real normed space E.

Let x_1, x_2 be in E. The midpoint $x = (x_1 + x_2)/2$ satisfies

$$\|x - x_1\| = \|x - x_2\| = \left\|\frac{x_1 - x_2}{2}\right\|. \tag{27}$$

But, for a general normed space, (27) does not characterize this midpoint (see Exercise 19 for a counterexample). In other words, the subset $M_0(x_1, x_2)$ of E, consisting of all x in E such that (27) holds, is not necessarily reduced to a point. We shall reduce the size of $M_0(x_1, x_2)$ to half and proceed recursively. We use the diameter Δ_0 of M_0,

$$\Delta_0 = \sup_{\substack{x' \in M_0 \\ x'' \in M_0}} \|x' - x''\|.$$

Naturally, if $\Delta_0 = 0$, $M_0(x_1, x_2)$ is reduced to $\{(x_1 + x_2)/2\}$. The diameter Δ_0 is always finite since, for any x in M_0, we have, from (27), the bound $\|x\| \leqslant \|(x_1 - x_2)/2\| + \|x_1\|$. Now define $M_1(x_1, x_2)$ as the subset ofl all x in M_0 such that, for each y in $M_0(x_1, x_2)$,

$$\|x - y\| \leqslant \frac{\Delta_0}{2}.$$

This subset $M_1(x_1, x_2)$ contains the midpoint $(x_1 + x_2)/2$. To show this, let x be in $M_0(x_1, x_2)$ and take $x' \in E$, such that $x + x' = x_1 + x_2$, that is $(x_1 + x_2)/2$ is the midpoint of x and x'. From $x' - x_1 = x_2 - x$ and $x' - x_2 = x_1 - x$, we also get that x' is in $M_0(x_1, x_2)$. Due to the construction of x', we get that $\|x - (x_1 + x_2)/2\| (= \|x' - (x_1 + x_2)/2\|)$ is exactly half of $\|x - x'\|$. As this last norm is majorized by Δ_0, we have established that $(x_1 + x_2)/2$ is in $M_1(x_1, x_2)$.

For $n \geqslant 1$, we define $M_{n+1}(x_1, x_2)$ recursively as the subset of all x in $M_n(x_1, x_2)$ for which $y \in M_n(x_1, x_2)$ implies $\|x - y\| \leqslant \Delta_n/2$, where Δ_n is the diameter of $M_n(x_1, x_2)$.

When there is no danger of misunderstanding, we write M_n for $M_n(x_1, x_2)$.

We prove by induction for all $n \geqslant 1$ that $(x_1 + x_2)/2 \in M_n$ and $x' = -x + (x_1 + x_2) \in M_{n-1}$ for all $x \in M_{n-1}$. We have proved these statements for $n = 1$. Suppose that they are true for n. Let $x \in M_n$ and $y \in M_{n-1}$. Since $-y + (x_1 + x_2) \in M_{n-1}$, we have

$$\|x' - y\| = \| -y + (x_1 + x_2) - x\| \leqslant \frac{\Delta_{n-1}}{2}$$

and so, indeed, $x' \in M_n$. Furthermore,

$$\left\|x - \frac{x_1 + x_2}{2}\right\| = \frac{\|x - x'\|}{2} \leqslant \frac{\Delta_n}{2},$$

which implies $(x_1 + x_2)/2 \in M_{n+1}$, as asserted.

From $\Delta_n \leqslant \Delta_0/2^n$, we deduce that $M(x_1, x_2) = \bigcap_{n\geqslant 0} M_n(x_1, x_2)$ is necessarily reduced to the midpoint $(x_1 + x_2)/2$.

Now let $f: E \to F$ be a surjective isometry. Equation (26) shows that f is a bijection. Let x, y be in E. For all $n \geqslant 0$ construct, in F, the subset $M_n(f(x), f(y))$. It coincides with $f(M_n(x, y))$ because f is an isometry and because of the way $\{M_n\}$ has been constructed. Therefore we deduce (25) by using $f(E) = F$:

$$\left\{\frac{f(x)+f(y)}{2}\right\} = \bigcap_{n\geqslant 0} M_n(f(x), f(y)) = \bigcap_{n\geqslant 0} f(M_n(x, y))$$
$$= f\left(\bigcap_{n\geqslant 0} M_n(x, y)\right) = \left\{f\left(\frac{x+y}{2}\right)\right\},$$

which concludes the proof of Proposition 10.

From (26), the mapping f is continuous. Putting $y = 0$ in (25), we get

$$f\left(\frac{x}{2}\right) = \frac{f(x)}{2} + \frac{f(0)}{2}. \tag{28}$$

Using (28) in (25), we get

$$\frac{f(x+y)}{2} + \frac{f(0)}{2} = \frac{f(x)}{2} + \frac{f(y)}{2},$$

that is,

$$f(x+y) = f(x) + f(y) - f(0)$$

and, with $g(x) = f(x) - f(0)$, the Cauchy equation

$$g(x+y) = g(x) + g(y) \tag{29}$$

(same argument as in the proof of Theorem 4.9). With f, the function g is also continuous. This continuity yields

$$g(\lambda x) = \lambda g(x) \quad \text{for all } x \text{ in } E, \text{ all } \lambda \text{ in } \mathbb{R}.$$

Only here was it essential to have *real* normed spaces. We may state now the geometric result which we have proved.

Theorem 11. *Every surjective isometry between two real normed spaces is an affine transformation (the composition of a homogeneous linear transformation and of a translation).*

Theorem 11 and Proposition 10 are no longer true, in general, if f is not onto (surjective) as we shall see in the proof of the next theorem. However, if we restrict ourselves to strictly convex spaces, we have a similar characterization in this case, too, and also a characterization of the strictly convex spaces themselves. The idea is that in such a strictly convex space, the set $M_0(x, y)$ always reduces to a point.

Theorem 12. *A normed space $(F,\|\cdot\|)$ is strictly convex if, and only if, for every normed space $(E,\|\cdot\|)$, every isometry $f:E\to F$ satisfies* (25).

Proof. First, we suppose that $(F,\|\cdot\|)$ is not strictly convex. We use a,b, two elements of F, for which $\|a\|=\|b\|=1$, $a\neq b$ and $\|a+b\|=\|a\|+\|b\|$. As in the proof of Theorem 9, using here simply $g(x)=x$, we define an $f:\mathbb{R}\to F$ by

$$f(x)=\begin{cases} xa & \text{if } |x|\leqslant 1\\ a+(x-1)b & \text{if } x\geqslant 1\\ -a+(x+1)b & \text{if } x\leqslant -1\end{cases}.$$

The computations made in the proof of Theorem 9 yield

$$\|f(x)+f(y)\|=\|f(x+y)\|=|x+y|.$$

From $f(0)=0$ we easily deduce that $f(x)=-f(-x)$. Therefore

$$\|f(x)-f(y)\|=\|f(x-y)\|=|x-y|.$$

Thus f is an isometry from $\mathbb{R}$ into F. However, f is not additive, as we have seen previously in the proof of Theorem 9. We can see by taking $x=2$ and $y=0$ that f does not satisfy (25). (The function f, which we constructed, is not surjective, even on the space generated by a and b. This function yields a counterexample, showing that the onto hypothesis is essential in Proposition 10.)

Conversely, let $(E,\|\cdot\|)$ be a normed space and suppose that $(F,\|\cdot\|)$ is strictly convex. Let $f:E\to F$ be an isometry. We compute

$$\left\|f\left(\frac{x+y}{2}\right)-f(x)\right\|=\left\|\frac{x+y}{2}-x\right\|=\left\|\frac{y-x}{2}\right\|$$
$$=\tfrac{1}{2}\|f(x)-f(y)\|.$$

In the same way,

$$\left\|f\left(\frac{x+y}{2}\right)-f(y)\right\|=\tfrac{1}{2}\|f(x)-f(y)\|$$
$$=\left\|f\left(\frac{x+y}{2}\right)-f(x)\right\|.$$

with $u=f\left(\frac{x+y}{2}\right)-f(y)$, $v=-\left(f\left(\frac{x+y}{2}\right)-f(x)\right)$, we get

$$\|u\|=\|v\|=\left\|\frac{u+v}{2}\right\|.$$

As $(F,\|\cdot\|)$ is strictly convex, this relation implies $u=v$. Therefore

$$2f\left(\frac{x+y}{2}\right)=f(x)+f(y),$$

as asserted. (Notice that, in Theorem 12, the spaces can be over the field of real or complex numbers, or over more general fields.)

We mention here that Andalafte and Blumenthal (1964) deal with metric characterizations of Banach and euclidean spaces by other means.

We shall now study the case where E and F are spaces of continuous functions.

Let X be a compact Hausdorff space and, as at the beginning of Section 1, $C_R(X)$ is the space of all real valued continuous functions on X, equipped with the uniform norm. The constant functions equal to 1 or 0 are denoted by $\mathbf{1}$ and 0, respectively. The space $C_R(X)$ is not strictly convex if X is not reduced to one point.

Theorem 13. *Let X and Y be two compact Hausdorff spaces. Let $P: C_R(X) \to C_R(Y)$ be a bijection mapping the zero function in $C_R(X)$ to the zero function in $C_R(Y)$. If P is an isometry, then, for all f and all g in $C_R(X)$, we have*

$$P(\mathbf{1})P(fg) = Pf \cdot Pg. \tag{30}$$

Proof. As $P(0) = 0$, using Theorem 11, the bijection P is linear and $\|Pf\| = \|f\|$ for all f in $C_R(X)$. By the Riesz representation theorem, the topological dual of $C_R(Y)$ is the set $M(Y)$ of all (bounded) Radon measures on Y. The adjoint operator P^* of P acts from $M(Y)$ into $M(X)$. Since $\|P\| = 1$, also $\|P^*\| = 1$. In fact, for any μ in $M(Y)$,

$$\|P^*\mu\| = \|\mu\|. \tag{31}$$

To prove (31), we just use the definition of $\|P^*\mu\|$:

$$\|P^*\mu\| = \sup_{\substack{\|f\|=1 \\ f \in C_R(X)}} |(f, P^*\mu)| = \sup_{\substack{\|f\|=1 \\ f \in C_R(X)}} |(Pf, \mu)|.$$

But P is both a bijection and an isometry. Therefore

$$\|P^*\mu\| = \sup_{\substack{\|g\|=1 \\ g \in C_R(Y)}} |(g, \mu)| = \|\mu\|.$$

The adjoint P^* is an isometry from $M(Y)$ into $M(X)$. It is also a bijection because P is a surjection from $C_R(X)$ onto $C_R(Y)$. This bijective isometry transforms the extreme points of the unit ball $M_1(Y)$ into the extreme points of $M_1(X)$, the unit ball of $M(X)$. But we have already proved (Proposition 4) that the extreme points μ of $M_1(X)$, for example, satisfy the functional equation

$$\mu(\mathbf{1})\mu(fg) = \mu(f)\mu(g) \quad \text{for all } f, g \text{ in } C_R(X). \tag{32}$$

Thus for all extreme points μ of $M_1(Y)$,

$$P^*\mu(1)P^*\mu(fg) = P^*\mu(f)P^*\mu(g) \quad \text{for all } f, g \text{ in } C_R(X),$$

or

$$(P(1), \mu)(P(fg), \mu) = (Pf, \mu)(Pg, \mu).$$

We also saw (Proposition 4 and Lemma 5) that all the Dirac measures δ_y ($y \in Y$) are extreme points of the unit ball $M_1(Y)$. Therefore, for all f, g in $C_R(X)$, we have

$$P(1)P(fg) = Pf \cdot Pg. \tag{30}$$

which concludes the proof of Theorem 13.

We have already given (in Proposition 6) the general form of all bounded linear operators $P: C(X) \to C(Y)$ satisfying (30) if, for each $y \in Y$, there exists an $f \in C(X)$ such that $Pf(y) \neq 0$. When P is a bijection from $C_R(X)$ onto $C_R(Y)$, as required in Theorem 13, this last condition is satisfied and we may easily adapt Proposition 6. We get, for $y \in Y$, $f \in C_R(X)$,

$$Pf(y) = \alpha(y) f(\phi(y)), \tag{33}$$

where $\alpha: Y \to \mathbb{R}$ is a continuous and nonvanishing function and where $\phi: Y \to X$ is continuous. Because of the symmetry of the roles played by P and P^{-1} under the assumptions of Theorem 13, we also get, for all x in X, and all g in $C_R(Y)$,

$$P^{-1}g(x) = \alpha'(x) g(\psi(x)), \tag{34}$$

where $\alpha': X \to \mathbb{R}$ is a continuous, nonvanishing function and where $\psi: X \to Y$ is continuous. From (33) and (34) we deduce that

$$g(y) = \alpha(y)\alpha'(\phi(y)) g(\psi \circ \phi(y)). \tag{35}$$

With $g = 1$ in (35), we get $\alpha(y)\alpha'(\phi(y)) = 1$. Then, as $C_R(Y)$ separates points of Y, we deduce that $\psi \circ \phi$ is the identity on Y. By symmetry, we get that $\phi \circ \psi$ is the identity on X. In other words, X and Y are homeomorphic compact spaces. Moreover, because P (and so P^{-1}) is an isometry, we deduce that $|\alpha| \leqslant 1$ and $|\alpha'| \leqslant 1$. As $\alpha(y)\alpha'(\phi(y)) = 1$, we finally get $|\alpha| = |\alpha'| = 1$.

We may say, using the terminology given at the end of Section 1, that a bijective isometry from $C_R(X)$ onto $C_R(Y)$, mapping the zero function in $C_R(X)$ to the zero function in $C_R(Y)$, is a nice operator.

As a consequence of what we have proved after Theorem 13, we obtain the Banach–Stone theorem:

Theorem 14. *There exists an isometrical bijection between $C_R(X)$ and $C_R(Y)$ if, and only if, X and Y are homeomorphic compact Hausdorff spaces.*

10.5 A topology on the set of all solutions of a functional equation: the Bohr group

In this section we make use of topology on function spaces, in particular of weak star topology. The reader will not need the results of this section in subsequent sections and chapters.

In Section 5.3, we investigated the spectrum $\mathscr{B}$ of the complex algebra $\mathscr{P}$ of all trigonometric polynomials on $\mathbb{R}$. We shall find more properties of $\mathscr{B}$. In Proposition 5.9 we proved that $\mathscr{B}$ is the set of all those solutions of the functional equation (5.26) which satisfy (5.27). It is convenient to introduce a new function space here.

We define $\mathscr{A}$ as the closure for the uniform norm of the algebra $\mathscr{P}$ in the space $C_b(\mathbb{R})$ of all complex valued bounded continuous functions on $\mathbb{R}$. An element of $\mathscr{A}$ is called an *almost periodic Bohr function*. Such a function f can be obtained by taking the uniform limit of a sequence f_n of trigonometric polynomials, that is,

$$\lim_{n\to\infty}\left(\sup_{x\in\mathbf{R}}|f(x)-f_n(x)|\right)=0. \tag{36}$$

Therefore f is a continuous complex valued function. As examples of almost periodic Bohr functions, we have $f(x)=\sin x$, or $f(x)=\sin x+\cos 2^{1/2}x$, or $f(x)=\sum_{n=0}^{\infty}1/2^n\sin\lambda_n x$ where $\{\lambda_n\}$ is any sequence of real numbers. In order to study $\mathscr{A}$, just as it was for $\mathscr{P}$, it is the best to investigate its spectrum. Recall that the *spectrum* of $\mathscr{A}$ is the set of all complex valued functions $\phi:\mathscr{A}\to\mathbb{C}$ for which

(i) ϕ is linear: $\quad\phi(\lambda f+\mu g)=\lambda\phi(f)+\mu\phi(g)\quad(\lambda,\mu\in\mathbb{C};f,g\in\mathscr{A})$,

(ii) ϕ is multiplicative: $\quad\phi(fg)=\phi(f)\phi(g)\quad(f,g\in\mathscr{A})$,

(iii) $\phi(1)=1$,

and

(iv) $|\phi(f)|\leqslant\|f\|_\infty=\sup_{x\in\mathbf{R}}|f(x)|\quad(f\in\mathscr{A})$.

It is easy to check that the spectrum $\mathscr{B}$ of $\mathscr{P}$ coincides with the spectrum of $\mathscr{A}$. We may now use what we proved in Section 5.3 (Proposition 5.9). There exists a bijection between $\mathscr{B}$ and the set $\mathscr{C}$ of all characters Φ where $\Phi:\mathbb{R}\to\mathbb{C}$. This bijection $b:\mathscr{B}\to\mathscr{C}$ is defined by

$$\phi(e_\lambda)=\Phi(\lambda)\quad(\lambda\in\mathbb{R}), \tag{37}$$

where e_λ is the complex exponential function: $e_\lambda(x)=e^{i\lambda x}$. We define $\Gamma=\{z\mid z\in\mathbb{C},|z|=1\}$. Then $\mathscr{C}$ is the set of all solutions $\Phi:\mathbb{R}\to\Gamma$ of the functional equation

$$\Phi(\lambda+\mu)=\Phi(\lambda)\Phi(\mu)\quad(\lambda,\mu\in\mathbb{R}). \tag{38}$$

Our purpose is to introduce a topological structure on $\mathscr{C}$ or on $\mathscr{B}$, which, added to the group structure, provides a topological compact group.

Recall that the mapping 'hat' $h:x\to\hat{x}$, where $\hat{x}(f)=f(x)$ for all f in $\mathscr{A}$, defines a mapping from $\mathbb{R}$ into $\mathscr{B}$. Among all characters of $\mathbb{R}$ belonging to $\mathscr{C}$, the characters $b(\hat{x})$ are continuous. Therefore, using Theorem 5.4, we deduce that the set of all $b(\hat{x})$, when x runs through $\mathbb{R}$, is a *proper* subset of $\mathscr{C}$.

On $\mathscr{B}$, we have defined a binary operation $*$ by

$$\phi_1*\phi_2(e_\lambda)=\phi_1(e_\lambda)\phi_2(e_\lambda)\quad(\lambda\in\mathbb{R}). \tag{39}$$

We proved that the *Bohr group* $(\mathscr{B},*)$ is an abelian group and that the mapping hat is an injective homomorphism from $(\mathbb{R},+)$ into $(\mathscr{B},*)$.

On $\mathscr{C}$, there exists a natural binary operation $\circledast$ which we may define by

$$\Phi_1\circledast\Phi_2(\lambda)=\Phi_1(\lambda)\Phi_2(\lambda)\quad(\lambda\in\mathbb{R}).$$

It is easy to see that $(\mathscr{B},*)$ and $(\mathscr{C},\circledast)$ are isomorphic abelian groups (with the isomorphism $b:\mathscr{B}\to\mathscr{C}$). Indeed,

$$\begin{aligned}b(\phi_1*\phi_2)(\lambda)&=\phi_1*\phi_2(e_\lambda)\\&=\phi_1(e_\lambda)\cdot\phi_2(e_\lambda)\\&=b(\phi_1)(\lambda)\cdot b(\phi_2)(\lambda).\end{aligned}$$

Thus

$$b(\phi_1*\phi_2)=b(\phi_1)\circledast b(\phi_2).$$

The set $\mathscr{B}$ is a subset of the unit ball of the topological dual $\mathscr{A}^*$ of the Banach algebra $\mathscr{A}$. On this dual, a usual topology is the weak star topology. A basis of neighbourhoods of zero for this topology is given by

$$V(\varepsilon,f_1,\ldots,f_n)=\{\phi\mid\phi\in\mathscr{A}^*,|\phi(f_k)|<\varepsilon\quad(k=1,\ldots,n)\},$$

where $\varepsilon>0$ and $f_1,\ldots,f_n$ is a finite family in $\mathscr{A}$. The Theorem of Alaoglu and Bourbaki (see Dunford–Schwartz 1958, pp. 423–4) states that the unit ball of $\mathscr{A}^*$ in the weak star topology is a compact Hausdorff space. Therefore $\mathscr{B}$ is relatively compact when equipped with the weak star topology. Let ϕ_i be a net in $\mathscr{B}$ converging to some ϕ in $\mathscr{A}^*$ (with respect to the weak star topology). In particular, for each f in $\mathscr{A}$, $\phi_i(f)$ converges to $\phi(f)$. Therefore it is easy to see that ϕ satisfies (i), (ii), (iii) and (iv). Thus ϕ belongs to $\mathscr{B}$. In other words, $\mathscr{B}$ is closed with respect to weak star topology. Since it is also relatively compact, $\mathscr{B}$ is a compact Hausdorff space.

Proposition 15. *The Bohr group $(\mathscr{B},*)$ is a compact Hausdorff topological group with respect to the weak star topology.*

Proof. $\mathscr{B}$ is a Hausdorff space and we have proved that $\mathscr{B}$ is compact. Moreover, $(\mathscr{B},*)$ is an abelian group. To get Proposition 15, we have to link the two structures by proving the continuity of the following mappings:

$$(\phi_1,\phi_2)\mapsto\phi_1*\phi_2:\mathscr{B}\times\mathscr{B}\to\mathscr{B}, \tag{40}$$

(here the product $\mathscr{B} \times \mathscr{B}$ is equipped with the product topology) and

$$\phi \mapsto \bar{\phi}: \mathscr{B} \to \mathscr{B}, \tag{41}$$

where $\bar{\phi}$ is the inverse of ϕ with respect to $*$ (and $\bar{\phi}(e_\lambda) = \overline{\phi(e_\lambda)}$, as we have seen in Section 5.3). Using (39) and the definition of the weak star topology, the continuity of the two mappings follow easily.

There exists a natural topology also for the set $\mathscr{C}$ of characters. In fact $\mathscr{C}$ can be viewed as a subset of the set $C_b(\mathbb{R})$ of all continuous bounded complex valued functions on $\mathbb{R}$. We take then on $C_b(\mathbb{R})$ the topology of pointwise convergence. A basis of neighbourhoods of 0 for this topology over $C_b(\mathbb{R})$ is given by

$$W(\varepsilon, x_1, \ldots, x_n) = \{\Phi \mid \Phi \in C_b(\mathbb{R}), |\Phi(x_k)| < \varepsilon \quad (k = 1, \ldots, n)\},$$

where $\varepsilon > 0$ and $x_1, x_2, \ldots, x_n$ is a finite family of $\mathbb{R}$. The set $\mathscr{C}$ is closed in $C_b(\mathbb{R})$ equipped with this Hausdorff topology as can be seen by using (38). We see also that $(\mathscr{C}, \circledast)$ is a topological group. We notice how easily the weak star topology or the topology of pointwise convergence can be used to prove that the sets of solutions for functional equations of the form (38) are closed.

Now we prove that the mapping $b: \mathscr{B} \to \mathscr{C}$ is a *homeomorphism* for the two topologies we described for $\mathscr{B}$ and for $\mathscr{C}$. Because of the topological group structure of $\mathscr{B}$ and $\mathscr{C}$, it is enough to prove the continuity of the mapping b at $\hat{0}$ and the continuity of the mapping b^{-1} at $b(\hat{0}) = \mathbf{1}$. As before, $\mathbf{1}$ denotes the function identically equal to 1. It is a character, the neutral element of $(\mathscr{C}, \circledast)$. On the other hand, b is a bijection, $\mathscr{B}$ is a compact Hausdorff and $\mathscr{C}$ a Hausdorff space. Therefore it is enough to prove the continuity of b at $\hat{0}$.

Let $W'(\varepsilon, \lambda_1, \ldots, \lambda_n)$ be a given neighbourhood of $\mathbf{1}$ in the topology of pointwise convergence in $C_b(\mathbb{R})$, that is,

$$W'(\varepsilon, \lambda_1, \ldots, \lambda_n) = \{\Phi \mid \Phi \in C_b(\mathbb{R}), |\Phi(\lambda_k) - 1| < \varepsilon \quad (k = 1, 2, \ldots, n)\} \cap \mathscr{C}.$$

In order to prove the continuity of b at $\hat{0}$, it is enough to show that there exists in $\mathscr{B}$ an open neighbourhood V' of $\hat{0}$ for the weak star topology on $\mathscr{A}^*$ and

$$b(V' \cap \mathscr{B}) \subset W'.$$

With $b(\phi) = \Phi$, this inclusion means that $|\Phi(\lambda_k) - 1| < \varepsilon$ for all $\phi \in V'$. This inequality is satisfied if we choose for V' the following neighbourhood of $\hat{0}$:

$$V' = \{\phi \mid \phi \in \mathscr{B}, |\phi(e_{\lambda_k}) - \hat{0}(e_{\lambda_k})| < \varepsilon \quad (k = 1, 2, \ldots, n)\}.$$

For the proof of our next proposition, it is easier to use $\mathscr{C}$ and its topology. By Proposition 15, and because $\mathscr{B}$ and $\mathscr{C}$ are isomorphic as topological groups, we know that $(\mathscr{C}, \circledast)$ is a compact Hausdorff topological group. Let $\mathbb{R}^\wedge$ be the image of $\mathbb{R}$ in $\mathscr{B}$ through the mapping hat ($\mathbb{R}^\wedge = h(\mathbb{R})$).

Proposition 16. $\mathbb{R}^\wedge$ *is a dense subgroup of the Bohr group* $(\mathscr{B}, *)$ *equipped with the weak star topology.*

Proof. In terms of $(\mathscr{C}, \circledast)$, we have to prove that $b(\mathbb{R}^\wedge)$ is a dense subgroup with respect to the topology of the pointwise convergence on $\mathscr{C}$. We have already noticed that an element of $b(\mathbb{R}^\wedge)$ is a continuous character of $\mathbb{R}$. We have to show that any nonempty open subset $\mathscr{O}$ of $\mathscr{C}$ contains a continuous character of $\mathbb{R}$. Let Φ be an element of $\mathscr{O}$. There exists $\varepsilon > 0$ and a finite family $x_1, x_2, \ldots, x_n$ of points in $\mathbb{R}$ such that $\Phi \circledast W'(\varepsilon, x_1, x_2, \ldots, x_n)$ is a subset of $\mathscr{O}$. It is then enough to show the existence of a continuous character belonging to $\Phi \circledast W'(\varepsilon, x_1, x_2, \ldots, x_n)$. Let $z_i = \Phi(x_i)$ for $i = 1, 2, \ldots, n$. Suppose that, for n relatively prime integers $d_1, d_2, \ldots, d_n$, the following relation holds:

$$\sum_{j=1}^{n} d_j x_j = 0.$$

Using (38), we see that

$$\prod_{j=1}^{n} z_j^{d_j} = \Phi(0) = 1.$$

The conditions of Theorem 5.8 are satisfied with $\mu_j = x_j$ $(j = 1, \ldots, n)$. By that theorem there exists an x in $\mathbb{R}$ such that

$$|e_x(x_i) - \Phi(x_i)| < \varepsilon \quad (i = 1, 2, \ldots, n).$$

In other words the continuous character e_x belongs to the neighbourhood $\Phi \circledast W'(\varepsilon, x_1, \ldots, x_n)$, which ends the proof of the density of $b(\mathbb{R}^\wedge)$ in $\mathscr{C}$.

The mapping hat, from $\mathbb{R}$ into $\mathscr{B}$, is a continuous mapping when $\mathscr{B}$ is equipped with the weak star topology (because $\hat{x}(f) = f(x)$ for all f in $\mathscr{A}$ and all x in $\mathbb{R}$, and so the convergence of a sequence x_n in $\mathbb{R}$ implies the simple convergence of the functionals $\hat{x}_n$ over $\mathscr{A}$). In Section 5.3 we proved that this mapping hat is an injection (one-to-one) and a homomorphism from $(\mathbb{R}, +)$ into $(\mathscr{B}, *)$. However, it is important to notice that the mapping hat is *not* a homeomorphism from $\mathbb{R}$ onto its image $h(\mathbb{R}) = \mathbb{R}^\wedge$ which is a topological subspace of $\mathscr{B}$ equipped with the weak star topology (see Exercise 3).

Propositions 15 and 16 show that the Bohr group, with its compact topology, realizes a compactification of the locally compact group $(\mathbb{R}, +)$. We leave to the reader the task of proving that, up to an isomorphism of compact topological groups, the Bohr group is the 'smallest' topological group which is a compactification of $(\mathbb{R}, +)$ (see Exercise 5).

We shall now show how the Bohr group aids in understanding the structure of the algebra $\mathscr{A}$ of all almost periodic Bohr functions. Let $f \in \mathscr{A}$. For all $\phi \in \mathscr{B}$ and $\Phi \in \mathscr{C}$ where $\Phi = b(\phi)$, we define

$$F(\Phi) = \phi(f). \tag{42}$$

Clearly, $F:\mathscr{C}\to\mathbb{C}$. We define $\mathscr{D}$ to be the set of all such functions F arising from all f in $\mathscr{A}$. Because $\mathscr{C}$ and $\mathscr{B}$ are homeomorphic spaces with the mapping b, and because of the topologies we used for $\mathscr{B}$ and $\mathscr{C}$, it is clear that F is a continuous function on $\mathscr{C}$. This means that $\mathscr{D}$ is included in $C(\mathscr{C})$, the algebra of all complex valued continuous functions over the compact Hausdorff space $\mathscr{C}$. As ϕ belongs to the spectrum of $\mathscr{A}$, using (42) we have

$$F_1(\Phi)+F_2(\Phi)=\phi(f_1+f_2)\quad (f_1,f_2\in\mathscr{A}),$$
$$\lambda F(\Phi)=\phi(\lambda f)\quad (\lambda\in\mathbb{C}),$$
$$F_1(\Phi)F_2(\Phi)=\phi(f_1f_2)\quad (f_1,f_2\in\mathscr{A}).$$

These relations show that $\mathscr{D}$ is a subalgebra of $C(\mathscr{C})$.

We notice that $\phi(e_{-\lambda})=\Phi(-\lambda)=\overline{\Phi(\lambda)}=\overline{\phi(e_\lambda)}$. Thus $\phi(\bar{f})=\overline{\phi(f)}$ for every trigonometric polynomial f and, by extension, for every almost periodic Bohr function f. We deduce that, when F belongs to $\mathscr{D}$, its conjugate function $\bar{F}$ also belongs to $\mathscr{D}$ ($\bar{F}(\Phi)=\phi(\bar{f})$).

Suppose that $\Phi_1\neq\Phi_2$. Then $\phi_1\neq\phi_2$ and there exists some f in $\mathscr{A}$ such that $\phi_1(f)\neq\phi_2(f)$. Thus $F(\Phi_1)\neq F(\Phi_2)$. The subalgebra $\mathscr{D}$ separates points of $\mathscr{C}$. Moreover, the functions constant on $\mathscr{C}$ are elements of $\mathscr{D}$.

All conditions of the Stone–Weierstrass theorem are satisfied (see Dunford–Schwartz 1958, pp. 272–6), so $\mathscr{D}$ is a dense subalgebra of the uniform algebra $C(\mathscr{C})$. As usual in this chapter, $\mathscr{C}$ being compact, we impose upon $\mathscr{C}$ the topology of the uniform norm

$$\|F\|=\sup_{\Phi\in\mathscr{C}}|F(\Phi)|.$$

At this point it is natural to compare the uniform norm of f, an almost periodic Bohr function, and the uniform norm of F, as an element of $C(\mathscr{C})$ provided by (42).

Using property (iv) in the definition of the spectrum $\mathscr{B}$ of $\mathscr{A}$, we easily get an inequality between the two norms $\|F\|$ and $\|f\|$, namely

$$\|F\|\leqslant\|f\|.$$

Conversely, let f be in $\mathscr{A}$. For any integer n ($n>1$), there exists a real number x_n such that $|f(x_n)|\geqslant(1-1/n)\|f\|$. But we get $f(x_n)=\hat{x}_n(f)$ and $\hat{x}_n(f)=F(b(\hat{x}_n))$ by using (42) to link F and f. We have obtained

$$\|F\|\geqslant|F(b(\hat{x}_n))|\geqslant(1-1/n)\|f\|.$$

Letting n go to infinity, we deduce that

$$\|F\|\geqslant\|f\|.$$

So we have obtained, for all f in $\mathscr{A}$, with F defined by (42),

$$\|F\|=\|f\|. \tag{43}$$

We are now ready to prove the main result of this section:

Theorem 17. *The algebra of all almost periodic Bohr functions with the uniform norm is isometrically isomorphic to the algebra of all complex valued continuous functions over the Bohr group equipped with the weak star topology.*

Proof. As topological spaces, $\mathscr{B}$ and $\mathscr{C}$ are homeomorphic. The Banach algebras $C(\mathscr{B})$ and $C(\mathscr{C})$ are isometrically isomorphic (Theorem 14). Therefore, in order to prove Theorem 17 it is enough to prove that $C(\mathscr{C})$ and $\mathscr{A}$ are isometrically isomorphic.

The mapping $f \mapsto F$, where f and F are linked by (42), is a homomorphism of complex algebras from $\mathscr{A}$ into $C(\mathscr{C})$. By (43), it is also an isometry. But we have already proved that the image $\mathscr{D}$ of this homomorphism was a dense subalgebra of $C(\mathscr{C})$. As the isometrical image of $\mathscr{A}, \mathscr{D}$ is also a Banach subalgebra of $C(\mathscr{C})$, it then coincides with $C(\mathscr{C})$, which proves that $C(\mathscr{C})$ and $\mathscr{A}$ are isometrically isomorphic, and concludes the proof of Theorem 17.

We may combine Proposition 16 and Theorem 17 in order to obtain a nice characterization of an almost periodic Bohr function.

Theorem 18. *A function $f:\mathbb{R} \to \mathbb{C}$ is an almost periodic Bohr function if, and only if, f can be extended to a continuous function on the Bohr group equipped with the weak star topology.*

To be more precise, the mapping hat (which we have also denoted by h) realizes a continuous injection from $\mathbb{R}$ into the Bohr group $\mathscr{B}$ equipped with the weak star topology. By an extension of f we mean an extension of $f \circ h^{-1}$ from $h(\mathbb{R}) = \mathbb{R}^\wedge$ into $\mathscr{B}$.

Proof. Let f be an almost periodic Bohr function and let F be defined by (42). Let $G:\mathscr{B} \to \mathbb{C}$ be defined by $G(\phi) = F(\Phi)$ $(= \phi(f))$. Consider for some x in $\mathbb{R}$ the value of G at $\hat{x} = h(x)$. We get

$$G(\hat{x}) = \hat{x}(f) = f(x)$$

and

$$f(h^{-1}(\hat{x})) = G(\hat{x}).$$

The function F, which is continuous on $\mathscr{B}$, extends $f \circ h^{-1}$ to all elements of $\mathscr{B}$.

Conversely, suppose that G, a continuous complex valued function on $\mathscr{B}$, extends $f \circ h^{-1}$, where $f:\mathbb{R} \to \mathbb{C}$. By the continuity of G and Theorem 17, there exists an almost periodic Bohr function $g:\mathbb{R} \to \mathbb{C}$ and $G(h(x)) = g(x)$ for all x in $\mathbb{R}$. This result yields $g(x) = f(x)$ for all x in $\mathbb{R}$ and ends the proof of Theorem 18.

From Theorem 18 the classical properties of almost periodic Bohr functions follow easily. See Exercise 6 for an explanation of the expression 'almost periodic'. We end this section by proving, in a direct manner, the existence of a Radon measure on $\mathscr{B}$ which is invariant under translation of the group $(\mathscr{B}, *)$.

Theorem 19. *On the Bohr group $(\mathscr{B}, *)$ there exists a positive Radon measure of total mass 1 which is invariant under translation.*

Proof. Using Lemma 3 and the Riesz representation theorem, we see that a positive Radon measure μ of total mass 1 is a continuous linear form $\mu: C(\mathscr{B}) \to \mathbb{C}$ such that $\mu(1) = 1$ and $\|\mu\| = 1$. But Theorem 17 states that $C(\mathscr{B})$ and $\mathscr{A}$ are isometrically isomorphic. Therefore, to construct a positive Radon measure μ of total mass 1 is equivalent to finding a linear continuous function $Q_\mu: \mathscr{A} \to \mathbb{C}$, with $\|Q_\mu\| = 1$ and $Q_\mu(1) = 1$. For μ to be invariant under translation means that $\mu(F_\phi) = \mu(F)$ for all ϕ in $\mathscr{B}$ where $F_\phi(\psi) = F(\phi * \psi)$. The analogous property for Q_μ is that $Q_\mu(f_y) = Q_\mu(f)$ for all y in $\mathbb{R}$ where $f_y(x) = f(y + x)$. Therefore, in order to find μ as stated in Theorem 19, it is enough to find Q, invariant under translation on $\mathscr{A}$, linear, of norm 1, and such that $Q(1) = 1$.

Consider first

$$Q(f) = \lim_{T \to \infty} \frac{1}{2T} \int_{-T}^{T} f(y)\, dy$$

for all trigonometric polynomials f. The existence of this limit is easy to prove, as $Q(e_\lambda) = 0$ for all $\lambda \neq 0$ and $Q(1) = 1$. Moreover, $\|Qf\| \leqslant \|f\|$ and so $\|Q\| = 1$; Q can be extended, with the same properties, to all almost periodic functions f. It is not difficult to check that the extension is still given by the expression

$$Q(f) = \lim_{T \to \infty} \frac{1}{2T} \int_{-T}^{T} f(y)\, dy. \tag{44}$$

But $Q(f_x) = Q(f)$ for all x in $\mathbb{R}$. This concludes the proof of Theorem 19.

The measure with the properties given in Theorem 19 is unique (see Exercise 20). It is usually denoted by $d\Phi$ on $\mathscr{C}$ (and by $d\phi$ on $\mathscr{B}$). We write, using (42),

$$\lim_{T \to \infty} \frac{1}{2T} \int_{-T}^{T} f(y)\, dy = \int_{\mathscr{C}} F(\Phi)\, d\Phi. \tag{45}$$

(See Exercises 5, 7, 9 for more properties of the Bohr group.)

The existence of a positive Radon measure, invariant under translation, on the Bohr group $\mathscr{B}$ is not restricted to this group. It can be proved that every locally compact abelian topological group has a *Haar measure*, which is positive and invariant under translation. And the natural algebra to consider in such a group is the algebra $L^1(G)$. A function $f: G \to \mathbb{C}$ belongs to $L^1(G)$ if it is Haar measurable and integrable with respect to the Haar measure denoted by dx ($\int_G |f(x)|dx$, is finite, and we consider as equivalent two functions f and g for which $\int_G |f(x)-g(x)|dx = 0$). The space $L^1(G)$ with the norm $\|f\|_1 = \int_G |f(x)|dx$ is a Banach space. It is also a Banach algebra if we introduce the convolution of two functions in $L^1(G)$,

$$f*g(x) = \int_G f(x-y)g(y)\,dy,$$

which is defined almost everywhere (with respect to the Haar measure dx). Then Theorem 7.10 (and the use of a conditional Cauchy equation in the proof) can easily be generalized to the present situation (see Exercise 23), for any locally compact abelian group G instead of just $\mathbb{R}$. In this manner, the spectrum of the algebra $L^1(G)$ coincides with the set of all *continuous* characters of G.

Recall (Chapter 8) that a character χ on G is a complex valued function $\chi: G \to \mathbb{C}$ such that

$$\chi(x+y) = \chi(x)\chi(y) \quad (x, y \in G) \tag{46}$$

and

$$|\chi(x)| = 1 \quad (x \in G). \tag{47}$$

On the set $G^\wedge$ of all continuous characters of G a binary operation is easily defined, as we have seen, by

$$\chi_1 * \chi_2(x) = \chi_1(x)\chi_2(x) \quad (x \in G),$$

$(G^\wedge, *)$ is an abelian group and it is called the *dual group* of G.

As in the case of the Bohr group, two topologies on $G^\wedge$ can be examined. The first topology is the weak star topology on the spectrum of the convolution algebra $L^1(G)$. The other is the topology of the uniform convergence on compact subsets of G. As in the case of the Bohr group, and with similar arguments, it is possible to prove the identity of the two topologies. Then the dual group, equipped with the topology as described above, becomes a *locally compact* topological group.

We may investigate the link of these results with what we did for the Bohr group. In fact, we started from $\mathbb{R}$ equipped with the discrete topology. We denote it here by $\mathbb{R}_d$. It is a locally compact abelian group. Its dual group $\mathbb{R}_d^\wedge$ is the Bohr group. Moreover, the topology we used for the Bohr group is exactly the topology on this dual group. In this case, we get a compactness result instead of just local compactness (Exercise 22).

But the theory goes further. The dual group $G^\wedge$, with the natural topology imposed on it, also has a dual group $(G^\wedge)^\wedge$. A beautiful theorem of Pontryagin asserts that $(G^\wedge)^\wedge$ coincides with G as a group and as a topological space. As essential tool for a proof of Pontryagin's theorem is Bochner's theorem which we mentioned earlier in Chapter 8 (see Exercise 18). For all these results, fundamental to abstract commutative harmonic analysis, we refer to Hewitt–Ross 1963 (pp. 355–439); and Rudin 1962 (pp. 1–57).

10.6 Valuations on the fields of rational and of real numbers

From Corollary 3.8, it is easy to determine all $f:\mathbb{R}\to\bar{\mathbb{R}}_+$ satisfying

$$f(xy)=f(x)f(y)\quad (x,y\in\mathbb{R}). \tag{48}$$

There are fewer functions $f:\mathbb{R}\to\bar{\mathbb{R}}_+$ satisfying (48) and the following conditions

$$f(x)=0\quad \text{if and only if } x=0, \tag{49}$$

$$f(x_0)\neq 1\quad \text{for some } x_0\neq 0, \tag{50}$$

$$f(x+y)\leqslant f(x)+f(y)\quad \text{for all } x,y\in\mathbb{R}. \tag{51}$$

A function $f:\mathbb{R}\to\bar{\mathbb{R}}_+$ satisfying (48), (49), (50) and (51) is called a *nontrivial valuation* on $\mathbb{R}$. It is easy to find such valuations.

Proposition 20. *Let $0<r\leqslant 1$. Then $f(x)=|x|^r$ is a nontrivial valuation on $\mathbb{R}$.*

Proof. The only nontrivial fact to prove is that (51) is satisfied, that is,

$$|x+y|^r\leqslant |x|^r+|y|^r\quad (x,y\in\mathbb{R}). \tag{52}$$

For $r=1$, this is obvious. If $0<r<1$, then we write

$$\frac{|x|+|y|}{(|x|^r+|y|^r)^{1/r}}=\left(\frac{|x|^r}{|x|^r+|y|^r}\right)^{1/r}+\left(\frac{|y|^r}{|x|^r+|y|^r}\right)^{1/r}. \tag{53}$$

Since

$$\frac{|x|^r}{|x|^r+|y|^r}\leqslant 1,\quad \frac{|y|^r}{|x|^r+|y|^r}\leqslant 1,$$

and also $(1/r)\geqslant 1$, therefore

$$\left(\frac{|x|^r}{|x|^r+|y|^r}\right)^{1/r}\leqslant\frac{|x|^r}{|x|^r+|y|^r}$$

and

$$\left(\frac{|y|^r}{|x|^r+|y|^r}\right)\leqslant\frac{|y|^r}{|x|^r+|y|^r}.$$

Using (53) we get

$$|x|+|y|\leqslant (|x|^r+|y|^r)^{1/r}.$$

This concludes the proof of (52) and of Proposition 20.

If the domain of f is no longer $\mathbb{R}$ but a field F, we still call an $f:F\to\bar{\mathbb{R}}_+$ satisfying (48), (49), (50) and (51) for all x,y in F, a *nontrivial valuation.*

Surprisingly if, for instance, $F = \mathbb{Q}$ then there exist nontrivial valuations which are not restrictions of $f(x) = |x|^r$ $(0 < r \leqslant 1)$ to $\mathbb{Q}$. Such an example is the so-called *p-adic valuation*. Take an $\alpha \in]0, 1[$ and let p be a prime number. Every rational number $x \neq 0$ can be written in the form

$$x = p^n \frac{q}{r}, \tag{54}$$

where $q \in \mathbb{Z}^*$ and $r \in \mathbb{N}$ and q (and r) are not divisible by p. Then define

$$f_{\alpha,p}(x) = \alpha^n \quad \text{if } x = p^n \frac{q}{r}$$

and

$$f_{\alpha,p}(0) = 0.$$

The function $f: \mathbb{Q} \to \bar{\mathbb{R}}_+$ is well defined (if $p^n(q/r) = p^{n'}(q'/r')$, then $n = n'$).

Proposition 21. *The function $f_{\alpha,p}$ is a nontrivial valuation on $\mathbb{Q}$.*

Proof. The conditions (49) and (50) are obviously satisfied. As to (48), if $x = p^n(q/r)$ and $y = p^m(s/t)$, the $xy = p^{n+m}(qs/rt)$. But p does not divide q, r, s or t. Therefore p does not divide qs or rt. Thus

$$\begin{aligned} f_{\alpha,p}(xy) &= \alpha^{n+m} \\ &= \alpha^n \cdot \alpha^m \\ &= f_{\alpha,p}(x) f_{\alpha,p}(y). \end{aligned}$$

We have proved (48). It just remains to prove (51). We will prove a stronger result, namely

$$f_{\alpha,p}(x + y) \leqslant \max(f_{\alpha,p}(x), f_{\alpha,p}(y)). \tag{55}$$

Clearly (55) implies (51). In order to prove (55), first choose $x \neq 0$ in $\mathbb{Q}$ so that $f_{\alpha,p}(x) \leqslant 1$. We have

$$x = p^n \frac{q}{r},$$

with some $n \geqslant 0$, where p is not a divisor of q or of r. So

$$1 + x = 1 + p^n \frac{q}{r} = \frac{r + p^n q}{r}.$$

Since p does not divide the denominator r, we are certain that $f_{\alpha,p}(1 + x) = \alpha^{n'}$ with $n' \geqslant 0$. Thus $f_{\alpha,p}(1 + x) \leqslant 1$.

Now let x and y be arbitrary in $\mathbb{Q}$. If x (or y) is zero, then (55) is obvious. We may suppose, by symmetry, that $0 < f_{\alpha,p}(x) \leqslant f_{\alpha,p}(y)$. Therefore

$$f_{\alpha,p}\left(\frac{x}{y}\right) \leqslant 1.$$

By what we just proved,

$$f_{\alpha,p}\left(1+\frac{x}{y}\right)\leqslant 1.$$

Thus

$$f_{\alpha,p}(x+y)\leqslant f_{\alpha,p}(y)=\max(f_{\alpha,p}(x),f_{\alpha,p}(y)),$$

which concludes the proof of (55) and so the proof of Proposition 21.

With the aid of the function $f_{\alpha,p}$ we can easily introduce a metric on $\mathbb{Q}$ by the definition $d(x,y)=f_{\alpha,p}(x-y)$. The topology it imposes upon $\mathbb{Q}$, the so-called p-adic topology, is very different from the usual topology on $\mathbb{Q}$ coming from the metric $|x-y|$. For instance, $f_{\alpha,p}(n)\leqslant 1$ for every positive integer n. Thus $f_{\alpha,p}(nx)\leqslant f_{\alpha,p}(x)$ for every positive integer n and every rational number x. Thus the well known archimedean property of the absolute value, that for $x\neq 0$ there exists an integer n such that

$$|nx|>|x|,$$

is not satisfied by the p-adic valuation. A valuation which does not satisfy the archimedean property is called a *nonarchimedean valuation.*

It is well known that every metric space is a dense subspace of a complete metric space, its completed space. For instance, $\mathbb{R}$ with its usual topology, is the completed space of $\mathbb{Q}$ where the metric on $\mathbb{Q}$ is $d(x,y)=|x-y|$. Let us call $\mathbb{Q}_{\alpha,p}$ the completed space of $\mathbb{Q}$, where the metric on $\mathbb{Q}$ is $d(x,y)=f_{\alpha,p}(x-y)$. It is not difficult to prove that $\mathbb{Q}_{\alpha,p}$ and $\mathbb{Q}_{\tilde{\alpha},p}$, for $0<\alpha<1$ and $0<\tilde{\alpha}<1$, are isomorphic topological fields. So we denote the p-adic field simply by $\mathbb{Q}_p$.

For exploring arithmetic properties of integers, $\mathbb{R}$ and the $\mathbb{Q}_p$'s play similar (or complementary) roles. Therefore, the $\mathbb{Q}_p$ are important fields.

It is possible (see Bachman 1964, p. 35) to prove the following. For every x in $\mathbb{Q}_p$, there exist integers n, n_j in $\mathbb{Z}$ such that

$$x=\sum_{j=n}^{\infty} n_j p^j,$$

In fact n is such that $f_{\alpha,p}(x)=f_{\alpha,p}(p)^n$.

It is important, therefore, to show that no nontrivial valuations exist on $\mathbb{Q}$ other than $f_{\alpha,p}(x)$ and the valuations $f(x)=|x|^r$ with an $r\in]0,1]$. This was first proved by Ostrowski (1917):

Theorem 22. *A function $f:\mathbb{Q}\to\bar{\mathbb{R}}_+$ is a nontrivial valuation on $\mathbb{Q}$ if and only if either there exist an $\alpha\in]0,1[$ and a prime p such that $f(x)=f_{\alpha,p}(x)$ for every x in $\mathbb{Q}$ or there exists an $r\in]0,1]$ such that $f(x)=|x|^r$ for every x in $\mathbb{Q}$.*

Proof. In consequence of Propositions 20 and 21, it only remains to prove that a nontrivial valuation on $\mathbb{Q}$ must have either the form $f_{\alpha,p}(x)$ or the form $f(x)=|x|^r$.

Let f be a nontrivial valuation on $\mathbb{Q}$. We clearly have $f(1)=1$ and f is even. A *majorant* for $f(n)$, when $n\in\mathbb{Z}$, is n since

$$f(n)=f(1+1+\cdots+1)\leqslant|n|f(1)=|n|.$$

Let us find a better majorization. Choose integers $m>1$, $n>1$. We then write m in base n as

$$m=\alpha_0+\alpha_1 n+\cdots+\alpha_k n^k,$$

where k is a nonnegative integer and the integers α_i satisfy $0\leqslant\alpha_i<n$, $(i=1,\ldots,k)$. If $\alpha_k>0$, we get $n^k\leqslant m$, that is,

$$k\leqslant\frac{\log m}{\log n}. \tag{56}$$

Moreover,

$$f(m)\leqslant f(\alpha_0)+f(\alpha_1)f(n)+\cdots+f(\alpha_k)f(n)^k.$$

As $f(\alpha_i)\leqslant\alpha_i\leqslant n$, we obtain

$$f(m)\leqslant n(1+f(n)+\cdots+f(n)^k). \tag{57}$$

We distinguish three cases.

(i) Suppose first that for every integer $n>1$, we have $f(n)>1$. Then, from (57),

$$f(m)\leqslant n(k+1)f(n)^k.$$

Using (56) to eliminate k, we deduce that

$$f(m)\leqslant n\left(\frac{\log m}{\log n}+1\right)f(n)^{\log m/\log n}. \tag{58}$$

Replace m by m^l in (58), l being an integer, and take the lth root,

$$f(m)\leqslant n^{1/l}\left(\frac{l\log m}{\log n}+1\right)^{1/l}f(n)^{\log m/\log n}.$$

Thus

$$f(m)^{1/\log m}\leqslant\left(n^{1/l}\left(1+\frac{l\log m}{\log n}\right)^{1/l}\right)^{1/\log m}f(n)^{1/\log n}.$$

Let l tend to infinity. As

$$\lim_{l\to\infty}n^{1/l}=1\quad\text{and}\quad\lim_{l\to\infty}\left(1+\frac{l\log m}{\log n}\right)^{1/l}=1, \tag{59}$$

we finally obtain

$$f(m)^{1/\log m}\leqslant f(n)^{1/\log n}.$$

But m and n being arbitrary integers (greater than 1), the converse inequality is true as well. Thus there exists a central α such that

$$f(n)^{1/\log n} = \alpha.$$

As $f(n) > 1$, we have $\alpha > 1$. We write $\log \alpha = r$. Then $r > 0$ and

$$f(n) = \alpha^{\log n} = e^{r \log n} = n^r \ (n \geqslant 1).$$

But we must have

$$f(n) \leqslant n \quad (n \geqslant 1),$$

therefore $r \leqslant 1$. We have obtained that there exists an $r \in]0, 1]$

$$f(n) = n^r \quad (n \geqslant 1).$$

So, for any pair p, q of integers,

$$p^r = f(p) = f\left(q\frac{p}{q}\right) = f(q)f\left(\frac{p}{q}\right) = q^r f\left(\frac{p}{q}\right).$$

Finally, from (48) $f(-1)^2 = f(1) = 1$ but $f(-1) \geqslant 0$, so $f(-1) = 1$, $f(-x) = f(x)$, f is even. So we have that

$$f(x) = |x|^r \quad (x \in \mathbb{Q})$$

for some $r \in]0, 1]$.

(ii) Suppose now that $f(n) = 1$ for all $n \in \mathbb{Z}^*$. In the same way as previously, we deduce that $f(x) = 1$ for all $x \in \mathbb{Q}^*$. This is impossible since f is a nontrivial valuation.

(iii) Finally, we suppose that there exists some integer $n > 1$ such that $f(n) < 1$.

We first prove that $f(m) \leqslant 1$ for all integers $m \in \mathbb{Z}^*$. In fact, from (57) we get for a positive integer m

$$f(m) \leqslant n(k+1), \tag{60}$$

or, by (56),

$$f(m) \leqslant n\left(1 + \frac{\log m}{\log n}\right). \tag{61}$$

Replace m by m^l, where l is an integer and let l go to infinity. By (59) we obtain

$$f(m) \leqslant 1.$$

Let p be the smallest positive integer such that $f(p) < 1$. If this number p is a product of two smaller numbers q and r, then

$$f(p) = f(q)f(r) = 1,$$

which is impossible. Therefore, p is a prime number.

We need a better majorant for $f(n+m)$ than $f(n) + f(m)$. Suppose that n and m are positive integers. We compute $f(n+m)^k$ using the binomial theorem

$$f(n+m)^k \leqslant f(n)^k + f(k)f(n)^{k-1}f(m) + f\left(\frac{k(k-1)}{2}\right)f(n)^{k-2}f(m)^2 + \cdots + f(m)^k.$$

But $f(k) \leqslant 1$, $f(k(k-1)/2) \leqslant 1$ and so on, since the $\binom{k}{j}$ are positive. Thus

$$f(n+m)^k \leqslant (k+1)(\max(f(n), f(m)))^k.$$

As f is positive valued and $\lim_{k\to\infty}(k+1)^{1/k} = 1$, we finally obtain

$$f(n+m) \leqslant \max(f(n), f(m)). \tag{62}$$

Let m be a positive integer and written in the form $m = pq + r$ where p is the prime already defined, $q = [m/p]$ the quotient and r the remainder $(0 \leqslant r < p)$.

If $r > 0$, then $f(r) = 1$ and $f(pq) = f(p)f(q) < 1$ (as $f(p) < 1$). Thus $f(pq) < f(r)$. But

$$f(r) = f(pq + r - pq) \leqslant \max(f(pq + r), f(pq)).$$

We cannot have $\max(f(pq+r), f(pq)) = f(pq)$ since $f(r) > f(pq)$. So $\max(f(pq+r), f(pq)) = f(pq+r)$. Therefore and by (62),

$$1 = f(r) \leqslant f(pq + r) \leqslant \max(f(pq), f(r)) = 1.$$

We have proved that

$$f(m) = f(pq + r) = 1$$

when $r \neq 0$.

If $r = 0$, $f(m) < 1$. In other words, $f(m) < 1$ for an integer $m \neq 0$ if, and only if, p divides m. Otherwise $f(m) = 1$.

Write any x in $\mathbb{Q}^*$ in the form $p^n(q/r)$, where $q \in \mathbb{Z}^*$, $r \in \mathbb{N}$ are such that p divides neither q nor r. Then

$$f\left(p^n \frac{q}{r}\right) = f(p^n)\frac{f(q)}{f(r)} = f(p)^n.$$

Set $f(p) = \alpha > 1$. We have

$$f\left(p^n \frac{q}{r}\right) = \alpha^n.$$

Clearly $f(0) = 0$ and this concludes the proof of Theorem 22.

Exercises and further results

1. (Rădulescu 1980; Dhombres 1983.) Give the general representation of an operator $P: C_R(X) \to C_R(X)$ such that, for all f, g in $C_R(X)$, one gets
$$P(f+g) \geqslant Pf + Pg,$$
$$P(fg) \geqslant Pf \cdot Pg.$$
2. Let C be a proper convex cone with vertex at 0. Let H be a hyperplane not going through 0. Prove or disprove that a point x of $H \cap C$ is an extreme point of $H \cap C$ if, and only if, the ray going through x is an extreme ray.
3. Using the fact that $\mathbb{R}$ is a proper subset of the Bohr group $\mathscr{B}$, show that the

mapping hat $\wedge:\mathbb{R}\to\mathbb{R}^\wedge$ is not a homeomorphism (with respect to the topology which $\mathbb{R}^\wedge$ inherits as a subset of the set $\mathscr{B}$ equipped with the weak star topology).

4. Prove or disprove the following characterization of almost periodic Bohr functions. A continuous function $f:\mathbb{R}\to\mathbb{C}$ is an almost periodic Bohr function if, and only if, the set of all translates f_x of f is a relatively compact subset of $C_b(\mathbb{R})$, the space of all complex valued bounded continuous functions over $\mathbb{R}$, equipped with the topology of uniform convergence.
5*. Let $(\mathscr{B}, *)$ be the Bohr group of $\mathbb{R}$ equipped with the weak star topology and take the hat mapping $h:\mathbb{R}\to\mathscr{B}$. Show that, for any compact Hausdorff topological group G and any continuous homomorphism $\alpha:\mathbb{R}\to G$, there exists a continuous homomorphism $\beta:\mathscr{B}\to G$ such that $\alpha=\beta\circ h$, that is, the following diagram applies:

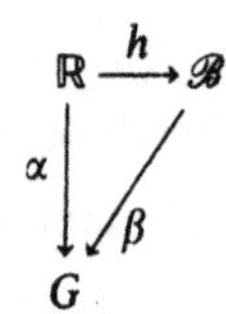

Prove or disprove that $(\mathscr{B}, *)$ is unique up to an isomorphism of compact Hausdorff topological groups.

6*. (Favard 1933, pp. 28–36, Weil 1940, pp. 133–4.) Prove the following characterisation of almost periodic functions (which explains the name). A continuous function $f:\mathbb{R}\to\mathbb{C}$ is an almost periodic Bohr function if, and only if, for every $\varepsilon>0$, there exists a relatively dense subset T of $\mathbb{R}$ such that $|f(x+t)-f(x)|\leqslant\varepsilon$ for all $x\in\mathbb{R}$, and all $t\in T$. (A subset T of $\mathbb{R}$ is said to be relatively dense if there exists an $L>0$ such that every interval of length L contains a point of T.)
7. Find all continuous $F:\mathscr{B}\to\Gamma$ (where Γ is the set of all complex numbers of modulus 1 and $\mathscr{B}$ the Bohr group equipped with the weak star topology) which are solutions of the functional equation

$$F(\phi*\psi)=F(\phi)F(\psi)\quad(\phi,\psi\in\mathscr{B}).$$

8. (Favard 1933, p. 36.) Using (44), show that, for an almost periodic Bohr function f, the subset Λ_f of all real λ for which $c_\lambda(f)=Q(fe_{-\lambda})\neq 0$ is at most countable. Prove $Q(|f|^2)=\sum|c_\lambda(f)|^2$ (Parseval's equality).
9. By noticing that $\mathbb{R}^\wedge$ is dense in the Bohr group equipped with the weak star topology and using an extension theorem for the functional equation $\Phi(x+y)=\Phi(x)\Phi(y)$, analogous to Theorem 7.5, give a proof for Theorem 5.8.
10*. We make use of the notations of Theorem 5.8. Consider the function

$$f(t)=\prod_{j=1}^{n}(1+\exp(i\mu_j t)\bar{z}_j).$$

which is a (continuous) almost periodic Bohr function. Obtain from Parseval's equality (Exercise 8) for $|f(t)|^{2p}$, where p is a positive integer, a proof for theorem 5.8. (This proof requires only (44) for a trigonometric polynomial and no further properties of almost periodic functions. Thus we have indeed another proof of Theorem 5.8 without circular reasoning).

11*. (Favard 1933, pp. 82–5.) Suppose $f:\mathbb{R}\to\mathbb{C}$ is an almost periodic Bohr function such that $c_\lambda(f)=0$ for all λ, $|\lambda|\leqslant\alpha$ where $\alpha>0$ (see Exercise 8 for the definition of $c_\lambda(f)$). Show that $\int_0^x f(t)dt$ is an almost periodic Bohr function.

12*. (Hewitt–Zuckerman 1950.) Let E be a real linear topological space and consider the set $E^\wedge$ of all continuous characters of the group $(E,+)$. Let E^* be the topological dual of E. Show that a function $\chi:E\to\mathbb{C}$ belongs to $E^\wedge$ if, and only if, there exists an f *in* E^* such that

$$\chi(x)=e^{if(x)}\quad(x\in\mathbb{R}).$$

Get from this result the dual group of $\mathbb{R}^n$.

13*. (Weil 1940, pp. 54–7.) Let E be a subset of positive Haar measure of a locally compact abelian group. Show that every additive function $f:G\to\mathbb{R}$, which is bounded from above on E, is continuous.

14. Prove or disprove that the trivial valuation is the only valuation of a finite field.

15. Let I be an integral domain and $g:I\to\bar{\mathbb{R}}_+$ such that

$$g(xy)=g(x)g(y)\qquad(x,y\in I),$$
$$g(x)=0\qquad\text{if and only if } x=0,$$
$$g(x+y)\leqslant g(x)+g(y)\quad(x,y\in I).$$

Prove or disprove that there exists a valuation f on the quotient field F of the integral domain I which extends g.

16. Let f be a *nonarchimedean valuation* on a field F (that is, a valuation on F such that $f(x+y)\leqslant\operatorname{Max}(f(x),f(y))$ for all x and y in F). Let U be the subset of all x in F for which $f(x)\leqslant 1$. Show that U is a ring with an identity. Prove or disprove that, if $x\notin U$, then $x^{-1}\in U$ (U is the valuation ring of F).

17. Prove or disprove that any valuation of a field with characteristic not equal to zero is nonarchimedean (for the definition, see the previous exercise).

18*. (Choquet 1969, pp. 246–64.) Let G be a locally compact abelian group. A function $f:G\to\mathbb{C}$ is positive definite if

$$\sum_{k=1}^{n}\sum_{i=1}^{n} z_i\bar{z}_k f(x_i-x_k)\geqslant 0,$$

for all integers $n\geqslant 1$, all complex numbers $z_1,\ldots,z_k$ and for all elements $x_1,\ldots,x_n$ of G (see Chapter 8).

(a) Show that the set P of all continuous positive definite functions on G is a convex cone which contains the continuous characters of G (elements of $G^\wedge$).

(b) Verify that the function $x\mapsto(1+|z|^2)f(x)+zf(x-a)+\bar{z}f(x+a)$ $(f\in P, a\in G, z\in\mathbb{C})$ is an element of P.

(c) Show that each extreme ray (cf. Section 2) of the cone P contains a function f satisfying the functional equation

$$f(x-a)=\lambda(a)f(x),$$

where $\lambda:G\to\mathbb{C}$.

(d) Prove that every extreme ray of P contains a continuous character ($\in G^\wedge$).

Then, by using the technique of Choquet described in Section 2, it is possible to write every continuous positive definite f in the form

$$f(x)=\int_{G^\wedge}\chi(x)d\mu(x),$$

where μ is a positive Radon measure on the dual group $G^\wedge$. The converse is easy (Bochner's theorem).

19. By using continuous $f:[0,1]\to[0,1]$ for which

$$\int_0^1|f(x)|dx=\int_0^1|1-f(x)|dx=\tfrac{1}{2},$$

show that, in $C[0,1]$ equipped with the integral norm $\|f\|=\int_0^1|f(x)|dx$, (27) does not characterize the middle points.

20. Prove that there exists exactly one positive Radon measure of total mass 1 which is invariant under translation on the Bohr group.

21. Without using a compactness argument, show that both $b:\mathscr{B}\to\mathscr{C}$ and its inverse are continuous.

22*. Consider the set S of all functions $f:\mathbb{R}\to\mathbb{C}$ which are zero outside a subset A_f, where A_f is at most countable. Define

$$\|f\|=\sum_{x\in\mathbb{R}}|f(x)|,$$

and

$$f*g(x)=\sum_{y\in\mathbb{R}}f(x-y)g(y),$$
$$(f+g)(x)=f(x)+g(x),$$
$$(\lambda f)(x)=\lambda f(x)$$

for $f,g\in S$, $\lambda\in\mathbb{C}$. Show that $(S,+,*,\|\cdot\|)$ is a Banach algebra. Prove or disprove that its spectrum (cf. Sections 5, 7.2) for the weak star topology can be identified with the compact Bohr group $\mathscr{B}$.

23. Let G be a locally compact abelian group and $L^1(G)$ its group algebra, which is a commutative Banach algebra. Prove or disprove that the spectrum of $L^1(G)$ (cf. Sections 5, 7.2) can be identified with the set $G^\wedge$ of all continuous characters on G.

24. (Kuczma 1961.) Let $f:\mathbb{R}^n\to E$ be an additive function, where $n\geqslant 1$ and E is a real Hilbert space. Suppose there exists an $a\neq 0$ in E and $\alpha>0$ such that

$$\langle f(x),a\rangle\leqslant\alpha\quad(x\in B(x_0,r)),$$

where $B(x_0,r)$ is a ball of radius r around x_0 on $\mathbb{R}^n$. Show that there exist $a_i\in\mathbb{R}$, $(i=1,2,\ldots,n)$ such that

$$\langle f(x)-\textstyle\sum x_ia_i,a\rangle=0$$

for all $x=(x_1,\ldots,x_n)\in\mathbb{R}^n$.

11

Characterizations of inner product spaces. An application to gas dynamics

Among normed spaces, the inner product spaces are particularly interesting. In this chapter, with minimal prerequisites from functional analysis, we use, in a natural way, various functional equations for the norm in order to characterize inner product spaces among normed spaces. We have to distinguish the real case from the complex case (and the case of quaternions). We conclude the chapter with an application of a conditional equation (orthogonal additivity) to the characterization of the Maxwell–Boltzmann distribution law for velocity in gas dynamics.

11.1 Quadratic functionals: a characterization of inner product spaces

As we have seen, in a real linear space E a norm is a function f from E into $\bar{\mathbb{R}}_+$, with the properties

$$f(x+y) \leqslant f(x)+f(y) \quad (x, y \in E), \tag{1}$$

$$f(\lambda x) = |\lambda| f(x) \quad (x \in E, \lambda \in \mathbb{R}), \tag{2}$$

and

$$f(x)=0 \quad \text{if and only if } x=0. \tag{3}$$

The subadditivity condition (1) is more difficult to handle than additivity. However, in some cases a form of additivity can be recovered for the function g defined by $g(x)=f(x)^2$. If so, the computations become much easier, with a geometric interpretation. We will now suppose E to be a *linear space over a commutative field F of characteristic* 0.

A function $g: E \to F$ is a *quadratic functional* if it satisfies the functional equation

$$g(x+y)+g(x-y)=2g(x)+2g(y); \tag{4}$$

and a function $S:E^2 \to F$ is *biadditive*, if

$$S(t_1+t_2,u)=S(t_1,u)+S(t_2,u) \quad \text{and} \quad S(u,t_1+t_2)=S(u,t_1)+S(u,t_2)$$
$$\text{for all } t_1,t_2,\ u\in E.$$

Our first result is the following.

Proposition 1. *The functional $g:E\to F$ is quadratic if, and only if, there exists a symmetric biadditive function $S:E^2\to F$ such that*

$$g(x)=S(x,x).$$

This S is unique.

Proof. If $g(x)=S(x,x)$, then

$$\begin{aligned} g(x+y)+g(x-y) &= S(x+y,x+y)+S(x-y,x-y)\\ &= 2S(x,x)+2S(y,y)\\ &= 2g(x)+2g(y). \end{aligned}$$

Conversely, let g be a quadratic functional. We define

$$S(x,y)=\tfrac{1}{4}(g(x+y)-g(x-y)). \tag{5}$$

By choosing $y=0$ in (4), we deduce that $g(0)=0$ and, by choosing $x=y$, we have $g(2x)=4g(x)$. Therefore

$$S(x,x)=\tfrac{1}{4}g(2x)=g(x).$$

We now prove the symmetry of S. Interchanging x and y in (4), we obtain

$$g(x-y)=g(y-x),$$

which shows that g is even. This and (5) imply the symmetry $S(x,y)=S(y,x)$ and

$$S(-x,y)=-S(x,y). \tag{6}$$

Thus S is odd in each variable and it only remains to show that S is additive in the first variable. We have

$$\begin{aligned} &4[S(x_1+x_2,y)+S(x_1-x_2,y)]\\ &\quad= g(x_1+x_2+y)+g(x_1-x_2+y)-(g(x_1+x_2-y)+g(x_1-x_2-y))\\ &\quad= 2g(x_1+y)+2g(x_2)-2g(x_1-y)-2g(x_2)\\ &\quad= 2(g(x_1+y)-g(x_1-y))\\ &\quad= 8S(x_1,y). \end{aligned}$$

Interchange x_1 and x_2 and subtract the equation thus obtained from the above in order to get

$$S(x_1-x_2,y)-S(x_2-x_1,y)=2S(x_1,y)-2S(x_2,y).$$

In view of (6), this reduces to

$$2S(x_1-x_2,y)=2S(x_1,y)-2S(x_2,y)$$

and, if we replace x_2 by $-x_2$ and take (6) into consideration again,

$$S(x_1 + x_2, y) = S(x_1, y) + S(x_2, y),$$

which is the required additivity in the first variable. We leave it to the reader to prove the uniqueness of S (Exercise 36).

Proposition 1 easily leads to a nice result concerning the geometry of normed spaces, a result proved by Jordan and von Neumann (1935). Recall that *a real normed linear space E is also a (real) inner product space* when its norm $\|\cdot\|$ is such that

$$\|x\|^2 = \langle x, x\rangle \quad (x \in E),$$

where $\langle , \rangle$ is an inner product on E, that is,

(i) $\langle x, y\rangle \in \mathbb{R}$ $\quad (x, y \in E)$,
(ii) $\langle x, x\rangle \geqslant 0$ $\quad$ with equality if and only if $x = 0$,
(iii) $\langle x, y\rangle = \langle y, x\rangle$ $\quad (x, y \in E)$,
(iv) $x \mapsto \langle x, y\rangle$ $\quad$ is linear in x for each given y in E.

Theorem 2. *If each two dimensional subspace of a real normed space $(E, \|\cdot\|)$ is an inner product space, then E itself is an inner product space.*

Proof. Let $(E, \|\cdot\|)$ have this property. We define $g: E \to \mathbb{R}$

$$g(x) = \|x\|^2 \quad (x \in E). \tag{7}$$

Let F be a two dimensional subspace of E. Then $(F, \|\cdot\|)$ is an inner product space and there exists an inner product $\langle , \rangle$ on F such that

$$g(x) = \langle x, x\rangle \quad (x \in F).$$

Since $\langle , \rangle$ is symmetric (iii) and biadditive ((iv) and (iii)), Proposition 1 shows that g satisfies (4) on F. But, since F may be any two dimensional subspace of E, we deduce that g satisfies (4) for all x and y in E. Proposition 1 then asserts the existence and uniqueness of a symmetric biadditive $S: E^2 \to \mathbb{R}$ such that

$$g(x) = S(x, x) \quad (x \in E).$$

The uniqueness applies also to F. So $\langle , \rangle$ is the restriction of S to F^2. It follows that S satisfies (i)–(iv). For instance, in order to show that S is homogeneous in the first variable, that is, that

$$S(\lambda x, y) = \lambda S(x, y) \quad (x, y \in E, \lambda \in \mathbb{R}), \tag{8}$$

(cf. (iv)) consider a two dimensional F, containing the given $x, y \in E$, with the inner product $\langle , \rangle$ on F. Then

$$S(\lambda x, y) = \langle \lambda x, y\rangle = \lambda\langle x, y\rangle = \lambda S(x, y),$$

that is, (8). This concludes the proof of Theorem 2.

This result may arouse our curiosity as to whether a similar result holds in complex spaces.

A *complex normed linear space is also a (complex) inner product space* when its norm $\|\cdot\|:E\to\mathbb{R}$ is such that

$$\|x\|^2=\langle x,x\rangle \quad (x\in E),$$

where $\langle,\rangle$ is a complex inner product on E, that is,

(i′)	$\langle x,y\rangle\in\mathbb{C}$	$(x,y\in E)$,
(ii′)	$\langle x,x\rangle\geqslant 0$	with equality if and only if $x=0$,
(iii′)	$\langle x,y\rangle=\overline{\langle y,x\rangle}$	for all x,y in E (*hermitian*),
(iv′)	$x\mapsto\langle x,y\rangle$	is linear in x for each given y in E.

Theorem 3. *If each two dimensional subspace of a complex normed space $(E,\|\cdot\|)$ is an inner product space, then E itself is an inner product space.*

Proof. Let $(E,\|\cdot\|)$ have this property. We define $g:E\to\mathbb{R}$ by (7). The space E can be viewed as a real normed linear space and every real subspace of dimension 2 is contained in a complex subspace of dimension 2. Therefore Theorem 2 yields that

$$g(x)=S(x,x),$$

where $S:E^2\to\mathbb{R}$, is defined by

$$S(x,y)=\tfrac{1}{4}(\|x+y\|^2-\|x-y\|^2) \tag{9}$$

and, in accordance with (5), satisfies (i), (ii), (iii) and (iv). Now we define $S':E^2\to\mathbb{C}$ by

$$S'(x,y)=S(x,y)-iS(ix,y). \tag{10}$$

We shall prove that $S'(x,y)$ is an inner product on the complex normed linear space E. The property (i′) is obvious. Let us compute $S'(x,x)$:

$$\begin{aligned}S'(x,x)&=S(x,x)-iS(ix,x)\\&=\|x\|^2-iS(ix,x).\end{aligned}$$

But using the definition in (9) of S, we get

$$\begin{aligned}S(ix,x)&=\tfrac{1}{4}(\|(i+1)x\|^2-\|(i-1)x\|^2)\\&=\frac{\|x\|^2}{4}(|i+1|^2-|i-1|^2)\\&=0.\end{aligned}$$

This proves (ii′).

Since S is additive in the first variable, we obtain

$$S'(x_1+x_2,y)=S'(x_1,y)+S'(x_2,y),$$

and so S' is additive in the first variable. Next, for $\lambda\in\mathbb{C}$ we compute $S'(\lambda x,y)$.

We have

$$S'(\lambda x, y) = S'((\lambda_1 + i\lambda_2)x, y) \quad \text{with } \lambda_1, \lambda_2 \in \mathbb{R}.$$

Using (10) we get

$$S'(\lambda x, y) = S((\lambda_1 + i\lambda_2)x, y) - iS((i\lambda_1 - \lambda_2)x, y).$$

Applying first the additivity of S and then its (real) homogeneity in the first variable we get

$$\begin{aligned} S'(\lambda x, y) &= S(\lambda_1 x, y) + S(i\lambda_2 x, y) - iS(i\lambda_1 x, y) + iS(\lambda_2 x, y) \\ &= \lambda_1 S(x, y) + \lambda_2 S(ix, y) - \lambda_1 iS(ix, y) + \lambda_2 iS(x, y) \\ &= \lambda_1 [S(x, y) - iS(ix, y)] + \lambda_2 [S(ix, y) + iS(x, y)]. \end{aligned}$$

Using (10) once more, we obtain

$$\begin{aligned} S'(\lambda x, y) &= \lambda_1 S'(x, y) + i\lambda_2 S'(x, y) \\ &= (\lambda_1 + i\lambda_2) S'(x, y). \end{aligned}$$

Finally,

$$S'(\lambda x, y) = \lambda S(x, y) \quad (x \in E, y \in E, \lambda \in \mathbb{C}).$$

We have thus proved (iv′).

We now compute $S'(y, x)$. Since S is real valued and (iii) holds, we get

$$\begin{aligned} S'(y, x) &= S(y, x) - iS(iy, x) \\ &= S(x, y) - iS(iy, x). \end{aligned}$$

Now, using (9) twice, we get

$$\begin{aligned} 4[S(iy, x) + S(ix, y)] &= \| iy + x \|^2 - \| iy - x \|^2 + \| ix + y \|^2 - \| ix - y \|^2 \\ &= \| iy + x \|^2 - \| i(x + iy) \|^2 + \| ix + y \|^2 - \| i(ix + y) \|^2 \\ &= \| iy + x \|^2 - \| x + iy \|^2 + \| ix + y \|^2 - \| ix + y \|^2 \\ &= 0. \end{aligned}$$

Thus

$$\overline{S'(y, x)} = S(x, y) - iS(ix, y) = S'(x, y),$$

which proves (iii′) and concludes the proof of Theorem 3.

In Theorem 2 (and in Theorem 3) we used particular solutions of (4) by choosing $g(x) = \| x \|^2$. The regularity of this function g yields the homogeneity of the function S in the first variable. But, in general, a quadratic functional is not necessarily homogeneous.

However, a mild regularity assumption on g as a solution of (4) generally assures the homogeneity. We call a function $g: E \to \mathbb{R}$ on a real linear space *ray-bounded* if, for any $x \neq 0$ and for any y in E, there exists a subset I_x of positive Lebesgue measure in $\mathbb{R}$ and a positive constant $M_{x,y}$ such that

$$|g(\lambda x + y)| \leqslant M_{x,y} \quad (\lambda \in I_x).$$

We have the following result.

Proposition 4. *A function $g:E \to \mathbb{R}$ is a ray-bounded quadratic functional on a real linear space E if and only if there exists a symmetric bilinear function $S:E^2 \to \mathbb{R}$ such that*

$$g(x) = S(x, x).$$

Proof. If S is a symmetric bilinear function, Proposition 1 shows that g is a quadratic functional. But g is ray-bounded, since

$$g(\lambda x + y) = \lambda^2 S(x, x) + 2\lambda S(x, y) + S(y, y).$$

Conversely, let g be a ray-bounded quadratic functional on E. Proposition 1 yields a symmetric and biadditive S such that

$$g(x) = S(x, x) \quad (x \in E).$$

We define $f:\mathbb{R} \to \mathbb{R}$ by $f(\lambda) = S(\lambda x, y)$ for a given pair of elements x, y in E, with $x \neq 0$. Clearly, f is additive. Since g is ray-bounded, and using (5), we have the following bound for $f(\lambda)$ for all λ in I_x:

$$|f(\lambda)| \leqslant \tfrac{1}{4}(|g(\lambda x + y)| + |g(\lambda x - y)|)$$
$$\leqslant \frac{M_{x,y} + M_{x,-y}}{4}.$$

Corollary 2.5 yields $f(\lambda) = \lambda f(1)$ and so $S(\lambda x, y) = \lambda S(x, y)$ which is all that remained to be proved in Proposition 4.

Instead of a regularity condition, one may just look for an algebraic condition for a quadratic functional so that the biadditive form S associated to it by (5) (as in Proposition 1) is bilinear. For example, one may want g to satisfy

$$g(\lambda x) = |\lambda|^2 g(x) \tag{11}$$

for all λ in the field F. We assume that the field F is a valued field and note that our final result depends upon F (cf. Kurepa 1964*a*, 1965*c*).

We first define a *sesquilinear* function $S:E^2 \to \mathbb{C}$ as a biadditive function satisfying $S(\lambda x, y) = \lambda S(x, y)$ and $S(x, \lambda y) = \bar{\lambda} S(x, y)$ for all $x, y \in E$, $\lambda \in \mathbb{C}$.

Theorem 5. *Let $g:E \to \mathbb{C}$ be a quadratic functional satisfying* (11) *where E is a complex linear space. Then there exists a sesquilinear function $S':E^2 \to \mathbb{C}$, such that*

$$S'(x, x) = g(x) \quad (x \in E).$$

Proof (following Vrbová 1973, with slight changes). We suppose first that g is a real valued function. Since the space E can be regarded as a real linear space, Proposition 1 provides a symmetric biadditive function $S:E^2 \to \mathbb{R}$ such that

$$g(x) = S(x, x).$$

Equation (11) gives now $S(\lambda x, \lambda x) = |\lambda|^2 S(x, x)$. From this and from the biadditivity and symmetry of S we get

$$\begin{aligned}|\lambda|^2[S(x,x)+2S(x,y)+S(y,y)] &= |\lambda|^2 S(x+y,x+y) = S[\lambda(x+y),\lambda(x+y)]\\ &= S(\lambda x,\lambda x)+2S(\lambda x,\lambda y)+S(\lambda y,\lambda y)\\ &= |\lambda|^2 S(x,x)+2S(\lambda x,\lambda y)+|\lambda|^2 S(y,y),\end{aligned}$$

and thus

$$S(\lambda x, \lambda y) = |\lambda|^2 S(x, y).$$

We prefer to write this as

$$S(\lambda x, y) = |\lambda|^2 S(x, y/\lambda) \quad (x, y \in E, \lambda \in \mathbb{C}^* = \mathbb{C}\setminus\{0\}). \tag{12}$$

Symmetry and biadditivity of S give $S(x, -y) = -S(x, y)$ and (12), with $\lambda = i$, yields

$$S(ix, y) = -S(iy, x). \tag{13}$$

As in the proof of Theorem 3, we define $S': E^2 \to \mathbb{C}$ by

$$S'(x, y) = S(x, y) - iS(ix, y).$$

The function S' is clearly biadditive. Since S is real valued and symmetric, (13) shows that S' is a hermitian function:

$$S'(y, x) = \overline{S'(x, y)}.$$

From (13), $S(ix, x) = 0$. Therefore,

$$S'(x, x) = g(x).$$

To complete the proof of Theorem 5 in the case where g is real valued, we have to show that S' is homogeneous in the first variable. Fix x, y in E and introduce the function $f: \mathbb{C} \to \mathbb{C}$ by

$$f(\lambda) := S'(\lambda x, y) - S'(x, \lambda y).$$

This function f is additive on $\mathbb{C}$. Moreover, (12) yields another functional equation for f:

$$f(\lambda) = -|\lambda|^2 f\left(\frac{1}{\lambda}\right) \quad (\lambda \in \mathbb{C}^*). \tag{14}$$

Indeed,

$$\begin{aligned}f(\lambda) &= S(\lambda x, y) - iS(\lambda i x, y) - [S(x, \lambda y) - iS(ix, \lambda y)]\\ &= |\lambda|^2\left[S\left(x, \frac{y}{\lambda}\right) - iS\left(ix, \frac{y}{\lambda}\right)\right] - |\lambda|^2\left[S\left(\frac{x}{\lambda}, y\right) - iS\left(i\frac{x}{\lambda}, y\right)\right]\\ &= |\lambda|^2\left[S'\left(x, \frac{y}{\lambda}\right) - S'\left(\frac{x}{\lambda}, y\right)\right]\\ &= -|\lambda|^2 f\left(\frac{1}{\lambda}\right).\end{aligned}$$

We solve (14) for an additive f. Let $\lambda_1 \in \mathbb{R}$. If $|\lambda_1| \leqslant 1$, there exists $\lambda_2 \in \mathbb{R}$ such that $\lambda = \lambda_1 + i\lambda_2$ and $|\lambda| = 1$. By (14), $f(\lambda) = -f(1/\lambda) = -f(\bar{\lambda})$ for this λ. This, the additivity of f, and (14) yield

$$\begin{aligned} f(\lambda_1) + f(i\lambda_2) = f(\lambda_1 + i\lambda_2) &= -f(\lambda_1 - i\lambda_2) \\ &= -f(\lambda_1) + f(i\lambda_2). \end{aligned}$$

Therefore $f(\lambda_1) = 0$ when $|\lambda_1| \leqslant 1$ and $\lambda_1 \in \mathbb{R}$.

If $|\lambda_1| > 1$, $f(\lambda_1) = -\lambda_1^2 f(1/\lambda_1) = 0$ from the result which we have just obtained. Therefore $f(\lambda_1) = 0$ for all $\lambda_1 \in \mathbb{R}$.

Let $\lambda_2 \in \mathbb{R}^* = \mathbb{R}\backslash\{0\}$. Equation (14) yields $f(i\lambda_2) = -\lambda_2^2 f(1/i\lambda_2) = \lambda_2^2(i/\lambda_2)$. But the solution of Exercise 2.10 shows that the additive function $\lambda_2 \mapsto f(i\lambda_2)$ is continuous. Therefore $f(i\lambda_2) = \lambda_2 f(i)$ for all $\lambda_2 \in \mathbb{R}$ (also for $\lambda_2 = 0$). Finally,

$$f(\lambda_1 + i\lambda_2) = \lambda_2 f(i) \quad (\lambda_1, \lambda_2 \in \mathbb{R}).$$

From the definition of f and from $f(\lambda_1) = 0$, we obtain

$$S'(\lambda_1 x, y) = S'(x, \lambda_1 y) \quad (\lambda_1 \in \mathbb{R}), \tag{15}$$

and, from $f(\lambda_2 i) = \lambda_2 f(i)$, we have

$$S'(\lambda_2 ix, y) - S'(x, \lambda_2 iy) = \lambda_2[S'(ix, y) - S'(x, iy)] \quad (\lambda_2 \in \mathbb{R}). \tag{16}$$

But

$$\begin{aligned} S'(ix, y) = S(ix, y) - iS(-x, y) &= S(ix, y) + iS(x, y) \\ &= i[S(x, y) - iS(ix, y)], \end{aligned}$$

so

$$S'(ix', y) = iS'(x, y). \tag{17}$$

We deduce from (17), since S' is hermitian, that

$$S'(x, iy) = -iS'(x, y). \tag{18}$$

Using (17) and (18), we get

$$S'(ix, y) - S'(x, iy) = 2iS'(x, y).$$

Since, from (15), $S'(\lambda_2 x, iy) = S'(x, \lambda_2 iy)$, we have

$$S'(\lambda_2 ix, y) - S'(x, \lambda_2 iy) = 2iS'(\lambda_2 x, y).$$

On the other hand, (16) gives

$$\begin{aligned} S'(\lambda_2 ix, y) - S'(x, \lambda_2 iy) &= \lambda_2[S'(ix, y) - S'(x, iy)] \\ &= 2i\lambda_2 S'(x, y). \end{aligned}$$

Therefore,

$$S'(\lambda_2 x, y) = \lambda_2 S'(x, y).$$

By this and (17) we have $S'(i\lambda_2 x, y) = iS'(\lambda_2 x, y) = i\lambda_2 S'(x, y)$ and

$$\begin{aligned} S'((\lambda_1 + i\lambda_2)x, y) &= S'(\lambda_1 x, y) + S'(i\lambda_2 x, y) \\ &= \lambda_1 S'(x, y) + i\lambda_2 S'(x, y) \\ &= (\lambda_1 + i\lambda_2)S'(x, y), \end{aligned}$$

which proves the homogeneity in the first variable. Since S' is hermitian, the proof of Theorem 5 is finished for the case when g is real valued.

We suppose now that g is complex valued and write $g(x) = g_1(x) + ig_2(x)$, where $g_1(x)$ and $g_2(x)$ are real. Both g_1 and g_2 satisfy (4) and (11). Hence there exist two sesquilinear (and hermitian) functions $S_1, S_2: E^2 \to \mathbb{C}$ such that

$$g_1(x) = S_1(x,x),$$
$$g_2(x) = S_2(x,x) \quad (x \in E).$$

If we use $S(x,y) = S_1(x,y) + iS_2(x,y)$, it is clear that S is sesquilinear (but, in general, no longer hermitian) and that

$$S(x,x) = S_1(x,x) + iS_2(x,y) = g_1(x) + ig_2(x)$$
$$= g(x).$$

This concludes the proof of Theorem 5.

On the other hand, an additive $f:\mathbb{R} \to \mathbb{R}$, satisfying (14), for all real $\lambda \in \mathbb{R}^*$, need not be continuous. Therefore, the situation is different in the real case.

Proposition 6. *Let E be a real linear space. Then there exists a symmetric and biadditive $S:E^2 \to \mathbb{R}$ such that $S(\lambda x, \lambda x) = |\lambda|^2 S(x)$ for all x in E, all λ in $\mathbb{R}$, which is not bilinear.*

We leave the proof of Proposition 6 to the reader (see Exercise 7).

In the case where the field F is the field of quaternions, an additive function $f:F \to F$, satisfying (14) for all λ in $F^* = F \backslash \{0\}$, is continuous (see Exercise 8).

Recall that the field of quaternions is the set of all

$$\lambda = \lambda_1 + \mathbf{i}\lambda_2 + \mathbf{j}\lambda_3 + \mathbf{k}\lambda_4,$$

where $\lambda_1, \lambda_2, \lambda_3$ and λ_4 are real numbers, with the two laws

$$\lambda + \lambda' = (\lambda_1 + \mathbf{i}\lambda_2 + \mathbf{j}\lambda_3 + \mathbf{k}\lambda_4) + (\lambda_1' + \mathbf{i}\lambda_2' + \mathbf{j}\lambda_3' + \mathbf{k}\lambda_4')$$
$$= (\lambda_1 + \lambda_1') + \mathbf{i}(\lambda_2 + \lambda_2') + \mathbf{j}(\lambda_3 + \lambda_3') + \mathbf{k}(\lambda_4 + \lambda_4'),$$
$$\lambda\lambda' = (\lambda_1\lambda_1' - \lambda_2\lambda_2' - \lambda_3\lambda_3' - \lambda_4\lambda_4') + \mathbf{i}(\lambda_1\lambda_2' + \lambda_2\lambda_1' + \lambda_3\lambda_4' - \lambda_4\lambda_3')$$
$$+ \mathbf{j}(\lambda_1\lambda_3' - \lambda_2\lambda_4' + \lambda_3\lambda_1' + \lambda_4\lambda_2') + \mathbf{k}(\lambda_1\lambda_4' + \lambda_2\lambda_3' - \lambda_3\lambda_2' + \lambda_4\lambda_1').$$

The conjugate of the quaternion $\lambda = \lambda_1 + \mathbf{i}\lambda_2 + \mathbf{j}\lambda_3 + \mathbf{k}\lambda_4$ is $\bar{\lambda} = \lambda_1 - \mathbf{i}\lambda_2 - \mathbf{j}\lambda_3 - \mathbf{k}\lambda_4$. Therefore,

$$\lambda\bar{\lambda} = \lambda_1^2 + \lambda_2^2 + \lambda_3^2 + \lambda_4^2.$$

The field of quaternions is a noncommutative field which is valued by

$$|\lambda|^2 = \lambda\bar{\lambda}.$$

Here too, if E is a linear space over F, a function $S:E^2 \to F$ is *sesquilinear*, if it is biadditive and

$$S(\lambda x, y) = \lambda S(x, y),$$
$$S(x, \lambda y) = \bar{\lambda} S(x, y) \quad (x, y \in E, \lambda \in F).$$

Proposition 7. *Let $g: E \to F$ be a quadratic functional satisfying* (11) *when E is a linear space over the field F of quaternions. There exists a sesquilinear function $S: E^2 \to F$ such that*

$$S(x, x) = g(x) \quad (x \in E).$$

Proof. We first suppose that g is real valued and, arguing as in the proof of Theorem 5, we have that there exists a symmetric and biadditive $S: E^2 \to \mathbb{R}$ such that

$$g(x) = S(x, x) \quad (x \in E).$$

Equation (11) gives

$$S(\lambda x, y) = |\lambda|^2 S\left(x, \frac{y}{\lambda}\right) \quad (x, y \in E)$$

for all nonzero quaternions λ. In the same way as in the proof of Theorem 5 (13), it follows that

$$S(\mathbf{i}x, y) = -S(\mathbf{i}y, x), \tag{19}$$

$$S(\mathbf{j}x, y) = -S(\mathbf{j}y, x), \tag{20}$$

and

$$S(\mathbf{k}x, y) = -S(\mathbf{k}y, x). \tag{21}$$

We define $S': E^2 \to F$ by

$$S'(x, y) = S(x, y) - \mathbf{i}S(\mathbf{i}x, y) - \mathbf{j}S(\mathbf{j}x, y) - \mathbf{k}S(\mathbf{k}x, y).$$

This is clearly a biadditive function. Moreover, S' is hermitian, that is,

$$S'(y, x) = \overline{S'(x, y)}. \tag{22}$$

Indeed,

$$\begin{aligned} S'(y, x) &= S(y, x) - \mathbf{i}S(\mathbf{i}y, x) - \mathbf{j}S(\mathbf{j}y, x) - \mathbf{k}S(\mathbf{k}y, x) \\ &= S(x, y) + \mathbf{i}S(\mathbf{i}x, y) + \mathbf{j}S(\mathbf{j}x, y) + \mathbf{k}S(\mathbf{k}x, y) \\ &= \overline{S'(x, y)}. \end{aligned}$$

Since $S(\mathbf{i}x, x) = S(\mathbf{j}x, x) = S(\mathbf{k}x, x) = 0$, we get

$$S'(x, x) = S(x, x) = g(x) \quad (x \in E).$$

Since S' is hermitian (22), in order to prove that S' is sesquilinear, it suffices to show that S' is linear in the first variable. We again introduce

$$f(\lambda) := S'(\lambda x, y) - S'(x, \lambda y).$$

As before, $f:F\to F$ is additive and satisfies the functional equation

$$f(\lambda)=-|\lambda|^2 f\left(\frac{1}{\lambda}\right)\quad(\lambda\in F^*=F\backslash\{0\}).$$

We leave it to the reader (Exercise 8) to prove that every additive solution of this equation is of the form

$$f(\lambda_1+\mathbf{i}\lambda_2+\mathbf{j}\lambda_3+\mathbf{k}\lambda_4)=\lambda_2 f(\mathbf{i})+\lambda_3 f(\mathbf{j})+\lambda_4 f(\mathbf{k}),$$

where $\lambda_1,\lambda_2,\lambda_3$ and λ_4 are real numbers. Therefore,

$$S'(\lambda_1 x,y)=S'(x,\lambda_1 y),$$
$$S'(\lambda_2\mathbf{i}x,y)-S'(x,\lambda_2\mathbf{i}y)=\lambda_2[S'(\mathbf{i}x,y)-S'(x,\mathbf{i}y)],$$
$$S'(\lambda_3\mathbf{j}x,y)-S'(x,\lambda_3\mathbf{j}y)=\lambda_3[S'(\mathbf{j}x,y)-S'(x,\mathbf{j}y)],$$

and

$$S'(\lambda_4\mathbf{k}x,y)-S'(x,\lambda_4\mathbf{k}y)=\lambda_4[S'(\mathbf{k}x,y)-S'(x,\mathbf{k}y)].$$

From $\mathbf{i}^2=\mathbf{j}^2=\mathbf{k}^2=-1$, we get

$$S'(\mathbf{i}x,y)=\mathbf{i}S'(x,y),\tag{23}$$

and similar equations with $\mathbf{i}$ replaced by $\mathbf{j}$, or $\mathbf{k}$. Since S' is hermitian, we also obtain

$$S'(x,\mathbf{i}y)=-\mathbf{i}S'(x,y),$$

and similar equations with $\mathbf{i}$ replaced by $\mathbf{j}$, or $\mathbf{k}$. Then, in the same way as in the proof of Theorem 5, we obtain

$$S'(\lambda_1 x,y)=\lambda_1 S'(x,y)\quad(\lambda_1\in\mathbb{R};x,y\in E).\tag{24}$$

Therefore, using the biadditivity, (23) and (24), we get

$$S'((\lambda_1+\mathbf{i}\lambda_2+\mathbf{j}\lambda_3+\mathbf{k}\lambda_4)x,y)=(\lambda_1+\mathbf{i}\lambda_2+\mathbf{j}\lambda_3+\mathbf{k}\lambda_4)S'(x,y).$$

This proves the linearity of S' in the first variable and completes the proof of Proposition 7 in the case where g is real valued.

When g takes its values in the quaternion field F, we write

$$g(x)=g_1(x)+\mathbf{i}g_2(x)+\mathbf{j}g_3(x)+\mathbf{k}g_4(x),$$

where $g_n(x)\in\mathbb{R}$ ($n=1,2,3$ and 4). By what we have just proved, $g_n(x)=S'_n(x,x)$, where $S'_n:E^2\to F$ ($n=1,2,3,4$) are sesquilinear functions. We define $S:E^2\to F$ by

$$S(x,y)=S'_1(x,y)+\mathbf{i}S'_2(x,y)+\mathbf{j}S'_3(x,y)+\mathbf{k}S'_4(x,y).$$

This S is sesquilinear and $S(x,x)=g(x)$, which concludes the proof of Proposition 7.

We conclude this section by solving a functional equation which conveniently combines (4) and (11) on quite general fields.

Theorem 8. *Let E be a linear space over a commutative field F. Suppose that*

F is of characteristic zero. Let $g:E\to F$ be such that $g(0)=0$. Then the function g satisfies

$$g(\lambda y+x)-g(\lambda y-x)=\lambda(g(y+x)-g(y-x)) \tag{25}$$

for all x, y in E and all λ in F if, and only if, there exists a symmetric and bilinear function $S:E^2\to F$ such that

$$g(x)=S(x,x) \quad (x\in E). \tag{26}$$

Proof. If (26) holds, then

$$g(\lambda y+x)-g(\lambda y-x)=4\lambda S(x,y)$$

and

$$\lambda(g(y+x)-g(y-x))=4\lambda S(x,y).$$

Therefore (25) is true.

Conversely, suppose that $g:E\to F$ satisfies (25) and $g(0)=0$. Our aim is to show that g is a quadratic functional. Because, once this is proved, Proposition 1 provides a symmetric biadditive function $S:E^2\to F$ such that $g(x)=S(x,x)$ for all x in E. With this,

$$S(\lambda y+x,\lambda y+x)-S(\lambda y-x,\lambda y-x)=2S(\lambda y,x),$$

and

$$S(y+x,y+x)-S(y-x,y-x)=2S(y,x).$$

Therefore (25) yields the homogeneity

$$S(\lambda y,x)=\lambda S(y,x)$$

of S in the first variable, and so S is symmetric and bilinear.

In order to prove that g is a quadratic functional, we first notice, by putting $\lambda=0$ into (25), that g is an even function. Next, we write (25) in the form

$$g(x+\lambda y)-g(x-\lambda y)=\lambda(g(x+y)-g(x-y)). \tag{27}$$

We fix x and y in E and define $f:F\to F$ by

$$f(\lambda)=g(x+\lambda y).$$

We prove that f satisfies a functional equation of the form (27) on F. Let λ, μ and ν be in F. We use (27):

$$\begin{aligned}\lambda(f(\mu+\nu)-f(\mu-\nu))&=\lambda(g(x+\mu y+\nu y)-g(x+\mu y-\nu y))\\&=g(x+\mu y+\lambda\nu y)-g(x+\mu y-\lambda\nu y)\\&=f(\mu+\lambda\nu)-f(\mu-\lambda\nu).\end{aligned}$$

Thus, indeed,

$$f(\mu+\lambda\nu)-f(\mu-\lambda\nu)=\lambda(f(\mu+\nu)-f(\mu-\nu)). \tag{28}$$

Take $\nu=1$ and $\mu=\lambda$ in (28),

$$f(2\lambda)-f(0)=\lambda(f(\lambda+1)-f(\lambda-1)). \tag{29}$$

But

$$f(\lambda+1)-f(\lambda-1)=f(1+\lambda)-f(1-\lambda)+f(1-\lambda)-f(\lambda-1).$$

Put $\nu=1$ and $\mu=1$ into (28),

$$f(1+\lambda)-f(1-\lambda)=\lambda(f(2)-f(0)),$$

then take $\mu=0$, $\nu=1$ and replace λ by $1-\lambda$ in (28),

$$\begin{aligned}f(1-\lambda)-f(\lambda-1)&=f(0+1-\lambda)-f(0-(1-\lambda))\\&=(1-\lambda)(f(1)-f(-1)).\end{aligned}$$

Therefore (29) gives

$$f(2\lambda)=\lambda^2(f(2)-f(0)-f(1)+f(-1))+\lambda(f(1)-f(-1))+f(0),$$

or

$$f(\lambda)=\frac{\lambda^2}{4}(f(2)-f(0)-f(1)+f(-1))+\frac{\lambda}{2}(f(1)-f(-1))+f(0). \tag{30}$$

But

$$\begin{aligned}f(2)-f(0)-f(1)+f(-1)&=g(x+2y)-g(x)-g(x+y)+g(x-y),\\f(1)-f(-1)&=g(x+y)-g(x-y)\end{aligned}$$

and $f(0)=g(x)$. Thus we get, from (30),

$$\begin{aligned}g(x+\lambda y)=&\frac{\lambda^2}{4}(g(x+2y)-g(x)-g(x+y)+g(x-y))\\&+\frac{\lambda}{2}(g(x+y)-g(x-y))+g(x).\end{aligned} \tag{31}$$

Adding (31) with $\lambda=1$ to (31) with $\lambda=-1$, we deduce that

$$g(x+2y)=3g(x+y)+g(x-y)-3g(x). \tag{32}$$

With $x=0$ in (32), using $g(0)=0$ and $g(-y)=g(y)$, we get

$$g(2y)=4g(y). \tag{33}$$

Using (33) and (31) with $x=0$, we get for all $\lambda\in F$,

$$g(\lambda y)=\lambda^2 g(y),\quad (y\in E). \tag{34}$$

When E is of dimension 1 (over F), then (34) easily leads to (4).

When E is of dimension larger than 1 (possibly infinite), let x and y be two linearly independent vectors of E. We consider $g(\alpha x+\beta y)$ for α and β in F. By using (31) twice we get

$$\begin{aligned}g(\alpha x+\beta y)&=P(\alpha)\beta^2+Q(\alpha)\beta+R(\alpha),\\g(\alpha x+\beta y)&=p(\beta)\alpha^2+q(\beta)\alpha+r(\beta).\end{aligned}$$

Since F is of characteristic zero, the functions P, Q and R are polynomials of degree at most 2 in α and, similarly, p, q and r are polynomials of degree at most 2 in β. Since g is even and $g(0)=0$, we deduce that, for some constants

A, B, C and D in F,

$$g(\alpha x + \beta y) = A\alpha^2\beta^2 + B\alpha\beta + C\alpha^2 + D\beta^2 \tag{35}$$

holds. But (34) implies $A = 0$. It is now easy to see that g is a quadratic functional on the linear space generated by x and y. Since x and y were arbitrarily chosen (but linearly independent on F), we deduce that g is a quadratic functional on E, which concludes the proof of Theorem 8.

11.2 Triangles in normed spaces: a second characterization of inner product spaces

We first interpret analytically the hypothesis that the lengths of the three sides of any proper triangle in a real or complex normed space $(E, \|\cdot\|)$ *determine the lengths of the* three *medians.* Since the triangle with sides $x, y, x - y$ has $\frac{1}{2}(x + y)$ as its median through the origin, this means, in particular, the existence of a function $L: \mathbb{R}_+^3 \to \mathbb{R}_+$ for which

$$\|x + y\| = L(\|x\|, \|y\|, \|x - y\|) \quad (x, y \in E). \tag{36}$$

(With y and $x - y$ or x and $x - y$ in place of x and y we get the existence of the other two medians, so these give the same condition (36).) When $(E, \|\cdot\|)$ is an inner product space, the lengths of the three sides of any proper triangle in E determine the lengths of the three medians. Indeed, $g(x) = \|x\|^2$ satisfies (4). Therefore,

$$\|x + y\| = (2\|x\|^2 + 2\|y\|^2 - \|x - y\|^2)^{1/2}.$$

We shall now use functional equations to establish the following characterization of inner product spaces (Lorch 1948):

Theorem 9. *Let $(E, \|\cdot\|)$ be a real or complex normed space such that the lengths of the three sides of any proper triangle in E determine the lengths of its medians. Then E is an inner-product space.*

Proof. We define $f(x) = \|x\|$. We call Γ the subset of all $\gamma \in \mathbb{R}$ such that, for all x, y in E with $f(x) = f(y)$, the equation $f(x + \gamma y) = f(\gamma x + y)$ holds. We prove that $\Gamma = \mathbb{R}$.

By the continuity of f in the norm topology, Γ is a closed subset of $\mathbb{R}$. This subset is not empty since, clearly, both 0 and 1 belong to Γ. Moreover, if $\gamma \in \Gamma$ and $\gamma \neq 0$, then $\gamma^{-1} \in \Gamma$. Similarly, if $\gamma \in \Gamma$, then $-\gamma \in \Gamma$. The set Γ has a very particular stability. Suppose γ_1 and γ_2 belong to Γ and take x, y in E such that $f(x) = f(y)$. Since $\gamma_1 \in \Gamma$, $f(\gamma_1 x + y) = f(x + \gamma_1 y)$ and, since $\gamma_2 \in \Gamma$, we deduce the following equality:

$$f(x + \gamma_1 y + \gamma_2(\gamma_1 x + y)) = f(\gamma_2(x + \gamma_1 y) + \gamma_1 x + y).$$

Thus

$$f((1 + \gamma_1\gamma_2)x + (\gamma_1 + \gamma_2)y) = f((\gamma_1 + \gamma_2)x + (1 + \gamma_1\gamma_2)y).$$

In other words, when γ_1 and γ_2 are in Γ, and $\gamma_1\gamma_2 \neq -1$, then

$$(\gamma_1+\gamma_2)/(1+\gamma_1\gamma_2)\in\Gamma.$$

Since $\tanh(u+v)=(\tanh u+\tanh v)/(1+\tanh u \tanh v)$ and $t\mapsto\frac{1}{2}\ln[(1+t)/(1-t)]$ is the inverse function of tanh, we define a new subset Γ' of $\mathbb{R}$ by $\Gamma'=\{\frac{1}{2}\ln[(1+\gamma)/(1-\gamma)]\,|\,\gamma\in\Gamma\cap]-1,+1[\}$.

This definition of Γ' has been chosen so that Γ' is a subgroup of $\mathbb{R}$. Since Γ is a closed subset, therefore Γ' is also a closed subgroup of $\mathbb{R}$. Let us prove that $0\ (\in\Gamma')$ is an accumulation point of Γ'. We will establish this by showing that 0 is an accumulation point of Γ and this will be proved if we prove that $1/n\in\Gamma$ for every positive integer n. Since Γ is stable under inversion (contains the reciprocal of every nonzero element), it is enough to show that $\mathbb{N}\subset\Gamma$. We already know that 0 and 1 belong to Γ. We proceed by induction and suppose that all positive integers up to $n-1$ belong to Γ. Suppose also that x and y are in E and $f(x)=f(y)$. Then

$$f(nx+y)=f(((n-1)x+y)+x).$$

Using (36), we deduce that

$$f(nx+y)=L(f((n-1)x+y),f(x),f((n-2)x+y)).$$

As $n-1, n-2\in\Gamma$ and $f(x)=f(y)$, we get

$$\begin{aligned}f(nx+y)&=L(f(x+(n-1)y),f(x),f(x+(n-2)y))\\&=L(f(x+(n-1)y),f(y),f(x+(n-2)y)).\end{aligned}$$

Using (36) a second time, we get

$$\begin{aligned}f(nx+y)&=f((x+(n-1)y)+y)\\&=f(x+ny).\end{aligned}$$

Also Γ' is dense and closed in $\mathbb{R}$. Therefore $\Gamma'=\mathbb{R}$. Thus $\Gamma\cap]-1,1[=]-1,1[$. But, since Γ is stable under inversion and since λ and $-\lambda$ belong to Γ, this proves that $\Gamma=\mathbb{R}$.

In other words, under the suppositions in Theorem 9, the norm $f(x)=\|x\|$ of E satisfies the conditional functional equation

$$f(x+\gamma y)=f(\gamma x+y)\quad(\gamma\in\mathbb{R})\quad\text{if } f(x)=f(y). \tag{37}$$

Theorem 9 is now a consequence of the following classical theorem first stated by Ficken 1944.

Theorem 10. *A real or complex normed space $(E,\|\cdot\|)$ is an inner product space if, and only if, for all $\gamma\in\mathbb{R}$ and for all $x,y\in E$ such that $\|x\|=\|y\|$, we have*

$$\|x+\gamma y\|=\|\gamma x+y\|. \tag{38}$$

Proof. If $(E,\|\cdot\|)$ is an inner product space, we have

$$\|x+\gamma y\|^2=\|x\|^2+2\gamma\,\mathrm{Re}\langle x,y\rangle+\gamma^2\|y\|^2$$

and

$$\|\gamma x + y\|^2 = \gamma^2 \|x\|^2 + 2\gamma \operatorname{Re}\langle x, y\rangle + \|y\|^2$$

for all $\gamma \in \mathbb{R}$. If $\|x\| = \|y\|$, then (38) holds which proves the 'only if' part in Theorem 10.

The proof of the 'if' part requires a more sophisticated procedure and we shall use functional equations. For convenience, we divide the proof into some lemmata and propositions which are interesting in themselves (in particular, Proposition 14, which gives a condition for continuous mappings to be quadratic functionals).

Proposition 11. *Let E be a real or complex topological linear space. Let a continuous function $f: E \to \mathbb{R}$ be such that*

(*i*) $f(x) > 0$ *for all* $x \neq 0$ *in* E *and* $f(0) = 0$,
(*ii*) $f(\lambda x) = |\lambda| f(x)$ *for* $x \in E$ *and* $\lambda \in \mathbb{R}$,

and

(*iii*) *f satisfies (37), that is,* $f(x + \gamma y) = f(\gamma x + y)$ *for all* x, y *in* E *with* $f(x) = f(y)$ *and for all* $\gamma \in \mathbb{R}$.

Then $f(x) = f(y) = f(\alpha x + (1-\alpha)y)$ *for a pair* x, y *in* E *and for an* $\alpha \in]0, 1[$, *if and only if* $x = y$.

Proof. We start from $f(x) = f(y) = f(\alpha x + (1-\alpha)y)$ for some x, y in E and some $\alpha \in]0, 1[$. We first want to prove by induction that, for all nonnegative integers, n, we have

$$f(x) = f(y) = f((n-\alpha)(x-y) - y). \tag{39}$$

Using (*ii*) with $\lambda = -1$, we see that (39) is true for $n = 0$. We suppose (39) true for some integer n. Using (37), we get, for all β in $\mathbb{R}$,

$$f((n-\alpha)(x-y) - y + \beta y) = f(\beta((n-\alpha)(x-y) - y) + y).$$

We choose $\beta = (n+1) - \alpha$. Then

$$f((n-\alpha)x) = f((n-\alpha)(n+1-\alpha)(x-y) + (\alpha - n)y).$$

Since α is not an integer, using (*ii*) and (*i*) we have

$$f(x) = f((n+1-\alpha)(x-y) - y),$$

which completes the inductive proof of (39).

Using (*i*), (*ii*), and (39) we have

$$\frac{f(x)}{n-\alpha} = f\left(x - y - \frac{y}{n-\alpha}\right).$$

Letting n tend to infinity and using the fact that f is continuous, we deduce that $f(x-y) = 0$. This yields $x = y$, using (*i*), and concludes the proof of Proposition 11.

Proposition 11 and the following Lemmata 12, 13 and Proposition 14 could be stated and proved on more general linear topological (including quaternionic) spaces (cf. Exercises 33, 34). For simplicity we restrict ourselves here to real and complex spaces.

Lemma 12. *Let E be a real or complex topological linear space and $f:E\to\mathbb{R}$ be a continuous function satisfying properties (i), (ii) and (iii) of Proposition* 11. *Suppose that $y\neq 0$. Let $r>0$. Then there exist at most two different real values t such that*

$$f(x+ty)=r.$$

Proof. Suppose to the contrary that

$$f(x+t_1y)=f(x+t_2y)=f(x+t_3y),$$

where t_1, t_2 and t_3 are three distinct real values. Without loss of generality, we may suppose that $X_2=x+t_2y$ lies in the segment $]X_1,X_3[=]x+t_1y,x+t_3y[$. Thus there exists a $\lambda\in]0,1[$ such that $X_2=\lambda X_1+(1-\lambda)X_3$. We have

$$f(\lambda X_1+(1-\lambda)X_3)=f(X_1)=f(X_3).$$

Proposition 11 implies that $X_1=X_3$, which is impossible since $y\neq 0$ and $t_1\neq t_3$. So Lemma 12 is proved.

Lemma 13. *Let E be a real or complex linear space and $f:E\to\mathbb{R}$ be a function satisfying properties (i), (ii) and (iii) of Proposition* 11. *Then f satisfies the functional equation*

$$f(x)f(y)f(x+y)=f(f(y)^2x+f(x)^2y)\quad (x,y\in E). \tag{40}$$

Proof. We first suppose that x and y are not zero. We write

$$f(x+y)=f\left[f(x)\frac{x}{f(x)}+f(y)\frac{y}{f(y)}\right].$$

From (*i*) and (*ii*) we get $f[x/f(x)]=f[y/f(y)]=1$. Using (37), and (*ii*) again, we obtain

$$f(x+y)=f\left[f(y)\frac{x}{f(x)}+f(x)\frac{y}{f(y)}\right], \tag{41}$$

or, using (*i*) and (*ii*) once more, we get (40).

Clearly, this equation is true for $x=0$ or for $y=0$. This proves (40) for all x,y in E.

Proposition 14. *Let E be a real or complex linear topological space and $f:E\to\mathbb{R}$ be a continuous function satisfying properties (i), (ii) and (iii) of Proposition* 11. *Then $g(x):=f(x)^2$ is a quadratic functional.*

Proof. Let x, y be in E. We first suppose that there exist $z \in E$ and $\lambda, \mu \in \mathbb{R}$ such that

$$x = \lambda z, \quad y = \mu z.$$

Using (*i*) and (*ii*), we compute

$$g(x+y) + g(x-y) = ((\lambda+\mu)^2 + (\lambda-\mu)^2)g(z),$$

and

$$2g(x) + 2g(y) = (2\lambda^2 + 2\mu^2)g(z).$$

Therefore g satisfies the functional equation (4) of quadratic functionals in this case.

Suppose now that x and y are linearly independent in E, regarded as a real linear space. Replace x by $x - y$ and y by $2y$ in (40) to get

$$f(x-y)f(2y)f(x+y) = f(g(2y)(x-y) + 2g(x-y)y),$$

or, with (*ii*),

$$f(x+y)f(x-y)f(y) = f(2g(y)x + (g(x-y) - 2g(y))y). \tag{42}$$

We may change y into $-y$ in (42) and, using (*ii*), we get

$$f(x+y)f(x-y)f(y) = f(2g(y)x + (2g(y) - g(x+y))y). \tag{43}$$

Now we write (41) in another way:

$$f(x+y) = f\left(\frac{g(y)x - g(x)y}{f(x)f(y)} + 2f(x)\frac{y}{f(y)}\right),$$

or

$$f(x+y) = f\left(\frac{2g(y)x - 2g(x)y}{f(x-y)f(y)}\frac{f(x-y)}{2f(x)} + 2f(x)\frac{y}{f(y)}\right). \tag{44}$$

But $f[(2g(y)x - 2g(x)y)/(f(x-y)f(y))] = f[2f(x)y/f(y)]$, because this equation is equivalent to $f(g(y)x - g(x)y) = f(x)f(y)f(x-y)$ which follows from (40) by changing y into $-y$. Applying (37) with $\gamma = f(x-y)/[2f(x)]$ to the right hand side of (44) yields

$$f(x+y) = f\left(\frac{2g(y)x - 2g(x)y}{f(x-y)f(y)} + f(x-y)\frac{y}{f(y)}\right)$$
$$= \frac{f(2g(y)x + (g(x-y) - 2g(x))y)}{f(x-y)f(y)},$$

or

$$f(x+y)f(x-y)f(y) = f(2g(y)x + (g(x-y) - 2g(x))y). \tag{45}$$

Changing y into $-y$ in (45) gives

$$f(x+y)f(x-y)f(y) = f(2g(y)x + (2g(x) - g(x+y))y). \tag{46}$$

Clearly, $2g(y)x$ and y are linearly independent vectors (in E considered as a real linear space). Comparing (42), (43), (45) and (46) and using Lemma 12, we see that of the four real numbers

$$A = g(x-y) - 2g(y), \qquad B = 2g(y) - g(x+y),$$
$$C = g(x-y) - 2g(x), \qquad D = 2g(x) - g(x+y),$$

at most two are distinct. We distinguish four cases.

(α) If $A = D$, then $g(x+y) + g(x-y) = 2g(x) + 2g(y)$.
(β) If $B = C$, then $g(x+y) + g(x-y) = 2g(x) + 2g(y)$.
(γ) If $A = B$ and $C = D$, then $g(x+y) + g(x-y) = 2g(x) + 2g(y)$.
(δ) If $A = C$ and $B = D$, then $g(x) = g(y)$ and we may distinguish two subcases here.

If $g(x+y) = g(x-y)$ then, in view of $g(x) = g(y)$ and of $g(x-y) \neq 0$ for $x \neq y$, we deduce from (42) (or (43)) that

$$f(x) = f(y) = f\left(\frac{2g(y)}{g(x-y)}x + \left(1 - \frac{2g(y)}{g(x-y)}\right)y\right).$$

Since x and y are linearly independent, Proposition 11 shows that we cannot have $2g(y) < g(x-y)$. But our conditions $g(x+y) = g(x-y)$ and $g(x) = g(y)$ are preserved if we replace x by $x+y$ and y by $x-y$. Therefore we also obtain the impossibility of $2g(x-y) < g(2y)$, that is, the impossibility of $g(x-y) < 2g(y)$. Therefore, we must have $g(x-y) = 2g(y)$. Thus $g(x+y) = g(x-y) = 2g(y) = 2g(x)$ and so

$$g(x+y) + g(x-y) = 2g(x) + 2g(y).$$

If $g(x+y) \neq g(x-y)$, then we may work with $X = x+y$ and $Y = x-y$ as we have done with x and y. But we have the hypothesis $g(X) \neq g(Y)$ which makes the case (δ), with X and Y in place of x and y, impossible. Therefore we have

$$g(X+Y) + g(X-Y) = 2g(X) + 2g(Y),$$

or

$$g(2x) + g(2y) = 2g\left(\frac{x+y}{2}\right) + 2g\left(\frac{x-y}{2}\right),$$

which gives

$$2g(x) + 2g(y) = g(x+y) + g(x-y).$$

We have thus established the validity of the functional equation of a quadratic functional in all possible cases, which concludes the proof of Proposition 14.

The *proof of the 'if' part of Theorem* 10 is now easy. Let $(E, \|\cdot\|)$ be a real or complex normed space and let $f(x) = \|x\|$. Suppose that (38) is true. Then $f: E \to \mathbb{R}$ is continuous and satisfies (*i*), (*ii*) and (*iii*). Proposition 14 shows that $g = f^2$ is a quadratic functional and the results of the first section of this chapter shows that the norm $\|\cdot\|$ comes from an inner product, which concludes the proof of Theorem 10.

Let $f:E\to\mathbb{R}$ be a function satisfying (40). Then $g=f^2$ satisfies

$$g(g(x)y+g(y)x)=g(x)g(y)g(x+y) \quad ((x,y)\in E^2). \tag{47}$$

It is rather surprising that under some restrictions on g, this functional equation has only quadratic functionals as solutions. We shall examine the case when $E=\mathbb{R}$, following Volkmann (1974).

Proposition 15. *Let $g:\mathbb{R}\to\mathbb{R}$ be a continuous function such that $g(0)=0$ and $g(x)>0$ for all $x\neq 0$. Suppose $\lim_{x\to\infty} g(x)=\infty$. Then g is a solution of the functional equation*

$$g(g(x)y+g(y)x)=g(x)g(y)g(x+y) \quad (x,y\in\mathbb{R}) \tag{48}$$

if, and only if, there exists a positive constant a such that

$$g(x)=ax^2 \quad (x\in\mathbb{R}).$$

Proof. Putting $y=-x$ into (48), and using $g(0)=0$, we get $g(g(-x)x-xg(x))=0$. But we have supposed that g is zero only at 0. Therefore $x(g(-x)-g(x))=0$ for all x in $\mathbb{R}$. Thus g is an even function.

We introduce two new functions F and h defined on $\mathbb{R}$ and on $\mathbb{R}^*$, respectively, by

$$F(x)=xg\left(\frac{x}{2}\right) \quad \text{and} \quad h(x)=\frac{g(x)}{x^2}.$$

With these two functions, (48) can be rewritten in the following manner (with x and y replaced by $x/2$):

$$h(F(x))=h(x) \quad (x\in\mathbb{R}^*). \tag{49}$$

We will first prove that h is constant outside a bounded interval. Because $\lim_{x\to\infty} g(x)=\infty$ and $g(0)=0$, using the definition of F, there is a largest nonnegative number X such that $F(X)=X$. Take any $x>X$ and notice that $F(x)>x$. Therefore there exists an x_1 such that $X\leqslant x_1\leqslant x$ and $F(x_1)=x$. In the same way, there exists x_2 such that $X\leqslant x_2\leqslant x_1$ and $F(x_2)=x_1$. We get, by induction, a nonincreasing sequence $\{x_n\}$, with $F(x_n)=x_{n-1}$ and $x_n\geqslant X$. This sequence necessarily has a limit x_∞. But, since g is a continuous function, so is F and thus $F(x_\infty)=x_\infty$. As X is the largest nonnegative number with this fixed point property, we obtain $x_\infty=X$. Using alternatingly (49) and the definition of the sequence $\{x_n\}$, we get

$$h(x)=h(F(x_1))=h(x_1)=h(F(x_2))=h(x_2)=\cdots$$
$$=h(x_n).$$

But h is a continuous function and the sequence $\{x_n\}$ converges to X. Thus

$$h(x)=h(X) \quad \text{for all } x\geqslant X.$$

Therefore

$$g(x)=h(X)x^2 \quad \text{for all } x\geqslant X. \tag{50}$$

Take an arbitrary positive Y. Let $y \geqslant Y > 0$. Noticing that $\lim_{x\to\infty}(g(x)y + g(y)x) = \infty$ and recalling that $\lim_{x\to\infty} g(x) = \infty$, we can find a (large) $x \geqslant X$, such that $g(x)y + g(y)x \geqslant X$ for all $y \geqslant Y$.

We may now apply (48) and (50) in order to get

$$h(X)[h(X)x^2y + g(y)x]^2 = h(X)x^2g(y)h(X)(x+y)^2.$$

It is possible to divide each term by $h(X)$ because $g(x) \neq 0$ when $x \neq 0$ and so $h(X) \neq 0$. We obtain for all $y \geqslant Y$

$$g(y)^2 - h(X)(x^2 + y^2)g(y) + h(X)^2x^2y^2 = 0.$$

This equation of the second degree in $g(y)$ must be true for all $y \geqslant Y$. The discriminant equals $h(X)^2(x^2 - y^2)^2$. Therefore we have two cases and the possible values of $g(y)$ are

$$g(y) = \frac{h(X)(x^2+y^2) + h(X)(x^2-y^2)}{2} \quad \text{and} \quad g(y) = \frac{h(X)(x^2+y^2) - h(X)(x^2-y^2)}{2},$$

that is, $h(X)x^2$ or $h(X)y^2$. The first value does not depend upon y. Since g is continuous, if, for some $y_0 > x$, $g(y_0) = h(X)x^2$ holds, then also $g(y) = h(X)x^2$ for all $y \geqslant x$. But this is impossible because, due to (49), $g(y) = h(X)y^2$ for $y > X$.

In the same way, if, for some $Y < y_0 < x$, $g(y_0) = h(X)x^2$, then $g(y) = h(X)x^2$ for all $Y \leqslant y \leqslant x$. The relation has to be true for all $Y > 0$. But this contradicts $\lim_{y\to 0} g(y) = g(0) = 0$. So, $g(y) = h(X)y^2$ for all $y \geqslant Y > 0$. Since $Y > 0$ was arbitrary and since $g(0) = 0$ and g is even, we can therefore deduce that $g(y) = h(X)y^2$ for all y in $\mathbb{R}$. This concludes the proof of Proposition 15.

Now the following can be proved (see Exercises 21, 22, 23).

Theorem 16. *A real or complex normed space $(E, \|\cdot\|)$ is an inner product space if, and only if, the function $g(x) = \|x\|^2$ satisfies the functional equation*

$$g(x)g(y)g(x+y) = g(xg(y) + yg(x)) \quad (x, y \in E). \tag{51}$$

Remark. From Theorem 16 it is also easy to give another proof of the 'if' part of Theorem 10: Let $(E, \|\cdot\|)$ be a real (or complex) normed space such that $\|\cdot\|$ satisfies (38) for the x, y in E with $\|x\| = \|y\|$ and for all $\gamma \in \mathbb{R}$. But we derived (47) from $g(x) = \|x\|^2$ in an inner product space by using only (38). Therefore g satisfies (51). Theorem 16 shows that $(E, \|\cdot\|)$ is then an inner product space. This concludes the proof of Theorem 10, using Theorem 16.

11.3 Orthogonal additivity

Let $(E, \langle , \rangle)$ be a real or complex inner product space of dimension at least 2. We define an orthogonality relation $\perp$ for two elements x and y in E, as usual, by

$$x \perp y \quad \text{if } \langle x, y \rangle = 0.$$

Our purpose is to solve the conditional Cauchy equation where the condition is $x \perp y$. Namely, we look for all $f: E \to \mathbb{R}$ such that

$$f(x+y) = f(x) + f(y) \quad \text{for all } x, y \text{ in } E \text{ with } x \perp y. \tag{52}$$

Such an f is called an *orthogonally additive* function on E.

Theorem 17. *Let $(E, \langle , \rangle)$ be a real inner product space of dimension at least 2. A continuous function $f: E \to \mathbb{R}$ is orthogonally additive if, and only if, there exists a real constant a and a continuous linear function $h: E \to \mathbb{R}$ such that*

$$f(x) = a\|x\|^2 + h(x) \quad (x \in E). \tag{53}$$

By linearity we mean, of course,

$$h(\lambda x + \mu y) = \lambda h(x) + \mu h(y) \quad \text{for all } \lambda, \mu \in \mathbb{R}, \quad x, y \in E.$$

Proof. Suppose first that f is of the form (53). The Pythagorean theorem shows that $\|x\|^2 + \|y\|^2 = \|x+y\|^2$ when $x \perp y$. Thus, for $x \perp y$,

$$\begin{aligned} f(x+y) &= a\|x+y\|^2 + h(x+y) \\ &= a\|x\|^2 + a\|y\|^2 + h(x) + h(y) \\ &= f(x) + f(y). \end{aligned}$$

Therefore f is orthogonally additive (and continuous).

Conversely, suppose that f is continuous and orthogonally additive. We decompose f into its even and odd parts,

$$f(x) = g(x) + h(x),$$

where $g(x) = (f(x) + f(-x))/2$ and $h(x) = (f(x) - f(-x))/2$. A simple computation shows that both g and h are continuous solutions of the conditional Cauchy equation (52). We intend to prove first that h is additive, then that g is a quadratic functional.

To prove the additivity of h, it is enough to show that h is additive on any two dimensional subspace H of E. Since E is an inner product space, in such a subspace H there exists a pair x, y of nonzero elements of H with $x \perp y$ and $(x+y) \perp (x-y)$. (Just choose $\|x\| = \|y\| = 1$ and $\langle x, y \rangle = 0$.) In fact, among the properties of an inner product space, our proof will use only the fact that, in each two dimensional subspace H, there exists a pair of nonzero elements x, y such that $x \perp y$ and $(x+y) \perp (x-y)$. We proceed to compute $h(2\lambda x)$ $(\lambda \in \mathbb{R})$:

$$h(2\lambda x) = h(\lambda(x+y) + \lambda(x-y))$$

by (52). Since $\lambda(x+y)\perp\lambda(x-y)$, $\lambda x\perp\lambda y$ and $\lambda x\perp-\lambda y$, we get

$$\begin{aligned}h(2\lambda x) &= h(\lambda(x+y))+h(\lambda(x-y))\\ &= h(\lambda x)+h(\lambda y)+h(\lambda x)+h(-\lambda y).\end{aligned}$$

But h is odd so

$$h(2\lambda x)=2h(\lambda x).$$

We also have $h(2\lambda y)=2h(\lambda y)$ because of the symmetric role played by x and y.

We prove by induction that $h(p\lambda x)=ph(\lambda x)$ and $h(p\lambda x)=ph(\lambda y)$ for all positive integers p. It is true for $p=1,2$. Suppose that this result is true for all integers p such that $0\leqslant p<n$. Then

$$h(\lambda nx+\lambda(n-2)y)=h(\lambda(n-1)(x+y)+\lambda(x-y)).$$

Since $\lambda(n-1)(x+y)\perp\lambda(x-y)$, by using the induction hypothesis we deduce that

$$\begin{aligned}h(\lambda nx+\lambda(n-2)y) &= h(\lambda(n-1)(x+y)+\lambda(x-y))\\ &= h(\lambda(n-1)x)+h(\lambda(n-1)y)+h(\lambda x)+h(-\lambda y)\\ &= (n-1)h(\lambda x)+h(\lambda x)+(n-1)h(\lambda y)-h(\lambda y),\end{aligned}$$

$$h(\lambda nx+\lambda(n-2)y)=nh(\lambda x)+(n-2)h(\lambda y). \tag{54}$$

But we may compute $h(\lambda nx+\lambda(n-2)y)$ in a different way using $\lambda nx\perp\lambda(n-2)y$:

$$h(\lambda nx+\lambda(n-2)y)=h(\lambda nx)+h(\lambda(n-2)y).$$

By the induction hypothesis, we obtain

$$h(\lambda nx+\lambda(n-2)y)=h(\lambda nx)+(n-2)h(\lambda y).$$

Comparing this with (54), we get

$$h(\lambda nx)=nh(\lambda x), \tag{55}$$

as asserted. Similarly,

$$h(\lambda ny)=nh(\lambda y).$$

We may take $\lambda=1/n$ in (55), so that

$$h\left(\frac{x}{n}\right)=\frac{1}{n}h(x),$$

and, using (55) once more, for all positive integers m,n, we get, as in Section 2.1 for Cauchy's equation,

$$h\left(\frac{m}{n}x\right)=\frac{m}{n}h(x).$$

Then the continuity of h on the space E (and therefore also on H) immediately yields that, for all λ in $\mathbb{R}$ and for the chosen x in H,

$$h(\lambda x)=\lambda h(x). \tag{56}$$

Similarly, we deduce that, for all λ in $\mathbb{R}$ and the chosen y in H,

$$h(\lambda y) = \lambda h(y). \tag{57}$$

We now compute $h(\lambda x + \mu y)$. By (52), (56) and (57), since $\lambda x \perp \mu y$, we obtain the linearity of h on H:

$$h(\lambda x + \mu y) = \lambda h(x) + \mu h(y).$$

We use the same technique for the even part g of f. As previously, let H be a two dimensional subspace of E and let x and y be nonzero elements of H such that $x \perp y$ and $(x+y) \perp (x-y)$. We write

$$\begin{aligned} g(2\lambda x) &= g(\lambda(x+y) + \lambda(x-y)) \\ &= g(\lambda(x+y)) + g(\lambda(x-y)) \\ &= g(\lambda x) + g(\lambda y) + g(\lambda x) + g(-\lambda y) \\ &= 2g(\lambda x) + 2g(\lambda y). \end{aligned}$$

Because of the symmetric roles of x and y, we also have

$$g(2\lambda y) = 2g(\lambda x) + 2g(\lambda y). \tag{58}$$

For all λ in $\mathbb{R}$, we obtain $g(2\lambda x) = g(2\lambda y)$ or, as λ is arbitrary, for x and y as chosen,

$$g(\lambda x) = g(\lambda y) \quad (\lambda \in \mathbb{R}). \tag{59}$$

Equations (58) and (59) give

$$g(2\lambda x) = 4g(\lambda x).$$

In the same way,

$$g(2\lambda y) = 4g(\lambda y).$$

We now prove by induction that

$$g(n\lambda x) = n^2 g(\lambda x)$$

and

$$g(n\lambda y) = n^2 g(\lambda y).$$

Suppose that, for all integers p with $1 \leqslant p < n$, we have

$$g(p\lambda x) = p^2 g(\lambda x) \quad \text{and} \quad g(p\lambda y) = p^2 g(\lambda y).$$

We compute $g(\lambda nx + \lambda(n-2)y)$ in two different ways

$$\begin{aligned} g(\lambda nx + \lambda(n-2)y) &= g(\lambda(n-1)(x+y) + \lambda(x-y)) \\ &= g(\lambda(n-1)(x+y)) + g(\lambda(x-y)) \\ &= g(\lambda(n-1)x) + g(\lambda(n-1)y) + g(\lambda x) + g(-\lambda y) \\ &= (n-1)^2 g(\lambda x) + (n-1)^2 g(\lambda y) + g(\lambda x) + g(\lambda y) \\ &= ((n-1)^2 + 1)g(\lambda x) + ((n-1)^2 + 1)g(\lambda y). \end{aligned}$$

Using (59),

$$g(\lambda nx + \lambda(n-2)y) = 2((n-1)^2 + 1)g(\lambda x). \tag{60}$$

But

$$g(\lambda nx + \lambda(n-2)y) = g(\lambda nx) + (n-2)^2 g(\lambda y),$$
$$g(\lambda nx + \lambda(n-2)y) = g(\lambda nx) + (n-2)^2 g(\lambda x). \tag{61}$$

Comparing (60) and (61) yields

$$g(\lambda nx) = [2((n-1)^2 + 1) - (n-2)^2]g(\lambda x),$$
$$g(\lambda nx) = n^2 g(\lambda x).$$

Similarly,

$$g(\lambda ny) = n^2 g(\lambda y),$$

which concludes our proof by induction.

From these two results we easily deduce that

$$g\left(\frac{m}{n}x\right) = \left(\frac{m}{n}\right)^2 g(x)$$

for all positive integers m and n. The continuity of g implies that

$$g(\lambda x) = \lambda^2 g(x) \quad (\lambda \in \bar{\mathbb{R}}_+)$$

and, as g is even,

$$g(\lambda x) = \lambda^2 g(x) \quad (\lambda \in \mathbb{R}). \tag{62}$$

Similarly, we get for y,

$$g(\lambda y) = \lambda^2 g(y) \quad (\lambda \in \mathbb{R}). \tag{63}$$

Now let λ, μ be real numbers and apply the orthogonal additivity using (59), (62) and (63):

$$g(\lambda x + \mu y) = g(\lambda x) + g(\mu y)$$
$$= \lambda^2 g(x) + \mu^2 g(y),$$

that is,

$$g(\lambda x + \mu y) = (\lambda^2 + \mu^2)g(x). \tag{64}$$

Let x' and y' be arbitrary elements of the two dimensional subspace H of E. Since x and y generate the subspace H, there exist real constants a, b, c and d such that

$$x' = ax + by,$$
$$y' = cx + dy.$$

Therefore, using (64), we get

$$g(x') + g(y') = (a^2 + b^2)g(x) + (c^2 + d^2)g(y).$$

But from (59),

$$g(x') + g(y') = (a^2 + b^2 + c^2 + d^2)g(x). \tag{65}$$

In a similar way, from (59) and (64),

$$g(x' + y') + g(x' - y') = [(a+c)^2 + (a-c)^2 + (b+d)^2 + (b-d)^2]g(x).$$

Thus

$$g(x'+y')+g(x'-y')=2(a^2+b^2+c^2+d^2)g(x). \tag{66}$$

Comparing (65) and (66), we see that g is a quadratic functional on H

$$g(x'+y')+g(x'-y')=2g(x')+2g(y') \quad (x',y'\in H).$$

Since g is continuous, there exists a bilinear $S:H\to\mathbb{R}$ such that

$$g(x)=S(x,x).$$

Since g is orthogonally additive, we see that $x\perp y$ implies $S(x,y)=0$. We now use the fact that H is an inner product space of dimension two. There exists an orthonormal basis (e_1,e_2) for the scalar product $\langle,\rangle$. Then, with

$$x=x_1e_1+x_2e_2$$

and

$$y=y_1e_1+y_2e_2,$$

we get

$$\langle x,y\rangle=x_1y_1+x_2y_2 \quad \text{and} \quad S(x,y)=x_1y_1S(e_1,e_1)+x_2y_2S(e_2,e_2),$$

since $S(e_1,e_2)=0$. But, since $\langle x,y\rangle=0$ implies $S(x,y)=0$, we must have $S(e_1,e_1)=S(e_2,e_2)=a\in\mathbb{R}$. Thus

$$g(x)=S(x,x)=a\|x\|^2.$$

This concludes the proof of Theorem 17.

In fact, it is possible to find all orthogonally additive functions on an inner product space without supposing continuity (see Exercise 25).

Conversely, orthogonally additive functions may be used to characterize inner product spaces.

Let E be a real normed space of dimension at least 2. We define an orthogonality relation $\perp$ for two elements x and y in E as follows:

$$x\perp y \quad \text{if } \|x+\lambda y\|\geqslant\|x\| \quad \text{for all } \lambda\in\mathbb{R}.$$

If $x\perp y$, then, for all μ,ν in $\mathbb{R}$, $\mu x\perp\nu y$. But it is difficult to say more about this relation $\perp$ in general. For instance, $x\perp y$ is not, in general, a symmetric relation. It has even been proved that, when the dimension of E is at least three, $x\perp y$ is a symmetric relation if and only if E is a real inner product space (James 1947). Therefore we will make an additional assumption concerning E. We say that E is *orthogonally rich* if it has at least dimension 3 and if, for any two dimensional subspace H of E, there exists a pair x,y of nonzero elements of H for which $x\perp y$ and $(x+y)\perp(x-y)$ (as above). Then, the following characterization can be established. We define the condition of orthogonal additivity as

$$f(x+y)=f(x)+f(y) \quad \text{if } x\perp y. \tag{67}$$

Theorem 18. *Let E be a real normed space which is orthogonally rich. Then*

E is an inner product space if and only if the condition of orthogonal additivity is not C-addundant.

Proof. If E is a inner product space, define $f(x)=\|x\|^2$. Then f is orthogonally additive (for the relation $x\perp y$ defined by $\langle x,y\rangle=0$), but is not an additive function.

Conversely, let $f:E\to\mathbb{R}$ be an orthogonally additive continuous function, in an orthogonally rich real normed space, which is not additive. Our proof of Theorem 17 has been organized so that it yields the unique decomposition

$$f=g+h,$$

where h is a linear function and g a quadratic functional. Because f is not additive, g is not identically zero. Therefore, using (64), we see that $g(x)\neq 0$ for all $x\neq 0$. So, possibly after changing g into $-g$, we have a continuous quadratic functional g on E which is strictly positive on E^*. Proposition 6 shows that

$$g(x)=S(x,x),$$

where S is a symmetric bilinear function on E. The strict positivity of g shows that $S(x,y)$ is an inner product on E. We now show that $x\perp y$ if and only if $S(x,y)=0$.

If $x\perp y$, then $f(x+y)=f(x)+f(y)$ (orthogonal additivity) and therefore $g(x+y)=g(x)+g(y)$, since g is the even part of f. But $g(x)=S(x,x)$. Thus $S(x,y)=0$.

Conversely, suppose that $S(x,y)=0$ (with $x,y\neq 0$) and let H be the two dimensional subspace of E generated by x and y. Because E is orthogonally rich, there exists y' in H^* such that $x\perp y'$. Therefore, by what we already proved, $S(x,y')=0$. But H is of dimension 2 and so y' is a scalar multiple of y. Thus $x\perp y=0$ as well.

The orthogonal relation defined by $S(x,y)=0$ and $x\perp y$ coincide. The first one is symmetric, and so is the second. And we already saw (James 1947) that in a real normed space of dimension 3, the symmetry of $x\perp y$ implies that the norm comes from an inner product. This concludes the proof of Theorem 18.

In fact, more can be proved. Theorem 18 remains true if we drop the continuity condition and consider any real normed space of dimension not less than 2 (see Exercise 37, and Szabó 1986).

11.4 An application to gas dynamics

We now proceed to an application to gas dynamics (cf. also Truesdell–Muncaster 1980, pp. 71–3, 88–9). An ideal gas consists of molecules occupying no volume, with a fixed and equal mass and without interactions except collisions in which molecules may exchange energy according to the usual laws of mechanics. Molecules in an ideal gas are independent in the sense that the probability of a molecule occupying a given position with a given energy is independent of the position and energy state of other molecules. We are looking for the density of the probability that a molecule has velocity v. In other words, if v_x, v_y, and v_z are the components of the velocity in some orthogonal basis, we look for $H(v)$ such that $H(v)\,dv_x dv_y dv_z$ is the probability that the components of the velocity of the molecule are between v_x and $v_x + dv_x$, v_y and $v_y + dv_y$, v_z and $v_z + dv_z$. For physical reasons, it is generally assumed that H is a positive even function and that H is continuous (but see Exercise 26).

In a collision between two molecules, the initial velocities v and w change into velocities v' and w' in such a way that

$$v' + w' = v + w \tag{68}$$

(conservation of momentum) and

$$\|v'\|^2 + \|w'\|^2 = \|v\|^2 + \|w\|^2 \tag{69}$$

(conservation of energy), where $\|\cdot\|$ denotes the usual euclidean norm on $E = \mathbb{R}^3$. We introduce the usual scalar product $\langle , \rangle$ so that

$$\langle v, v \rangle = \|v\|^2.$$

For the probability H we must have

$$H(v)H(w) = H(v')H(w').$$

We naturally assume that $0 < H(v) < 1$. Taking the logarithm of both sides, we get

$$h(v) + h(w) = h(v') + h(w'),$$

where $h(v) = \log H(v)$. Using (68), we get

$$h(v) + h(w) = h(v') + h(v + w - v').$$

From (68) and (69) we deduce that

$$\langle v - v', w - v' \rangle = 0.$$

We now fix $v' \in E$ and we define $V = v - v'$ and $W = w - v'$. Then

$$h(V + v') + h(W + v') = h(v') + h(V + W + v').$$

We define $f: E \to \mathbb{R}$ by

$$f(V) = h(V + v') - h(v').$$

Then we obtain the conditional Cauchy equation for f

$$f(V + W) = f(V) + f(W) \quad \text{if } \langle V, W \rangle = 0 \quad (V \in E, W \in E). \tag{70}$$

In other words, f is *orthogonally additive on E*. Theorem 17 shows that

$$f(V) = \beta \|V\|^2 + l(V),$$

where β is a real constant and $l:E \to \mathbb{R}$ is a (continuous) linear function, i.e. $l(V) = \langle V, t \rangle$ for some $t \in E$.

Going back to H, in the case when $v' = 0$, we have that, for some real constant α and for all $v \in E$,

$$H(v) = \exp(\alpha + \beta \| v \|^2 + \langle v, t \rangle).$$

But, since H is a probability, there are further restraints. In particular,

$$0 < H(v) < 1 \quad (v \in E)$$

yields $\beta \leqslant 0$. Moreover,

$$\iiint_E H(v) dv_x dv_y dv_z = 1.$$

The case $\beta = 0$ is impossible because it implies $t = 0$ and H cannot be a nonzero constant. Thus $\beta = -a$ with $a > 0$. Using the fact that H is even, we deduce that $t = 0$.

Thus we have derived the so-called Maxwell–Boltzmann distribution law for velocities in an ideal gas:

$$H(v) = A \exp(-a \| v \|^2). \tag{71}$$

Thermodynamics shows how to relate a to the absolute temperature T, to the mass m of a molecule and to the Boltzmann constant k via

$$a = \frac{m}{2kT}.$$

It should be noted that Maxwell (1860) used a different procedure to deduce the law (71). However, he too solved a functional equation. He too supposed H to be even and considered the density of the probability that the x-component of the speed of a molecule is between v_x and $v_x + dv_x$. For this probability, he wrote

$$F(v_x^2) dv_x.$$

Maxwell then made the strong assumption that the distribution laws along the three axis are independent. Thus

$$H(v) = F(v_x^2) F(v_y^2) F(v_z^2).$$

Moreover, he assumed complete isotropy of the gas, that is,

$$H(v) = G(\| v \|^2).$$

This led him to the functional equation

$$F(v_x^2) F(v_y^2) F(v_z^2) = G(v_x^2 + v_y^2 + v_z^2), \tag{72}$$

which he solved under the assumption that F is differentiable. However, it is easy to show that Lebesgue measurability of F is sufficient (Banach–Ruziewicz 1922; see also Exercise 24). Indeed, suppose that F and G are strictly positive. Take the logarithm of both sides of (72) and define the new functions f and g by

$$f(v_x^2) = \log F(v_x^2) \quad \text{and} \quad g(v_x^2) = \log G(v_x^2).$$

Since F is Lebesgue measurable, so is G (equation (72)), f and g. Then

$$f(v_x^2) + f(v_y^2) + f(v_z^2) = g(v_x^2 + v_y^2 + v_z^2). \tag{73}$$

We proceed similarly as in Section 4.3 for Pexider equations. We put $v_x = v_y = v_z = 0$ into (73) in order to get $3f(0) = g(0)$. Now take $v_y = v_z = 0$ in (73) and use $f(0) = g(0)/3$ to get

$$f(v_x^2) = g(v_x^2) - \frac{2g(0)}{3}. \tag{74}$$

Putting (74) into (73), we get

$$g(v_x^2) + g(v_y^2) + g(v_z^2) - 2g(0) = g(v_x^2 + v_y^2 + v_z^2).$$

With $k(v_x^2) := g(v_x^2) - g(0)$, this equation becomes

$$k(v_x^2) + k(v_y^2) + k(v_z^2) = k(v_x^2 + v_y^2 + v_z^2).$$

With $v_z = 0$, we obtain the additivity of k on $\bar{\mathbb{R}}_+$:

$$k(v_x^2) + k(v_y^2) = k(v_x^2 + v_y^2),$$

since $k(0) = 0$. By Corollary 2.9 there exists a real constant b such that

$$h(v_x^2) = bv_x^2.$$

Going back to F, we obtain with the convenient constants $A^{1/3}$ and a,

$$F(v_x^2) = A^{1/3} \exp(-av_x^2).$$

For H, the result is

$$\begin{aligned} H(v) = F(v_x^2)F(v_y^2)F(v_z^2) &= A \exp(-a(v_x^2 + v_y^2 + v_z^2)) \\ &= A \exp(-a\|v\|^2). \end{aligned}$$

This is (71) again, the Maxwell–Boltzmann distribution law of velocities.

Exercises and further results

(For Exercises 1–3, see Aczél 1965c.)

1. Find all continuous solutions $f: \mathbb{R} \to \mathbb{R}$ of the functional equation

$$f(x+y) + f(x-y) = 2f(x) + 2f(y) \quad (x, y \in \mathbb{R}).$$

2. Let G be an abelian group and H an abelian group divisible by 2 (every equation of the form $2y = a \in H$ has a unique solution $y \in H$). Show that the general solution $f: G \to H$ of

$$f(x+y) + f(x-y) = 2f(x) + 2f(y)$$

$(x, y \in G)$ is of the form $f(x) = S(x, x)$, where $S: G^2 \to H$ is an arbitrary symmetric biadditive function. Compare the result to Proposition 1.

3. Determine the general real solution $f:\mathbb{R}\to\mathbb{R}$ of the functional equation in the previous two exercises.
4. (Davison 1980.) Let F be a commutative ring and M a module over F. A function $f:M\to F$ is a quadratic form when $f(x)=B(x,x)$ for all x in M where B is a bilinear form $B:M^2\to F$. If F is a field with more than two elements and M a projective F-module (see, e.g., Lambek 1976, p. 82) show that the following functional equation characterizes quadratic forms:
$$f(ax+by)=af(x)+bf(y)-abf(x-y),$$
for all x,y in M and all a,b in F.
5. Let E be a real normed space of dimension greater than 2 and M a fixed one dimensional subspace of E, such that each two dimensional subspace L with $M\subset L\subset E$ is an inner product space. Prove or disprove that E has to be an inner product space too.
6. Let E be a real inner product space. Let $f:E\to E$ be additive. Suppose $x\mapsto\langle f(x),x\rangle$ is a continuous function. (This condition is, of course, weaker than the continuity of f.) Prove or disprove that f is a linear operator.
7. Let $d:\mathbb{R}\to\mathbb{R}$ be a nontrivial derivation (see Exercise 2.11). Show that $g:\mathbb{R}^2\to\mathbb{R}$ defined by
$$g(x,y)=ax^2+2bxy+cy^2+xd(y)-yd(x)$$
(a,b and c are real constants) is a real quadratic functional on $\mathbb{R}^2$ satisfying
$$g(\lambda x,\lambda y)=\lambda^2 g(x,y)\quad(\lambda\in\mathbb{R},x,y\in\mathbb{R}).$$
Use this example to prove Proposition 6.
8. Let F be the field of quaternions and let $f:F\to F$ be an additive function. Prove that f satisfies
$$f(\lambda)=-|\lambda|^2 f\left(\frac{1}{\lambda}\right)\quad(\lambda\in F^*=F\backslash\{0\})$$
if and only if
$$f(\lambda_1+\mathbf{i}\lambda_2+\mathbf{j}\lambda_3+\mathbf{k}\lambda_4)=\lambda_2 f(\mathbf{i})+\lambda_3 f(\mathbf{j})+\lambda_4 f(\mathbf{k})$$
where $\lambda_1,\lambda_2,\lambda_3,\lambda_4\in\mathbb{R}$.
9. Let F be a field. Show that derivations (see Exercise 2.11) are the only additive functions $f:F\to F$ such that
$$f(\lambda)=-\lambda^2 f\left(\frac{1}{\lambda}\right)\quad(\lambda\in F^*).$$
10. (Kurepa 1965c.) Let $g:E\to\bar{\mathbb{R}}_+$ be a quadratic fuctional on a real linear space E. Prove that $g(x)=f(x)^2$ where $f:E\to\mathbb{R}$ is additive if and only if
$$(g(x+y)-g(x-y))^2=16g(x)g(y)\quad(x,y\in E).$$

(For Exercises 11, 12, see Grząślewicz 1979a.)

11. Let E be a real or complex linear space and $g:E\to\mathbb{R}$ a function satisfying

(i) $\qquad (g(x+y)-g(x-y))^2=16g(x)g(y)\quad(x,y\in E),$

(ii) $\qquad g(x)\geqslant 0 \quad (x\in E).$

Show that g is a quadratic functional on E and use Exercise 10 to find its general form.

12. Show that a solution $g:\mathbb{R}\to\mathbb{R}$ of the functional equation

$$(g(x+y)-g(x-y))^2 = 16g(x)g(y) \quad (x, y\in\mathbb{R}),$$

such that

$$g(x) = x^4 g\left(\frac{1}{x}\right) \quad (x\in\mathbb{R}^*)$$

and

$$g(x) \geqslant 0 \qquad (x\in\mathbb{R}),$$

is of one of the following forms:

$$g(x) = g(1)x^2 \quad (x\in\mathbb{R}),$$

or

$$g(x) = d(x)^2 \quad (x\in\mathbb{R}).$$

Here $d:\mathbb{R}\to\mathbb{R}$ is a derivation (see Exercise 2.11).

13. (Rätz 1976*a*.) Let E be a real normed space and $g:E\to\mathbb{R}$ a quadratic functional. Suppose that, for some positive constants A, B,

(i) $|g(x)-g(y)| \leqslant A\|x\|\,\|y\|$ for all $x, y\in E$ with $\|x\| = \|y\|$,

and

(ii) $|g(x)| \leqslant B$ for all $x\in F$ where F is a given segment of E (that is, $F = \{z\,|\,z\in E,\ z = \lambda x_0 + (1-\lambda)y_0,\ 0\leqslant\lambda\leqslant 1,\ x_0\neq y_0,\ x_0, y_0\in E\}$).

Show that $g(x) = S(x,x)$ where S is a symmetric bilinear mapping from E^2 into $\mathbb{R}$.

14*. (Baker 1968.) Let E be a complex linear space and $g:E\to\mathbb{C}$ a quadratic functional. Suppose that, for all $\lambda\in\mathbb{C}$ and all $x\in E$, we have

$$|f(\lambda x)| = |\lambda^2 f(x)|.$$

Show that one of the following holds for all $\lambda\in\mathbb{C}$ and all $x\in E$:

(i) $|f(\lambda x)| = \lambda^2 f(x)$.
(ii) $f(\lambda x) = |\lambda|^2 f(x)$,
or
(iii) $f(\lambda x) = \bar{\lambda}^2 f(x)$.

15. Let $a(x, y)$ be the coefficient of λ^2 in equation (31). Prove that

(i) $g(x+y) + g(x-y) = 2a(x,y) + 2g(x)$,
(ii) $a(\lambda x, y) = a(x, y) \quad (\lambda\in F^*)$,
and
(iii) $a(0, y) = g(y)$.

16. (Gleason 1966.) Prove that Theorem 8 remains true when F is a commutative field which contains more than three elements and has characteristic different from 2.

17. Let E be a real linear space. Show that $g:E\to\mathbb{R}$ is a solution of the functional equation (27) for all $\lambda\in\mathbb{R}$, x and y in E if, and only if, g is of the form

$$g(x) = \alpha + \beta(x) + \gamma(x),$$

where $\alpha\in\mathbb{R}$, $\beta:E\to\mathbb{R}$ is linear and $\gamma:E\to\mathbb{R}$ is a quadratic fuctional such that $\gamma(x)=S(x,x)$ where $S:E^2\to\mathbb{R}$ is a symmetric bilinear function.

(For Exercises 18, 19, see Fischer–Mokanski 1981.)

18. Let E be a linear space over a commutative field of characteristic different from 2. Prove or disprove that, if $g:E\to F$ satisfies

$$g(x+\lambda y)+g(\lambda x-y)=g(\lambda x)+g(\lambda y)+g(x)+g(y)$$

for all $x,y\in E$, $A\in F$, then there exists a biadditive and symmetric $S:E^2\to F$ such that

$$g(x)=S(x,x)$$

and

$$S(\lambda x,y)=S(x,\lambda y)\quad(\lambda\in F;x,y\in E).$$

19. Prove or disprove that $g:\mathbb{R}\to\mathbb{R}$ satisfies

$$g(x+\lambda y)+g(\lambda x-y)=g(\lambda x)+g(\lambda y)+g(x)+g(y)$$

for all $x,y,\lambda\in\mathbb{R}$ if and only if $g(x)=f(x^2)$ where f is an additive function.

20. (Röhmel 1977.) Let E be a linear space over a commutative field of characteristic different from 2. Show that, if $g:E\to F$ satisfies

$$g(x+\lambda y)+g(x-\lambda y)=(1+\lambda^2)(g(x)+g(y))$$

for all $x,y\in E$, $\lambda\in F$, then there exists a bilinear symmetric $S:E^2\to F$ such that $g(x)=S(x,x)$ for all x in E.

(For Exercises 21 and 22, see Volkmann 1974.)

21*. Prove the following result. Let $g:\mathbb{R}^2\to\mathbb{R}$ be such that g is continuous, $g(0)=0$, $g(x)>0$ for all nonzero x in $\mathbb{R}^2$ and $\lim_{x\to\infty}g(x)=\infty$. If g satisfies the functional equation

$$g(g(x)y+g(y)x)=g(x)g(y)g(x+y)\quad(x,y\in\mathbb{R}^2),$$

then g is a quadratic functional in $\mathbb{R}^2$.

22. Let E be a real or complex normed space $(E,\|\cdot\|)$. Deduce from Exercise 21 that every continuous $g:E\to\mathbb{R}$ satisfying $g(0)=0$, $g(x)>0$ $(x\in E^*=E\backslash\{0\})$, $\lim_{\|x\|\to\infty}g(x)=0$, and

$$g(g(x)y+yg(x))=g(x)g(y)g(x+y)\quad(x,y\in E),$$

is a quadratic functional over E.

23. Using the result of Exercise 22, prove Theorem 16.

24. (Banach–Ruziewicz 1922.) Show that all solutions $F:\mathbb{R}\to\mathbb{R}$ and $G:\mathbb{R}\to\mathbb{R}$ of

$$F(x)F(y)F(z)=G(x^2+y^2+z^2)\quad(x,y,z\in\mathbb{R}),$$

which are not identically zero on $\mathbb{R}^*$, are of the form

$$F(x)=A\exp(a(x^2)$$

and

$$G(x)\doteq A^3e^{a(x)}$$

where $a:\mathbb{R}\to\mathbb{R}$ is an additive function and $A\in\mathbb{R}^*$.

25*. Let $(E,\langle\cdot,\cdot\rangle)$ be a real inner product space. Show that $f:E\to\mathbb{R}$ is orthogonally additive if and only if there exist additive functions $a:\mathbb{R}\to\mathbb{R}$ and $h:E\to\mathbb{R}$ such that

$$f(x)=a(\|x\|^2)+h(x)\quad (x\in E).$$

26. Using Exercise 25, show that, if H is even, it is enough to suppose that H is Lebesgue integrable in order to obtain the form (71) of Maxwell–Boltzmann for the law of distribution.
27. (Hyers 1941.) Let $f:\mathbb{R}\to\mathbb{R}$ be a function satisfying

$$|f(x+y)-f(x)-f(y)|<\varepsilon\quad (x,y\in\mathbb{R})$$

for some $\varepsilon>0$. Show that there exists an additive $g:\mathbb{R}\to\mathbb{R}$ such that

$$|g(x)-f(x)|\leqslant\varepsilon\quad (x\in\mathbb{R}).$$

(For Exercises 28 and 29, see Baker 1980*b*.)

28. Let $f:G\to\mathbb{C}$ where G is a semigroup. Suppose that $|f(xy)-f(x)f(y)|\leqslant\varepsilon$ for all x,y in G for some $\varepsilon>0$. Show that either $\|f\|$ is bounded over G or

$$f(xy)=f(x)f(y)\quad (x,y\in G).$$

29. Let $f:G\to\mathbb{C}$ where G is an abelian group. Suppose that $|f(x+y)+f(x-y)-2f(x)f(y)|\leqslant\varepsilon$ for all x,y in G. Show that either $f(x)\leqslant(1+(1+2\varepsilon)^{1/2})/2$ for all $x\in G$ or f satisfies d'Alembert's equation.
30. (Cholewa 1983.) Let $(G,+)$ be an abelian group divisible by 2 and let f be a complex-valued function on G. Suppose that

$$|f(x+y)f(x-y)-f(x)^2+f(y)^2|\leqslant\varepsilon$$

for an $\varepsilon>0$ and for all $x,y\in G$. Then f is either bounded or satisfies the sine functional equation (cf. (13.53))

$$f(x+y)f(x-y)=f(x)^2-f(y)^2\quad (x,y\in G).$$

31. Let $\mathscr{A}$ be a complex Banach algebra with a unit $\mathbf{1}$. For a given element x in $\mathscr{A}$ we define its resolvent $\rho(x)$ as the set of all complex numbers λ for which $x-\lambda\mathbf{1}$ has an inverse in $\mathscr{A}$. Define $f:\rho(x)\to\mathscr{A}$ by $f(\lambda)=(x-\lambda\mathbf{1})^{-1}$. Show that f is an analytic function on $\rho(x)$ and that $\rho(x)$ is an open unbounded proper subset of $\mathbb{C}$. Prove that f satisfies the functional equation

$$f(\lambda)-f(\mu)=(\lambda-\mu)f(\lambda)f(\mu).$$

32. We keep the notations of the previous exercise. Let $F(x)$ be the family of all complex valued functions which are analytic on some neighbourhood of the spectrum $\sigma(x)$, that is, on the complement of the resolvent $\rho(x)$. Let U be an open subset of $\mathbb{C}$ whose boundary ∂U consists of a finite number of rectifiable Jordan curves. We assume that ∂U is oriented in such a way that the index is 1 inside U and 0 outside $U\cup\partial U$. We suppose that $U\cup\partial U$ is contained in the domain of analyticity of some f in $F(x)$. We then define

$$f(x)=-\frac{1}{2i\pi}\int_{\partial U}f(\lambda)(x-\lambda I)^{-1}d\lambda.$$

Show that $f(x)$ does not depend on the subset U chosen to define $f(x)$. Using the functional equation in Exercise 31, prove that

$$fg(x)=f(x)g(x).$$

33. Let E be a linear space over a field F and let f be a nontrivial valuation over F (cf. Section 10.6). We say that (F,f) is complete if (F,d) with the distance $d(\lambda,\mu)=f(\lambda-\mu)$ is a complete metric space. We generalize the definition of a norm $\|\cdot\|_f$ on E by requiring that $\|\cdot\|_f$ be a mapping from E into $\bar{\mathbb{R}}_+$ such that

$$\|x\|_f=0 \quad \text{if and only if } x=0,$$
$$\|\lambda x\|_f=f(\lambda)\|x\|_f \quad (\lambda\in F, x\in E),$$
$$\|x+y\|_f\leqslant\|x\|_f+\|y\|_f \quad (x,y\in E).$$

We again say that $(E,\|\cdot\|_f)$ is complete if (E,D) with the distance $D(x,y)=\|x-y\|_f$ is a complete metric space.

If $E\neq\{0\}$ is finite dimensional over F, prove or disprove that (F,f) is complete if and only if $(E,\|\cdot\|_f)$ is complete.

34. Generalize Propositions 11, 14 and Lemmata 12, 13 to the case where E is a linear space over a field F with a nontrivial valuation f (cf. Exercise 33).

35. Let $E\neq\{0\}$ be a finite dimensional linear space over a field F and f be a nontrivial valuation on F. Two norms $\|\cdot\|_f^{(1)}$ and $\|\cdot\|_f^{(2)}$ on E are equivalent when there exist positive constants α,β such that

$$\alpha\|x\|_f^{(1)}\leqslant\|x\|_f^{(2)}\leqslant\beta\|x\|_f^{(1)} \quad (x\in E).$$

Prove or disprove that all norms on E are equivalent.

36. Prove the uniqueness of S in Proposition 1.

37. (Sundaresan 1972.) Prove that a real normed vector space of dimension at least 3 is orthogonally rich (as defined just before Theorem 18).

(For Exercises 38–42, see Rätz 1985.)

38. Let X be a real vector space of dimension at least 2 and $\perp$ a binary relation. $(X,\perp)$ is called an *orthogonality space* when

    (i) $x\perp 0$ and $0\perp x$ for every x in X.
    (ii) For nonzero x,y in X, $x\perp y$ implies that x and y are linearly independent.
    (iii) $x\perp y$ implies $\alpha x\perp\beta y$ for all α, $\beta\in\mathbb{R}$.
    (iv) For each 2 dimensional subspace P of X and for all $x\in P$, $\lambda\in\mathbb{R}_+$, there exists an $y\in P$ such that $x\perp y$ and $x+y\perp\lambda x-y$.

    Show that the orthogonality relation, defined before Theorem 18 by $x\perp y$ if $\|x+\lambda y\|\geqslant\|x\|$ for all λ in $\mathbb{R}$, is such that $(X,\perp)$ is an orthogonality space when X is a real normed vector space.

39*. Using the technique of the proof for Theorem 17, extend Theorem 17 to the case where $(E,\perp)$ is an orthogonality space (cf. Exercise 38).

40. Let $(E,\|\cdot\|)$ be a real inner product space and $f:E\to\mathbb{R}$ an orthogonally additive function, that is, $f(x+y)=f(x)+f(y)$ if $x\perp y$ $(x,y\in E)$. Suppose $f(x)\geqslant 0$ for all x in E. Show that $f(x)=c\|x\|^2$ for some c in $\bar{\mathbb{R}}_+$ and all x in E.

41. Let $(E,\|\cdot\|)$ be a real inner product space and $f:E\to\mathbb{R}$ an orthogonally additive function. Suppose $|f(x)|\leqslant c\|x\|$ for some $c\geqslant 0$, and all x in E. Show that f is a continuous linear form of E.

42. Let $(E,\|\cdot\|)$ be a real inner product space. Prove that E is complete if and only if, for every orthogonally additive function $f:E\to\mathbb{R}$, there exists an x_0 in E such that $f(x_0)\leqslant f(x)$ for all x in E.

43. Let E be a real vector space and $f:E\to\mathbb{R}$ an odd function such that $f(x+y)=f(x)+f(y)$ whenever x and y are linearly independent. Does it follow that f is additive?
44. (Cholewa 1984*a*.) Let G be an abelian group and E be a Banach space. Let $f:G\to E$ such that
$$\|f(x+y)+f(x-y)-2f(x)-2f(y)\|\leqslant\delta$$
for all $x, y\in G$ and some $\delta\geqslant 0$. Show that there exists a function $g:G\to E$ such that
$$g(x+y)+g(x-y)=2g(x)+2g(y)$$
(g is a *quadratic functional*) and
$$\|f(x)-g(x)\|\leqslant\frac{\delta}{2}\quad(x\in G).$$
45. Let E be a real linear space. Suppose that $f:E\to\mathbb{R}$, is continuous and satisfies
$$f(\lambda x)=\lambda^2 f(x)\quad\text{for all } x\in E,\quad \lambda\in\mathbb{R}\tag{1}$$
and, for some $\mu\in]0,1[$, depending on x and y,
$$f(\mu x+(1-\mu)y)-\mu f(x)-(1-\mu)f(y)=\mu(1-\mu)f(x-y).\tag{2}$$
Prove or disprove that f is a quadratic functional (cf. Exercise 44).
46. (Heuvers 1985.) Let E be a real linear space and $f:E\to\mathbb{R}$. Suppose that
$$f(x+\lambda y)+f(\lambda x-y)=f(\lambda x)+f(\lambda y)+f(x)f(y)$$
for all x, y in E and all λ in $\mathbb{R}$. Show that
$$g(x,y)=\frac{f(x+y)-f(x-y)}{y}$$
is symmetric, biadditive, and satisfies $g(\lambda x,y)=g(x,\lambda y)$ for all x, y in E and all λ in $\mathbb{R}$ and
$$f(x)=g(x,x).$$

12

Some related equations and systems of equations. Applications to combinatorics and Markov processes

In this chapter we will see an example of the fact that essentially the same system of functional equations has several applications in different fields of mathematics.

In (5.29) we encountered an infinite system of functional equations for infinitely many unknown functions $(H_2, H_3, \ldots)$. Another such system arises in probability theory, more exactly Markov processes.

The problem is simply the following (Jánossy–Rényi–Aczél 1950; Redheffer 1953; Aczél 1966*c*, pp. 111–16, 1977*a*; Schmidt 1974, etc., cf. Hadwiger 1945). Determine the probability $p_k(t)$ that an event happens k times $(k = 0, 1, 2, \ldots)$ during a time interval of length t (>0) under the following two conditions. *Homogeneity*, which is implicit in the notation $p_k(t)$, means that the probability depends only upon the length t of the time interval, not upon its endpoints; and the *Markov condition* means that the numbers of events, happening in two time intervals which have at most one point in common, are independent. The p_k $(k = 0, 1, 2, \ldots)$ being probabilities of a complete system, we have, of course,

$$\sum_{k=0}^{\infty} p_k(t) = 1 \quad (t \in \mathbb{R}_+) \tag{1}$$

and

$$0 \leqslant p_k(t) \leqslant 1 \quad (k = 0, 1, 2, \ldots; t \in \mathbb{R}_+). \tag{2}$$

The Markov condition, on the other hand, implies

$$p_k(t+u) = \sum_{j=0}^{k} p_j(t) p_{k-j}(u) \quad (k = 0, 1, 2, \ldots, t, u \in \mathbb{R}_+). \tag{3}$$

because an event can then happen exactly k times during the time interval $t+u$ if it happens j times during t and $k-j$ times during u and, as j runs through $0, 1, 2, \ldots, k$, this exhausts all possibilities. This (remember the independence in the Markov condition) means exactly that (3) is satisfied.

We will take (1) and (2) into consideration later, but first we concentrate upon the system (3) of functional equations, mentioning also a few problems from combinatorics which lead to similar equations (there are others in geometry, etc., see Aczél–Vrănceanu 1972).

In combinatorics, Rota and Mullin (1969) have introduced polynomials P_k (in one variable and of kth degree, $k = 0, 1, 2, \ldots$) of 'binomial type', that is, satisfying

$$P_k(t+u) = \sum_{j=0}^{k} \binom{k}{j} P_j(t) P_{k-j}(u) \quad (k = 0, 1, 2, \ldots; t, u \in \mathbb{R}). \tag{4}$$

(Many polynomials, important in combinatorics, are of the binomial type, see, e.g., Rota–Mullin 1969; Brown 1973.) We will determine, more generally, *all functions* P_k $(k = 0, 1, 2, \ldots)$ *which satisfy* (4) and, in particular, all solutions *bounded on a set of positive measure*. If we introduce

$$p_k = P_k/k! \quad (k = 0, 1, 2, \ldots), \tag{5}$$

it is clear that these p_k satisfy (3) for all real t, u (not just for positive ones). Also in combinatorics the *Bell polynomials* are of importance (Bell 1934; Riordan 1946, 1958). The kth Bell polynomial is a polynomial of kth degree in k variables. They satisfy a system of equations similar to (4) but for functions $F_k: \mathbb{R}^k \to \mathbb{R}$ $(k = 1, 2, \ldots)$:

$$\begin{aligned} F_k(t_1 + u_1, \ldots, t_k + u_k) = {} & F_k(t_1, \ldots, t_k) + F_k(u_1, \ldots, u_k) \\ & + \sum_{j=1}^{k-1} \binom{k}{j} F_j(t_1, \ldots, t_j) F_{k-j}(u_1, \ldots, u_{k-j}) \\ & (t_j, u_j \in \mathbb{R}; j = 1, 2, \ldots; k = 1, 2, \ldots). \end{aligned} \tag{6}$$

If we formally introduce another variable t_0, in which all F_k are constant and a 'function' F_0 which is identically 1, then (6) goes over into

$$\begin{aligned} F_k(t_0 + u_0, t_1 + u_1, \ldots, t_k + u_k) = {} & \sum_{j=0}^{k} \binom{k}{j} F_j(t_0, t_1, \ldots, t_j) F_{k-j}(u_0, u_1, \ldots, u_{k-j}) \\ & \cdot (t_i, u_i \in \mathbb{R}; i = 0, 1, 2, \ldots, k; k = 0, 1, 2, \ldots), \end{aligned} \tag{7}$$

which is even more similar to (4).

With some care, all these systems can be solved simultaneously in the following way. Let us take, instead of infinite, just a finite number, say $m + 1$, of functions and equations, but let the variables be n-dimensional vectors, n being fixed (again we may take elements of more general algebraic structures). That is, for the unknown functions $q_k: \mathbb{R}^n \to \mathbb{R}$ $(k = 0, 1, 2, \ldots, m)$, we consider the system (cf. (3), (5))

$$q_k(\mathbf{x} + \mathbf{y}) = \sum_{j=0}^{k} q_j(\mathbf{x}) q_{k-j}(\mathbf{y}) \quad (k = 0, 1, 2, \ldots, m), \tag{8}$$

where $\mathbf{x}$, $\mathbf{y}$ are in $\mathbb{R}^n$ or in $\mathbb{R}^n_+$. As in Sections 4.1 and 4.4, all results are similar for $\mathbb{R}^n$ and $\mathbb{R}^n_+$ so, for a while, we deal with $\mathbb{R}^n$. We solve one equation at a time, beginning with $k=0$ and ending with $k=m$.

For $k=0$ we get

$$q_0(\mathbf{x}+\mathbf{y})=q_0(\mathbf{x})q_0(\mathbf{y}) \quad (\mathbf{x},\mathbf{y}\in\mathbb{R}^n). \tag{9}$$

By Proposition 4.4, either

$$q_0(\mathbf{x})=0 \quad (\mathbf{x}\in\mathbb{R}^n), \tag{10}$$

or

$$q_0(\mathbf{x})=\exp(f_0(\mathbf{x})) \quad (\mathbf{x}\in\mathbb{R}^n), \tag{11}$$

where f_0 satisfies (4.10), that is,

$$f_0(\mathbf{x}+\mathbf{y})=f_0(\mathbf{x})+f_0(\mathbf{y}) \quad (\mathbf{x},\mathbf{y}\in\mathbb{R}^n). \tag{12}$$

Now we write (8) for $k=1$:

$$q_1(\mathbf{x}+\mathbf{y})=q_0(\mathbf{x})q_1(\mathbf{y})+q_1(\mathbf{x})q_0(\mathbf{y}) \quad (\mathbf{x},\mathbf{y}\in\mathbb{R}^n). \tag{13}$$

In case (10), equation (13) would give $q_1(\mathbf{z})=0$ for all $\mathbf{z}\in\mathbb{R}^n$ and, similarly, from (8) by induction,

$$q_k(\mathbf{x})=0 \quad \textit{for all } \mathbf{x}\in\mathbb{R}^n \quad \textit{and} \quad k=0,1,2,\ldots,m. \tag{14}$$

This is certainly a solution of (8) but we will disregard it for the time being. If, on the other hand, we substitute (11) into (13) and multiply both sides by (cf. (12))

$$\exp(-f_0(\mathbf{x}+\mathbf{y}))=\exp(-f_0(\mathbf{x}))\exp(-f_0(\mathbf{y})),$$

we get, for the function f_1 defined by

$$f_1(x)=\exp(-f_0(\mathbf{x}))q_1(\mathbf{x}), \tag{15}$$

the functional equation

$$f_1(\mathbf{x}+\mathbf{y})=f_1(\mathbf{x})+f_1(\mathbf{y}) \quad (\mathbf{x},\mathbf{y}\in\mathbb{R}^n), \tag{16}$$

that is, (4.10) again. So, from (15),

$$q_1(\mathbf{x})=\exp(f_0(\mathbf{x}))f_1(\mathbf{x}) \quad (\mathbf{x}\in\mathbb{R}^n), \tag{17}$$

where f_0, f_1 satisfy (for $j=0,1$) the Cauchy equation

$$f_j(\mathbf{x}+\mathbf{y})=f_j(\mathbf{x})+f_j(\mathbf{y}) \quad (\mathbf{x},\mathbf{y}\in\mathbb{R}^n). \tag{18}$$

Now we substitute (11) and (17) into (8) for $k=2$

$$q_2(\mathbf{x}+\mathbf{y})=\exp(f_0(\mathbf{x}))q_2(\mathbf{y})+\exp(f_0(\mathbf{x})+f_0(\mathbf{y}))f_1(\mathbf{x})f_1(\mathbf{y})+\exp(f_0(\mathbf{y}))q_2(\mathbf{x}).$$

We introduce at this time the function f_2 by

$$f_2(\mathbf{x})=\exp(-f_0(\mathbf{x}))q_2(\mathbf{x})-\frac{f_1(\mathbf{x})^2}{2}$$

and get

$$f_2(\mathbf{x}+\mathbf{y})=f_2(\mathbf{x})+f_2(\mathbf{y}),$$

that is, (18) for $j = 2$ and

$$q_2(\mathbf{x}) = \exp(f_0(\mathbf{x}))\left[f_2(\mathbf{x}) + \frac{f_1(\mathbf{x})^2}{2}\right],$$

where f_0, f_1, f_2 all satisfy (18).

By induction (see Exercise 1), one proves

$$q_k(x) = \exp(f_0(\mathbf{x})) \sum_{r_1+2r_2+\cdots+kr_k=k} \prod_{j=1}^{k} \frac{f_j(\mathbf{x})^{r_j}}{r_j!} \quad \text{for all } k = 0, 1, 2, \ldots, m \tag{19}$$

where all $f_j: \mathbb{R}^n \to \mathbb{R}$ satisfy (18) ($j = 0, 1, 2, \ldots, m$), the summation being on all k-tuples of nonnegative integers $r_1, r_2, \ldots, r_k$ such that $r_1 + 2r_2 + 3r_3 + \cdots + kr_k = k$. (For $k = 0$ this gives (11), since $0! = 1$.)

On the other hand, the functions given by (19), with arbitrary f_j ($j = 0, 1, 2, \ldots, m$) satisfying (18), always satisfy (8). So we have the following:

Proposition 1. *The general solutions $q_k: \mathbb{R}^n \to \mathbb{R}$ ($k = 0, 1, \ldots, m$) of* (8) *are given by* (14) *and* (19), *where $f_j: \mathbb{R}^n \to \mathbb{R}$ ($j = 0, 1, \ldots, m$) are arbitrary functions satisfying* (18).

The result is similar for $\mathbb{R}^n_+$. We notice that we have solved the system (8) of equations successively, first for $k = 0$, then the pair with $k = 0$ and $k = 1$, then the triple with $k = 0$, $k = 1$, and $k = 2$, and so on. So there is no problem in extending the result to infinitely many equations and functions.

Theorem 2. *The general solutions $q_k: \mathbb{R}^n \to \mathbb{R}$ or $\mathbb{R}^n_+ \to \mathbb{R}$ ($k = 0, 1, 2, \ldots$) of the infinite system*

$$q_k(\mathbf{x} + \mathbf{y}) = \sum_{j=0}^{k} q_j(\mathbf{x}) q_{k-j}(\mathbf{y}) \quad (k = 0, 1, 2, \ldots) \tag{20}$$

of equations, for all $\mathbf{x}, \mathbf{y} \in \mathbb{R}^n$ or $\in \mathbb{R}^n_+$, are given by $q_k(\mathbf{x}) \equiv 0$ and by

$$q_k(\mathbf{x}) = \exp(f_0(\mathbf{x})) \sum_{r_1+2r_2+\cdots+kr_k=k} \prod_{j=1}^{k} \frac{[f_j(\mathbf{x})]^{r_j}}{r_j!}$$
$$(k = 0, 1, \ldots; x \in \mathbb{R}^n \text{ or } x \in \mathbb{R}^n_+) \tag{21}$$

where the f_j ($j = 0, 1, 2, \ldots$) are arbitrary solutions of (18) (*for $\mathbb{R}^n$ or $\mathbb{R}^n_+$*).

In particular,

Corollary 3. *The general solutions $p_k: \mathbb{R}_+ \to \mathbb{R}$ of* (3) *are given by $p_k(t) \equiv 0$ and by*

$$p_k(t) = \exp(f_0(t)) \sum_{r_1+2r_2+\cdots+kr_k=k} \prod_{j=1}^{k} \cdot \frac{[f_j(t)]^{r_j}}{r_j!} \quad (k = 0, 1, 2, \ldots; t \in \mathbb{R}_+), \tag{22}$$

where the $f_j: \mathbb{R}_+ \to \mathbb{R}$ are arbitrary solutions of Cauchy's fundamental equation

$$f_j(t + u) = f_j(t) + f_j(u) \quad (j = 0, 1, 2, \ldots). \tag{23}$$

Note that until now no regularity conditions were supposed. Of course, with aid of Theorem 2.8, the general solutions in the class of functions, bounded on a set of positive measure, in $\mathbb{R}_+^n$ or $\mathbb{R}^n$ can also be easily determined.

Also, in view of (5), we have the following:

Corollary 4. *The general solutions* $P_k:\mathbb{R}\to\mathbb{R}$ *of* (4) *are given by* $P_k(t)\equiv 0$ *for all* $k\geqslant 0$ *and*

$$P_k(t)=\exp(f_0(t))k!\sum_{r_1+2r_2+\cdots+kr_k=k}\prod_{j=1}^{k}\frac{f_j(t)^{r_j}}{r_j!}\quad(k=0,1,2,\ldots;t\in\mathbb{R}),$$

where the f_j $(j=0,1,2,\ldots)$ *satisfy* (23). *In particular, the general solutions of* (4), *bounded on a set of positive measure are given by* $P_k(t)\equiv 0$ *and by*

$$P_k(t)=\exp(c_0t)\,k!\sum_{r_1+2r_2+\cdots+kr_k=k}\prod_{j=1}^{k}\frac{(c_jt)^{r_j}}{r_j!}\quad(k=0,1,2,\ldots;t\in\mathbb{R}),\tag{24}$$

where the c_j $(j=0,1,2,\ldots)$ *are arbitrary real constants.*

Notice that the expressions which follow $\exp(c_0t)$ in (24) are *polynomials in one variable of kth degree. These are the most general polynomials of binomial type.* So, *the most general functions of binomial type, bounded on a set of positive measure, differ from the general polynomials of binomial type only by a common exponential factor.*

Let us look now at the system (1), (2) and (3) which was our starting point. By (2), all p_k $(k=0,1,2,\ldots)$ are bounded functions on $\mathbb{R}$ and thus (cf. (11), (15), (18)) $f_j(t)=c_jt$ $(t\in\mathbb{R})$ with real constants $c_j(j=0,1,\ldots)$. Also, since $p_0(t)\leqslant 1$ for $t>0$, we have $c_0\leqslant 0$. One can also prove $c_k\geqslant 0$ $(k=1,2,\ldots)$ from $p_k(t)\geqslant 0$. Finally, with some calculations (see Exercise 2 or, e.g., Jánossy–Rényi–Aczél 1950), we derive $c_0=-\sum_{j=1}^{\infty}c_j$ (including the convergence of the series on the right hand side), from (1). Equation (1) excludes also the solution $p_k(t)\equiv 0$ $(t\in\mathbb{R}_+,k=0,1,\ldots)$. So we have proved the following.

Theorem 5. *The general solution of* (1), (2) *and* (3) *is given by*

$$p_k(t)=\exp\left(-t\sum_{j=1}^{\infty}c_j\right)\sum_{r_1+2r_2+\cdots+kr_k=k}\prod_{j=1}^{k}\frac{(c_jt)^{r_j}}{r_j!}\quad(k=0,1,2,\ldots;t\in\mathbb{R}_+),\tag{25}$$

where $\sum_{j=1}^{\infty}c_j$ *is convergent, but otherwise the* c_j $(j=1,2,\ldots)$ *are arbitrary nonnegative constants. If the 'rarity condition'*

$$\lim_{t\to 0+}\frac{1-p_0(t)}{p_1(t)}=1\tag{26}$$

is added (*for short time intervals it is approximately as probable that the event will happen once, as it is that it will happen at least once), then* (25) *reduces to*

$$p_k(t) = e^{-ct}\frac{(ct)^k}{k!} \quad (k = 0, 1, 2, \ldots; t \in \mathbb{R}_+) \text{ (}c\text{ a nonnegative constant).} \quad (27)$$

(It is easy to see that (25) satisfies (1), (2), (3), that (25) and (26) imply (27) (Exercise 3) and that (1), (2), (3) and (26) are satisfied by (27).)

Probability distributions described by (25) are called *compound* (or *composed*) *Poisson distributions*, while (27) describes the (ordinary) *Poisson distribution*. Compound Poisson distributions can be 'composed' from ordinary ones, see Exercise 4 or Jánossy–Rényi–Aczél 1950.

The systems (3), (4), or (20) can be solved also by aid of *generating functions* defined by

$$f(t, z) = \sum_{k=0}^{\infty} p_k(t) z^k = \sum_{k=0}^{\infty} \frac{P_k(t)}{k!} z^k \quad \text{or} \quad f(\mathbf{x}, z) = \sum_{k=0}^{\infty} q_k(\mathbf{x}) z^k. \quad (28)$$

Such generating functions satisfy Cauchy's exponential equations (cf. Chapters 3 and 4) depending on a parameter z,

$$f(t + u, z) = f(t, z) f(u, z) \quad \text{and} \quad f(\mathbf{x} + \mathbf{y}, z) = f(\mathbf{x}, z) f(\mathbf{y}, z),$$

respectively, which are easy to solve, but either convergence has to be proved in (28) (see Exercise 5 or Jánossy–Rényi–Aczél 1950; Redheffer 1953; Schmidt 1974) or the formal calculus of (not necessarily convergent) power series solutions of functional equations has to be developed.

If, however, we wanted to solve (6) or (7), both the method of generating functions and of Theorem 2 would lead to vectors of infinite dimensions. But we can avoid this by the use of Proposition 1. In this Proposition, we would just choose $n = m$ or $n = m + 1$. We could then restrict q_k to k or $k + 1$ variables by similarly restricting the f_j appearing in the solutions of (8) according to Proposition 1. Since this can be done for *every m*, we have the following:

Theorem 6. *The general solutions* $F_k: \mathbb{R}^{k+1} \to \mathbb{R}$ *of* (7) *are given by*

$$F_k(t_0, t_1, \ldots, t_k) = k! \exp(f_0(t_0)) \sum_{r_1 + 2r_2 + \cdots + kr_k = k} \prod_{j=1}^{k} \frac{[f_j(t_0, t_1, \ldots, t_j)]^{r_j}}{j!}$$
$$\cdot (t_i \in \mathbb{R}, i = 0, 1, \ldots, k; k = 0, 1, 2, \ldots), \quad (29)$$

where the f_j *are arbitrary solutions of the Cauchy equation*

$$f_j(t_0 + u_0, t_1 + u_1, \ldots, t_j + u_j) = f_j(t_0, t_1, \ldots, t_j) + f_j(u_0, u_1, \ldots, u_j)$$
$$(t_i, u_i \in \mathbb{R}; i = 0, 1, 2, \ldots, j; j = 0, 1, 2, \ldots). \quad (30)$$

In particular, the general solutions of (7), *where each* F_k *is bounded on a set*

S_k of positive measure, are given by

$$F_k(t_0, t_1, \ldots, t_k) = k! \exp(c_{00}t_0) \sum_{r_1 + 2r_2 + \cdots + kr_k = k} \prod_{j=1}^{k} \frac{1}{r_j!}\left(\sum_{i=0}^{j} c_{ji}t_i\right)^{r_j}$$

$$(t_i \in \mathbb{R};\ c_{ji} \text{ real constants};\ i = 0, 1, \ldots, j;\ j = 1, 2, \ldots, k;\ k = 0, 1, 2, \ldots). \quad (31)$$

For (6), *the most general solution and the general solution bounded on a set of positive measure, can be obtained from* (29), (30) *and from* (31), *by choosing all f_j independent of t_0 or $c_{j0} = 0$ $(j = 0, 1, 2, \ldots)$, respectively.*

Clearly, the last mentioned functions will be all polynomials, of which the classical Bell polynomials are special cases and it happens that they all can be obtained from Bell polynomials by linear transformations (Aczél–Vrănceanu 1972).

Exercises and further results

(For Exercises 1–4, see Jánossy–Rényi–Aczél 1950.)

1. Prove (19) by induction.
2. Show that (1) and (2) combined with (22) and (23) give

$$f_j(t) = c_j t \quad (j = 0, 1, 2, \ldots),$$

$c_0 \leqslant 0$, $c_j \geqslant 0$ $(j = 1, 2, \ldots)$, $\sum_{j=1}^{\infty} c_j$ convergent and $c_0 = -\sum_{j=1}^{\infty} c_j$, that is, (25).

3. Show that (25) and (26) imply (27).
4. Let $\xi_j(t)$ be independent Poisson-distributed random variables, the mean value of $\xi_j(t)$ being $c_j t$, that is, the probability of $\xi_j(t) = k$ is $\exp(-c_j t)(c_j t)^k/k!$. Let

$$\xi(t) = \sum_{j=1}^{\infty} j\xi_j(t),$$

that is, ξ is the sum of infinitely many ordinary Poisson processes where an event in the jth process is the simultaneous occurrence of j events in the original process. Show that the probability of $\xi(t) = k$ is given by (25).

5. (Redheffer 1953.) Suppose that $\{p_k(t)\}$ satisfy (1), (2) and (3). Show that

$$\sum_{k=0}^{\infty} p_k(t)z^k$$

is convergent for $|z| < 1$. Denote the sum by $f(t, z)$ and show that f satisfies

$$f(t + u, z) = f(t, z)f(u, z).$$

By solving this equation give another proof of Theorem 5.

6. Prove Theorem 5 for $\bar{\mathbb{R}}_+$ instead of $\mathbb{R}_+$.
7. From (1), (2), (3) and (26) on $\bar{\mathbb{R}}_+$ deduce that the p_k $(k = 0, 1, \ldots)$ are differentiable on $\bar{\mathbb{R}}_+$ and that

$$p_k'(t) = c[p_{k-1}(t) - p_k(t)] \quad (k = 1, 2, \ldots; t \geqslant 0)$$

for some nonnegative constant c and use this recursive formula to prove the Poisson distribution formula (27) on $\bar{\mathbb{R}}_+$.

(For Exercises 8–10, see Aczél 1966c, pp. 224–6, 359–64.)

8. Give the general solution of the system

$$\pi_n(s,u) = \sum_{k=0}^{n} \pi_k(s,t)\pi_{n-k}(t,u) \quad (a \leqslant s \leqslant t \leqslant u; n = 0,1,2,\ldots)$$

of functional equations under the supposition

$$\pi_0(s,t) > 0 \quad (a \leqslant s \leqslant t),$$

in analogy to Corollary 3.

9. Denote the matrix $\|p_{ij}(s,t)\|$ of nth order by $\mathbf{F}(s,t)$ $(a \leqslant s \leqslant t)$. Suppose that $\det \mathbf{F}(a,t) \neq 0$ for all $t \geqslant a$. Under these conditions, determine the general solution of

$$\mathbf{F}(s,t)\mathbf{F}(t,u) = \mathbf{F}(s,u) \quad (a \leqslant s \leqslant t \leqslant u).$$

10. Specify those solutions of the previous exercise which also satisfy

$$\sum_{j=1}^{n} p_{ij}(a,t) = 1 \quad (t \geqslant a; i = 1,2,\ldots,n),$$

and use the result to find another way to solve Exercise 8.

11. (Aczél 1977*b*.) Let G be a divisible abelian group ($nx = y$ has unique solutions $x \in G$ for all $y \in G$, $n \in \mathbb{N}$). Determine the general solutions $S_k: \mathbb{N} \to G$ $(k = 0,1,2,\ldots)$ of

$$S_k(m+n) = S_k(m) + \sum_{j=0}^{k} \binom{k}{j} m^j S_{k-j}(n) \quad \left(m,n \in \mathbb{N}; \binom{k}{0} = 1; k = 0,1,2,\ldots\right).$$

12. (Aczél 1980*a*.) Let R be a divisible commutative ring. Determine the general solutions $T_k: \mathbb{N} \to R$ of the system

$$T_k(m+n) = T_k(m) + \sum_{j=0}^{k=1} \binom{k}{j} T_{k-j}(n) m^j h^j \quad (k,m,n \in \mathbb{N})$$

of functional equations, where h is a fixed element of R.

13. (Aczél 1977*b*, 1980*a*.) By specifying $S_k(1) = 1$ $(k = 0,1,2,\ldots)$ in Exercise 11, and $T_k(1) = ka^k$ $(k = 1,2,\ldots)$ in Exercise 12, find formulas for

$$\sum_{i=1}^{n} i^k \quad \text{and} \quad \sum_{i=1}^{n} [a + (i-1)h]^k$$

respectively.

13

Equations for trigonometric and similar functions

We will now consider general solutions of systems of equations and then of single equations which are satisfied by trigonometric and related (e.g. hyperbolic, exponential) functions. D'Alembert's equation (cf. Chapter 8) will be among the latter. We will solve these equations, without any regularity condition, by reducing them to Cauchy's complex exponential equations (5.8) and (5.9). D'Alembert's equation will be considered also for Banach algebras.

As we will see, some of the equations characterize these (say, trigonometric) functions. The equations represent some of their most essential, important properties, well known from trigonometry, and have a long history beginning with d'Alembert (1769).

First we consider the 'addition theorems' of the trigonometric and hyperbolic functions. Both pairs

$$f = \sin, \quad g = \cos \tag{1}$$

and

$$f = \sinh, \quad g = \cosh \tag{2}$$

satisfy the functional equation

$$f(x+y) = f(x)g(y) + g(x)f(y), \tag{3}$$

while (1) also satisfies

$$g(x+y) = g(x)g(y) - f(x)f(y) \tag{4}$$

and (2) satisfies

$$g(x+y) = g(x)g(y) + f(x)f(y) \tag{5}$$

for all real or complex x, y. For convenience, and since we will consider (first) complex valued f, g (or ranges in other *quadratically closed* commutative *fields*, that is, where $k^2 = c$ has a solution k in the field, whatever c we choose in that field), we combine (4) and (5) into the more general form

$$g(x+y) = g(x)g(y) + k^2 f(x)f(y), \tag{6}$$

where k is a complex constant.

We now solve (cf., e.g., Aczél 1958) the system composed of the two functional equations (3) and (6). The variables x, y are in $\mathbb{R}$ or in $\mathbb{C}$ or in any other group or semigroup (cf. Section 2.1), in some cases even in a groupoid (cf. Section 4.3). We multiply (3) by k and add or subtract it from (6) obtaining

$$h_j(x+y) = h_j(x)h_j(y) \quad (j = 1, 2) \tag{7}$$

for

$$h_1(x) = g(x) + kf(x) \quad \text{and} \quad h_2(x) = g(x) - kf(x) \tag{8}$$

respectively. From (8) we can recover g and f, given h_1 and h_2. For g we have

$$g(x) = \frac{h_1(x) + h_2(x)}{2}. \tag{9}$$

As to f, we have to distinguish two cases. If $k \neq 0$, we get

$$f(x) = \frac{h_1(x) - h_2(x)}{2k}. \tag{10}$$

If $k = 0$, then

$$g(x) = h_1(x) = h_2(x). \tag{11}$$

Now, as we have seen in Chapters 3 and 5 (proofs of Theorems 3.5, 5.3), if G is a group, h_1 is either everywhere or nowhere 0. If $h_1 = 0$, then

$$g(x) = 0 \tag{12}$$

and also, from (3),

$$f(x) = 0. \tag{13}$$

If h_1 is nowhere 0, divide (3) by (7) ($j = 1$), taking (11) into consideration and get, for the function l, defined by

$$l(x) = \frac{f(x)}{h_1(x)}, \tag{14}$$

the equation

$$l(x+y) = l(x) + l(y). \tag{15}$$

From (14) we have

$$f(x) = l(x)h_1(x). \tag{16}$$

Notice that (11) and (16) contain (12) and (13) (with $h_1 = 0$). We have proved the following:

Theorem 1. *The general solutions $f, g: G \to F$ of the system (3), (6) on G^2, where*

G is a group and F a commutative quadratically closed field with characteristic different from 2, are given by (9), (10), if $k \neq 0$ and by (11), (16) if $k = 0$. Here h_1, h_2 and l are arbitrary solutions of (7) and (15).

One sees indeed immediately that all these pairs of functions satisfy (3) and (6), if (7) and (15) hold. In the case of $k \neq 0$, Theorem 1 remains true if one supposes only that G is a groupoid (rather than a group).

On cartesian squares of semigroups, the equation (3) *alone* yields as much as (3) and (6) together (see Vincze 1962*b*, for the case of abelian semigroups; Chung–Kannappan–Ng 1985, for semigroups in general).

Proposition 2. *Let G be a semigroup and F a field. Let $f, g: G \to F$ satisfy* (3) *on G^2, and $f(x) \not\equiv 0$. Then there exists a constant A in F such that*

$$g(x+y) = g(x)g(y) + Af(x)f(y) \tag{17}$$

holds on G^2.

We have excluded the case $f \equiv 0$, as (3) would then be satisfied for an arbitrary g.

Proof. Since (3) and associativity on G are the only properties we have, we will use these repeatedly. In order to make full use of the associativity, introducing a third variable z will prove convenient:

$$\begin{aligned} f(x+y)g(z) + g(x+y)f(z) &= f((x+y)+z) \\ &= f(x+(y+z)) \\ &= f(x)g(y+z) + g(x)f(y+z), \end{aligned}$$

that is,

$$\begin{aligned} f(x)g(y)g(z) + g(x)f(y)g(z) + g(x+y)f(z) = f(x)g(y+z) &+ g(x)f(y)g(z) \\ &+ g(x)g(y)f(z), \end{aligned}$$

or

$$[g(x+y) - g(x)g(y)]f(z) = f(x)[g(y+z) - g(y)g(z)]. \tag{18}$$

Since $f(x) \not\equiv 0$, there exists a z_0 such that $f(z_0) \neq 0$, and so, with $h(y) := [g(y+z_0) - g(y)g(z_0)]f(z_0)^{-1}$,

$$g(x+y) - g(x)g(y) = f(x)h(y). \tag{19}$$

Putting this back into (18), we get

$$f(x)h(y)f(z) = f(x)f(y)h(z). \tag{20}$$

With $A = h(z_0)f(z_0)^{-1} \in F$, (20) gives, for all y in G,

$$h(y) = f(y)A,$$

so that (19) indeed goes over into (17). (Equation (20) also shows that A commutes with $f(z)$ for all $z \in G$.)

Here we have used the definition of field which does not suppose commutativity (what is sometimes called 'skew field' or 'sfield').

Note 1. When F is not commutative, it is important to write the right hand side of (3) in the given order and not as $f(x)g(y)+f(y)g(x)$.

Note 2. In Proposition 2, we can weaken the hypothesis that F is a field, and just suppose F is a ring (Exercise 2).

Equation (17) is the same as (6) ($A=k^2$) in a quadratically closed field. With the help of Theorem 1, Proposition 2 yields:

Theorem 3. *If G is a group and F a quadratically closed commutative field with characteristic different from 2, the general solutions of* (3) *are given by*

$$f\equiv 0, \quad g \text{ arbitrary},$$

or (9) *and* (10) *or* (11) *and* (16) *where k ($\neq 0$) is an arbitrary element of F; $h_1, h_2: G\to F$ are arbitrary solutions of* (7) *and $l: G\to F$ is an arbitrary solution of* (15).

(We have already noticed that Theorem 1 was valid on a groupoid instead of a group at least when $k\neq 0$. But Proposition 2 requires G to be a semigroup. In Theorem 3 we suppose that G is a group, because that is what was supposed in Theorem 1 when $k=0$. See, however, Exercise 10 for generalizations.)

Theorem 3, Proposition 5.2 and Theorem 5.4 also give the following (with $b=1/k$ in the first solution):

Corollary 4. *The functions $f, g: \mathbb{R}\to\mathbb{C}$ given in the following list and only these are solutions of* (3) *for all x, y in $\mathbb{R}$ in the class of functions measurable on a proper interval I of $\mathbb{R}$:*

$$f(x)=b\frac{e^{cx}-e^{dx}}{2}, \quad g(x)=\frac{e^{cx}+e^{dx}}{2}; \quad f(x)=b\frac{e^{cx}}{2}, \quad g(x)=\frac{e^{cx}}{2};$$

$$f(x)=e^{cx}bx, \quad g(x)=e^{cx},$$

$f\equiv 0$ *and g an arbitrary function, measurable on I,*

where b, c, d are arbitrary complex constants.

Corollary 5. *The general solutions $f, g: \mathbb{R}\to\mathbb{C}$ of* (3) *and* (5) *or* (3) *and* (4), *in the class of functions measurable on an interval, are given by*

$$f(x)=e^{ax}\sinh bx, \quad g(x)=e^{ax}\cosh bx; \quad f(x)=\frac{e^{cx}}{2}, \quad g(x)=\frac{e^{cx}}{2};$$

$$f(x)=-\frac{e^{cx}}{2}, \quad g(x)=\frac{e^{cx}}{2}; \quad f(x)=g(x)=0$$

and

$$f(x)=e^{ax}\sin bx,\quad g(x)=e^{ax}\cos bx;\quad f(x)=\frac{e^{cx}}{2}i,\quad g(x)=\frac{e^{cx}}{2};$$

$$f(x)=-\frac{e^{cx}}{2}i,\quad g(x)=\frac{e^{cx}}{2};\quad f(x)=g(x)=0,$$

respectively (a, b, c *arbitrary complex constants*).

We made use here of

$$\sinh p=\frac{e^{p}-e^{-p}}{2},\quad \cosh p=\frac{e^{p}+e^{-p}}{2},$$

$$\sin p=\frac{e^{pi}-e^{-pi}}{2i},\quad \cos p=\frac{e^{pi}+e^{-pi}}{2}. \tag{21}$$

The 'subtraction theorems' are stronger in the sense that their solutions contain fewer constants (and functions) which can be arbitrarily chosen. On the other hand, we have to deal here with a group (in order that $x-y$ should make sense) rather than a semigroup as in Proposition 2. The functions (1) and (2) satisfy also

$$f(x-y)=f(x)g(y)-g(x)f(y) \tag{22}$$

and (1) satisfies

$$g(x-y)=g(x)g(y)+f(x)f(y), \tag{23}$$

while (2) satisfies

$$g(x-y)=g(x)g(y)-f(x)f(y). \tag{24}$$

The domains are (cartesian squares of) *groups* (G^2) and the values of g and f are complex or in a *quadratically closed commutative field F divisible by* 2 (that is, all equations of the form $2u=v$ have unique solutions). Again we unite (23) and (24) into

$$g(x-y)=g(x)g(y)-k^2f(x)f(y). \tag{25}$$

(Some of what follows is based on Wilson 1919.)

We now solve the system (22), (25) of functional equations for functions $f,g:G\to F$. By interchanging x and y in (22) and (25), one sees that f is odd and g is even:

$$f(-z)=-f(z),\quad g(-z)=g(z). \tag{26}$$

Whenever (26) is satisfied, equations (22) and (25) go over into

$$f(x+y)=f(x)g(y)+g(x)f(y)$$

and

$$g(x+y)=g(x)g(y)+k^2f(x)f(y),$$

that is, (3) and (6).

Let us now examine the consequences of (26) upon the solutions (9), (10) and (11), (16) of (3) and (6). As to (9) and (10), that is, when $k \neq 0$ we get, from (26),

$$h_1(-x) + h_2(-x) = h_1(x) + h_2(x)$$
$$h_1(-x) - h_2(-x) = -h_1(x) + h_2(x)$$

and, by adding them,

$$h_2(x) = h_1(-x).$$

So we have

$$g(x) = \frac{h_1(x) + h_1(-x)}{2}, \quad f(x) = \frac{h_1(x) - h_1(-x)}{2k} \tag{27}$$

as solutions if $k \neq 0$. Indeed (27), with an arbitrary solution h_1 of (7) satisfies (22) and (25) when $k \neq 0$.

If $k = 0$, we have (11) and (16), that is, omitting the subscript on h,

$$g(x) = h(x), \quad f(x) = l(x)h(x), \tag{28}$$

with $h(-x) = h(x)$. Indeed (28), with an arbitrary solution of (15) and with an arbitrary solution h of the system

$$h(x + y) = h(x)h(y) \quad \text{and} \quad h(-x) = h(x), \tag{29}$$

satisfies (22) and (25) with $k = 0$. So we have proved the following.

Theorem 6. *The general solutions of the system* (22), (25), *or equivalently of* (3), (6) *and* (26), *when the domain of the equations is the cartesian square of a group G and the ranges of the functions are in a quadratically closed commutative field divisible by* 2, *are given by* (27) *if* $k \neq 0$ *and by* (28) *if* $k = 0$. *Here* h_1, l *and* h *are arbitrary solutions of* (7), (15) *and* (29) *respectively.*

The system (29) is quite particular. Suppose that G is a group divisible by 2 (here we need only that, for any z in G, an equation of the form $2x = x + x = z$ has a solution $x \in G$, which we denote by $x = z/2$; it need not be unique). Then $h(z) = h(2x) = h(x)h(x) = h(x)h(-x) = h(0)$ by (29), so that h is constant over G. But $h(0) = h(0)^2$ and so $h(0) = 0$ or $h(0) = 1$ (in the field F).

Proposition 7. *If, in Theorem* 6, *G is divisible by* 2, *the solutions for* $k = 0$ *are given by* $h \equiv 0$ *or* $h \equiv 1$ *in* (28), *l being an arbitrary solution of* (15).

However, there exist solutions of (29) which are not identically 0 or 1 if the group is not divisible by 2. Take, for instance, the group of integers under addition and define $h(n) = (-1)^n$. This satisfies (29) and is not constant. We have also the following:

Corollary 8. *The general complex valued solutions of the system (22), (26), for x, y, z in a group, are given by $f \equiv 0$, g an arbitrary even function, by (27) and by (28), where $k(\neq 0)$ is an arbitrary constant and h_1, l and h are arbitrary solutions of (7), (15) and (29), respectively. If the group is divisible by 2 then $h(x) \equiv 1$ in (28).*

Proof. In view of (26), replacing y by $-y$ in (22) gives (3) and then Proposition 2 gives (6) so Theorem 6 yields the result as stated. (For (28), if G is divisible by 2, we have omitted the case where $h(x) \equiv 0$, because this solution is contained in $f = 0$, g arbitrary even.)

Corollary 9. *The general solutions $f, g: \mathbb{R} \to \mathbb{C}$ of (22), (25), in the class of functions measurable on an interval, are given by*

$$f(x) = \frac{1}{k} \sinh cx, \quad g(x) = \cosh cx \quad \text{if } k \neq 0, \tag{30}$$

by

$$f(x) = cx, \quad g(x) = 1 \quad \text{if } k = 0, \tag{31}$$

and by

$$f(x) = g(x) = 0 \quad \text{in both cases,} \tag{32}$$

(c is an arbitrary complex constant).

Just to see how this works, we now determine the *real valued* solutions $f, g: \mathbb{R} \to \mathbb{R}$ of (22) and (23). In this case (in (25)) $k = i$ (so (31) does not come up) and both $f(x)$ and $g(x)$ have to be real. Of course, (32) is real valued. As to (30), we write $c = \alpha + \beta i$. Since, from (21),

$$\cosh pi = \cos p \quad \text{and} \quad \sinh pi = i \sin p, \tag{33}$$

and since (2) satisfies (3),

$$\begin{aligned} f(x) &= -i \sinh cx = -i \sinh(\alpha x + \beta x i) \\ &= -i \sinh(\alpha x) \cosh(\beta x i) - i \cosh(\alpha x) \sinh(\beta x i) \\ &= \cosh(\alpha x) \sin(\beta x) - i \sinh(\alpha x) \cos(\beta x) \end{aligned}$$

will be real if, and only if, the (real) coefficient of i is identically 0, that is, when $\alpha = 0$. Thus

$$f(x) = \sin \beta x \quad \text{and} \quad g(x) = \cos \beta x \tag{34}$$

and we have the following:

Corollary 10. *The general solutions $f, g: \mathbb{R} \to \mathbb{R}$ of (22) and (23) in the class of functions measurable on an interval are given by (32) and (34) (β an arbitrary real constant).*

In the same way, as we have treated (3) alone (without (6)), we will investigate the solutions of the equation

$$f(x-y)=f(x)g(y)-g(x)f(y) \tag{22}$$

alone. The domain of the equation is this time the cartesian square of an *abelian* group G. The functions f and g are supposed to take values in a quadratically closed commutative field F, divisible by 2. Equation (22) implies that f is odd (exchange x and y in (22) using the commutativity of F):

$$f(-x)=-f(x), \tag{35}$$

but we do not know whether g is even or not. So we split g into an even and an odd part

$$g=g_e+g_o, \tag{36}$$

where

$$g_e(x):=\frac{g(x)+g(-x)}{2}, \quad g_o(x):=\frac{g(x)-g(-x)}{2},$$

$$g_e(-x)=g_e(x), \quad g_o(-x)=-g_o(x). \tag{37}$$

Substituting (36) into (22), we get

$$f(x-y)=f(x)g_e(y)+f(x)g_o(y)-g_e(x)f(y)-g_o(x)f(y). \tag{38}$$

Replacing x by $-x$ and y by $-y$ we get, in view of (35) and (37), and since in abelian groups $-x+y=y-x=-(x-y)$,

$$-f(x-y)=-f(x)g_e(y)+f(x)g_o(y)+g_e(x)f(y)-g_o(x)f(y). \tag{39}$$

First we take care of the odd function g_o by adding the two equations (38) and (39). We obtain, after division by 2 (which, by supposition, can be done in F)

$$f(x)g_o(y)=g_o(x)f(y). \tag{40}$$

Now, either $f(x)\equiv 0$, and then (22) is satisfied whatever g is, which gives the solution

$$f=0, \quad g \text{ arbitrary}, \tag{41}$$

or there exists a z_0 such that $f(z_0)\neq 0$. Thus $f(z_0)^{-1}$ exists. We will denote it by c:

$$c=f(z_0)^{-1}. \tag{42}$$

Putting $y=z_0$ into (40) and writing $d=g_o(z_0)c$, we get

$$g_o(x)=df(x), \tag{43}$$

so that we will know g_o, once we have determined f.

We take care of the even function g_e by subtracting (39) from (38) (and dividing by 2). We obtain

$$f(x-y)=f(x)g_e(y)-g_e(x)f(y). \tag{44}$$

As we see from (35), (37) and (44), f and g_e satisfy (3) and (26). We already saw the consequences of (26) on the solutions of (3) just before stating Theorem 6 (cf. also Proposition 2).

We get

$$f(x) = a\frac{h_1(x) - h_1(-x)}{2}, \quad g_e(x) = \frac{h_1(x) + h_1(-x)}{2} \tag{45}$$

($a = 1/k \neq 0$, but (45) is a solution also if $a = 0$)

$$f(x) = l(x)h(x), \quad g_e(x) = h(x) \tag{46}$$

where h_1, l and h satisfy (7), (15) or (29) respectively.

In view of (36), (43) and writing $b := da$, we get, from (45),

$$f(x) = a\frac{h_1(x) - h_1(-x)}{2}, \quad g(x) = \frac{h_1(x) + h_1(-x)}{2} + b\frac{h_1(x) - h_1(-x)}{2}. \tag{47}$$

From (46) we get

$$f(x) = l(x)h(x), \quad g(x) = (1 + dl(x))h(x). \tag{48}$$

Now (47) and (48) satisfy (22) whenever h_1, l and h satisfy (7), (15) or (29) respectively. Thus we have proved the following:

Theorem 11. *Let F be a quadratically closed commutative field divisible by 2 and let G be an abelian group. The general solutions $f: G \to F$ of the functional equation* (22) *are given by* (41), (47) *and* (48) *where a, b, d are arbitrary elements in F and h_1, l, h arbitrary solutions of* (7), (15) *and* (29) *respectively.*

If G is divisible by 2 then, in view of Proposition 7, we have $h(x) \equiv 1$ in (48), that is,

$$f(x) = l(x), \quad g(x) = 1 + dl(x). \tag{49}$$

Corollary 12. *The general complex valued solutions of the functional equation* (22) *on the cartesian square G^2 of an abelian group G divisible by 2 are given by* (41), (47) *and* (49) *where a, b and d are complex constants.*

A mild regularity assumption on a function f satisfying (22) leads to the following:

Corollary 13. *In the class of complex valued functions of a real variable, measurable on a proper interval I, the general solution of* (22) *on $\mathbb{R}^2$ are given by*

$$f(x) = a\sinh cx, \quad g(x) = \cosh cx + b\sinh cx, \tag{50}$$

$$f(x) = ax, \quad g(x) = 1 + bx, \tag{51}$$

and by

$$f(x) = 0, \quad g \text{ an arbitrary function, measurable on } I \tag{52}$$

(a, b, c arbitrary complex constants).

For treatment of (4), (5), (6), (23), (24) or (25) alone, see among others, Wilson 1919; Vietoris 1944; Vincze 1962*b*; Aczél 1966*c*, pp. 176–80; Chung–Kannappan–Ng 1985.

The remarkable equation

$$f(x+y)f(x-y)=f(x)^2-f(y)^2 \tag{53}$$

just characterizes the sin, sinh (and identity) functions up to a multiplicative constant. (The function $f=\sin$ indeed satisfies (53) (as do sinh and the identity function) *not* because, as you sometimes hear from students,

$$\sin(x\pm y)=\sin x\pm\sin y \quad \text{and} \quad (\sin x+\sin y)(\sin x-\sin y)=\sin^2 x-\sin^2 y,$$

but because

$$\sin(x\pm y)=\sin x\cos y\pm\cos x\sin y$$

and

$$(\sin x\cos y+\cos x\sin y)(\sin x\cos y-\cos x\sin y)=\sin^2 x-\sin^2 y.)$$

We will first investigate the functional equation (53) for an $f:G\to F$, where G is an abelian group divisible by 2 and F a *ring with a unit* 1. We will suppose F to be divisible by 2.

We notice that $f(x)\equiv 0$ is a trivial solution of (53). We temporarily discard this solution. More exactly, *we will suppose that there exists a* z_0 *in* G *such that the inverse of* $f(z_0)$ *exists in* F:

$$f(z_0)^{-1}=c. \tag{54}$$

We may write (53) as

$$f(t)f(u)=f\left(\frac{t+u}{2}\right)^2-f\left(\frac{t-u}{2}\right)^2, \tag{55}$$

because the equations

$$x+y=t, \quad x-y=u$$

have solutions $x=(t+u)/2$, $y=(t-u)/2$, since the domain of f is an abelian group divisible by 2 (again only the existence of such x,y is needed, not their uniqueness). Putting $-u$ instead of u into (55) and comparing the equation thus obtained with (55), we get

$$f(t)f(-u)=f\left(\frac{t-u}{2}\right)^2-f\left(\frac{t+u}{2}\right)^2=-f(t)f(u). \tag{56}$$

With $t=z_0$ and (54) we obtain

$$f(-u)=-f(u), \tag{57}$$

that is, f *is odd*.

We also try to reduce (53) to (22). In order to do that, we have to define a g which, together with f, satisfies (22). This can be done as follows:

$$g(x) := \frac{f(x+z_0) - f(x-z_0)}{2} f(z_0)^{-1}. \tag{58}$$

(This makes sense because $f(z_0)$ has an inverse by supposition, cf. (54). By taking $f = \sin$, we see why g has been chosen that way. This g also happens to be even since, from (58) and (57),

$$g(-x) = \frac{-f(x-z_0) + f(x+z_0)}{2} f(z_0)^{-1} = g(x).)$$

By putting into (55) $u = z_0$ and $t = x + y$ or $t = x - y$, we get

$$f(x+y)f(z_0) = f\left(\frac{x+y+z_0}{2}\right)^2 - f\left(\frac{x+y-z_0}{2}\right)^2$$

and

$$f(x-y)f(z_0) = f\left(\frac{x-y+z_0}{2}\right)^2 - f\left(\frac{x-y-z_0}{2}\right)^2$$

respectively. Adding these up and using (55) again with $t = x$, $u = y + z_0$ or $t = x$, $u = y - z_0$ we have, by (58),

$$[f(x+y) + f(x-y)]f(z_0) = f(x)f(y+z_0) - f(x)f(y-z_0) = 2f(x)g(y)f(z_0).$$

Again we interchange x and y and use (57)

$$[f(x+y) - f(x-y)]f(z_0) = 2f(y)g(x)f(z_0).$$

If we subtract this from the previous equation and divide by $2f(z_0)$ (cf. (54)) we obtain, as required,

$$f(x-y) = f(x)g(y) - f(y)g(x).$$

This equation coincides with (22) when F is *commutative*. It seems that no careful study has been made in the noncommutative case. But anyway, equation (22) was solved in Theorem 11 just in the case where F is a *commutative* and quadratically closed *field*. In view of (57) we obtain the following:

Theorem 14. *Let F be a quadratically closed commutative field divisible by 2 and let G be an abelian group divisible by 2. The general solutions $f: G \to F$ of (53) are given by*

$$f(x) = a\frac{h(x) - h(-x)}{2} \quad \textit{and} \quad f(x) = l(x)$$

where h, l are arbitrary solutions of (7) and (15) and a is an arbitrary element of F.

Corollary 15. *The general solutions $f: \mathbb{R} \to \mathbb{C}$ or $F: \mathbb{R} \to \mathbb{R}$ of (53) in the class* *of functions measurable on a proper interval are given by*

$$f(x) = a\sinh cx, \quad f(x) = cx,$$

and

$$f(x) = a\sinh cx, \quad f(x) = a\sin cx, \quad f(x) = cx,$$

where a and c are complex or real constants, respectively.

We return now, as promised in Section 8, to d'Alembert's equation (8.1). We can determine the general solution on the cartesian square of an abelian group G (in particular, on $\mathbb{C}^2$ and $\mathbb{R}^2$) *without supposing divisibility by* 2 and without reduction to (22) and (25). We suppose $g: G \to \mathbb{C}$ ($\mathbb{C}$ may be replaced by other quadratically closed fields, see Exercise 4, or by more general algebraic structures, see Exercise 5). So we deal with

$$g(x+y) + g(x-y) = 2g(x)g(y) \quad (x, y \in G). \tag{59}$$

The following proof is due to Baker (1980*b*):

As in Section 8, we temporarily disregard the trivial solution

$$g(x) = 0 \quad (x \in G) \tag{60}$$

and notice that, for all other solutions,

$$g(0) = 1, \quad g(-y) = g(y) \quad (y \in G). \tag{61}$$

(Put $y = 0$, then $x = 0$ into (59).) In a similar fashion as we have introduced an auxiliary function g by (58) in order to solve (53), here we introduce another function, this time f, as

$$f(x) := g(x+p) - g(x-p) \tag{62}$$

(note that for $g = \cos$ this gives $f = a\sin$) in order that the function h, defined by

$$h(x) = g(x) + kf(x), \tag{63}$$

satisfies

$$h(x+y) = h(x)h(y) \quad (x, y \in G) \tag{64}$$

(cf. (7)). We will determine p and k later so that the reasons for their particular choice will be apparent.

We use (63), (62), (59) and (61) and calculate

$$\begin{aligned} h(x)h(y) &= g(x)g(y) + k[f(x)g(y) + g(x)f(y)] + k^2 f(x)f(y) \\ &= g(x)g(y) + k[g(x+p)g(y) - g(x-p)g(y) + g(x)g(y+p) \\ &\quad - g(x)g(y-p)] + k^2[g(x+p)g(y+p) - g(x-p)g(y+p) \\ &\quad - g(x+p)g(y-p) + g(x-p)g(y-p)] \\ &= \tfrac{1}{2}[g(x+y) + g(x-y)] + (k/2)[g(x+y+p) + g(x-y+p) \\ &\quad - g(x+y-p) - g(x-y-p) + g(x+y+p) + g(x-y-p) \\ &\quad - g(x+y-p) - g(x-y+p)] + (k^2/2)[g(x+y+2p) \end{aligned}$$

$$+g(x-y)-g(x+y)-g(x-y-2p)-g(x+y)-g(x-y+2p)$$
$$+g(x+y-2p)+g(x-y)] = [\tfrac{1}{2}+k^2(g(2p)-1)]g(x+y)$$
$$+[\tfrac{1}{2}-k^2(g(2p)-1)]g(x-y)+kf(x+y).$$

In order to satisfy (64), the right hand side of the last equation should be $h(x+y)=g(x+y)+kf(x+y)$. This is satisfied exactly if

$$k^2(g(2p)-1)=\tfrac{1}{2}.$$

Such a k exists,

$$k=\pm(2g(2p)-2)^{-1/2},$$

if there exists a p such that

$$g(2p)-1\neq 0 \tag{65}$$

and that is how we choose p and k in (62) and (63). From (61) and (62) we also have

$$f(-x)=-f(x), \tag{66}$$

so that, from (63),

$$h(-x)=g(x)-kf(x)$$

and, combined with (63),

$$g(x)=\frac{h(x)+h(-x)}{2} \quad (x\in G). \tag{67}$$

Only if

$$g(2x)=1 \quad \text{for all } x\in G \tag{68}$$

does there exist no p satisfying (65). Since we do not suppose that the group G is divisible by 2, this does not imply $g(x)\equiv 1$. With aid of (59) and (61) we see that we have (68) exactly when

$$2g(x)^2=g(2x)+1=2, \quad \text{that is,} \quad g(x)^2=1.$$

In other words,

$$\text{for any } x\in G, \quad \text{either } g(x)=1 \quad \text{or} \quad g(x)=-1, \quad \text{that is,} \quad g(G)=\{1,-1\}. \tag{69}$$

We show that, in this case too, g is of the form (67) with (64): If $g(x)g(y)=1$ then, by (59), $g(x+y)+g(x-y)=2$ and, by (69) $g(x+y)=g(x-y)=1$, so that

$$g(x+y)=g(x)g(y) \quad (x,y\in G). \tag{70}$$

If, on the other hand, $g(x)g(y)=-1$ then, again by (59) and (69), $g(x+y)+g(x-y)=-2$, $g(x+y)=g(x-y)=-1$, and (70) holds again. Since (61) implies $g(x)=[g(x)+g(-x)]/2$, we can define in this case $h(x):=g(x)$ and have both (67) (by definition) and (64) (by (70)).

The trivial solution (60) is also of this form ($h=0$). We have proved the following:

Theorem 16. *The general complex valued solutions of d'Alembert's functional equation* (59) *on the cartesian square of an abelian group, are given by*

$$g(x)=\frac{h(x)+h(-x)}{2},$$

where h is an arbitrary solution of (64).

Corollary 17. *The general solutions* $g:\mathbb{R}\to\mathbb{C}$ *of* (59) *in the class of functions measurable on a proper interval are given by*

$$g(x)=0,$$

and by

$$g(x)=\cosh cx, \tag{71}$$

where c is an arbitrary complex constant.

Indeed, all these functions satisfy (59).

Again we also determine the (measurable) real solutions. Writing $c=\alpha+\beta i$ in (71) and remembering that (2) satisfies (5), we get, with (33),

$$\begin{aligned}g(x)=\cosh(\alpha x+\beta xi)&=\cosh(\alpha x)\cosh(\beta xi)+\sinh(\alpha x)\sinh(\beta xi)\\&=\cosh(\alpha x)\cos(\beta x)-i\sinh(\alpha x)\sin(\beta x).\end{aligned}$$

This is real for all x if $\alpha=0$ or $\beta=0$. So we have the following:

Corollary 18. *The general solutions* $g:\mathbb{R}\to\mathbb{R}$ *of* (59), *measurable on a proper interval, are given by*

$$g(x)=0,\quad g(x)=\cosh\alpha x\quad \textit{and}\quad g(x)=\cos\beta x$$

(α, β *arbitrary real constants*).

This corollary (cf. Theorem 8.1) shows that d'Alembert's equation (59) characterizes the cos and cosh functions (alone).—See Exercise 4 for a generalization of Theorem 16.

In the last decades, an increasing interest developed for d'Alembert's equation in the case when the range of f is in a space of linear operators. We conclude this chapter by giving some results in this direction. Here we need integration of functions defined on $\mathbb{R}$ and with values in a Banach space (see Dunford–Schwartz 1963, pp. 145–55). We wish to determine those (continuous) $f:\mathbb{R}\to\mathscr{A}$ which satisfy

$$f(x+y)+f(x-y)=2f(x)f(y)\quad (x,y\in\mathbb{R}) \tag{72}$$

and

$$f(0) = 1, \tag{73}$$

where $\mathscr{A}$ is an algebra (over the real or the complex field), with an identity 1.

In the special case when $\mathscr{A}$ is the complex field $\mathbb{C}$, Theorem 16 provides the general solution of (72) and (73). If we suppose also that f is continuous, then, by Corollary 17,

$$f(x) = \cosh bx \quad (x \in \mathbb{R})$$

for some $b \in \mathbb{C}$. We can write $b = a^{1/2}$, since we are in $\mathbb{C}$, and so

$$f(x) = \sum_{n=0}^{\infty} \frac{a^n x^{2n}}{(2n)!} \quad (n \in \mathbb{R}). \tag{74}$$

Even when Theorem 16 and Corollary 17 do not apply for some general algebras $\mathscr{A}$, we may get the same representation (74), at least if $\mathscr{A}$ is a Banach algebra and f is continuous.

Theorem 19. *Let $\mathscr{A}$ be a Banach algebra with an identity. A continuous $f: \mathbb{R} \to \mathscr{A}$ satisfies* (72) *and* (73) *if and only if there exists an $a \in \mathscr{A}$ such that*

$$f(x) = \sum_{n=0}^{\infty} \frac{a^n x^{2n}}{(2n)!}. \tag{75}$$

Proof. The series in (75) always converges in a Banach algebra. It is then easy to verify that a function f of the form (75) is a continuous solution of (72), which satisfies (73).

Conversely, let $f: \mathbb{R} \to \mathscr{A}$ be a continuous solution of (72) satisfying (73). We start from

$$\lim_{t \to 0} \frac{1}{t} \int_0^t f(x) dx = 1.$$

There exists an $x_0 \in \mathbb{R}^*$ such that for $y_0 = (1/x_0) \int_0^{x_0} f(x) dx$ we have

$$\| 1 - y_0 \| < 1.$$

We therefore notice that $z_0 := \sum_{n=0}^{\infty} (1 - y_0)^n \in \mathscr{A}$ and $z_0 y_0 = y_0 z_0 = 1$. Integrate (72) from 0 to x_0 with respect to y:

$$\int_0^{x_0} f(x+y) dy + \int_0^{x_0} f(x-y) dy = 2f(x) \int_0^{x_0} f(y) dy.$$

Changes of variables give

$$\int_x^{x+x_0} f(t) dt - \int_x^{x-x_0} f(t) dt = 2f(x) \int_0^{x_0} f(t) dt.$$

Thus

$$f(x) = \frac{1}{2} \left[\int_x^{x+x_0} f(t) dt - \int_x^{x-x_0} f(t) dt \right] \frac{z_0}{x_0}. \tag{76}$$

Since f is continuous, the right hand side of (76) is differentiable. Consequently, the left hand side is differentiable too. So f is differentiable on $\mathbb{R}$. By differentiating (76) we obtain

$$f'(x)=\tfrac{1}{2}\left[f(x+x_0)-f(x-x_0)\right]\frac{z_0}{x_0}. \tag{77}$$

The right hand side of (77) is differentiable, so f is twice differentiable. We differentiate (72) twice with respect to y and substitute $y=0$ in order to obtain

$$f''(x)=f(x)a,$$

where $a=f''(0)$. However, (72) with $x=0$ and (73) imply that f is even and so, from (72),

$$f(x)f(y)=f(y)f(x).$$

Differentiating this equation twice with respect to y and putting $y=0$, we get $f(x)a=af(x)$. By (77), $f'(0)=0$, so we have

$$f'(s)=a\left(\int_0^s f(t)dt\right).$$

But from this and from $f(0)=1$, we obtain

$$f(x)=1+a\int_0^x\left[\int_0^s f(t)dt\right]ds.$$

Changing the order of integration gives

$$f(x)=1+a\int_0^x (x-s)f(s)ds. \tag{78}$$

Now we apply (78) repeatedly till we get

$$f(x)=1+a\frac{x^2}{2!}+\cdots+\frac{a^n x^{2n}}{2n!}+\frac{a^{n+1}}{(2n+1)!}\int_0^x (x-s)^{2n+1}f(s)ds.$$

The last term tends to zero as $n\to\infty$. Therefore we have proved that f has the form (75), which concludes the proof of theorem 19.

Remark 1. If there exists a $b\in\mathscr{A}$ such that $b^2=a$, then (75) can be written in the form

$$f(x)=\cosh bx,$$

where $\cosh t=\sum_{n=0}^{\infty} t^{2n}/(2n)!$ for any t in $\mathscr{A}$.

Remark 2. See Exercise 16 for the case where $\mathbb{R}$ is replaced by $\mathbb{R}_+$.

Suppose that X is a Banach space and take for $\mathscr{A}$ the algebra $\mathscr{L}(X)$ of bounded linear operators from X into X. With an argument quite similar to the one used for Theorem 2.8, one may prove the following (see Baker 1971*b*).

Let $f:\mathbb{R}\to\mathcal{L}(X)$ *be a solution of* (72) *and* (73). *Suppose that, for every* x *in* X, *the function* $g:\mathbb{R}\to\bar{\mathbb{R}}_+$, *defined by*

$$g(t) = \| f(t)(x) \|,$$

is measurable. Then f *is continuous at* 0 (that is, $\lim_{x\to 0} f(x) = 1$, in the norm topology of $\mathcal{L}(X)$).

When the continuity of f has been obtained, Theorem 19 gives the general form of f. It can then be proved that

$$a = \lim_{x\to 0} \frac{2}{x^2}(f(x) - 1),$$

where the limit is in the norm topology of $\mathcal{L}(X)$.

We refer to Lutz 1981 for the links between the d'Alembert equation in this setting and the differential equation

$$y''(t) + Ax(t) = 0.$$

Exercises and further results

1. Try to find a generalization of Theorem 1 when G is a groupoid and F a ring, in the case where k has a (multiplicative) inverse in F. What can be said if $k = 0$?
2. Let G be a semigroup and R be a ring with a unit. Let $f, g:G\to R$ satisfy (3) on G^2. Suppose there is a $z_0 \in G$ for which $f(z_0)^{-1}$ exists in R. Can Proposition 2 be extended to this situation?
3. Let G be an abelian group and F be a commutative field of characteristic different from 2. Let I be the set of all square roots of 1 in F (that is, $x \in I$ if $x^2 = 1$, where 1 is the unit of F). Prove or disprove that $g:G\to I$ is a solution of the d'Alembert equation if, and only if, g satisfies the Cauchy exponential equation.

4*. Show that Theorem 16 holds when the values of g are in a quadratically closed commutative field of characteristic different from 2.

5*. Let $g:G\to B$ where G is an abelian group and B a real (or complex) Banach algebra with a unit e ($\| e \| = 1$). Suppose that there exist z_0 and t_0 in G such that $\| g(z_0) - e \| < 1$ and $\| g(2t_0) \| < 1$. Determine the general solution of the d'Alembert functional equation for all x, y in G.

6*. Prove that Theorem 16 remains true when $\mathbb{C}$ is replaced by a quadratically closed field of characteristic different from 2 and when, instead of supposing that G is abelian, G is assumed to be a nilpotent group, all elements of which are of odd order.

7. (Aczél 1966c, p. 138.) Let $f:G\to F$ (G a group, F a field, both commutative and divisible by 2) be a solution of (53). Show that the definition

$$g(x) := \left(\frac{f(x + z_0) - f(x - z_0)}{2}\right) f(z_0)^{-1}$$

(equation (58)) is independent of the choice of z_0 as long as $f(z_0) \neq 0$.

8. Let G be an abelian group, divisible by 2, and let K be a commutative field, divisible by 2 and quadratically closed. Determine all solutions $f:G\to K$ of the

functional equation

$$f(x)^2+f(y)^2=f(x+y)f(x-y)+1 \quad (x,y\in G).$$

9. Let $A\in\mathbb{R}$. Determine the general solution of the functional equation

$$f(x+y+A)-f(x-y+A)=2f(x)f(y) \quad (x,y\in\mathbb{R}),$$

with aid of the general solution of the d'Alembert equation.

10. Find the general solutions $f,g:G\to R$ of (3), where G is a semigroup, R a ring with unit and divisible by 2, the constant k has an inverse in R and commutes with all values of f and g, and there exists a $z_0\in G$ such that $f(z_0)$ has an inverse $f(z_0)^{-1}$ in R and $(g(2z_0)-g(z_0)^2)(f(z_0)^{-1})^2=k^2$.
11. Derive Theorem 8.2 from Theorem 16 of the present chapter.
12. Let $f:\mathbb{R}\to\mathbb{R}$ satisfy the inequality

$$f(x+y)f(x-y)\leqslant f(x)^2-f(y)^2 \quad (x,y\in\mathbb{R}).$$

Show, without regularity assumption for f, that the inequality has to be an equality. Generalize the result to $f:G\to F$ where G is an abelian group and F a conveniently chosen (how?) ordered field.

13. Let $f:\mathbb{R}\to\mathbb{R}$ satisfy the inequality

$$f(x+y)f(x-y)\geqslant f(x)^2-f(y)^2 \quad (x,y\in\mathbb{R}).$$

Are there solutions for which strict inequality holds for all real x,y?

14*. Let $f:G\to F$ where G is a group and F a field of characteristic different from 2. Find the general solution of the functional equation

$$2f(xy)=f(x)f(y)+f(y)f(x).$$

Compare the result to that of Exercise 6.10.

15. (Kannappan 1968*a*.) Let G be a group. Show that, if $f:G\to\mathbb{C}$ satisfies

$$f(xy)+f(xy^{-1})=2f(x)f(y) \quad (x,y\in G)$$

and

$$f(xyz)=f(xzy) \quad (x,y,z\in G),$$

then

$$f(x)=\tfrac{1}{2}[g(x)+g(x)^{-1}] \quad (x\in G),$$

where g satisfies

$$g(x+y)=g(x)g(y).$$

16*. (Baker 1971*b*.) Let $\mathscr{A}$ be a Banach algebra and let $C:\mathbb{R}_+\to\mathscr{A}$ satisfy

$$C(t+s)+C(t-s)=2C(t)C(s) \quad (t>s>0).$$

Suppose that $\lim_{t\to 0}C(t)$ exists and denote it by j. Prove that $j^2=j$ and that there exist a,b in $\mathscr{A}$ such that

$$ja=aj=a, \quad bj=b, \quad jb=0,$$

and

$$C(t)=\left(j+\frac{at^2}{2!}+\frac{a^2t^4}{4!}+\cdots\right)+b\left(tj+\frac{t^3a}{3!}+\frac{t^5a^2}{5!}+\cdots\right)$$

for all $t>0$.

17. (Lutz 1981.) Let $f:\mathbb{R}\to\mathscr{L}(X)$ where X is a Banach space and $\mathscr{L}(X)$ the algebra of all bounded linear operators from X into X. Suppose that $f(x+y)+f(x-y)=2f(x)f(y)$ $(x,y\in\mathbb{R})$, $f(0)=I$ and that $g:\mathbb{R}\to X$ with $g(t)=f(t)(x)$ is continuous. Prove or disprove that

$$\|f(x)\|\leqslant Me^{\omega|x|}$$

for some $\omega>0$ and some $M\geqslant 1$.

14

A class of equations generalizing d'Alembert and Cauchy Pexider-type equations

The purpose of this chapter is to study systematically generalizations of the Cauchy, d'Alembert and other trigonometric equations, thus extending results of the Chapters 2, 3, 8 and 13, and to show how regularity conditions concerning the solutions can be subsequently relaxed. We have just seen in (13.76) how a functional equation can improve the regularity of its solutions (see also (5.19)). This will be a basic tool of this chapter.

The equations (2.1), (3.14), (4.25), (4.35), (8.1), (12.3), (12.4), (13.3), (13.4), (13.5), (13.6), (13.22), (13.23), (13.24), (13.25), and later (15.2), (15.6), (15.9), are special cases of the very general equation

$$f(x+y)+g(x-y)=\sum_{k=1}^{n} f_k(x)g_k(y). \tag{1}$$

As Speiser, Whitehouse and Berg 1978 show, this equation has significant applications in signal processing. Notice that the left hand side of (1) is the general solution of the wave equation $u_{xx}=u_{yy}$, while the right hand side is the inner product of the vectors $\mathbf{f}(x)=(f_1(x),\dots,f_n(x))$ and $\mathbf{g}(x)=(g_1(x),\dots,g_n(x))$.

We suppose first that $f_k:E\to\mathbb{R}$, $g_k:E'\to\mathbb{R}$ $(k=1,\dots,n)$, $f:E+E'\to\mathbb{R}$ and $g:E-E'\to\mathbb{R}$ where E and E' are nonempty sets to be specified later. On the right hand side of (1), we have a certain amount of freedom in the sense that, if, for instance, $f_1,\dots,f_n$ are linearly dependent (one of them is a linear combination of the others), then the number of terms can be reduced. So we may suppose without loss of generality that the n functions $f_1,\dots,f_n$ are *linearly independent* on their domain E of definition; that is, if $c_1,\dots,c_n$ in $\mathbb{R}$ are such that

$$\sum_{k=1}^{n} c_k f_k(x)=0 \quad (x\in E),$$

then $c_1=c_2=\cdots=c_n=0$.

A similar hypothesis is made for the n functions $g_1,\dots,g_n$. The following lemma (Curtiss 1910, Vincze 1962*a*) has a basic role in what follows.

Lemma 1. *The n functions $f_1,f_2,\dots,f_n:E\to\mathbb{R}$ are linearly dependent if, and only if,*

$$\begin{vmatrix} f_1(x_1) & f_1(x_2) & \cdots f_1(x_{n-1}) & f_1(x_n) \\ f_2(x_1) & f_2(x_2) & \cdots f_2(x_{n-1}) & f_2(x_n) \\ \cdots & \cdots & \cdots & \cdots \\ f_{n-1}(x_1) & f_{n-1}(x_2) & \cdots f_{n-1}(x_{n-1}) & f_{n-1}(x_n) \\ f_n(x_1) & f_n(x_2) & \cdots f_n(x_{n-1}) & f_n(x_n) \end{vmatrix} = 0 \tag{2}$$

for all n-tuples $(x_1,x_2,\dots,x_n)\in E^n$.

Proof. The linear dependence of $f_1,f_2,\dots,f_n$ on E means that there exist real constants $c_1,c_2,\dots,c_n$, not all 0, such that

$$c_1f_1(x)+c_2f_2(x)+\cdots+c_nf_n(x)=0 \quad (x\in E). \tag{3}$$

By putting $x=x_k$ in (3), for $k=1,2,\dots,n$, we get a linear system of n equations in $c_1,c_2,\dots,c_n$. Since there exists a nontrivial solution, the determinant of this system is zero, that is, we have (2).

We prove now by induction on n that, conversely, (2) implies the linear dependence of $f_1,f_2,\dots,f_n$ on E. The statement is obviously true for $n=1$ where both linear dependence and (2) mean just $f(x)=0$ for all x in E.

Suppose that it is true for $n-1$ and that (2) holds for n. Then there are two cases. Either

$$\begin{vmatrix} f_1(x_1) & f_1(x_2) \cdots\cdots f_1(x_{n-1}) \\ f_2(x_1) & f_2(x_2) \cdots\cdots f_2(x_{n-1}) \\ \cdots & \cdots \\ f_{n-1}(x_1) & f_{n-1}(x_2) \cdots\cdots f_{n-1}(x_{n-1}) \end{vmatrix} = 0 \text{ for all } x_1,x_2,\dots,x_{n-1} \text{ in } E, \tag{4}$$

in which case, by the induction hypothesis, $f_1,f_2,\dots,f_{n-1}$ are linearly dependent and, so much the more, $f_1,f_2,\dots,f_{n-1},f_n$ are also linearly dependent. Or (4) does not hold, that is, there exist points $a_1,a_2,\dots,a_{n-1}$ in E such that

$$c_n := \begin{vmatrix} f_1(a_1) & f_1(a_2) & ,\dots\dots\dots, f_1(a_{n-1}) \\ f_2(a_1) & f_2(a_2) & ,\dots\dots\dots f_2(a_{n-1}) \\ \cdots & \cdots & \cdots \\ f_{n-1}(a_1) & f_{n-1}(a_2) & ,\dots\dots\dots, f_{n-1}(a_{n-1}) \end{vmatrix} \neq 0.$$

Put $x_k=a_k$ $(k=1,2,\dots,n-1)$ and $x_n=x$ into (2) and develop the determinant in (2) with respect to the last column. Then (2) goes over into an equation of the form (3) with $c_n\neq 0$, so that $f_1,f_2,\dots,f_n$ are again linearly dependent, and Lemma 1 is proved.

In a way, this result is analogous to the theorems linking the linear dependence of differentiable or integrable functions to the disappearance of the Wronski and Gram determinants, respectively (cf. Curtiss 1910). It can also be considered as a statement about the general solution of the functional equation (2).

We will occasionally denote the left hand side of (2), the so-called Casorati determinant, by

$$\det f_k(x_j) = \begin{vmatrix} f_1(x_1) & f_1(x_2) \cdots\cdots f_1(x_n) \\ f_2(x_1) & f_2(x_2) \cdots\cdots f_2(x_n) \\ \cdots & \cdots \\ f_n(x_1) & f_n(x_2) \cdots\cdots f_n(x_n) \end{vmatrix}. \tag{5}$$

In what follows, we first solve (1) under strong regularity assumptions concerning all the unknown functions, $f, g, f_1, \ldots, f_n, g_1, \ldots, g_n$, and using functions defined on subsets of $\mathbb{R}$. Then we shall reduce the regularity conditions for the unknown functions which are solutions of (1), but still remain in $\mathbb{R}$. Finally, we will explore some possible generalizations when the domains of definitions are no longer subsets of $\mathbb{R}$, but are in abelian groups.

Accordingly, we first suppose the following:

Hypothesis H. *The $f_k: E \to \mathbb{R}$ and $g_k: E' \to \mathbb{R}$ $(k = 1, \ldots, n)$, $f: E + E' \to \mathbb{R}$, and $g: E - E' \to \mathbb{R}$ have derivatives of all orders.*

Proposition 2. *Let E and E' be two nonempty open intervals in $\mathbb{R}$. We suppose that $0 \in E'$, that the hypothesis H is satisfied, that $(f_1, \ldots, f_n)$ are linearly independent (on E) and $(g_1, \ldots, g_n)$ are linearly independent (on E') and that $f_1, \ldots, f_n; g_1, \ldots, g_n, f$ and g satisfy the functional equation (1). Then the f_k and the g_k $(k = 1, \ldots, n)$ satisfy systems of explicit second order homogeneous linear differential equations with constant coefficients and without first order terms, while $f + g$ and $f' - g'$ satisfy linear homogeneous differential equations of order 2n with constant coefficients and with all terms of even order.*

Proof. Indeed, differentiating (1) twice with respect to x, we get

$$\sum_{k=1}^{n} f_k''(x) g_k(y) = f''(x + y) + g''(x - y). \tag{6}$$

Differentiating (1) twice with respect to y, we get

$$\sum_{k=1}^{n} f_k(x) g_k''(y) = f''(x + y) + g''(x - y). \tag{7}$$

Since $g_1, \ldots, g_n$ are linearly independent, using Lemma 1, there exist points $y_1, \ldots, y_n$ in E' such that

$$\det g_k(y_j) \neq 0. \tag{8}$$

Using (6) and (7) with $y = y_j$, for $j = 1, 2, \ldots, n$, we obtain a system of n

equations

$$\sum_{k=1}^{n} f_k''(x)g_k(y_j) = \sum_{k=1}^{n} f_k(x)g_k''(y_j) \quad (j = 1, \ldots, n). \tag{9}$$

With (9) we get a linear system of equations for $f_k''(x)$ $(k = 1, 2, \ldots, n)$, whose determinant is different from zero by (8). Thus (9) can be solved and each $f_k''(x)$ is a linear combination, with constant coefficients, of the $f_k(x)$ $(k = 1, 2, \ldots, n)$:

$$f_k''(x) = \sum_{m=1}^{n} a_{k,m} f_m(x) \quad (k = 1, 2, \ldots, n). \tag{10}$$

in a similar way we prove that the g_k satisfy an explicit linear system of n differential equations of second order with constant coefficients and without first order terms,

$$g_k''(y) = \sum_{m=1}^{n} b_{k,m} g_m(y) \quad (k = 1, 2, \ldots, n), \tag{11}$$

as asserted in Proposition 2.

As to f and g, put $y = 0$ into (1) (remember that we have supposed $0 \in E'$) and define $\phi: E \to \mathbb{R}$ by

$$\phi = f + g.$$

We get, with $c_{0,k} = g_k(0)$, the equation

$$\phi(x) = \sum_{k=1}^{n} c_{0,k} f_k(x). \tag{12_0}$$

Differentiating (12_0) twice, we have

$$\phi''(x) = \sum_{k=1}^{n} c_{0,k} f_k''(x),$$

or, using (10), with

$$c_{1,k} = \sum_{l=1}^{n} c_{0,l} a_{l,k} \quad (k = 1, 2, \ldots, n),$$

$$\phi''(x) = \sum_{k=1}^{n} c_{1,k} f_k(x). \tag{12_1}$$

Similarly,

$$\phi^{(2j)}(x) = \sum_{k=1}^{n} c_{j,k} f_k(x) \quad (j = 0, 1, \ldots, n-1, n). \tag{12_j}$$

In particular,

$$\phi^{(2n-2)}(x) = \sum_{k=1}^{n} c_{n-1,k}\, f_k(x), \tag{12_{n-1}}$$

$$\phi^{(2n)}(x) = \sum_{k=1}^{n} c_{n,k}\, f_k(x), \tag{12_n}$$

for all x in E.

We now distinguish two cases. If

$$\det c_{j,k} = 0 \quad (j = 0, 1, \ldots, n-1; k = 1, 2, \ldots, n),$$

then there exist $b_0, b_1, \ldots, b_{n-1}$, not all 0, such that

$$\sum_{j=0}^{n-1} b_j c_{j,k} = 0 \quad (k = 1, 2, \ldots, n).$$

Therefore, from (12_0), $(12_1), \ldots, (12_{n-1})$, we deduce that, for all x in E,

$$\sum_{j=0}^{n-1} b_j \phi^{(2j)}(x) = 0.$$

If, on the other hand,

$$\det c_{j,k} \neq 0 \quad (j = 0, 1, \ldots, n-1; k = 1, 2, \ldots, n),$$

then the f_k $(k = 1, 2, \ldots, n)$ can be determined from (12_0), $(12_1), \ldots, (12_{n-1})$ as linear combinations (with constant coefficients) of $\phi, \phi'', \ldots, \phi^{(2n-2)}$ so that (12_n) goes over into

$$\phi^{(2n)}(x) = \sum_{j=0}^{n-1} b_j \phi^{(2j)}(x) \quad (x \in E).$$

In both cases ϕ satisfies a homogeneous linear differential equation of order $2n$ with constant coefficients (not all 0), where all terms are of even order.

We now define $\psi(x) = f'(x) - g'(x)$ $(x \in E)$. If we differentiate (1) with respect to y and substitute $y = 0$ we get, with $d_{0,k} = g_k'(0)$,

$$\psi(x) = \sum_{k=0}^{n} d_{0,k} f_k(x)$$

and, differentiating this equation twice, four times, ..., $2n$ times, we get for ψ the same result as above for ϕ. This concludes the proof of Proposition 2.

To find the solutions of the linear systems (10) and (11) for the functions $f_1, \ldots, f_n$ or $g_1, \ldots, g_n$ is a routine task in the theory of differential equations (though it can be rather lengthy). The same is true for the differential equations satisfied by ϕ or ψ. Notice that f and g are easily determined once ϕ and ψ are known.

We leave it as an exercise (cf. Exercise 1) for the reader to find, when $E = E' = \mathbb{R}$, all C^∞ solutions of the functional equation

$$f(x+y) + g(x-y) = f_1(x)g_1(y) + f_2(x)g_2(y), \tag{13}$$

where (f_1, f_2) and (g_1, g_2) are linearly independent on $\mathbb{R}$.

Both in Proposition 2, regarding (1), and in this example, regarding (13), the condition C^∞ could be replaced by C^2 for f_k, g_k and by C^{2n+1} (resp. C^5) for f and g. But we will eliminate all differentiability conditions anyway.

So we shall now considerably weaken the hypothesis H, which we made about the unknown functions f, g, f_k, and g_k $(k = 1, 2, \ldots, n)$. To do this, we shall first show that the f_k and the g_k can be expressed by convenient combinations of shifts of f and g.

Naturally we maintain our hypothesis of linear independence of the f_k on E (as well as the linear independence of the g_k on E'). Then Lemma 1 yields $y_1,\ldots,y_n$ in E' such that $\det g_k(y_j) \neq 0$. Substituting these y_j for y in (1), we get n linear equations which can be uniquely solved for $f_1(x), f_2(x),\ldots,f_n(x)$. In other words, for $x \in E$,

$$f_k(x) = \sum_{j=1}^{n} \alpha_{k,j}(f(x+y_j) + g(x-y_j)), \tag{14}$$

where $\alpha_{k,j}$, $k=1,\ldots,n$ and $j=1,\ldots,n$ are real constants. In a similar way,

$$g_k(y) = \sum_{j=1}^{n} \beta_{k,j}(f(x_j+y) + g(x_j-y)), \tag{15}$$

where $\beta_{k,j}$, $k=1,\ldots,n$ and $j=1,\ldots,n$ are real constants.

Substitute $u=x+y$ into (1). We get

$$f(u) = \sum_{k=1}^{n} f_k(u-y)g_k(y) - g(u-2y). \tag{16}$$

Similarly, by substituting $v=x-y$ into (1), we get

$$g(v) = \sum_{k=1}^{n} f_k(x)g_k(x-v) - f(v+2y). \tag{17}$$

Suppose first that $f: E+E' \to \mathbb{R}$ and $g: E-E' \to \mathbb{R}$ are continuous functions where E and E' are nonempty open intervals of $\mathbb{R}$. Then (14) and (15) show that $f_k: E \to \mathbb{R}$ and $g_k: E' \to \mathbb{R}$ are continuous functions. Let us integrate (1) over $[x_0, t]$ where x_0 is a fixed point in E and where $t \in E$:

$$\int_{x_0}^{t} f(x+y)dx + \int_{x_0}^{t} g(x-y)dx = \sum_{k=1}^{n} F_k(t)g_k(y),$$

where $F_k(t) = \int_{x_0}^{t} f_k(x)dx$. We may also write, after introducing the new variables $u=x+y$, $v=x-y$,

$$\int_{x_0+y}^{t+y} f(u)du + \int_{x_0-y}^{t-y} g(v)dv = \sum_{k=1}^{n} F_k(t)g_k(y). \tag{18}$$

Now, the n functions $F_1, F_2,\ldots,F_n$ are linearly independent on E. If not, there would exist real constants $\alpha_1, \alpha_2,\ldots,\alpha_n$, not all zero, such that

$$\sum_{k=1}^{n} \alpha_k F_k(t) = 0 \quad (t \in E).$$

As the f_k are continuous functions, we get, by differentiation,

$$\sum_{k=1}^{n} \alpha_k f_k(x) = 0 \quad (x \in E),$$

which contradicts the supposed linear independence of the f_k on E. Therefore, using Lemma 1, we can find some $t_1, \ldots, t_n \in E$ such that $\det F_k(t_j) \neq 0$. We can now solve (18) and obtain for some real constants $\gamma_{k,j}$ $(k = 1, \ldots, n; j = 1, \ldots, n)$

$$g_k(y) = \sum_{j=1}^{n} \gamma_{k,j} \left(\int_{x_0+y}^{t_j+y} f(u)du + \int_{x_0-y}^{t_j-y} g(v)dv \right).$$

From this we deduce that g_k is continuously differentiable on E'.

Let $y_o \in E'$ so that $]y_o - \delta, y_o + \delta[\subset E'$ for some $\delta > 0$. This is possible as E' is an open interval of $\mathbb{R}$. We may integrate (16) with respect to y on $[y_o, t]$ where $t \in]y_o - \delta, y_o + \delta[$. We get

$$(t - y_0)f(u) = \sum_{k=1}^{n} \int_{y_0}^{t} f_k(u-y)g_k(y)dy - \int_{y_0}^{t} g(u-2y)dy$$

or

$$(t - y_0)f(u) = \sum_{k=1}^{n} \int_{u-t}^{u-y_0} f_k(v)g_k(u-v)dv - \tfrac{1}{2} \int_{u-2t}^{u-2y_0} g(w)dw. \tag{19}$$

We have proved that the f_k are continuous and that the g_k are continuously differentiable $(k = 1, \ldots, n)$, and we know that g is continuous. Therefore the right hand side of (19) is a continuously differentiable function of u and, consequently, so is f. Here $u \in E + E'$.

In a similar way, using (17), we can also prove that g is a continuously differentiable function on $E - E'$. Using (14) and (15), we deduce that the f_k and the g_k $(k = 1, \ldots, n)$ are continuously differentiable functions on E and E' respectively. Now we may differentiate (1) with respect to x to get

$$f'(x+y) + g'(x-y) = \sum_{k=1}^{n} f'_k(x)g_k(y). \tag{20}$$

The linear independence of the $\{f'_k\}$ does not follow directly from that of the $\{f_k\}$ but, as mentioned at the beginning of this chapter, the right hand side of (20) can be replaced by a sum (possibly with fewer terms) of the same form

$$f'(x+y) + g'(x-y) = \sum_{j=1}^{m} \phi_j(x)\psi_j(y) \tag{21}$$

with both $\{\phi_j\}$ and $\{\psi_j\}$ linearly independent and each ϕ_j and ψ_j $(j = 1, \ldots, m)$ continuous on E and E' respectively.

Now we may apply to (21) what we have previously obtained for (1). Therefore f' and g' are continuously differentiable functions. Thus f and g are twice continuously differentiable on their domains of definition. So are the f_k and the g_k because of (14) and (15). By induction, one clearly sees that f, g, f_k and g_k $(k = 1, \ldots, n)$ have derivatives of all orders on their respective domains of definition. We have proved the following:

Proposition 3. *Proposition 2 remains valid when we replace hypothesis H by the following: The functions $f: E + E' \to \mathbb{R}$ and $g: E - E' \to \mathbb{R}$ are continuous.*

But we can go further in weakening the regularity conditions imposed upon f and g. Let us suppose that f and g are locally Lebesgue integrable functions on their respective domains of definition. We will need only the most basic facts about such functions. Equations (14) and (15) show that the f_k and the g_k ($k = 1,\ldots,n$) are also locally Lebesgue integrable functions (on E and E', respectively).

Let χ_ε be a nonnegative continuous function whose support S is a subset of $]-\varepsilon, +\varepsilon[$ where $\varepsilon > 0$ and such that $\int_S \chi_\varepsilon(x)dx = 1$. By multiplying (1) by $\chi_\varepsilon(t)$ where x in (1) is replaced by $x + t$, and integrating, we get

$$\int_S f(x+t+y)\chi_\varepsilon(t)dt + \int_S g(x+t-y)\chi_\varepsilon(t)dt = \sum_{k=1}^{n} \Phi_k(x)g_k(y), \tag{22}$$

where $\Phi_k(x) = \int_S f_k(x+t)\chi_\varepsilon(t)dt$.

We define

$$\Phi(x) = \int_S f(x+t)\chi_\varepsilon(t)dt, \quad \Psi(y) = \int_S g(y+t)\chi_\varepsilon(t)dt.$$

These functions Φ and Ψ are continuous. However, their domains of definition are not $E + E'$ or $E - E'$ but subintervals of $E + E'$ and $E - E'$, the extremities of which have been shifted by ε to the inside of $E + E'$ (and $E - E'$). Of course, we choose ε small enough so that the remaining intervals are nonempty. Then we get

$$\Phi(x+y) + \Psi(x-y) = \sum_{k=1}^{n} \Phi_k(x)g_k(y). \tag{23}$$

Now there exist $x_1,\ldots,x_n$ in E so that $\det f_k(x_j) \neq 0$. By choosing ε small enough, we get $\det \Phi_k(x_j) \neq 0$ as well. Therefore the $\Phi_1,\ldots,\Phi_n$ are linearly independent functions. What we have already proved (Proposition 3) then shows that the g_k are continuous functions. Now, ε being arbitrarily small, this proves the continuity of the g_k on E'. In a similar fashion we prove the continuity of the f_k on E. Thus (19) shows that f is a continuous function. We may obtain the continuity of g in a similar fashion. Proposition 3 completes the proof of the following:

Theorem 4. *Let E and E' be two nonempty open intervals in $\mathbb{R}$ and $0 \in E'$. Suppose that $f: E + E' \to \mathbb{R}$ and $g: E - E' \to \mathbb{R}$ are locally Lebesgue integrable, that both $f_1,\ldots,f_n$ and $g_1,\ldots,g_n$ are linearly independent and that they satisfy*

$$f(x+y) + g(x-y) = \sum_{k=1}^{n} f_k(x)g_k(y) \quad (x \in E, y \in E').$$

Then the f, g, f_k, g_k are C^∞ and the f_k and the g_k ($k = 1,\ldots,n$) satisfy systems of explicit second order homogeneous linear differential equations with constant coefficients without first order terms, while $f + g$ and $f' - g'$ satisfy linear *homogeneous differential equations of order 2n with constant coefficients and with all terms of even order.*

Remark 1. The technique of the proof we used to get Proposition 3 (and Theorem 13.19) goes back to Andrade (1900) and was used later on for various kinds of functional equations of the form (1). See Levi–Civita 1913; Satô 1928; Aczél 1966*c* pp. 190–6. In Aczél–Chung 1982, a weaker version of Theorem 4 is proved under the additional assumption of L-independence for the f_k and the g_k. L-independence of the f_k means that

$$\sum_{k=1}^{n} a_k f_k(x) = 0 \text{ almost everywhere on } E$$

implies $a_k = 0$, $k = 1,\ldots,n$.

However Járai (1984) has proved that L-independence can be replaced by linear independence and that in Theorem 4 the hypothesis of local Lebesgue integrability for f and g can be replaced by Lebesgue measurability (see Exercise 4). His proof relies heavily on a very general theorem (Járai 1979) concerning regularity for functional equations of the form (16) (and of more general forms).

Remark 2. The technique we have used here can be easily adapted to treat the case of a more general functional equation than (1), namely

$$\sum_{l=1}^{m} h_l(\alpha_l x + \beta_l y) = \sum_{k=1}^{n} f_k(x) g_k(y) \quad (x \in E, y \in E'). \tag{24}$$

In Aczél-Chung 1982, the following theorem is proved (Exercise 15):

Theorem 5. *Let E and E′ be nonempty open intervals in* $\mathbb{R}$, $\alpha_l \beta_l \neq 0$ *and* $\alpha_l/\beta_l \neq \alpha_j/\beta_j$ *for* $j \neq l = 1,\ldots,m$. *Suppose that the functions* h_l *are locally integrable on* $\alpha_l E + \beta_l E'$ $(l = 1,\ldots,m)$ *and both* $\{f_1,\ldots,f_n\}$ *and* $\{g_1,\ldots,g_n\}$ *are linearly independent on E or E′, respectively, and that they satisfy* (24). *Then the functions* h_l, f_k, g_k *are* C^∞ *on their respective intervals. Moreover, the* f_k *and* g_k $(k = 1,\ldots,n)$ *satisfy systems of explicit homogeneous linear differential equations of mth order with constant coefficients and, if* 0 *is both in E and E′, then the* $h_l^{(m-1)}(t)$ *are linear combinations of the* $f_k^{(s)}(t/\alpha_l)$ *and also of the* $g_k^{(s)}(t/\beta_l)$ $(k = 1,\ldots,n; l = 1,\ldots,m; s = 0, 1,\ldots,m-1)$.

The reader may notice not only the similarity but also the difference between Theorems 4 and 5. There is no analogue in Theorem 5 to the absence of first order terms stated in Theorem 4. This returns if we go over from real to complex domains and have $\alpha_l = 1$, $\beta_l = \varepsilon^l$ $(l = 1,\ldots,m)$, where $\varepsilon = \exp(2\pi i/m)$ is an mth root of unity: then the derivatives of orders $2,\ldots,m-1$ are missing in the systems of differential equations for f_k and g_k. In this case we have also an analogue of the last statement in Theorem 4: the functions

$\phi_j = \sum_{l=1}^{m} \varepsilon^{jl} f_l^{(j)}$ satisfy homogeneous linear differential equations of mn-th order with constant coefficients, having only terms of pmth order ($p = 0, 1, \ldots, n$); see Exercise 16.

Remark 3. The proof for regularization, which we gave in Theorem 4, can easily be adapted to the case where $\mathbb{R}$ is replaced by a locally compact abelian group, where E and E' are open subsets and where local Lebesgue measurability is replaced by local Harr measurability. However, the conclusion with differential equations is no longer valid and an alternative approach has to be used. As one may guess, the alternative approach is to replace differential equations by difference type functional equations.

If G is an abelian group, a function $f: G \to \mathbb{C}$ is called a *polynomial of degree at most n* if

$$\Delta_y^{n+1} f(x) = 0 \quad (x, y \in G),$$

where $\Delta_y f(x) = f(x+y) - f(x)$ and $\Delta_y^{k+1} f(x) = \Delta_y(\Delta_y^k f)(x)$ ($k = 1, 2, \ldots, n$).

For abelian topological groups G, a function $f: G \to \mathbb{C}$ is called an *exponential polynomial,* if

$$f = \sum_{i=1}^{n} m_i P_i,$$

where P_i is a continuous complex-valued polynomial on G and if $m_i: G \to \mathbb{C}$ is continuous and satisfies

$$m_i(x+y) = m_i(x) m_i(y) \quad ((x, y) \in G).$$

The following theorem (Székelyhidi 1981*c*; cf. McKiernan 1977*a*, *b*) holds:

Theorem 6. *Let G be a locally compact abelian group divisible by 2. Suppose f, g, f_k, g_k ($k = 1, \ldots, n$) are Haar measurable complex functions on G for which (1) is valid for all x, y in G. Then f and g are exponential polynomials.*

For further results we refer the reader to McKiernan 1977*a*, *b*; Penney–Rukhin 1979; Székelyhidi 1981*c*; and Rukhin 1982; and to the references in these papers.

Exercises and further results

1. (Aczél–Chung 1982.) Determine the general locally Lebesgue integrable solutions of equation (13).
2. By specializing the result of the previous exercise determine the general locally Lebesgue integrable solution of

$$f(x+y) = f_1(x) g_1(y) + f_2(x) g_2(y) \quad (x, y \in \mathbb{R}).$$

3. Same problem as in Exercise 2, this time for the equation

$$f(x+y) = f_1(x) g_1(y) + f_2(x).$$

4*. (Járai 1984, 1986.) Prove that Theorem 4 remains valid when f and g are not supposed to be Lebesgue integrable, only measurable.

5. Is Proposition 2 still valid when E and E' are two nonempty *closed* intervals in $\mathbb{R}$?

6. Same question as in the preceding exercise, this time when E and E' are nonempty open subsets of $\mathbb{R}$ (not necessarily intervals).

(For Exercises 7–8, see Kaczmarz 1924.)

7. Show that, if $\phi, f:\mathbb{R}\to\mathbb{R}$ satisfy the functional equation

$$f(x)+f(x+y)=\phi(y)f\left(x+\frac{y}{2}\right) \quad (x,y\in\mathbb{R}), \tag{25}$$

then ϕ satisfies the functional equation

$$\phi(x+y)+\phi(x-y)=\phi(x)\phi(y) \quad (x,y\in\mathbb{R}).$$

8. Find all measurable solutions of the functional equation (25) in the preceding exercise.

9. Find all Lebesgue integrable solutions $f:\mathbb{R}\to\mathbb{R}$ of the functional equation

$$f(x+y+A)-f(x-y-A)=2f(x)f(y) \quad (x,y\in\mathbb{R}),$$

where A is a real constant (cf. Exercise 13.9)

10. (E.g. Guilmard 1972.) Show that every locally integrable solution $f:\mathbb{R}\to\mathbb{C}$ of the functional equation

$$f(x)^2-f(y)^2=f(x+y)f(x-y) \quad (x,y\in\mathbb{R})$$

is continuous.

11*. (Haruki 1984c.) Prove the following theorem: A meromorphic function $f:\mathbb{C}\to\mathbb{C}$ is a solution of the functional equation

$$(f(x+y)+f(x-y))(f(x)^2+f(y)^2)=2f(x)f(y)(f(x+y)f(x-y)+1) \quad (x,y\in\mathbb{C}),$$

if, and only if,

$$f(x)=\mathrm{cn}(\alpha x,k),$$

where α, k are arbitrary complex numbers and cn denotes Jacobi's elliptic cosine function.

12. (Vincze 1978b.) Let G be an abelian group divisible by 2 and let F be a field of characteristic zero. Determine the general solutions $f,g:G\to F$ of

$$f(x+y)g(x-y)=f(x)g(y)g(x)g(-y) \quad (x,y\in G).$$

(For Exercises 13–14, see Horinouchi–Kannappan 1971.)

13. Let G be a commutative integral domain and F be a commutative field of characteristic different from 2. Show that, if there exists a not identically zero additive $f:G\to F$ such that

$$f(xy)=\alpha(x)f(y)+\alpha(y)f(x) \quad (x,y\in G), \tag{26}$$

then $\alpha:G\to F$ is additive, not identically zero and

$$\alpha(x^3)-3\alpha(x)\alpha(x^2)+2\alpha(x)^3=0. \tag{27}$$

Does (26) have a nontrivial additive solution for every additive α satisfying (27)?

14. Let G be a commutative integral domain and F a commutative field of characteristic different from 2. Prove or disprove that there exists an additive

and not identically zero $f:G\to F$ such that

$$f(xy)=\alpha(x)f(y)+\beta(y)f(x)\quad(x,y\in G),$$

where $\alpha,\beta:G\to F$ are not identically equal, if, and only if, α and β are additive and satisfy

$$\alpha(x^2)-\alpha(x)^2=\beta(x^2)-\beta(x)^2.$$

15. Prove Theorem 5.
16. Prove the last statement in Remark 2.
17. (Vincze 1960*b*, 1963*a*.) Determine the general solutions $f,f_1,f_2,g_1,g_2:G\to F$ (G is an abelian group, F a commutative quadratically closed field divisible by 2) of

$$f(x+y)=f_1(x)g_1(y)+f_2(x)g_2(y).$$

This is the equation in Exercise 2, but here no regularity conditions are supposed and the domain is more general.

15

A further generalization of Pexider's equation. A uniqueness theorem. An application to mean values.

In this chapter, we first study more general functional equations of the Pexider type than those in Section 4.3. We keep our arguments in a quite general algebraic setting. The main feature is the proof of a typical uniqueness theorem in the theory of functional equations. A direct application of this versatile theorem is given to the representation of classical means, like arithmetic and quasiarithmetic means. Among these, which shall be studied more thoroughly in Chapter 17 (where we will also apply this uniqueness theorem again), we will consider in particular the arithmetic, the geometric, the exponential means and the root-mean-power.

A common generalization of Pexider's equations (4.25)

$$k(x+y) = g(x) + h(y) \tag{1}$$

and

$$k(x+y) = g(x)l(y)$$

is

$$k(x+y) = g(x)l(y) + h(y). \tag{2}$$

Here, as in Section 4.3, we consider functions defined on an abelian (commutative) groupoid N (written additively) with a neutral element 0, but with values in a field F. This equation is again an example of one equation determining more (four) unknown functions. Putting $y = 0$ into (2) we get

$$k(x) = g(x)l(0) + h(0). \tag{3}$$

If $l(0) = 0$ then, by (3), k is constant. This case can be easily handled, and we exclude it here. So $l(0) \neq 0$ and, from (3),

$$g(x) = \frac{k(x) - h(0)}{l(0)}. \tag{4}$$

Putting this back into (2) and introducing

$$\phi(y) = \frac{l(y)}{l(0)}, \quad \psi(y) = h(y) - h(0)\frac{l(y)}{l(0)}, \tag{5}$$

we get

$$k(x+y) = k(x)\phi(y) + \psi(y) \quad (x, y \in N). \tag{6}$$

It is equation (6) which has important applications (see, for instance, the proof of Theorem 8 below) and, when it is solved, we can determine l, h and g using (5) and (4), respectively (Exercise 1). We can reduce (6) further by substituting $x = 0$ this time:

$$k(y) = k(0)\phi(y) + \psi(y). \tag{7}$$

By subtracting this eqution from (6) and denoting

$$\kappa(x) = k(x) - k(0), \tag{8}$$

we get

$$\kappa(x+y) = \kappa(x)\phi(y) + \kappa(y) \quad (x, y \in N). \tag{9}$$

Notice that there are four unknown functions in (2), three in (6) and just two in (9).

We distinguish two cases in (9). First take

$$\phi(y) = 1 \quad \text{for all } y \in N. \tag{10}$$

Then (9) goes over into Cauchy's equation and

$$\kappa(x) = a(x) \quad (x \in N),$$

where $a: N \to F$ is an arbitrary additive function. From (8) and (7), with $k(0) = b$ (constant) we have now

$$\phi(x) = 1, \quad k(x) = a(x) + b, \quad \psi(x) = a(x) \quad (x \in N). \tag{11}$$

If (10) does not hold, then there exists a y_0 such that

$$\phi(y_0) \neq 1. \tag{12}$$

Interchanging x and y in (9), so that we can profit from the commutativity of N, we get

$$\kappa(x)\phi(y) + \kappa(y) = \kappa(x+y) = \kappa(y+x) = \kappa(y)\phi(x) + \kappa(x),$$

or

$$\kappa(x)[\phi(y) - 1] = \kappa(y)[\phi(x) - 1].$$

Here we put in the y_0 for which (12) holds and write $c = \kappa(y_0)/[\phi(y_0) - 1]$ (constant) in order to get

$$\kappa(x) = c[\phi(x) - 1]. \tag{13}$$

We have now two subcases. If $c = 0$, then $\kappa(x) = 0$ on N and ϕ can be arbitrary (cf. (9)). Writing again $k(0) = b$, we get from (7) and (8)

$$k(x) = b, \quad \phi \text{ arbitrary}, \quad \text{and} \quad \psi(y) = b[1 - \phi(y)] \text{ on } N. \tag{14}$$

If, on the other hand, $c \neq 0$ then, substituting (13) into (9), we get

$$\phi(x + y) = \phi(x)\phi(y) \quad \text{for all } x, y \in N.$$

Thus ϕ is a homomorphism from N into the multiplicative structure of the field F. By (13), (8) and (7) we get, with $b := k(0) - c$ this time,

$$\phi(x) = e(x), \quad k(x) = ce(x) + b, \quad \psi(x) = b[1 - e(x)] \quad (x \in N), \tag{15}$$

where e is an arbitrary solution of

$$e(x + y) = e(x)e(y). \tag{16}$$

It is easy to check that under these circumstances (11), (14) and (15) always satisfy (6). We have proved the following (cf., e.g., Aczél 1966*c*, pp. 148–53, 1969*a* pp. 20–2):

Theorem 1. *The general solutions* $k, \phi, \psi : N \to F$ *(*N *an abelian groupoid with a neutral element,* F *a field) of* (6) *on* N^2 *are given by* (11), (14) *and* (15) *where* b, c *are arbitrary constants and* e, a *are arbitrary solutions of* (16) *and of* $a(x + y) = a(x) + a(y)$ $(x, y \in N)$, *respectively.*

Corollary 2. *The general solutions of* (6) *on* $\bar{\mathbb{R}}_+^2$ *in the class of functions* $k, \phi, \psi : \bar{\mathbb{R}}_+ \to \mathbb{R}$, *where* k *is strictly monotonic, are given by*

$$\phi(x) = 1, \quad k(x) = ax + b, \quad \psi(x) = ax, \tag{17}$$

and

$$\phi(x) = e^{ax}, \quad k(x) = ce^{ax} + b, \quad \psi(x) = b[1 - e^{ax}], \tag{18}$$

where $a \neq 0$, $c \neq 0$, *but otherwise* a, b, c *are arbitrary real constants.*

Next we deal with a very simple functional equation already encountered as (10.25) – it can also be considered as a special case of (1) – the Jensen equation (so called, because it is the equality case of the Jensen inequality (9.20), (10.13)),

$$f\left(\frac{x + y}{2}\right) = \frac{f(x) + f(y)}{2} \quad [(x, y) \in I^2], \tag{19}$$

in which I is an arbitrary (open, closed, half-closed, finite or infinite) interval. The general continuous solution can be found for instance in a way similar to that applied in Chapter 8 to (8.1), but we know also immediately that it is given by

$$f(x) = ax + b \quad (x \in I), \quad a, b \text{ arbitrary real constants}, \tag{20}$$

since, by (19) and the continuity, f has to be both convex and concave on I and is therefore affine. Similarly, we can prove the following about the 2-place Jensen equation ((19) is the 1-place Jensen equation).

Proposition 3. *The general continuous solution of*

$$G\left(\frac{x+y}{2},\frac{u+v}{2}\right)=\frac{G(x,u)+G(y,v)}{2}\quad (x,y,u,v\in I)$$

is given by

$$G(x,u)=ax+bu+c\quad (x,u\in I);\quad a,b,c \text{ arbitrary real constants.}$$

Remark. If $0\in I$ then it is particularly easy to find the general solution of the 1- or 2- (or n-) place Jensen equations. For instance for (19), $y=0$ gives $f(x/2)=[f(x)+f(0)]/2$ so $A(x)=f(x)-f(0)$ satisfies $A(x/2)=A(x)/2$ and, by (19), also $A[(x+y)/2]=[A(x)+A(y)]/2$. Thus

$$A(x+y)=2A\left(\frac{x+y}{2}\right)=A(x)+A(y),$$

and *the general solution of* (19) (with $0\in I$, but this condition can be eliminated, cf. also Theorem 4.12) *is given by* $f(x)=A(x)+b$, *where* A *is an arbitrary additive function and* b *is a constant.*

But here (19) will serve as an example of a *uniqueness theorem* on functional equations (Aczél 1964*a*), which has been followed by many others (Aczél–Hosszú 1965; Howroyd 1969, 1970*b*; Miller 1970*a*; Ng 1970; Paganoni Marzegalli 1971*a*, *b*; Paganoni 1972*b*, 1974*a*, 1976; Mills 1974, etc.).

Theorem 4. *Let* $f_1,f_2:I\to\mathbb{R}$ *be continuous solutions of the equation*

$$f[F(x,y)]=H[f(x),f(y),x,y]\quad [(x,y)\in I^2],\tag{21}$$

where I *is an (open, closed, half-open, finite or infinite) interval. Suppose that* $F:I^2\to I$ *is continuous and internal that is,*

$$\min(x,y)<F(x,y)<\max(x,y)\quad \text{if } x\neq y\quad (x,y\in I),\tag{22}$$

and that either $u\mapsto H(u,v,x,y)$ *or* $v\mapsto H(u,v,x,y)$ *are injections (if* $H(t_1,v,x,y)=H(t_2,v,x,y)$ *for fixed* v,x,y *or if* $H(u,t_1,x,y)=H(u,t_2,x,y)$ *for fixed* u,x,y, *then* $t_1=t_2$*). Further, let* $\gamma,\delta\in I$, $\gamma<\delta$, *and*

$$f_1(\gamma)=f_2(\gamma)\quad \text{and}\quad f_1(\delta)=f_2(\delta).\tag{23}$$

Then

$$f_1(x)=f_2(x)\quad \text{for all } x\in I.\tag{24}$$

Proof. We divide the proof of (24) into three steps. We first show that

(i) $$f_1(x)=f_2(x)\quad \text{for all } x\in[\gamma,\delta].$$

Indeed, in view of (23), suppose that there exists an $x_0\in]\gamma,\delta[$ such that $f_1(x_0)\neq f_2(x_0)$. Write

$$C=\sup\{x\,|\,f_1(x)=f_2(x),\gamma\leqslant x<x_0\}$$

and

$$D=\inf\{x\,|\,f_1(x)=f_2(x),x_0<x\leqslant\delta\}.$$

By these definitions and by the continuity of f_1 and f_2

$$C < x_0 < D, \quad f_1(C) = f_2(C), \quad f_1(D) = f_2(D), \tag{25}$$

but

$$f_1(x) \neq f_2(x) \quad \text{for all } x \in]C, D[. \tag{26}$$

This leads to a contradiction, since both f_1 and f_2 satisfy (21) and, with (25), we get

$$f_1[F(C,D)] = H[f_1(C), f_1(D), C, D] = H[f_2(C), f_2(D), C, D] = f_2[F(C,D)]. \tag{27}$$

But, by the internality (22),

$$F(C,D) \in]C, D[,$$

so that (27) contradicts (26), and (i) is proved.

The remaining two steps are to show that

(ii) $f_1(x) = f_2(x)$ from the left end of I (say α, which may also be $-\infty$) up to γ and

(iii) $f_1(x) = f_2(x)$ from δ up to the right end of I (say ε, which may also be ∞).

The two proofs are completely similar, so we prove only (ii). Also we suppose that it is $u \mapsto H(u, v, x, y)$ which is injective. In view of (i), define

$$A = \inf\{t \mid f_1(x) = f_2(x) \text{ for all } x \in [t, \delta]\} \tag{28}$$

(notice the difference compared to the definition of D in (i)). By (i), $A \leqslant \gamma < \delta$. We will prove that $A = \alpha$, the left end of I. If this were not so, there would exist an increasing sequence

$$x_n \to A \quad (x_n \in I), \tag{29}$$

such that

$$f_1(x_n) \neq f_2(x_n) \quad (n = 1, 2, \ldots). \tag{30}$$

Take an m such that

$$F(x_m, \delta) \in]A, \delta]. \tag{31}$$

(There is such an m because, if we had

$$F(x_n, \delta) \leqslant A \quad \text{for all } n = 1, 2, \ldots,$$

then, F being continuous, with $n \to \infty$ (see (29))

$$F(A, \delta) \leqslant A$$

would hold, contrary to the strict inequalities (22).) Applying (21) again, we get, in view of (28), (31) and (23),

$$\begin{aligned} H[f_1(x_m), f_1(\delta), x_m, \delta] = f_1[F(x_m, \delta)] &= f_2[F(x_m, \delta)] \\ &= H[f_2(x_m), f_2(\delta), x_m, \delta] \\ &= H[f_2(x_m), f_1(\delta), x_m, \delta]. \end{aligned}$$

Since $u \mapsto H(u, v, x, y)$ is injective (one-to-one), this would imply

$$f_1(x_m) = f_2(x_m),$$

in contradiction to (30). This contradiction proves (ii) and, since the proof of (iii) goes the same way, we have proved Theorem 4.

This also proves, as a very special case, that the general continuous solution of (19) is given by (20). Indeed, in this case,

$$F(x, y) = \frac{x + y}{2},$$

which is both continuous and internal. Also

$$u \mapsto H(u, v, x, y) = \frac{u + v}{2}$$

is an injection ('one-to-one'). Now take $\gamma, \delta \in I$ so that $\gamma < \delta$. Let f be an arbitrary continuous solution of (19). Then the function defined by

$$f_1(x) = \frac{f(\delta) - f(\gamma)}{\delta - \gamma} x + \frac{\delta f(\gamma) - \gamma f(\delta)}{\delta - \gamma}$$

is affine (that is, of the form (20)) and

$$f_1(\gamma) = f(\gamma), \quad f_1(\delta) = f(\delta).$$

So, by Theorem 4,

$$f(x) = f_1(x) = ax + b \quad \text{for all } x \in I,$$

as asserted. Of course, (20) does satisfy (19).

Now we give another application of Theorem 4, in the theory of quasi-arithmetic mean values.

Let $k: \bar{\mathbb{R}}_+ \to \mathbb{R}$ be a continuous and strictly monotonic function (injection) with the inverse k^{-1}. The range of k is an interval (also $\bar{\mathbb{R}}_+$ may be replaced by other infinite or finite intervals). Then

$$M(x, y) = k^{-1}\left(\frac{k(x) + k(y)}{2}\right) \quad (x, y \in \bar{\mathbb{R}}_+) \tag{32}$$

is a quasi-arithmetic mean. The arithmetic mean $[k(x) = x]$, the exponential mean $(k(x) = e^{cx}, c \neq 0)$, the root-mean-square $(k(x) = x^2)$, the root-mean-power $(k(x) = x^c$ for $c > 0$; if $\bar{\mathbb{R}}_+$ is replaced by $\mathbb{R}_+$ or a subinterval of $\mathbb{R}_+$, then, also, the root-mean-powers with $c < 0$, including the harmonic mean with $k(x) = 1/x)$ and the geometric mean $(k(x) = \log x)$ are examples of quasi-arithmetic means. The function k in (32) is, however, not uniquely determined. If M has a representation of the form (32) with k, then it has the same representation with $\phi k + \psi$, where $\phi \neq 0$ and ψ are real constants. We now prove that these are the only functions with this property.

Corollary 5. *We have*

$$k^{-1}\left(\frac{k(x)+k(y)}{2}\right)=f^{-1}\left(\frac{f(x)+f(y)}{2}\right) \quad (x,y\in\bar{\mathbb{R}}_+) \tag{33}$$

for two continuous, strictly monotonic functions $f, k:\bar{\mathbb{R}}_+\to\mathbb{R}$, *if and only if there exist constants* $\phi\neq 0$, ψ *such that*

$$f(x)=\phi k(x)+\psi \quad \textit{for all } x\in\bar{\mathbb{R}}_+. \tag{34}$$

We prove a stronger result. In order to formulate it, we rewrite (33), with (32) as

$$f[M(x,y)]=\frac{f(x)+f(y)}{2} \tag{35}$$

or, with

$$g(x):=k^{-1}(f(x)), \quad \text{that is,} \quad f(x)=k(g(x)), \tag{36}$$

as

$$g[M(x,y)]=k^{-1}\left(\frac{kg(x)+kg(y)}{2}\right)=M[g(x),g(y)] \quad (x,y\in\bar{\mathbb{R}}_+).$$

In order that (36) makes sense, we should have $f(\bar{\mathbb{R}}_+)\subseteq k(\bar{\mathbb{R}}_+)$. Both are intervals, since f and k are continuous. Equation (33) remains valid if f is replaced by $\alpha f+\beta$ $(\alpha\neq 0)$ and by appropriate choice of the constants α and β, one can indeed make the range of this new function over the nonnegative reals a subset of the range of k. While it is clear from (36) that g is continuous and strictly monotonic, we will suppose only continuity and replace $\bar{\mathbb{R}}_+$ by an arbitrary (finite or infinite, open, half-open or closed) interval I. So (with M replaced by F) we look for continuous functions $g:I\to I$ such that

$$F(g(x),g(y))=g(F(x,y)) \quad \text{for all } x,y \text{ in } I, \tag{37}$$

where F denotes the quasi-arithmetic mean

$$F(x,y)=k^{-1}\left(\frac{k(x)+k(y)}{2}\right) \quad (x,y\in I) \tag{38}$$

for some continuous and strictly monotonic $k:I\to\mathbb{R}$. Note that $F(x,y)\in I$, so that (37) makes sense. The range of k will be denoted, as before, by $k(I)$.

Proposition 6. *Let* a, b *be such that* $ak(I)+b$ *is a subset of* $k(I)$. *The functions*

$$g(x)=k^{-1}(ak(x)+b) \tag{39}$$

are the only continuous solutions of (37) *with* (38).

Clearly, Corollary 5 follows from Proposition 6 (compare (39) to (36) in order to get (34), in view of the strict monotonicity of f in Corollary 5).

Proof of Proposition 6. Equation (37), with (38), is satisfied by the functions of the form (39).

On the other hand, equation (37) is of the form (21). Since F as defined by (38) is continuous and internal (satisfies (22)) and $u \mapsto H(u,v,x,y) = F(u,v)$ is injective, the conditions of Theorem 4 are satisfied.

If g is constant, then (39) is satisfied with $a=0$. If g is not constant, then there exist $\gamma, \delta \in I$ such that $\gamma \neq \delta$ and $g(\gamma) \neq g(\delta)$. Let us write $h(x) := k[g(x)]$. As k is strictly monotonic, $h(\gamma) \neq h(\delta)$. Define g_1 by

$$g_1(x) = k^{-1}\left(\frac{h(\delta)-h(\gamma)}{k(\delta)-k(\gamma)}k(x) + \frac{k(\delta)h(\gamma)-k(\gamma)h(\delta)}{k(\delta)-k(\gamma)}\right) \quad (x \in I). \tag{40}$$

(We have $k(\gamma) \neq k(\delta)$ because k is strictly monotonic.) So $g_1(\gamma) = g(\gamma)$ and $g_1(\delta) = g(\delta)$. Also g and g_1 are both continuous and both satisfy (37). Thus, by Theorem 4,

$$g(x) = g_1(x) \quad \text{for all } x \in I$$

and, since g_1, as defined in (40), has the form

$$g_1(x) = k^{-1}(ak(x) + b),$$

we have proved

$$g(x) = k^{-1}(ak(x) + b) \quad (x \in I)$$

(a, b constants), that is (39), and have finished the proof of Proposition 6.

As a generalization of a special case of Proposition 6, we prove the following:

Proposition 7. *The solutions* $g: \bar{\mathbb{R}}_+ \to \bar{\mathbb{R}}_+$ *of*

$$\left(\frac{g(x)^2 + g(y)^2}{2}\right)^{1/2} = g\left[\left(\frac{x^2+y^2}{2}\right)^{1/2}\right] \tag{41}$$

are of the form $(ax^2+b)^{1/2}$ *with* $a \geqslant 0$, $b \geqslant 0$.

Note. As usual, $t^{1/2} \geqslant 0$ for *all* $t \in \bar{\mathbb{R}}_+$. So equation (41) makes sense only if $g(z) \geqslant 0$ for all $z \in \bar{\mathbb{R}}_+$. That is why we have supposed $g(\bar{\mathbb{R}}_+) \subseteq \bar{\mathbb{R}}_+$.

Proof. If g is assumed to be continuous, Proposition 6 leads to $g(x) = (ax^2+b)^{1/2}$ under the condition that $ax + b \in \bar{\mathbb{R}}_+$ for all $x \in \bar{\mathbb{R}}_+$. This clearly implies $a \geqslant 0$ and $b \geqslant 0$.

However, continuity for g is a consequence of the functional equation (41) itself and can be seen as follows. Equation (41) is nothing but (37) with

$$F(x,y) = \left(\frac{x^2+y^2}{2}\right)^{1/2},$$

that is, the root-mean square. So here we have (38) with $k(x) = x^2$. The range of k is an interval $J = k(\bar{\mathbb{R}}_+)$. (We do not write $J = (\bar{\mathbb{R}}_+)^2$ in order to avoid confusion with $\bar{\mathbb{R}}_+^2 = \bar{\mathbb{R}}_+ \times \bar{\mathbb{R}}_+$.) Define $h: k(\bar{\mathbb{R}}_+) \to \bar{\mathbb{R}}_+$ by $h = k \circ g \circ k^{-1}$.

Clearly, $x' = k(x)$, $y' = k(y)$ will be in the interval J and (37) becomes

$$\frac{h(x') + h(y')}{2} = h\left(\frac{x' + y'}{2}\right) \quad \text{for all } x', y' \in J. \tag{42}$$

Equation (42) is the Jensen equation. On the other hand, 0 belongs to J, since $k(0) = 0$. Therefore $h - h(0)$ satisfies the Cauchy equation on J^2 (see the Remark after Proposition 3). But the values of h are also in $\bar{\mathbb{R}}_+$, so $h - h(0)$ is bounded from below. Therefore, Theorem 2.1 and Corollary 2.5 lead to the continuity of h. This proves the continuity of g, since $h = k \circ g \circ k^{-1}$, which concludes the proof of Proposition 7.

A similar proposition can be proved for other root-mean-powers (Exercise 10) or the exponential mean (indeed, for all quasi-arithmetic means (38), see Exercise 12). It can easily be extended to quasi-arithmetic means of n variables

$$M(x_1, \ldots, x_n) = k^{-1}\left(\frac{k(x_1) + \cdots + k(x_n)}{n}\right).$$

Conversely, if we have a function $f: I \to \mathbb{R}^n$, we may ask about the existence of a continuous and strictly monotonic k such that $f(x) = k^{-1}(ak(x) + b)$, that is,

$$k \circ f = ak + b \tag{43}$$

for some real constants $a(\neq 0)$, b. This is equivalent to asking whether f leaves some quasi-arithmetic mean invariant. The equation (43) has been the subject of a large number of works (see Kuczma 1968). As it is a functional equation in a single variable, we shall give no details here.

Next we determine which quasi-arithmetic means are translative (homogeneous with respect to addition, cf. Chapter 20), that is,

$$M(x + t, y + t) = M(x, y) + t \quad \text{for all } x, y, t \in \bar{\mathbb{R}}_+ \tag{44}$$

for (32), or

$$k^{-1}\left(\frac{k(x + t) + k(y + t)}{2}\right) = k^{-1}\left(\frac{k(x) + k(y)}{2}\right) + t \quad (x, y, t \in \bar{\mathbb{R}}_+). \tag{45}$$

Keeping t fixed for the moment and introducing the (continuous, strictly monotonic) function f_t by

$$f_t(x) = k(x + t), \tag{46}$$

its inverse will be given by

$$f_t^{-1}(z) = k^{-1}(z) - t,$$

so that (45) goes over into

$$k^{-1}\left(\frac{k(x)+k(y)}{2}\right)=f_t^{-1}\left(\frac{f_t(x)+f_t(y)}{2}\right)\quad (x,y\in\bar{\mathbb{R}}_+).$$

By Corollary 5 (34),

$$f_t(x)=\phi(t)k(x)+\psi(t)$$

(the 'constants' ϕ, ψ now depend upon t, since for every t we have a separate equation of the form (33)). In view of (46), this is

$$k(x+t)=k(x)\phi(t)+\psi(t)\quad (x,t\in\bar{\mathbb{R}}_+),$$

that is, equation (6). The conditions of Corollary 2 are satisfied, so we have either

$$k(x)=cx+b\quad (c\neq 0)\quad \text{or}\quad k(x)=ae^{cx}+b\quad (ac\neq 0).$$

Substituting these into (32), and verifying that arithmetic and exponential means are translative, we have proved the following:

Theorem 8. *The arithmetic and the exponential means,*

$$\frac{x+y}{2}\quad \textit{and}\quad \frac{1}{c}\log\left(\frac{e^{cx}+e^{cy}}{2}\right)\quad (c\neq 0),$$

and only these are translative quasi-arithmetic means, that is, satisfy both (32) *and* (44).

It can be proved similarly (on $\mathbb{R}_+$ or a subinterval) that the *gometric mean and the root-mean-power,*

$$(xy)^{1/2}\quad \textit{and}\quad \left(\frac{x^c+y^c}{2}\right)^{1/c}.$$

and only these are homogeneous quasi-arithmetic means (32). Here we mean *homogeneity of degree one with respect to multiplication,*

$$M(xu,yu)=uM(x,y)\quad (x,y,u\in\mathbb{R}_+).$$

Thus *the arithmetic is the only quasi-arithmetic mean which is both translative and homogeneous* (that is, homogeneous with respect to both addition and multiplication). The same also holds for n variables. But in our case of two-place means, a result similar to the last assertion can be proved in an elementary way on $\mathbb{R}$ (and on more general structures) *without the quasi-arithmeticity supposition.*

Proposition 9. *The function $M:\mathbb{R}^2\to\mathbb{R}$ has the properties*

$$M(x+t,y+t)=M(x,y)+t\quad \textit{and}\quad M(xu,yu)=M(x,y)u\quad (x,y,t,u\in\mathbb{R}, u\neq 0)\tag{47}$$

if, and only if, there exists a constant q such that

$$M(x,y)=(1-q)x+qy \quad (x,y\in\mathbb{R}). \tag{48}$$

If M is symmetric, then $q=\frac{1}{2}$.

Proof. From (47), with $M(0,1)=q$, we have

$$M(x,y)=M[0+x,(y-x)+x]=M(0,y-x)+x=M(0,1)(y-x)+x$$
$$=(1-q)x+qy \quad \text{for } y\neq x$$

and, with $x=y=0$, $M(0,0)=M(0u,0u)=M(0,0)u$ for all $u\neq 0$, thus $M(0,0)=0$ and

$$M(t,t)=M(0,0)+t=t.$$

The rest of Proposition 9 is obvious.

Again, the same result could also be proved in the same way on more general algebraic structures than the reals (cf. Exercise 19). Notice that, had we supposed $M(xu,yu)=M(x,y)u$ (for $x,y\in\mathbb{R}$ but) *only for* $u>0$, then instead of (48) only

$$M(x,y)=\begin{cases}(1-q)x+qy & \text{if } y\geqslant x\\ px+(1-p)y & \text{if } y<x\end{cases}$$

would hold (p,q arbitrary constants). Even if M is supposed to be symmetric, we have only $p=q$, but not necessarily $q=\frac{1}{2}$.

Exercises and further results

1. (Cf. Aczél 1969*a*, pp. 20–2.) Determine all solutions $g,h,k,l:N\to F$ of equation (2), where N is a commutative groupoid with neutral element and F is a commutative field.

(For Exercises 2–6, cf. Janardan 1978.)

2. Let M denote the exponential mean

$$M(x,y)=\frac{1}{\alpha}\log\left(\frac{e^{\alpha x}+e^{\alpha y}}{2}\right),$$

where $\alpha\neq 0$ is a constant. Find all solutions $f:\mathbb{R}_+\to\mathbb{R}$ of the functional equation

$$f(x)f(y)=f(M(x,y)) \quad (x,y\in\mathbb{R}_+) \tag{49}$$

which are bounded on a subset of positive measure of $\mathbb{R}_+$.

3. Same question as in the previous exercise, but M is now an arbitrary quasi-arithmetic mean on a real interval I and $f:I\to\mathbb{R}$,

4. Solve the Pexider analogue

$$f(x)g(y)=h(M(x,y)) \quad (x,y\in\mathbb{R}_+)$$

of the functional equation (49) in Exercise 2.

5*. Using Exercise 4, prove the following result: A nondegenerate nonnegative random variable X has the Weibull distribution

$$P(X \leqslant x) = 1 - \exp(-\lambda x^{\alpha}) \quad (x \geqslant 0)$$

$(\lambda > 0, \alpha > 0)$ if, and only if,

$$P(X > M(x, y) > x) = P(X > y).$$

6. Using the notation in Exercise 2, solve for all $x_i \in \mathbb{R}_+$, $y_i \in \mathbb{R}_+$ $(i = 1, \ldots, n)$

$$f(x_1, \ldots, x_n) f(y_1, \ldots, y_n) = f(M(x_1, y_1), \ldots, M(x_n, y_n)),$$

where $f: \mathbb{R}_+^n \to \mathbb{R}$.

7. Let G and F be two abelian groups divisible by 2. For $f: G \to F$, show that the following two functional equations are equivalent:

$$f(x + y) + f(x - y) = 2f(x) \quad (\forall x, y \in G)$$

and

$$f\left(\frac{x + y}{2}\right) = \frac{f(x) + f(y)}{2}.$$

8. Determine the general continuous solutions $f: \mathbb{R} \to \mathbb{R}$ of

$$f(x + y) + f(x - y) = 2f(x),$$

using the result of Exercise 7.

9*. (Aczél 1966c, pp. 53–57.) Let $I \subseteq \mathbb{R}$ be a nondegenerate interval and $\circ$: $I^2 \to I$. Prove that the equation

$$f(x + y) = f(x) \circ f(y) \quad (x, y \in \mathbb{R})$$

has a nonconstant continuous solution $f: I \to \mathbb{R}$ if, and only if, I is open and $(I, \circ)$ is a continuous group.

(For Exercises 10–18, see Dhombres 1982b.)

10. Let $r \neq 0$ be a real constant. Find all solutions $g: \bar{\mathbb{R}}_+ \to \bar{\mathbb{R}}_+$ of

$$\left(\frac{g(x)^r + g(y)^r}{2}\right)^{1/r} = g\left(\left(\frac{x^r + y^r}{2}\right)^{1/r}\right) \quad (x, y \in \bar{\mathbb{R}}_+).$$

Solve the same problem if $\bar{\mathbb{R}}_+$ is replaced by $\mathbb{R}$ but r is a positive odd integer.

11. Find all solutions $g: \mathbb{R}_+ \to \mathbb{R}_+$ of

$$g((xy)^{1/2}) = \frac{2g(x)g(y)}{g(x) + g(y)} \quad (x, y \in \mathbb{R}_+).$$

12*. Let I be a proper subinterval of $\mathbb{R}$. Let M be a quasi-arithmetic mean defined on I by a continuous injection $f: I \to \mathbb{R}$:

$$M(x, y) = f^{-1}\left(\frac{f(x) + f(y)}{2}\right) \quad (x, y \in I).$$

A function $g: I \to I$ is *M-affine* if

$$g(M(x, y)) = M(g(x), g(y)) \quad (x, y \in I)$$

and it is *f-affine* if, for some $A \neq 0$, a in $\mathbb{R}$,

$$f(g(x)) = Af(x) + a \quad (x \in I).$$

Find a condition on $f(I)$, necessary and sufficient for all M-affine functions to be f-affine.

13. Find all solutions $g:\mathbb{R}_+ \to \mathbb{R}_+$ of

$$g\left(\frac{2xy}{x+y}\right)=\frac{2g(x)g(y)}{g(x)+g(y)} \quad (x, y\in\mathbb{R}_+).$$

14. Find all solutions $g:\bar{\mathbb{R}}_+ \to \mathbb{R}$ of

$$g(x)g(y)=g((xy)^{1/2})^2 \quad (x, y\in\bar{\mathbb{R}}_+).$$

15*. Let I be a nonempty open convex subset of $\mathbb{R}^2$ $(n>1)$. Suppose that $f:I\to I$ is a homeomorphism and define

$$M(x, y)=f^{-1}\left(\frac{f(x)+f(y)}{2}\right).$$

Find a condition on I, necessary and sufficient for all solutions $g:I\to I$ of

$$g(M(x, y))=M(g(x), g(y)) \quad (x, y\in I)$$

to be of the form

$$g(x)=Af(x)+a \quad (x\in I),$$

where A is an $n\times n$ matrix and $a\in\mathbb{R}^n$.

16. Find all solutions $f:\bar{\mathbb{R}}_+ \to \bar{\mathbb{R}}_+$ of

$$F\left(\frac{x^2+y^2}{2}\right)^{1/2}=\frac{F(x)+F(y)}{2} \quad (x, y\in\bar{\mathbb{R}}_+).$$

17*. Find all solutions $F:\mathbb{R}_+ \to \mathbb{R}_+$ of

$$F((xy)^{1/2})=\frac{F(x)+F(y)}{2} \quad (x, y\in\mathbb{R}_+).$$

18. Find all solutions $g:\bar{\mathbb{R}}_+ \to \mathbb{R}_+$ of

$$g\left(\frac{x+y}{2}\right)=\frac{2g(x)g(y)}{g(x)+g(y)} \quad (x, y\in\bar{\mathbb{R}}_+).$$

19. In Proposition 9, replace $\mathbb{R}$ by a ring R. Under what restriction on R will the statement remain true?

20. Let $(E, \|\cdot\|)$ be a real normed space. Let $A:\bar{\mathbb{R}}_+^2 \to \bar{\mathbb{R}}_+$ be of the form

$$A(x, y)=f^{-1}(f(x)+f(y)),$$

where $f:\bar{\mathbb{R}}_+ \to \bar{\mathbb{R}}_+$ is strictly increasing, continuous and $f(0)=0$. Let $M:\bar{\mathbb{R}}_+^2 \to \bar{\mathbb{R}}_+$ be a quasi-arithmetic mean

$$M(x, y)=g^{-1}\left(\frac{g(x)+g(y)}{2}\right),$$

where $g:\bar{\mathbb{R}}_+ \to \bar{\mathbb{R}}_+$ is strictly monotonic and continuous. Suppose that we have

$$M(\|u+v\|^2, \|u-v\|^2)=A(\|u\|^2, \|v\|^2) \quad \text{for all } u, v\in E.$$

Show that the norm $\|\cdot\|$ comes from an inner product.

21. (Cf. Aczél 1984*d*.) Determine all solutions of (cf. (6))

$$f(xy) = f(x)g(y) + h(y)$$

(i) on $\mathbb{R}_+^2$, (ii) on $[1, \infty[^2$ and (iii) on I^2 where $1 \in I \subset \mathbb{R}_+$ and I is an interval.

22*. (Young 1987; Aczél 1987*a*, *b*.) For the utility $u(x)$ of an income x, let the real valued continuous function u, defined (i) on $I = \mathbb{R}_+$ or (ii) on $I = [1, \infty[$. The 'sacrifice' $u(x) - u(y)$ (y being the income decreased, say, by taxation), 'when equal is scale-invariant' if

$$u(x) - u(y) = u(x') - u(y') \Rightarrow u(rx) - u(ry) = u(rx') - u(ry') \quad (x, y, x', y' \in I).$$

In both cases (i) and (ii), determine all continuous nonconstant functions $u: I \to \mathbb{R}$ satisfying this implication.

16

More about conditional Cauchy equations. Applications to additive number theoretical functions and to coding theory

This chapter is entirely devoted to conditional Cauchy equations which were introduced in Chapter 6 and already studied in Chapter 7. In the first section, by using a fixed point theorem and also the order structure of $\mathbb{R}$, and by adding a strong regularity condition on the unknown function, we deduce the addundancy of some conditions on rather thin sets. In the second and third sections, we use group decomposition and then proceed to an application to additive number theory. An analogous equation in the fourth section yields an application to information theory, for mean codeword lengths. In the two last sections, we come back to additive number theory, proving the basic result that logarithms are the only monotonic functions defined on integers which transform a product into a sum. We extend this and similar results in various ways.

16.1 Expansions of the Cauchy equation from curves

Thus far, for conditional Cauchy equations, we have been dealing with rather 'large' subsets Z (see Chapters 6 and 7). Under some additional regularity assumptions on f, we may proceed to far smaller Z. A typical result can be obtained in the plane by using for Z some curve (Dhombres 1979, pp. 3.32–3.37). We recall from Section 7.1 that $Z \subset \mathbb{R}^2$ is addundant in a class of functions if the restriction of the Cauchy equation from $\mathbb{R}^2$ to Z changes neither the domain nor the form of its general solution in the class.

Theorem 1. *Let $g: \mathbb{R} \to \mathbb{R}$ and $h: \mathbb{R} \to \mathbb{R}$ be two continuous functions such that $h(0) = g(0) = 0$ and $g(x)h(x) > 0$ for $x \neq 0$. Let $h + g$ be a bijection from $\mathbb{R}$ onto $\mathbb{R}$. Then*

$$Z = C_{g,h} = \{(g(x), h(x)) \mid x \in \mathbb{R}\}$$

(a curve) is addundant in the class of all functions differentiable at 0.

Proof. The conditional Cauchy equation relative to $C_{g,h}$ can be written as

$$f(g(x)+h(x))=f(g(x))+f(h(x)) \quad \text{for all } x \text{ in } \mathbb{R}. \tag{1}$$

Using $H(x)=g(x)+h(x)$, where $H:\mathbb{R}\to\mathbb{R}$ is a bijection, and $y=H(x)$, the equation (1) is transformed into

$$f(y)=f(g[H^{-1}(y)])+f(y-g[H^{-1}(y)]) \quad \text{for all } y \text{ in } \mathbb{R}. \tag{2}$$

It is nicer to transform (2) into a functional equation somewhat similar to Jensen's equation (15.19). We define

$$G(y)=g(H^{-1}(y)) \quad \text{and} \quad F(y)=\begin{cases} f(y)/y & \text{for } y\neq 0 \\ f'(0) & \text{for } y=0. \end{cases}$$

(remember that f is in the class of functions differentiable at 0 and $f(0)=0$) in order to get

$$yF(y)=G(y)F[G(y)]+[y-G(y)]F[y-G(y)].$$

If we define α by $\alpha(y)=G(y)/y$ for $y\neq 0$ and $\alpha(0)=1$, then we get

$$F(y)=\alpha(y)F[G(y)]+[1-\alpha(y)]F[y-G(y)]. \tag{3}$$

We note the following consequences for G and α.

If $H(z)>0$ then, from our hypotheses, we get $0<g(z)<H(z)$, which yields $0<G(y)<y$ for all $y>0$.

In the same way, we deduce $y<G(y)<0$ for all $y<0$. Thus $0<\alpha(y)<1$ for $y\neq 0$ and, by definition, $\alpha(0)=1$. We see that $G(y)=y$, that is, $\alpha(y)=1$ if and *only if* $y=0$. Our choice of $F(0)=f'(0)$ was such that F is continuous at 0, and our choice of $\alpha(0)=1$ assures that (3) holds also at $y=0$. Therefore Theorem 1 will be a consequence of the following lemma:

Lemma 2. *Let $\alpha:\mathbb{R}\to\mathbb{R}$ be such that $0\leqslant\alpha(y)\leqslant 1$ and let $G:\mathbb{R}\to\mathbb{R}$ be continuous and satisfy*

$$0<G(y)<y \quad \text{for } y>0,$$
$$0>G(y)>y \quad \text{for } y<0.$$

Then every solution $F:\mathbb{R}\to\mathbb{R}$ of (3), which is continuous at 0, is constant.

The *proof* of this lemma is a nice and easy interplay between continuity and the kind of convexity represented by (3). This equation shows that $F(G(y))$ and $F(y-G(y))$ cannot both be greater than $F(y)$ and cannot both be smaller than $F(y)$. Thus, we start from an arbitrarily chosen positive number y_0. We construct a sequence $\{y_n\}$ $(n=0,1,2,\ldots)$ as follows. If y_n is known, y_{n+1} is either $G(y_n)$ or $y_n-G(y_n)$, the choice being made in such a way that $F(y_{n+1})\leqslant F(y_n)$.

We also construct a sequence $\{y'_n\}$ by the opposite choice: We start from $y'_0=y_0$. If y'_n is known, y'_{n+1} is that one of the two values $G(y'_n)$ or $y'_n-G(y'_n)$

for which $F(y'_{n+1}) \geqslant F(y'_n)$. The sequence $\{y_n\}$ is nonincreasing and bounded from below by zero:

$$0 \leqslant y_{n+1} \leqslant \max(G(y_n), y_n - G(y_n)) \leqslant y_n.$$

Similarly, we find that $\{y'_n\}$ is nonincreasing and bounded below. The sequence $\{y_n\}$ then has a limit y_∞ which, by the continuity of G, satisfies the inequalities

$$y_\infty \leqslant \max(G(y_\infty), y_\infty - G(y_\infty)) \leqslant y_\infty.$$

This yields either $y_\infty = G(y_\infty)$ or $G(y_\infty) = 0$ and, by the conditions on G, we get $\lim_{n\to\infty} y_n = y_\infty = 0$ in both cases. The same result applies to the sequence $\{y'_n\}$, so that $\lim_{n\to\infty} y'_n = 0$. Going back to y_n and y'_n, we may write

$$F(y_n) \leqslant F(y_0) \leqslant F(y'_n).$$

Both $\{F(y_n)\}$ and $\{F(y'_n)\}$ converge to $F(0)$ because of the continuity of F at 0, so $F(y_0) = F(0)$, proving that F is constant, at least on $\mathbb{R}_+$, since y_0 was arbitrarily chosen on this set. But the same argument works for $y_0 < 0$, using an inequality of the form $y_n \leqslant \min(G(y_n), y_n - G(y_n)) \leqslant y_{n+1} < 0$. This concludes the proof of the Lemma 2, and thus also the proof of Theorem 1.

We also get

Corollary 3. *All solutions, differentiable at* 0, *of the functional equation*

$$f(2x) = f(x + \sin x) + f(x - \sin x) \tag{4}$$

are given by $f(x) = ax$ *(a arbitrary).*

The following corollary was obtained by Zdun 1972 in connection with a proof of Lemma 2.

Corollary 4. *Let Z be the graph of an increasing continuous function $h: \mathbb{R} \to \mathbb{R}$ with $h(0) = 0$. Then Z is addundant for the class of all functions differentiable at* 0.

Otherwise stated, if $h: \mathbb{R} \to \mathbb{R}$ is an increasing continuous function with $h(0) = 0$ and if $f'(0)$ exists, then the only solution $f: \mathbb{R} \to \mathbb{R}$ of

$$f(x + h(x)) = f(x) + f(h(x)) \tag{5}$$

is $f(x) = ax$ for some a in $\mathbb{R}$. For the *proof* it is enough to show that for $g(x) = x$ the hypotheses of Theorem 1 are fulfilled. One has to prove that $x + h(x)$ is a continuous bijection from $\mathbb{R}$ onto $\mathbb{R}$. But $\lim_{x\to-\infty}(x + h(x)) = -\infty$, and in the same way $\lim_{x\to\infty}(x + h(x)) = \infty$. Moreover, $x \mapsto x + h(x)$ is increasing and continuous.

Curves such as $C_{g,h}$ in Theorem 1 are not very oscillatory and we would like to obtain at least the regular solutions of (5) for more general continuous h, and not only for the increasing ones. Let us define the cone $C_1 = \{(x, y) | x = y = 0 \text{ or } 0 < y < -x \text{ or } 0 > y > -x\}$ in the plane $\mathbb{R}^2$ (see Figure 13).

Theorem 5. *If Z is the graph of a continuous $h: \mathbb{R} \to \mathbb{R}$ with $h(0) = 0$ and if Z is in the cone C_1, then Z is addundant for the class of odd functions differentiable at 0.*

Proof. With $H(x) = -h(x)$, we may write (5) as

$$f(x) = f(x - H(x)) - f(-H(x)) = f(x - H(x)) + f(H(x)).$$

In the same way as in Theorem 1 and under the same kind of conditions, we can reduce this equation to (3), using our hypothesis that Z is in C_1. Lemma 2 yields the conclusion.

Clearly, the cone C_1 can be replaced by its reflection through the line $y = x$ without changing the conclusion of Theorem 5.

The existence of a derivative at 0 for f is essential for the validity of Theorems 1 and 5. If we just require that f be continuous, then the general solution of an equation like (1) or (5) depends upon an arbitrary (continuous) function and the set Z is no longer addundant for this class of continuous functions. This can be shown by a very simple example where we choose $h(x) = x$ in (5). We then get

$$f(2x) = 2f(x) \quad (x \in \mathbb{R}). \tag{6}$$

Let ϕ be a continuous real valued function on $[\frac{1}{2}, 1]$ such that $2\phi(\frac{1}{2}) = \phi(1)$.

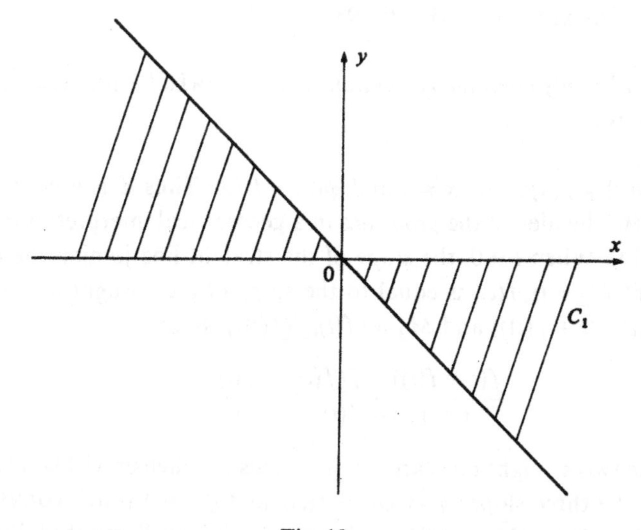

Fig. 13

We define

$$f(0)=0,$$
$$f(x)=\phi(x)\quad \text{for } x\in]\tfrac{1}{2},1],$$

and

$$f(x)=2^n\phi\left(\frac{x}{2^n}\right)\quad \text{for } x\in]2^{n-1},2^n],$$

$$f(x)=2^{-n}\phi(2^n x)\quad \text{for } x\in]2^{-n-1},2^{-n}]\quad (n\in\mathbb{N}).$$

The function f as constructed above clearly satisfies (6) on $[0,\infty[$ and is continuous there. For example, at the point 1 we get

$$f(1+)=\lim_{x\to 1+} f(x)=\lim_{x\to 1+} 2\phi\left(\frac{x}{2}\right)=2\phi(\tfrac{1}{2}+)=2\phi(\tfrac{1}{2})=\phi(1)=f(1)=f(1-)$$

and, as to the point 0, with $M=\max_{x\in]1/2,1]}|\phi(x)|$, we notice that $|f(x)|\leqslant 2Mx$ if $x>0$ so that $f(0+)=0=f(0)$. With another continuous ψ on $[-1,-\frac{1}{2}]$, satisfying $2\psi(-\frac{1}{2})=\psi(-1)$, we may construct a continuous solution of (6) depending upon the almost arbitrarily chosen continuous functions ϕ and ψ. However, if we require f to be differentiable at 0, it is easy to see directly or as a consequence of Theorem 1, that $\psi(x)=\phi(x)=ax$ for some constant $a\in\mathbb{R}$.

It is still an open problem to solve (5) with just a continuity assumption, when h is equal to the unknown function f, that is,

$$f(x+f(x))=f(x)+f(f(x))\quad (x\in\mathbb{R}). \tag{7}$$

There are solutions of the form $x\mapsto \sup(0,\alpha x)$ or $x\mapsto \inf(0,\alpha x)$. More generally, all continuous idempotent ($f[f(x)]\equiv f(x)$) solutions of (7) are known (cf. Section 19.3, in particular (19.41)). Also, the continuous solutions of (7) for which $f'(0)$ exists are known (Forti 1983).

Theorem 6. *The only continuous solutions of* (7), *for which* $f'(0)$ *exists, are the linear functions* $f(x)=ax$.

We define $\phi(x)=f(x)/x$ for $x\neq 0$ and $\phi(0)=f'(0)$. Thus ϕ is a continuous function on $\mathbb{R}$. The idea of the *proof* lies in a geometrical interpretation of (7) (see Figure 14). When $t\neq 0$, the slope of the straight line joining the origin and the point $M_1=(t,f(t))$ is equal to the slope of the straight line joining $M_2=(t+f(t),f(t+f(t)))$ and $M_3=(f(t),f(f(t)))$, since

$$\frac{f(t+f(t))-f(f(t))}{t+f(t)-f(t)}=\frac{f(t)}{t}.$$

Therefore the two straight lines are parallel. This parallelism yields inequalities between the three slopes $\phi(t)$, $\phi(t+f(t))$, and $\phi(f(t))$ which correspond to the three lines $0M_1$, $0M_2$ and $0M_3$. These inequalities will give the following.

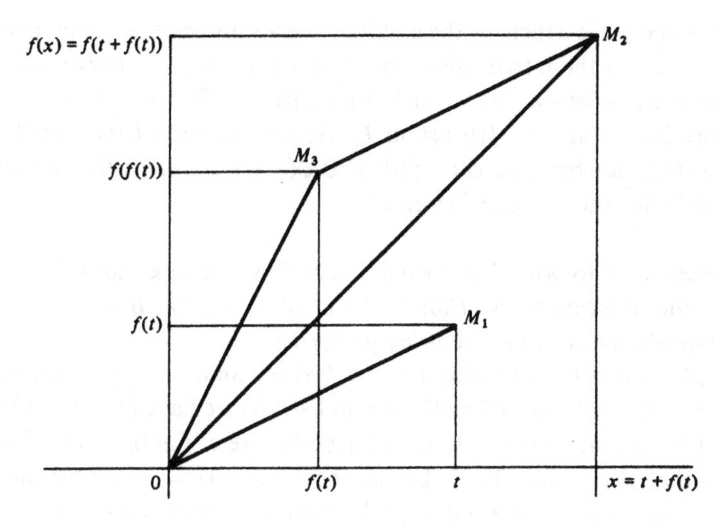

Fig. 14

Lemma 7. *Let $f:\mathbb{R}\to\mathbb{R}$ be a continuous solution of* (7), *differentiable at* 0. *Take an $x\neq 0$ for which $f(x)\neq 0$, $x+f(x)\neq 0$ and, if $(x+f(x))/x<0$, then $(-x+f(-x))/(-x)\leqslant 0$. Define*

$$I_x=\{t\mid t\in\mathbb{R}, 0<|t|<|x|\}\quad \textit{and}\quad \phi(x)=\frac{f(x)}{x}.$$

Then there exist y,z in I_x such that

$$\phi(y)\leqslant\phi(x)\leqslant\phi(z).$$

Proof. We define

$$I'_x=\{t\mid t\in I_x, tx>0\}.$$

Let $x\neq 0$, $f(x)\neq 0$, and $x+f(x)\neq 0$. Suppose there exists a $t\in I'_x$ so that $t+f(t)=x$. Then $f(t)\in I'_x$. Because of the parallelism, as explained earlier, $\phi(x)$ lies between $\phi(t)$ and $\phi(f(t))$. Therefore, depending on the values $\phi(t)$ and $\phi(f(t))$, we may take either $y=t$, $z=f(t)$ or $y=f(t)$, $z=t$, as illustrated in Figure 14 (in the case $x>0$).

On the other hand, if, for all $t\in I'_x$, we have $t+f(t)\neq x$, then $t+f(t)$ remains for all $t\in I'_x$ on the same side of x as 0, because f is a continuous function, and $f(0)=0$. By continuity, the sign of $f(x)$ is opposite to that of x (the suppositions in Lemma 7 exclude the case $f(x)=0$). Moreover $[x+f(x)]/x<1$. Now we examine two cases.

If $[x+f(x)]/x>0$, we see that $x+f(x)\in I'_x$ while $f(x)\in I_x$. Using the parallelism, as explained earlier, we see that $\phi(x)$ lies between $\phi(x+f(x))$ and $\phi(f(x))$. Therefore, depending on the values $\phi(x+f(x))$ and $\phi(f(x))$, we may take either $y=x+f(x)$, $z=f(x)$ or $y=f(x)$, $z=x+f(x)$.

If $[x+f(x)]/x<0$ then, by the hypothesis made in Lemma 7, x lies between 0 and $f(-x)$. By the intermediate value theorem, since f is continuous and $f(0)=0$, there exists a $t\in I'_{-x}$ such that $f(t)=x$. But $t+f(t)=t+x\in I'_x$. Therefore both t and $t+f(t)$ are in I_x. Again, because of the parallelism, $\phi(x)=\phi(f(t))$ lies between $\phi(t+f(t))$ and $\phi(t)$ which yields the required y and z and ends the proof of Lemma 7.

We proceed now to *prove Theorem 6*. From (7) we deduce that $f(0)=0$. We first consider the case where $f'(0)\neq 0$ and $f'(0)\neq -1$. *Let B be the set of all positive numbers b with the following properties*:

For all $x\neq 0$ in $[-b,b]$, $f(x)\neq 0$, $x+f(x)\neq 0$ and, if $[x_0+f(x_0)]/x_0<0$ for some $x_0\in[-b,b]$, then for all $x\neq 0$ in $[-b,b]$ we have $[x+f(x)]/x<0$.

As $f'(0)\neq 0$, and $f'(0)\neq -1$, using the derivative of f at 0, we may find an $\varepsilon>0$ so that B contains $]0,\varepsilon]$. Let us show that $f(x)=xf'(0)$ for all x *in* $[-b,b]$ when $b\in B$. Let $x\neq 0$ in $[-b,b]$. Define Y as the subset of all $y\neq 0$ in $[-b,b]$ such that

$$\phi(y)\leqslant\phi(x).$$

Lemma 7 yields that $Y\neq\varnothing$. We prove by contradiction that 0 is a cluster point of Y. If not, with $c=|\inf Y|>0$, we apply Lemma 7 to c, using the continuity of f to get $\phi(c)\leqslant\phi(x)$. Then we deduce the existence of some y in Y with $|y|<c$. This contradicts the definition of c.

As 0 is a cluster point of Y, using the continuity of ϕ at 0, we get

$$\phi(x)\geqslant\lim_{\substack{y\in Y\\ y\to 0}}\phi(y)=f'(0).$$

In a similar way, we introduce the set Z of all $z\neq 0$ in $[-b,b]$ for which $\phi(z)\geqslant\phi(x)$. And, in the same way, we obtain

$$\phi(x)\leqslant\lim_{\substack{z\in Z\\ z\to 0}}\phi(z)=f'(0).$$

Therefore $\phi(x)=f'(0)$ or $f(x)=f'(0)x$ for all x in $[-b,b]$. We shall now prove that $B=\mathbb{R}_+$, which is enough to conclude the proof of Theorem 6 in the case $f'(0)\neq 0$ and $f'(0)\neq -1$. Suppose for contradiction that $\bar{b}=\sup_{b\in B}b<\infty$. By the continuity of f, we deduce that $f(x)=f'(0)x$ for all x in $[-\bar{b},\bar{b}]$. This continuity of f easily yields the existence of an $\eta>0$ such that $\bar{b}+\eta\in B$. This contradicts the definition of $\bar{b}$ and yields $B=\mathbb{R}_+$ as required.

Next, suppose $f'(0)=0$. *Let B' be the set of all positive numbers b with the following properties*:

For all $x\neq 0$ in $[-b,b]$, we have $x+f(x)\neq 0$ and, if $[x_0+f(x_0)]/x_0<0$ for some x_0 in $[-b,b]$, then $[x+f(x)]/x<0$ for all $x\neq 0$ in $[-b,b]$.

As previously, we prove that B' is not empty. Suppose that for some x_0

in $[-b,b]$ we have $f(x_0)\neq 0$. We use again the set Y of all $y\neq 0$ in $[-b,b]$ such that $\phi(y)\leqslant\phi(x)$. Due to Lemma 7, Y is not empty. Either $\bar{Y}\cap\ker f\neq\emptyset$ which then gives $\phi(x_0)\geqslant 0$ or $\bar{Y}\cap\ker f=\emptyset$. We see in this last case, as previously, that 0 is a cluster point of Y, which gives also $\phi(x_0)\geqslant 0$. Now using Z instead of Y, we deduce that $\phi(x_0)\leqslant 0$. Thus $\phi(x_0)=0$, which contradicts $f(x_0)\neq 0$. Therefore $f(x)=0$ for all x in $[-b,b]$ where $b\in B'$. In the same way as was done for B, we prove that B' is unbounded. Thus $B'=\mathbb{R}_+$ and $f(x)=0$ for all x in $\mathbb{R}$.

Finally, we suppose that $f'(0)=-1$. Define B'' as the set of all positive numbers b such that $|[x+f(x)]/x|<\frac{1}{2}$ for all $x\neq 0$ in $[-b,b]$. The set B'' is not empty. If, for some $x_0\in[-b,b]$, where $b\in B''$, we have $[f(x_0)+x_0]/x_0>0$, since $f(x_0)\neq 0$ (a consequence of $|[x_0+f(x_0)]/x_0|<\frac{1}{2}$), we may apply Lemma 7 to get that there exists a $z\in I_x$ with

$$\phi(z)\geqslant\phi(x_0).$$

Let Z, as before, be the set of all $z\neq 0$ in $[-b,b]$, such that $\phi(z)\geqslant\phi(x_0)$. For no z in Z can we have $[f(z)+z]/z<0$ because this would give $-1>\phi(z)\geqslant\phi(x_0)>-1$, a contradiction. Therefore, as before, we prove that 0 is a cluster point of Z and so $\phi(x_0)\leqslant -1$. But this contradicts our hypothesis $[f(x_0)+x_0]/x_0>0$. The only remaining possibility is that $[f(x)+x]/x\leqslant 0$ for all x in $[-b,b]$. Lemma 7 can be applied and we deduce, as before, that, for all x in $[-b,b]$, we have $\phi(x)=-1$, that is, $f(x)=-x$. From our definition of B'' and in view of the continuity of f, we see that B'' is unbounded. Finally, $f(x)=-x$ for all $x\in\mathbb{R}$. This concludes the proof of Theorem 6.

The result of Theorem 6 can also be stated in the following way. Let $\mathscr{X}$ map the function $g:\mathbb{R}\to\mathbb{R}$ into its graph. Then $\mathscr{X}$ is additively expandable (cf. Section 7.1) for the class of continuous functions differentiable at 0.

The technique developed for the proof of Theorem 6 brings some results, similar to Theorem 1, where we require less from g and h, but more from the class of functions f which are considered ($f'(0)$ exists and f is continuous in a neighbourhood of zero; see Exercises 9 and 10).

16.2 Cylindrical conditions

We shall now turn to another kind of conditional Cauchy equation, the so-called 'cylinder' case (Dhombres–Ger 1978). In this section and the next, we will use again notions from group theory such as liftings and canonical epimorphisms and, later in this section, some basic concepts of measure theory and topology (such as Lebesgue measure and Baire property).

Proposition 8. *Let Y be a nonempty subset of an abelian group G and denote by G_0 the subgroup of G generated by Y. Let F be an abelian group. A mapping*

$f:G \to F$ is a solution of a conditional Cauchy equation relative to $Z = G \times Y$, that is,

$$f(x+y) = f(x) + f(y) \quad (x \in G,\ y \in Y) \tag{8}$$

if, and only if, f can be written in the form

$$f(x) = g(x - \xi[\pi(x)]) + h(\pi(x)), \tag{9}$$

where $h: G/G_0 \to F$ is a function with $h(0) = 0$, $g: G_0 \to F$ is an additive function, $\xi: G/G_0 \to G$ is a lifting relative to G_0 such that $\xi(0) = 0$ and $\pi: G \to G/G_0$ is the canonical epimorphism.

Proof. Recall that a lifting $\xi: G/G_0 \to G$, for the quotient group G/G_0, is a mapping such that $\pi \circ \xi$ is the identity on G/G_0.

First let $f:G \to F$ be a mapping given by (9) with the stated properties. For $y \in Y$ and $x \in G$, one gets

$$\begin{aligned} f(x+y) &= g(x + y - \xi[\pi(x)]) + h(\pi(x)) \\ &= g(x - \xi[\pi(x)]) + g(y) + h(\pi(x)) \\ &= f(x) + g(y) \\ &= f(x) + g(y - \xi[\pi(y)]) + h(\pi(y)) \\ &= f(x) + f(y). \end{aligned}$$

Conversely, let $f:G \to F$ be a solution of (8) (that is, a $G \times Y$-additive function). The restriction $g: G_0 \to F$ of f is clearly additive on G_0. Consider now the subset H of all y in G for which

$$f(x+y) = f(x) + f(y)$$

for all x in G. We prove that H is a subgroup and contains G_0. Let $y \in Y$ and take $x = 0$ in (8). We deduce that $f(y) = f(0) + f(y)$ and so $f(0) = 0$. Therefore $0 \in H$. Let $y \in H$. Then $0 = f(0) = f(-y + y) = f(-y) + f(y)$ which yields $f(-y) = -f(y)$. Now we compute that, for $x \in G$,

$$f(x) = f(x - y + y) = f(x - y) + f(y),$$

and so

$$f(x) + f(-y) = f(x) - f(y) = f(x - y).$$

Therefore $-y \in H$. Let $y_1, y_2 \in H$. Then $y_1 + y_2 \in H$, as can be seen from

$$\begin{aligned} f(x + y_1 + y_2) &= f(x + y_1) + f(y_2) \\ &= f(x) + f(y_1) + f(y_2) \\ &= f(x) + f(y_1 + y_2). \end{aligned}$$

Now we select a lifting $\xi: G/G_0 \to G$ such that $\xi(0) = 0$. As $x - \xi[\pi(x)]$ belongs to G_0 and therefore also to H, we deduce that

$$f(x) = f(\xi[\pi(x)]) + f(x - \xi[\pi(x)]).$$

Using the definition of g, this yields

$$f(x) = g(x - \xi[\pi(x)]) + f(\xi[\pi(x)]). \tag{10}$$

We may define $h: G/G_0 \to F$ by

$$h(\tilde{x}) = f(\xi(\tilde{x})) \quad (\tilde{x} \in G/G_0).$$

In particular, $h(0) = f(\xi(0)) = f(0) = 0$. Then (10) reads as

$$f(x) = g(x - \xi[\pi(x)]) + h(\pi(x)),$$

which concludes the proof of Proposition 8.

Notes. In [8] we have a *cylinder type functional equation.* If there exists a lifting $\xi: G/G_0 \to G$, which is also a homomorphism, then the abelian group G can be identified with the direct product $G_0 \oplus (G/G_0)$. With this identification, we may write any x in G in the form

$$x \oplus x'',$$

where $x' \in G_0$, $x'' \in G/G_0$. Then the general solution $f: G \to F$ of (8) can be written as

$$f(x) = g(x') + h(x''),$$

where $g: G_0 \to F$ is additive and $h: G/G_0 \to F$ is a mapping with $h(0) = 0$.

From Proposition 8, and the definitions of Section 7.1, we deduce the following:

Corollary 9. *Let $Y \subset \mathbb{R}$ be a subset of positive Lebesgue measure. Then $\mathbb{R} \times Y$ is addundant.*

Proof. Because of Theorem 2.6, the subgroup G_0 generated by Y is $\mathbb{R}$ itself. Thus $\mathbb{R}/G_0$ is reduced to one point 0, $\xi(0) = 0$ and $h(0) = 0$. Thus we see from Proposition 8 that $f(x) = g(x)$ for all x in $\mathbb{R}$. Since g is additive, so is f.

Corollary 9 can be generalized to any locally compact and connected topological group if we replace the positive Lebesgue measure for Y by a nonzero Haar measure (see Exercise 3).

An analogue of Corollary 9 can be obtained if we replace the condition regarding the measure of Y by a topological one. We make use of the following classical definition.

A *meagre subset* of a topological space is a countable union of subsets A_n for which the closure $\bar{A}_n$ has empty interior.

A subset Y of a topological space G has the *Baire property* if there exists a nonempty open subset $\mathcal{O}$ of G such that $\mathscr{C}Y \cap \mathcal{O}$ is meagre, where $\mathscr{C}Y$ is the complement of Y in G.

Theorem 10. *Let G be a connected and locally compact abelian group. Let F*

be an abelian group. Suppose that the subset Y of G has the Baire property and is not meagre. Then any $f: G \to F$ satisfying

$$f(x+y) = f(x) + f(y) \quad \textit{for all } x \in G,\ y \in Y$$

is additive on G^2.

In other words (Section 7.1), $G \times Y$ is expandable.

Proof. Let $Y \subset G$ be as stated. There exists a nonempty open subset $\mathcal{O}$ of G such that $\mathcal{O} \cap \mathscr{C}Y$ is meagre. Let $x \in \mathcal{O}$ and define $\mathscr{V} = \mathcal{O} - x = \{y - x \mid y \in \mathcal{O}\}$ $Z = Y - x$. Clearly, $\mathscr{V}$ is a neighbourhood of 0 and, for any t in $\mathscr{V}$, we define $\mathscr{V}_t = \mathscr{V} \cap (t - \mathscr{V})$. This subset $\mathscr{V}_t$ is open since $\mathscr{V}$ is open and $\mathscr{V}_t$ is not empty since $t \in \mathscr{V}_t$. Moreover, from

$$\mathscr{V}_t \cap \mathscr{C}Z \subset \mathscr{V} \cap \mathscr{C}Z = (\mathcal{O} \cap \mathscr{C}Y) - x$$

and

$$\mathscr{V}_t \cap \mathscr{C}(t - Z) \subset (t - \mathscr{V}) \cap (t - \mathscr{C}Z) = t - (\mathscr{V} \cap \mathscr{C}Z),$$

we deduce that $\mathscr{V}_t \cap \mathscr{C}Z$ and $V_t \cap \mathscr{C}(t - Z)$ are meagre. So the union $\mathscr{V}_t \cap \mathscr{C}(Z \cap (t - Z))$ is also meagre.

Suppose we are able to prove that $\mathscr{V}_t$ is not meagre. Then the subset $\mathscr{V}_t \cap (Z \cap (t - Z))$ cannot be meagre. This strong result implies a seemingly far simpler one, namely that $Z \cap (t - Z)$ is not empty. This is all what we need to conclude the proof of Theorem 10. Indeed, if this is true, then for every t in $\mathscr{V}$ there exist $z_0 \in Z$ and $y_0 \in Z$ with

$$y_0 = t - z_0.$$

With $y_0 = -x + y$, $y \in Y$ and $z_0 = -x + z$, $z \in Y$, we have $t + 2x = y + z$. This yields $Y + Y \supset \mathscr{V} + 2x$. Thus $Y + Y$ contains a nonempty open subset of G. So the subgroup G_0 generated by Y is open. But G_0 must also be closed because, if x belongs to the closure of G_0, there exists y in the subgroup G_0 such that $x - y$ belongs to G_0, this last subset being a neighborhood of 0. But then $x = (x - y) + y$ belongs to G_0 and so $G_0 = \bar{G}_0$. However, G is supposed to be connected. Therefore $G_0 = G$. As in the proof of Corollary 9, we deduce that f is additive on G (with values in F).

The only thing which remains to be proved is that V_t is not meagre. This follows from a general result of Baire which we prove for completeness' sake.

Lemma 11. *A nonempty open subset of a locally compact space cannot be meagre.*

Proof. We want to prove that $\bigcup_{n=1}^{\infty} A_n$ does not cover a nonempty open subset of a locally compact space G whenever $\bar{A}_n$ has no interior ($n \geqslant 1$). We use $\mathcal{O}_n = \mathscr{C}\bar{A}_n$, an open and dense subset of G. We define $X = \bigcap_{n=1}^{\infty} \mathcal{O}_n$. To prove our lemma it is enough to show that X is dense in G.

Let $\mathcal{O}$ be a nonempty open subset of G. We shall prove that $X \cap \mathcal{O} \neq \varnothing$. We define $\mathscr{V}_0 = \mathcal{O}$ and, by induction, $\mathscr{V}_n$ as a nonempty open subset of G for which the closure $\bar{\mathscr{V}}_n$ is compact in $\mathscr{V}_{n-1} \cap \mathcal{O}_n$. This definition is possible in the nth step, since G is locally compact and $\mathscr{V}_{n-1} \cap \mathcal{O}_n \neq \varnothing$ because of the density of $\mathcal{O}_n$ in G. By a compactness argument, we see that $\mathscr{V} = \bigcap_{n=1}^{\infty} \mathscr{V}_n$ is not empty. But $\varnothing \neq \mathscr{V} \subset \bar{\mathscr{V}}_n \subset \mathscr{V}_{n-1} \subset \mathcal{O} \cap \mathcal{O}_n$. Finally, $\mathcal{O} \cap X \neq \varnothing$.

In the rest of this chapter we deal with equations satisfied by certain functions which are used in number theory, and apply some methods of the first two sections of this chapter.

16.3 Additive number theoretical functions and related equations

Number theorists have for a long time considered *additive number theoretical functions*, i.e., functions $f:\mathbb{N} \to \mathbb{R}$ such that, for all relatively prime numbers m, n in $\mathbb{N}$, we get

$$f(mn) = f(m) + f(n), \quad \text{where } (m, n) = 1. \tag{11}$$

This equation is a typical conditional equation. A classical example is the Euler function which associates to any integer n the number of its prime divisors (different from 1). In particular, the Euler function is equal to 1 on any prime number. We can construct a family of solutions of (11) as follows: let the values of a function $f:\mathbb{N} \to \mathbb{R}$ be arbitrarily given on the subset of all prime numbers and suppose $f(1) = 0$. We use the unique representation of an arbitrary positive integer n as a product of powers of prime numbers

$$n = p_1^{k_1} \cdots p_m^{k_m}$$

to define f everywhere on $\mathbb{N}$:

$$f(n) = k_1 f(p_1) + \cdots + k_m f(p_m).$$

Then $f:\mathbb{N} \to \mathbb{R}$ clearly satisfies (11) and in fact satisfies more:

$$f(mn) = f(m) + f(n) \quad \text{for all } m, n \text{ in } \mathbb{N}. \tag{12}$$

The set of all solutions of (12) for $f:\mathbb{N} \to \mathbb{R}$, that is, the set of all *totally additive number theoretical functions*, coincides with the set of the above extensions of all real valued functions defined on the subset of all prime numbers. For (12) this subset plays the role of a Hamel basis in the case of the usual additive Cauchy equation (Section 2.2). The example of the Euler function shows that (12) has fewer solutions than (11).

We shall further restrict the set of solutions by looking at functions

$f:[1,\infty[\to\mathbb{R}$ such that

$$f(xn)=f(x)+f(n) \quad \text{for all } x \text{ in } [1,\infty[\text{ and all } n \text{ in } \mathbb{N}. \tag{13}$$

Indeed, (13) can be turned into a conditional Cauchy equation of cylinder type on $G=\mathbb{R}_+$, an abelian group under multiplication, and $F=\mathbb{R}$, the usual abelian group under addition. To see this, we extend f to $\mathbb{R}_+$ by

$$F(x):=f(nx)-f(n) \quad \text{for some integer } n \text{ such that } nx\geqslant 1.$$

To prove that the definition of F makes sense, we notice that, if $nx\geqslant 1$ and $mx\geqslant 1$, then

$$f(nx)+f(m)=f(nmx)=f(mx)+f(n).$$

Moreover, as $f(1)=0$, $F(x)=f(x)$ for all $x\geqslant 1$. But $F:\mathbb{R}_+\to\mathbb{R}$ satisfies the functional equation

$$F(xn)=F(x)+F(n) \quad \text{for all } x \text{ in } \mathbb{R}_+ \text{ and all } n \text{ in } \mathbb{N}. \tag{14}$$

For the proof, suppose that m is an integer such that $mx\geqslant 1$. Then

$$F(x)+F(n)=f(mx)-f(m)+f(n)$$

and

$$F(xn)=f(mnx)-f(m).$$

But $f(mnx)=f(mx)+f(n)$ by (13), which yields (14).

Therefore we can determine all solutions of (13) by applying Proposition 8 to F, which satisfies (14) with $G=\mathbb{R}_+$ (and the multiplication as the operation), $Y=\mathbb{N}$, $G_0=\mathbb{Q}_+$ and $F=\mathbb{R}$. Most of the solutions have a weird behaviour. For instance, the restriction of every (discontinuous) solution of $f(xy)=f(x)+f(y)$ (see (3.1), (3.4)) to $[1,\infty[$ satisfies (13). (We will deal with a similar equation at the end of the next section.) As usual, some regularity assumption for f excludes those weird solutions. The technique in the following Proposition 12 is typical of such reductions and that is why we present it here, even though a direct proof of Proposition 12 is not difficult either (see e.g. Aczél–Daróczy 1975, pp. 22–3).

Proposition 12. *A monotonic function $f:[1,\infty[\to\mathbb{R}$ is a solution of the conditional Cauchy equation* (13) *if and only if there exists a real constant α such that $f(x)=\alpha\log x$ for all x in $[1,\infty[$.*

Proof. The functions $x\mapsto\alpha\log x$ are clearly monotonic solutions of (13). To prove the converse, we may, without loss of generality, suppose f to be nondecreasing (else replace f by $-f$). We first notice that the extension F of f, as previously defined, is also monotonic. (If $0<x\leqslant x'$, there exists an n such that $nx\geqslant 1$. Therefore $F(x)=f(nx)-f(n)\leqslant f(nx')-f(n)=F(x')$ if f is nondecreasing.) As F satisfies a conditional Cauchy equation of cylinder type, we apply Proposition 8 in order to get

$$F(x)=g\left(\frac{x}{\xi(\pi(x))}\right)+h(\pi(x)), \tag{15}$$

where π is the canonical epimorphism from $\mathbb{R}_+$ onto $\mathbb{R}_+/\mathbb{Q}_+$, and where $\xi:\mathbb{R}_+/\mathbb{Q}_+ \to \mathbb{R}_+$ is a lifting relative to $\mathbb{Q}_+$, $g:\mathbb{Q}_+ \to \mathbb{R}$ a homomorphism ($g(xy) = g(x) + g(y)$, for all x, y in $\mathbb{Q}_+$) and $h = \mathbb{R}_+/\mathbb{Q}_+ \to \mathbb{R}$ some function with $g(\xi(\pi(1))) = h(\pi(1))$. Take $x = \beta y$ where β is a rational number $0 < \beta \leqslant 1$. So $x \leqslant y$ and $F(x) \leqslant F(y)$. But $\pi(x) = \pi(y)$. Therefore

$$g\left(\frac{\beta y}{\xi(\pi(y))}\right) \leqslant g\left(\frac{y}{\xi(\pi(y))}\right),$$

which yields, as g is a homomorphism on $\mathbb{Q}_+$, that $g(\beta) \leqslant 0$. In the same way, we can prove that $g(\beta) \geqslant 0$ for a rational $\beta \geqslant 1$. We extend g to all of $\mathbb{R}_+$ by

$$G(x) = g(x) \quad \text{if } x \in \mathbb{Q}_+.$$

$$G(x) = \sup_{\substack{y<x \\ y \in \mathbb{Q}_+}} g(y) \quad \text{if } x \in \mathbb{R}_+ \backslash \mathbb{Q}_+.$$

A limit argument gives, for all (x, y) in $\mathbb{R}^2_+$,

$$G(xy) = G(x) + G(y). \tag{16}$$

Corollary 3.2 yields, for some α,

$$G(x) = \alpha \log x, x \in \mathbb{R}_+,$$

with a constant α. Using (15), we get, for x in $\mathbb{R}_+$,

$$F(x) = \alpha \log x + h(\pi(x)) - \alpha \log(\xi[\pi(x)]).$$

we define $\zeta(\pi(x)) = h(\pi(x)) - \alpha \log(\xi[\pi(x)])$ and will prove that $\zeta(\pi(x)) = 0$ for all x in $\mathbb{R}_+$.

Indeed, let $x \neq y$ in $\mathbb{R}_+$, and consider positive rational numbers γ, δ such that $\gamma x > y$ and $\delta x < y$. Since $\pi(\gamma x) = \pi(x) = \pi(\delta x)$,

$$F(\gamma x) \geqslant F(y) \quad \text{yields } \zeta(\pi(x)) - \zeta(\pi(y)) \geqslant \alpha \log \frac{y}{\gamma x}$$

$$F(\delta x) \leqslant F(y) \quad \text{yields } \zeta(\pi(x)) - \zeta(\pi(y)) \leqslant \alpha \log \frac{y}{\delta x}.$$

We may choose a sequence $\{\gamma_n\}$ so that $\lim_{n\to\infty}(y/\gamma_n x) = 1$ and a sequence $\{\delta_n\}$ so that $\lim_{n\to\infty}(y/\delta_n x) = 1$. We obtain

$$\zeta(\pi(x)) = \zeta(\pi(y)).$$

Since $x \neq y$ are arbitrary in $\mathbb{R}_+$, we get $\zeta(\pi(x)) = \zeta(\pi(1)) =$ constant for all x in $\mathbb{R}_+$. But from (14) this constant is 0. So $\zeta(\pi(x)) \equiv 0$, which concludes the proof of Proposition 12.

16.4 An application to mean codeword lengths

Proposition 12 can be applied to information theory (see e.g. Aczél–Daróczy 1975, pp. 145–47). Here we discuss the application of an equation similar to (13) in *coding theory.*

The Shannon entropy (5.30) (cf. (5.31)) has the important property that it is the (greatest) lower bound for the average codeword length in the following sense.

A code associates codewords of lengths $k_1, k_2, \ldots, k_n$ to n messages with probabilities (frequencies) $p_1, p_2, \ldots, p_n$, where $\sum_{j=1}^n p_j = 1$. A code has an 'alphabet' consisting of a number D ($\geqslant 2$), of symbols (in the most often used binary codes $D = 2$ and the symbols are represented by 0 and 1). The codewords are strings of such symbols, the codeword lengths are the lengths of these strings. A code is uniquely decipherable (or uniquely decodable) if, given a sequence of codewords written without spacing (without separating the consecutive codewords), the original sequence of messages, which were coded into these codewords, can be uniquely determined. While this seems to be a rather qualitative condition, it is equivalent (see, e.g., Gallager 1968, pp. 47–9) to a quantitative condition, the Kraft inequality

$$\sum_{j=1}^{n} D^{-k_j} \leqslant 1, \tag{17}$$

in the following sense. For every uniquely decipherable code the Kraft inequality (17) is satisfied. On the other hand, if (17) is satisfied for D and $k_1, k_2, \ldots, k_n$, then there *exists* a uniquely decipherable code with D symbols and with codewords of lengths $k_1, k_2, \ldots, k_n$.

The *average codeword length* is

$$\sum_{j=1}^{n} p_j k_j \tag{18}$$

and the source (or noiseless) coding theorem (see, e.g., Gallager 1968, p. 51; Aczél–Daróczy 1975, pp. 46–7) states that, for every uniquely decipherable code, the following inequality holds between the average codeword length and the Shannon entropy:

$$\sum_{j=1}^{n} p_j k_j \geqslant -\sum_{j=1}^{n} p_j \log_D p_j,$$

where $\log_D$ is the logarithm with base D (so we take $c = -1/\log D$ in (5.30)). Moreover, for given $p_1, p_2, \ldots, p_n$ and D, there exist positive integers $\tilde{k}_1, \tilde{k}_2, \ldots, \tilde{k}_n$ which satisfy (17) and

$$\sum_{j=1}^{n} p_j \tilde{k}_j < -\sum_{j=1}^{n} p_j \log_D p_j + 1.$$

Rényi (1960, 1965, 1970) introduced the following *entropies of rank a*

$$H_n^{(a,D)}(p_1, p_2, \ldots, p_n) = \frac{a+1}{a} \log_D \left(\sum_{j=1}^{n} p_j^{1/(a+1)} \right) \quad (a \neq 0).$$

These are exponential means (cf. Theorem 15.8) of the $-\log_D p_j$ with weight p_j and exponent $a/(a+1)$ while the Shannon entropies are arithmetic means of the same variables with the same weights. (Actually Rényi used base 2 and the parameter $1/(a+1)$ and called these entropies of order $1/(a+1)$, but the above form, while more complicated, is more convenient for our purposes.) It is easy to see that the Shannon entropy is the limiting case as a tends to 0:

$$\lim_{a\to 0} H_n^{(a,D)}(p_1,p_2,\ldots,p_n) = -\sum_{j=1}^{n} p_j \log_D p_j.$$

Campbell (1965) has proved that an analogue of the above coding theorem exists for $a > -1$ $(a \neq 0)$: In every uniquely decipherable code

$$\frac{1}{a}\log_D\left(\sum_{j=1}^{n} p_j D^{ak_j}\right) \geqslant H_n^{(a,D)}(p_1,p_2,\ldots,p_n)$$

and, for given $p_1,p_2,\ldots,p_n$ and D, there exist positive integers $\tilde{k}_1,\tilde{k}_2,\ldots,\tilde{k}_n$ which satisfy (17) and

$$\frac{1}{a}\log_D\left(\sum_{j=1}^{n} p_j D^{a\tilde{k}_j}\right) < H_n^{(a,D)}(p_1,p_2,\ldots,p_n) + 1.$$

The quantity on the left hand side of these inequalities is the *a-average codeword length*. Again, the average codeword length (18) is the limit of the a-average codeword length as $a \to 0$.

One may still think that the a-average codeword length has been introduced just in order that the last two inequalities should hold. However, it has other advantages. One usually tries to keep the average codeword length (18) small (and the source coding theorem shows that it cannot be made smaller than the Shannon entropy, but it can get within one unit – bit – of the Shannon entropy) for the following reason. In an economical code, frequently occurring messages (with large p_j) should have short codewords (small k_j), while rare messages (p_j small) may have long codewords (k_j large). This is nicely balanced in (18) – but also in the a-average codeword length

$$\frac{1}{a}\log_D\left(\sum_{j=1}^{n} p_j D^{ak_j}\right) \tag{19}$$

if $a > 0$. However, the average codeword length (18) can still be reasonably small even if the codeword lengths for *very infrequent* messages are *very large* which is not economical and often not even possible because of the bound (buffer) of capacity of the encoder (cf. Humblet 1981). It is easy to see that the limit of (19) as $a \to \infty$ is the largest k_j, independent of p_j as long as $p_j \neq 0$ (Hardy–Littlewood–Pólya 1934, p. 15), so, if a is large enough, then very large k_j's will contribute very strongly to the a-average codeword length, even for small p_j, thus keeping it from being small.

On the other hand, both (18) and (19) are *quasilinear mean codeword lengths*. This means that both are of the form (cf. (15.32) and (17.16))

$$M(p_1,p_2,\ldots,p_n;k_1,k_2,\ldots,k_n)$$
$$=\psi^{-1}\left[\sum_{j=1}^{n}p_j\psi(k_j)\right]\left(\sum_{j=1}^{n}p_j=1,p_j\geqslant 0,k_j\in\mathbb{N},j=1,2,\ldots,n\right), \qquad (20)$$

where ψ is a continuous and strictly monotonic function ($\psi(x)=x$ in (18) and $\psi(x)=D^{ax}$ in (19)). We note that in (20) ψ is applied to positive integers, so it would seem more reasonable to require strict monotonicity just on $\mathbb{N}$. But ψ^{-1} should be defined for $\sum_{j=1}^{n}p_j\psi(k_j)$, which changes continuously with the p_j, so we suppose ψ^{-1} to be defined and continuous on an interval. Accordingly, we suppose that $\psi:[1,\infty[\to\mathbb{R}$ is continuous. Further, it is easy to check that both (18) and (19) (for all a) are *additive*, that is, for these M,

$$M(p_1q_1,\ldots,p_1q_m,p_2q_1,\ldots,p_2q_m,\ldots,p_nq_1,\ldots,p_nq_m;$$
$$k_1+l_1,\ldots,k_1+l_m,k_2+l_1,\ldots,k_n+l_1,\ldots,k_n+l_m)$$
$$=M(p_1,p_2,\ldots,p_n;k_1,k_2,\ldots,k_n)+M(q_1,\ldots,q_m;l_1,\ldots,l_m)$$
$$\left(\sum_{j=1}^{n}p_j=\sum_{i=1}^{m}q_i=1;\ \sum_{j=1}^{n}D^{-k_j}\leqslant 1,\ \sum_{i=1}^{m}D^{-l_i}\leqslant 1;\ p_j\geqslant 0;\ q_i\geqslant 0;\right.$$
$$\left.k_j,\ l_i\in\mathbb{N};\ j=1,\ldots,n;\ i=1,\ldots,m\right) \qquad (21)$$

This important property means that, if we have two sets of messages (with probabilities $p_1,\ldots,p_n$ and $q_1,\ldots,q_m$) which are independent (each message of the first set is independent of each in the second set), and the coding of such pairs of messages is done by simply joining the two codewords, then the mean codeword length will also be the sum of the mean codeword lengths belonging to the two sets of messages. (It is easy to see that, if $\{k_1,\ldots,k_n\}$ and $\{l_1,\ldots,l_m\}$ satisfy the Kraft inequality (17) for the same D, then so do $\{k_1+l_1,\ldots,k_1+l_m,\ldots,k_n+l_1,\ldots,k_n+l_m\}$). This property is useful, for instance, to sharpen the source coding theorem, by reducing the 1 bit span between its two inequalities to an arbitrarily small quantity (see, for instance, Gallager 1968, p. 51; Aczél-Daróczy 1975, pp. 48–9).

We will prove now that (18) *and* (19) *are the only additive quasilinear mean codeword lengths*. Campbell (1966) has proved a similar theorem under stronger conditions; it first appeared in the present form in Aczél 1974*a* (cf. also Aczél–Daróczy 1975, pp. 165–72).

Theorem 13. *A quasilinear mean codeword length* (20) *with strictly monotonic and continuous* ψ *is additive* (21) *if, and only if,*

$$\psi(x)=cx+b \quad (x\in[1,\infty[),$$

or

$$\psi(x)=cD^{ax}+b \quad (x\in[1,\infty[),$$

where $a \neq 0$, $c \neq 0$ but otherwise a, b, c are arbitrary constants (and $D \geqslant 2$ a fixed integer as in (21)).

Proof. As mentioned before, the 'if' part can be easily checked. For the 'only if' part it will be enough to suppose ((20) and) (21) for fixed m, n; for the sake of simplicity we take $m = n = 2$. In the latter case the Kraft inequalities (17) are automatically satisfied because $D \geqslant 2$, $k_1 \geqslant 1$, $k_2 \geqslant 1$, $l_1 \geqslant 1$, $l_2 \geqslant 1$. If we put (20) into (21) with $m = n = 2$, we get

$$\psi^{-1}[p_1q_1\psi(k_1+l_1)+p_1q_2\psi(k_1+l_2)+p_2q_1\psi(k_2+l_1)+p_2q_2\psi(k_2+l_2)]$$
$$=\psi^{-1}[p_1\psi(k_1)+p_2\psi(k_2)]+\psi^{-1}[q_1\psi(l_1)+q_2\psi(l_2)]$$
$$(p_1+p_2=q_1+q_2=1;\ p_1p_2,q_1,q_2\geqslant 0;\ k_1,k_2,l_1,l_2\in\mathbb{N}), \tag{22}$$

or, in the special case $l_1 = l_2 = l$,

$$\psi^{-1}[p_1\psi(k_1+l)+p_2\psi(k_2+l)]=\psi^{-1}[p_1\psi(k_1)+p_2\psi(k_2)]+l$$
$$(p_1+p_2=1;\ p_1,p_2\geqslant 0;\ k_1,k_2,l\in\mathbb{N}).$$

This equation is very similar to (15.45) but the method of solution will be somewhat different because the variables are restricted to positive integers. On the positive side, p_1 is allowed to vary. However, the proofs also have similarities. We fix l for the time being, define

$$\psi_l(x)=\psi(x+l)\quad(x\in[1,\infty[,\ l\in\mathbb{N}),$$

and get, with $p_1 = p$ (and so $p_2 = 1 - p$),

$$\psi_l^{-1}[p\psi_l(k_1)+(1-p)\psi_l(k_2)]=\psi^{-1}[p\psi(k_1)+(1-p)\psi(k_2)]$$
$$(p\in[0,1];\ k_1,k_2\in\mathbb{N}), \tag{23}$$

in analogy to (15.33). Of course, with ψ, ψ_l is also continuous and strictly monotonic. In analogy to Corollary 15.5, we prove the following:

Lemma 14. *The general continuous and strictly monotonic solution ψ_l:$[1,\infty[\to\mathbb{R}$ of* (23) *is given by*

$$\psi_l(x)=A\psi(x)+B\quad(x\in[1,\infty]),$$

where $A \neq 0$, but otherwise A and B are arbitrary constants.

Proof. Choose $k_2 = 1, k_1 > 1$ in (23). With

$$a_1:=\psi_l(1),\quad a_2:=\psi_l(k_1)-\psi_l(1)\neq 0,\quad (a_2+a_1=\psi_l(k_1)),$$
$$b_1:=\psi(1),\quad b_2:=\psi(k_1)-\psi(1)\neq 0,$$

we get

$$\psi_l^{-1}(a_2p+a_1)=\psi^{-1}(b_2p+b_1)\quad(p\in[0,1]).$$

From the definition of a_1, a_2 and from the continuity of ψ_l we see that $y = a_2p + a_1$ assumes all values between $\psi_l(1)$ and $\psi_l(k_1)$. So we have

$$\psi_l^{-1}(y)=\psi^{-1}(c_2y+c_1)\quad\text{for all } y\in[\psi_l(1),\psi_l(k_1)]$$

(we write $[C,D]$ even if $C>D$) and

$$\psi_l(x) = A\psi(x) + B$$

for all $x\in[1,k_1]$. But k_1 can be arbitrarily large, so this equation holds for all $x \geq 1$ (in analogy to (15.34)), and the Lemma is proved.

In order to finish the *proof of Theorem* 13, we let now l vary again. Then A and B may depend upon l and we get, taking the definition $\psi_l(x) = \psi(x+l)$ into consideration,

$$\psi(x+l) = A(l)\psi(x) + B(l) \quad (x\in[1,\infty[; l\in\mathbb{N}). \tag{24}$$

This equation is similar to (15.6) (cf. the proof of Theorem 15.8), but the restricted domain makes the solution different. Actually we will need, at one point in the proof, the stronger equation (22) in order to get the analogues of (15.17) and (15.18) (as shown in Aczél 1974*a*, cf. Aczél–Daróczy 1975, pp. 171–72, this is indeed indispensable; see Exercise 8).

Again we distinguish two cases:

(i) If $A(l) \equiv 1$ then, with integer $x = k$, we have

$$\psi(k+l) = \psi(k) + B(l)$$

and, because of the symmetry of the left hand side,

$$\psi(k) + B(l) = \psi(l) + B(k),$$

that is, $B(l) = \psi(l) + \gamma$, so that (24) goes over into

$$\psi(x+l) = \psi(x) + \psi(l) + \gamma \quad (x\in[1,\infty[; l\in\mathbb{N}).$$

This equation is similar to (13).

(ii) On the other hand, if $A(l) \not\equiv 1$, then there exists a k_0 such that

$$A(k_0) \neq 1.$$

From (24) we obtain

$$\psi(x+k+l) = A(l)\psi(x+k) + B(l) = A(k)A(l)\psi(x) + A(l)B(k) + B(l).$$

Again, the left hand side is symmetric in k and l, so

$$A(l)B(k) + B(l) = A(k)B(l) + B(k).$$

We put into this equation $k = k_0$ and get, with $\delta = B(k_0)/[A(k_0)-1]$,

$$B(l) = \delta A(l) - \delta.$$

So (24) becomes

$$\psi(x+l) = A(l)[\psi(x) + \delta] - \delta,$$

or with $x = k\in\mathbb{N}$ and, again by symmetry,

$$A(l)[\psi(k) + \delta] = \psi(k+l) + \delta = A(k)[\psi(l) + \delta].$$

As ψ is strictly monotonic, $\psi(k) \not\equiv -\delta$, and so

$$A(l) = \alpha[\psi(l) + \delta]$$

($\alpha = A(k_1)/[\psi(k_1) + \delta]$ where $\psi(k_1) \neq -\delta$). So, in this second case, (24) is transformed into

$$\psi(x + l) = \alpha\psi(x)\psi(l) + \alpha\delta\psi(x) + \alpha\delta\psi(l) + \alpha\delta^2 - \delta.$$

In both cases we got equations of the form

$$\psi(x + l) = \alpha\psi(x)\psi(l) + \beta\psi(x) + \beta\psi(l) + \gamma$$

(in the first case (i) $\alpha = 0$, $\beta = 1$, in the second (ii) $\alpha \neq 0$, $\beta = \alpha\delta$, $\gamma = \alpha\delta^2 - \delta$).

This transforms (22), which is now needed, into

$$\begin{aligned}\psi^{-1}(\alpha[p_1\psi(k_1) + p_2\psi(k_2)][q_1\psi(l_1) + q_2\psi(l_2)] + \beta[p_1\psi(k_1) + p_2\psi(k_2)]\\ + \beta[q_1\psi(l_1) + q_2\psi(l_2)] + \gamma)\\ = \psi^{-1}[p_1\psi(k_1) + p_2\psi(k_2)] + \psi^{-1}[q_1\psi(l_1) + q_2\psi(l_2)]\\ (p_1 + p_2 = q_1 + q_2 = 1;\ p_1, p_2, q_1, q_2 \geqslant 0;\ k_1, k_2, l_1\, l_2 \in \mathbb{N}).\end{aligned}$$

If $k_1 = l_1 = 1, k_2 = l_2 = n \in \mathbb{N}, p_1 = p, p_2 = 1 - p, q_1 = q, q_2 = 1 - q\ (p, q \in [0, 1])$, then $u = p\psi(1) + (1 - p)\psi(n)$ and $v = q\psi(1) + (1 - q)\psi(n)$ assume all values between $\psi(1)$ and $\psi(n)$ as p and q run through $[0, 1]$. So

$$\psi^{-1}(\alpha uv + \beta u + \beta v + \gamma) = \psi^{-1}(u) + \psi^{-1}(v) \quad \text{for all } u, v \in [\psi(1), \psi(n)],$$

and thus

$$\psi(x + y) = \alpha\psi(x)\psi(y) + \beta\psi(x) + \beta\psi(y) + \gamma$$

for all $x, y \in [1, n]$ and, since we can choose n as large as we want, for all $x, y \in [1, \infty[$. (We have used this trick before, when proving Lemma 14.)

We had either (i) $\alpha = 0$, $\beta = 1$ or (ii) $\alpha \neq 0$, $\beta = \alpha\delta$, $\gamma = \alpha\delta^2 - \delta$. In the first case we have

$$f(x + y) = f(x) + f(y) \quad (x, y \in [1, \infty[)$$

for

$$f(x) = \psi(x) + \gamma,$$

in the second,

$$g(x + y) = g(x)g(y) \quad (x, y \in [1, \infty[)$$

for

$$g(x) = \alpha[\psi(x) + \delta],$$

($\alpha \neq 0$, so g is also strictly monotonic.) These are conditional Cauchy equations with $Z = [1, \infty[^2$ as domain. (For g we have Cauchy's exponential equation, but it behaves like Cauchy's equation (7.1) because g is nowhere 0 since $g(x_0) = 0$ would imply $g(x_0 + y) = 0$, contrary to the strict monotonicity of g.) Using the notation of Section 7.1, here $\Pi_1(Z) = \Pi_2(Z) = t(Z)$. By our remark there (after (7.12)), there exist unique extensions which satisfy the same equations on $\mathbb{R}^2$. Finally, since with ψ, f and g are also strictly monotonic.

$$f(x) = cx \quad (c \neq 0),$$

$$g(x) = e^{Cx} \quad (C \neq 0).$$

In the first case this gives

$$\psi(x) = cx + \gamma,$$

in the second,

$$\psi(x) = \frac{1}{\alpha}e^{Cx} - \delta = cD^{ax} + b$$

($c = 1/\alpha \neq 0$, $b = -\delta$, $a = C/\ln D$; D given), which concludes the proof of Theorem 13.

Remark. In (i) we arrived at the equation

$$B(x + l) = B(x) + B(l) \quad (x \in [1, \infty[, l \in \mathbb{N}).$$

This has solutions different from $B(x) = cx$, and as irregular or regular (but still unusual) ones as we want them to be. Indeed, *we can choose B arbitrarily on* $[1, 2[$ and *then continue by*

$$B(x) = B(x - k) + kB(1) \quad \textit{for} \quad x \in [k + 1, k + 2[.$$

It is easy to see that *this algorithm gives the general solution.* We can get from it even *the general continuous and strictly monotonic solution by choosing B on* $[1, 2[$ *continuous, strictly monotonic and such that* $\lim_{x \to 2-} B(x) = 2B(1)$ (but *otherwise arbitrary*) *on* $[1, 2[$ (cf. Exercise 8 and Aczél–Daróczy 1975, pp. 171–2).

16.5 Totally additive number theoretical functions and their generalizations

Proposition 12 solves (13) for monotonic functions under a condition which is of a cylinder type $Z = [1, \infty[\times \mathbb{N}$ (and we had the same Z in Theorem 13 for the Cauchy equation). In fact, this set can be replaced by a discrete square set $Z = \mathbb{N}^2$ (see (12)) as proved by Erdös (1946) (with (25) supposed for relative primes m, n; cf. (11) and Theorem 17 below).

Theorem 15. *A monotonic function* $f: \mathbb{N} \to \mathbb{R}$ *is a solution of the functional equation*

$$f(mn) = f(m) + f(n) \quad \textit{for all } m, n \textit{ in } \mathbb{N} \tag{25}$$

if, and only if, there exists a constant α *such that* $f(n) = \alpha \log n$ *for all n in* $\mathbb{N}$.

As mentioned before, a function satisfying (25) is called a *totally additive number theoretical function.*

Many proofs and generalizations were given for Theorem 15 (see, e.g. Exercise 6; and Erdös 1957; Kátai 1967; Máté 1967; Wirsing 1969, 1979, 1980). Those generalizations replace the assumption of monotonicity for f by less stringent conditions and/or require that (25) is valid only for relatively prime m and n. Here we shall give a different generalization (Dhombres 1979, pp. 4.59–4.68), showing that the particular set of numbers $\mathbb{N}$ does not play a serious role in Theorem 15. We use the terminology of Section 7.1 to state the following:

Theorem 16. *Let* Γ *be a subsemigroup of the additive group* $(\mathbb{R}, +)$. *The set* Γ^2 *is quasi-expandable in the class of monotonic functions.*

This theorem means that, if $f:\mathbb{R}\to\mathbb{R}$ is a monotonic solution of

$$f(x+y)=f(x)+f(y) \quad (x\in\Gamma, y\in\Gamma), \tag{26}$$

then there exists a monotonic quasi-extension of f to all of $\mathbb{R}$. Using Corollary 2.5 and the definition of a quasi-extension, this reduces to the existence of two real numbers α and β such that, for all x in Γ,

$$f(x)=\alpha x+\beta. \tag{27}$$

Let us first deduce Theorem 15 from Theorem 16. We use $\Gamma=\log N=\{0,\log 2,\log 3,\ldots,\log n,\ldots\}$ which is a subsemigroup of the additive group $(\mathbb{R}, +)$. Let $f:\mathbb{N}\to\mathbb{R}$ be a monotonic solution of (25). Then $f_1:\Gamma\to\mathbb{R}$ is defined by $f_1(\log n)=f(n)$. Clearly, f_1 satisfies (26) and is monotonic. Thus $f(n)=f_1(\log n)=\alpha\log n+\beta$. Putting this into (25) we get $\beta=0$ (the same follows from $0\in\Gamma$, cf. Section 7.1). So $f(n)=\alpha\log n$ for all n in $\mathbb{N}$.

Let us now *prove Theorem* 16.

Let $\Delta=\Gamma-\Gamma=\{z|z=x-y, x\in\Gamma, y\in\Gamma\}$ and, given a monotonic Γ^2-additive function f, define $g:\Delta\to R$ as in (2.4) by

$$g(z)=f(x)-f(y) \quad \text{for } z=x-y. \tag{28}$$

In order that this definition make sense, we again prove that two different representations of z as difference of elements of Γ give the same value for g. If $z=x-y=x'-y'$, where x, y, x' and y' are in Γ, then $x+y'=y+x'$ and so $f(x)+f(y')=f(x+y')=f(y+x')=f(y)+f(x')$, which yields $g(x-y)=g(x'-y')$.

Let z, z' be in Δ and let us compute $g(z+z')$. If $z=x-y$, $z'=x'-y'$, $z+z'=(x+x')-(y+y')$ then, as in Theorem 2.1,

$$g(z+z')=f(x+x')-f(y+y')=f(x)+f(x')-f(y)-f(y')=g(z)+g(z').$$

Therefore g is Δ^2-additive.

Suppose $z>z'$, where z and z' are in Δ. We get $z=x-y>z'=x'-y'$ or $x+y'>y+x$. So $f(x+y')\geqslant f(y+x')$, since we may suppose, without loss of generality, that f is nondecreasing. Thus we get $g(z)=f(x)-f(y)\geqslant g(z')=f(x')-f(y')$ and the function g is also nondecreasing on Δ. Clearly, Δ is a subgroup of $\mathbb{R}$, namely the subgroup generated by Γ.

Suppose first that $\Delta=\mathbb{Z}x_0$ for some x_0 in $\mathbb{R}$. Then $g(nx_0)=ng(x_0)$ for all

$\mathbb{Z} := \{0, \pm 1, \pm 2, \ldots,\}$. If $x_0 = 0$, $g(0) = f(0) = 0$, $\Gamma = \{0\}$, and an additive extension of f is the function identically equal to 0. If $x_0 \neq 0$, then $g(x) = \alpha x$ for all x in Δ (with $\alpha = g(x_0)/x_0$). Therefore, for some x_1 in Γ and for all x in Γ, $f(x) = \alpha(x - x_1) + f(x_1) = \alpha x + \beta$. But f is Γ^2-additive, which implies $2\beta = \beta$ or $\beta = 0$. Then, on Γ and so much the more on $\Gamma + \Gamma$, f is the restriction of an additive, monotonic function. We have proved the quasi-expandability in this simple case.

Suppose now $\Delta = \mathbb{Z}x_0 + \mathbb{Z}y_0 = \{z \mid z = mx_0 + ny_0, m \in \mathbb{Z}, n \in \mathbb{Z}\}$, where x_0, y_0 are positive numbers, independent over $\mathbb{Z}$. In order that g be Δ^2-additive we must have

$$g(nx_0 + my_0) = n\alpha_0 + m\beta_0 \quad \text{for all } n, m \text{ in } \mathbb{Z},$$

with $\alpha_0 = g(x_0)$ and $\beta_0 = g(y_0)$. It only remains to use the fact that g is nondecreasing, which means that

$$nx + my_0 \geqslant 0 \quad \text{implies } n\alpha_0 + m\beta_0 \geqslant 0. \tag{29}$$

Since x_0 and y_0 are strictly positive, therefore $\alpha_0 \geqslant 0$ and $\beta_0 \geqslant 0$. But it is easy to check that either $\alpha_0 = \beta_0 = 0$ and so $g \equiv 0$ on Δ or $\alpha_0 > 0$ and $\beta_0 > 0$. We have again quasi-expandability in the first case. In the second case, let $\alpha = \alpha_0/x_0 > 0$ and $\beta = \beta_0/\alpha > 0$. We get

$$nx_0 + my_0 \geqslant 0 \quad \text{implies } nx_0 + m\beta \geqslant 0. \tag{30}$$

First we prove that, for any real $a > 0$, there exists an $n \in \mathbb{Z}$ and an $m \in \mathbb{N}$ such that $0 < nx_0 - my_0 < a$. For the purpose of contradiction, let $0 < b = \inf(nx_0 - my_0)$, where the infimum is taken over all n in $\mathbb{Z}$, and all m in $\mathbb{N}$, such that $nx_0 - my_0 > 0$. The set of all $n'x_0 + m'y_0$, with n', m' in $\mathbb{Z}$, is a subgroup of $\mathbb{R}$, which does not reduce to $z_0\mathbb{Z}$, as x_0 and y_0 are independent over $\mathbb{Z}$. Its closure must then be $\mathbb{R}$. Let us choose ε with $0 < 2\varepsilon < b$. There exist $n \in \mathbb{Z}$, $m \in \mathbb{N}$ such that $b \leqslant nx_0 - my_0 < b + \varepsilon$ and $n' \in \mathbb{Z}$, $m' \in \mathbb{Z}$ such that $\varepsilon < n'x_0 + m'y_0 < 2\varepsilon$. But, due to our hypothesis, m' must belong to $\mathbb{N}$. Thus the number z, where $z = (nx_0 - my_0) - (n'x_0 + m'y_0) = (n - n')x_0 - (m + m')y_0$, is such that $0 < z < b$ and $(m + m') \in \mathbb{N}$. This contradicts the definition of b and proves our result. Clearly, the role played by $\pm x_0$ and $\pm y_0$ is symmetric and in the same way we get $n \in \mathbb{Z}$, $m \in \mathbb{N}$ such that $0 < nx_0 + my_0 < a$.

Let us now prove that (30) yields $\beta = y_0$. Suppose for the purpose of contradiction that $\beta > y_0$. There exist $n \in \mathbb{Z}$, $m \in \mathbb{N}$ such that

$$0 < nx_0 - my_0 < \beta - y_0, \tag{31}$$

which implies

$$0 < nx_0 - (m - 1)y_0 < \beta.$$

Using (30) and the first inequality in (31), we get $nx_0 - m\beta \geqslant 0$, therefore

$$(m - 1)\beta < (m - 1)y_0. \tag{32}$$

We have supposed $m \geqslant 1$ and cannot have $m = 1$ because of the strict inequality in (32). Thus we obtain a contradiction to the hypothesis $\beta > y_0$. A similar contradiction would arise if we were to suppose $\beta < y_0$, this time using $nx_0 + my_0$ ($n \in \mathbb{Z}$ but $m \in \mathbb{N}$). For all x in Δ, we have obtained $g(x) = \alpha x$ and so, for all x in Γ and some β in $\mathbb{R}$,

$$f(x) = \alpha x + \beta.$$

We easily deduce from the Γ^2-additivity of f that $\beta = 0$ and so we have proved the result of Theorem 16 in this second case also.

In the general case of a subgroup Δ, we first define $\Lambda = \{z \mid z = x/n, x \in \Delta, n \in \mathbb{N}\}$ Clearly, Λ is the divisible subgroup generated by Γ. We also define, as in the proof of Theorem 7.4, $h: \Lambda \to \mathbb{R}$ by

$$h(z) = h\left(\frac{x}{n}\right) := \frac{g(x)}{n}.$$

This h is well defined, since $z = x/n = y/m$, $x, y \in \Delta$, $n, m \in \mathbb{N}$ implies $mx = ny$ and so $mg(x) = ng(y)$. Moreover, g is Λ^2-additive since, with $z = x/n$, $z' = y/n$, and $x, y \in \Delta$, $n, m \in \mathbb{N}$, we get

$$h(z + z') = h\left(\frac{mx + ny}{mn}\right) = \frac{g(mx + ny)}{mn} = \frac{mg(x) + ng(y)}{mn}$$

$$= \frac{g(x)}{n} + \frac{g(y)}{m} = h(z) + h(z').$$

In the same way, along with g, h is also nondecreasing and extends g to all of Λ. Let $\{x_i\}_{i \in I}$ be a Hamel basis for the divisible subgroup Λ (see Exercise 20). Using the same argument as in the case where $\Delta = x_0\mathbb{Z} + y_0\mathbb{Z}$, but with a subgroup generated by two distinct elements of the Hamel basis, we deduce the existence of an $\alpha \in \mathbb{R}$ such that $h(x_i) = \alpha x_i$ for all i in the index set I. Therefore, for any x in Λ, $h(x) = \alpha x$. As a consequence, $f(x) = \alpha x + \beta$ for all x in Γ. But the Γ^2-additivity of f yields $\beta = 0$ and therefore concludes the proof of Theorem 16.

Remark 1. Theorem 15 remains true if we restrict the validity of (25) to all relative prime numbers n and m, that is, if we deal with *additive number theoretical functions.* This is easy to understand from our proof of Theorem 15, which makes use of Theorem 16. In this case, with $f_1(\log n) = f(n)$, f_1 satisfies (26) for $x = \log n$, $y = \log m$ and $(n, m) = 1$. Now the subset H of all $y \in \mathbb{R}$ such that

$$f_1(x + y) = f_1(x) + f_1(y) \quad (x \in \Gamma = \log \mathbb{N}) \tag{33}$$

is clearly a subsemigroup of $\mathbb{R}$. This subsemigroup H contains $\log P$ where P is the subset of $\mathbb{N}$ consisting of all prime numbers. Since $\log P$ generates $\log \mathbb{N}$ as a subsemigroup, we deduce that H contains $\log \mathbb{N}$. By symmetry, we are back to (26) with a subsemigroup Γ or $\mathbb{R}$. We state the following:

Theorem 17. *A monotonic function $f:\mathbb{N}\to\mathbb{R}$ is a solution of*

$$f(mn)=f(n)+f(m) \quad \text{for all relative prime } m,n \tag{34}$$

if, and only if, $f(n)=\alpha\log n$ for all $n\in\mathbb{N}$ and some real $\alpha\in\mathbb{R}$.

Remark 2. The fact that f is monotonic is not very important. What is relevant is to have a class $\mathscr{F}$ of functions which is preserved both under the transition from Γ to Δ and from Δ to Λ in the proof of Theorem 16, and such that the additive functions f on $\mathbb{R}$ in the class $\mathscr{F}$ are of the form $f(x)=dx$. (Cf. Exercise 6 for another such class.)

It is certainly possible to generalize Theorem 16 to an ordered archimedean group G instead of $\mathbb{R}$. Another, more interesting, generalization would be to replace the monotonicity of f by something applicable to $\mathbb{R}^2$, for example, in order to get rid of the use of the order properties of $\mathbb{R}$. To achieve this goal, which is still an open problem, an idea would be to determine, starting from any subsemigroup Γ of $\mathbb{R}$, the maximal class C of functions such that the set Γ^2 is C-quasi-expandable. A different point of view is to give up the symmetry involved in the consideration of the set Γ^2 and replace it by $X\times Y$, where X and Y have some kind of algebraic independence. To study this more thoroughly, it is best to suppose X to be a semigroup generated by an element x_0 ($X=x_0\mathbb{N}$), and Y a semigroup generated by another element y_0 ($Y=y_0\mathbb{N}$). We shall suppose $x_0>0$, $y_0>0$. An $(X\times Y)$-additive function $f:\mathbb{R}\to\mathbb{R}$ thus satisfies

$$f(nx_0+my_0)=f(nx_0)+f(my_0) \quad \text{for all } n,m \text{ in } \mathbb{N}. \tag{35}$$

This $X\times Y$-subset is not quasi-expandable. Let, for example, $\phi:\mathbb{R}\to\mathbb{R}$ be a periodic function of period x_0 such that $\phi(0)=0$, and let $\psi:\mathbb{R}\to\mathbb{R}$ be a periodic function of period y_0 such that $\psi(0)=0$. Define $f=\phi+\psi$. We verify, for all n,m in $\mathbb{N}$,

$$\begin{aligned} f(nx_0+my_0) &= \phi(nx_0+my_0)+\psi(nx_0+my_0)=\phi(my_0)+\psi(nx_0)\\ &= \phi(my_0)+\psi(my_0)+\phi(nx_0)+\psi(nx_0)\\ &= f(nx_0)+f(my_0). \end{aligned}$$

With convenient choices of ϕ and ψ, f is not additive on $x_0\mathbb{N}+y_0\mathbb{N}$ and thus $X\times Y$ is not quasi-expandable.

There is no quasi-expandability even for the class of continuous functions, or for monotonic functions. In this last case, however, we may find all monotonic $(X\times Y)$-additive functions (Pisot–Schoenberg 1964). Let us begin with a simpler result:

Proposition 18. *Let x_0, y_0 be positive numbers which are supposed to be independent over $\mathbb{Z}$. Let $X=x_0\mathbb{N}$ and $Y=y_0\mathbb{N}$. Suppose that $f:\mathbb{R}\to\mathbb{R}$ is a* *monotonic function which is $(X \times Y)$-additive. Then there exist a real constant α and two bounded periodic functions*

$$\phi:\mathbb{R}\to\mathbb{R}, \quad \textit{of period } x_0, \quad \textit{with } \lim_{x\to\infty}\frac{\phi(x)}{x}=0,$$

$$\psi:\mathbb{R}\to\mathbb{R}, \quad \textit{of period } y_0, \quad \textit{with } \lim_{x\to\infty}\frac{\psi(x)}{x}=0,$$

such that, for all z in $t(X \times Y)=\{x+y\,|\,x\in X, y\in Y\}\subset x_0\mathbb{N}+y_0\mathbb{N}$,

$$f(z)=\alpha z+\phi(z)+\psi(z). \tag{36}$$

Proof. The crucial step is to prove that, for this f, $\lim_{x\to\infty}(f(x)/x)$ exists and is finite, where the limit is taken over the x in $x_0\mathbb{N}+y_0\mathbb{N}$. Without loss of generality we may suppose $x_0>y_0$. For each positive integer m, there exists at least one integer m' such that

$$mx_0<m'y_0<(m+1)x_0.$$

Therefore, for all positive integers n,

$$nx_0+mx_0<nx_0+m'y_0<nx_0+(m+1)x_0.$$

Without loss of generality we may suppose f to be nondecreasing, so

$$f(nx_0+mx_0)\leqslant f(nx_0)+f(m'y_0)\leqslant f(nx_0)+f((m+1)x_0),$$
$$f(nx_0)+f(mx_0)\leqslant f(nx_0)+f(m'y_0)=f(nx_0+m'y_0)\leqslant f((n+m+1)x_0),$$

which yields

$$f(nx_0)+f((m-1)x_0)\leqslant f((n+m)x_0)\leqslant f(nx_0)+f((m+1)x_0).$$

We now apply this inequality with $n+m,\dots,n+(h-1)m$ instead of n:

$$f((n+m)x_0)+f((m-1)x_0)\leqslant f((n+2m)x_0)$$
$$\leqslant f((n+m)x_0)+f((m+1)x_0),$$

....

....

$$f((n+(h-1)m)x_0)+f((m-1)x_0)\leqslant f((n+hm)x_0)$$
$$\leqslant f((n+(h-1)m)x_0)+f((m+1)x_0).$$

Thus, if we add all inequalities, we get, by successive cancellations,

$$f(nx_0)+hf((m-1)x_0)\leqslant f((n+hm)x_0)\leqslant f(nx_0)+hf((m+1)x_0).$$

Let N be any positive integer and first fix m. There exist integers h and n $(0\leqslant n<m)$ such that $N=hm+n$. Therefore

$$\frac{f(nx_0)}{N}+\frac{h}{N}f((m-1)x_0)\leqslant\frac{f(Nx_0)}{N}\leqslant\frac{f(nx_0)}{N}+\frac{h}{N}f((m+1)x_0).$$

Let N go to infinity. We use $\lim_{N\to\infty}h/N=1/m$. From the left inequality, we deduce

$$\underline{\lim}_{N\to\infty} \frac{f(Nx_0)}{N} \geqslant \frac{f((m-1)x_0)}{m} \tag{37}$$

and, from the right inequality,

$$\overline{\lim}_{N\to\infty} \frac{f(Nx_0)}{N} \leqslant \frac{f((m+1)x_0)}{m}.$$

So $\underline{\lim}_{N\to\infty} f(Nx_0)/N$ and $\overline{\lim}_{N\to\infty}(f(Nx_0))/N$ are finite. Further, from (37),

$$\frac{m}{m-1} \underline{\lim}_{N\to\infty} \frac{f(Nx_0)}{N} \geqslant \frac{f((m-1)x_0)}{m-1}.$$

Taking $\overline{\lim}_{m\to\infty}$ of both sides we get

$$\begin{aligned} \underline{\lim}_{N\to\infty} \frac{f(Nx_0)}{N} &\geqslant \overline{\lim}_{m\to\infty} \frac{f((m-1)x_0)}{m-1} \\ &= \overline{\lim}_{N\to\infty} \frac{f(Nx_0)}{N}. \end{aligned}$$

So

$$\underline{\lim}_{N\to\infty} \frac{f(Nx_0)}{N} = \overline{\lim}_{N\to\infty} \frac{f(Nx_0)}{N} = \lim_{N\to\infty} \frac{f(Nx_0)}{N} =: \alpha$$

exists and is finite. Now let $x = nx_0 + my_0 > 0$ $(n \geqslant 1, m \geqslant 1)$. There exists an integer N such that

$$Nx_0 \leqslant nx_0 + my_0 < (N+1)x_0.$$

Thus

$$\frac{f(Nx_0)}{(N+1)x_0} \leqslant \frac{f(nx_0 + my_0)}{nx_0 + my_0} \leqslant \frac{f((N+1)x_0)}{Nx_0}.$$

Therefore the middle term has, as $x = nx_0 + my_0 \to \infty$, the (finite) limit α:

$$\lim_{\substack{x\to\infty \\ (x = nx_0 + my_0)}} \frac{f(x)}{x} = \alpha.$$

We now define $g(x) = f(x) - \alpha x$ for all x in $\mathbb{R}$. Recall that x_0 and y_0 are independent over $\mathbb{Z}$. It is then possible to define, with much freedom, a function $\phi: \mathbb{R} \to \mathbb{R}$, of period x_0, such that ϕ coincides with g on $y_0\mathbb{N}$. Further, we get $\lim_{x\to\infty}(\phi(x)/x) = 0$ by taking ϕ bounded on the subset of points of $[0, x_0[$ which are not equal to some ny_0, $n \in \mathbb{N}$, modulo x_0. Similarly, we define a function $\psi: \mathbb{R} \to \mathbb{R}$, of period y_0, coinciding with g on $x_0\mathbb{N}$ and bounded on $[0, y_0[$. Then $\lim_{x\to\infty}[\psi(x)/x] = 0$. Moreover,

$$\begin{aligned} g(nx_0 + my_0) &= g(nx_0) + g(my_0) = \psi(nx_0) + \phi(my_0) \\ &= \psi(nx_0 + my_0) + \phi(nx_0 + my_0). \end{aligned}$$

In other words $f(x) = \alpha x + \phi(x) + \psi(x)$ for all x in $x_0\mathbb{N} + y_0\mathbb{N}$, that is, for all x in $t(X \times Y)$, which concludes the proof of Proposition 18.

To get the general monotonic solution for $(X \times Y)$-additivity, we now have to look for more properties of ϕ and ψ. In fact, let us suppose that $f: \mathbb{R} \to \mathbb{R}$ is defined everywhere by (36), where α is an arbitrary real number and ϕ, ψ possess the properties described in Proposition 18. We have already noticed that f is $(X \times Y)$-additive. However, it need not be monotonic in general. A necessary and sufficient condition for f to be nondecreasing (reverse the inequality for nonincreasing) is that the following inequality be valid for all x, y in $\mathbb{R}$, $x \neq y$:

$$-\alpha \leqslant \frac{\phi(y) - \phi(x)}{y - x} + \frac{\psi(y) - \psi(x)}{y - x}.$$

If we were to impose

$$\inf_{\substack{x, y \in \mathbb{R} \\ x \neq y}} \frac{\phi(y) - \phi(x)}{y - x} = -\beta; \quad \inf_{\substack{x, y \in \mathbb{R} \\ x \neq y}} \frac{\psi(y) - \psi(x)}{y - x} = -\gamma$$

and $\alpha \geqslant \beta + \gamma$, then (36) defines a nondecreasing $(X \times Y)$-additive function. These are sufficient conditions. However, it so happens that these conditions are also necessary. We just have to be more precise since ϕ and ψ are arbitrarily chosen and $(X \times Y)$-additivity says something about $x_0\mathbb{N} + y_0\mathbb{N}$ only. It can be proved that, if f is a monotonic $(X \times Y)$-additive function, then ϕ and ψ are completely determined on $x_0\mathbb{Z} + y_0\mathbb{N}_0$ and $x_0\mathbb{N}_0 + y_0\mathbb{Z}$, respectively, with $\mathbb{N}_0 = \mathbb{N} \cup \{0\}$, and there exist finite β, γ such that

$$-\beta = \inf_{\substack{x \neq y \\ x, y \in x_0\mathbb{Z} + y_0\mathbb{N}_0}} \frac{\phi(y) - \phi(x)}{y - x}; \quad -\gamma = \inf_{\substack{x \neq y \\ x, y \in x_0\mathbb{N}_0 + y_0\mathbb{Z}}} \frac{\psi(y) - \psi(x)}{y - x}.$$

Moreover, $\alpha \geqslant \beta + \gamma$. See Pisot–Schoenberg 1964.

16.6 Further equations for number theoretical functions

We end the chapter with some more sophisticated cases of number theoretical functions.

Let G, F be abelian groups and, for an integer $k \geqslant 2$, let X_i be nonempty subsets of G with $i = 1, 2, \ldots, k$. We look for $f: G \to F$ such that

$$f\left(\sum_{i=1}^{k} x_i\right) = \sum_{i=1}^{k} f(x_i) \tag{38}$$

for all $x_i \in X_i$. We write

$$\mathop{\times}_{i=1}^{k} X_i = \{(x_1, \ldots, x_k) \mid x_i \in X_i; i = 1, \ldots, k\}.$$

For $k=2$, the domain (condition) is a 'rectangle'. The expansibility of $\times_{i=1}^{k} X_i$ is easily defined but the most interesting definition is that of $\mathscr{C}$-quasi-expansibility: $\times_{i=1}^{k} X_i$ is $\mathscr{C}$-quasi-expandable if, for any $f:G\to F$ belonging to a class $\mathscr{C}$ of functions, and satisfying (38), there exists $g:G\to F$, additive and belonging to the class $\mathscr{C}$, such that, for all $x_i \in X_i$, we get

$$g\left(\sum_{i=1}^{k} x_i\right)=f\left(\sum_{i=1}^{k} x_i\right),$$

(notice the slight difference as compared to the definitions in Section 7.1). We shall only investigate cases where $X_i = a_i\mathbb{N}_0$ with $\mathbb{N}_0 := \mathbb{N}\cup\{0\}$ and where $G=F=\mathbb{R}$, with monotonic functions. It can be proved that $\times_{i=1}^{k} a_i\mathbb{N}_0$ or even $\times_{i=1}^{k} a_i\mathbb{Z}$ is not quasi-expandable. The following result is not difficult to prove and settles the case of nonindependent $x_{0,i}$'s over $\mathbb{Z}$:

Proposition 19. *Let $a_1, a_2, \ldots, a_k$ be k positive integers where $k\geq 2$. Suppose this set is relatively prime. Let α_i be the greatest common divisor of $(a_1, a_2, \ldots, a_{i-1}, a_{i+1}, \ldots, a_k)$. An $f:\mathbb{R}\to\mathbb{R}$ is $(\times_{i=1}^{k} a_i\mathbb{N}_0)$-additive if, and only if, there exists a real number α and periodic functions $\phi_i:\mathbb{R}\to\mathbb{R}$, of period α_i, with $\phi_i(0)=0$ $(i=1,\ldots,k)$ such that, for all $x\in\sum_{i=1}^{k} a_i\mathbb{N}_0$,*

$$f(x)=\alpha x+\sum_{i=1}^{k}\phi_i(x).$$

However, if $k\geq 3$, the situation here is different from Proposition 18: monotonic solutions to (38) are the usual linear ones on $\sum_{i=1}^{k} a_i\mathbb{N}_0$. We prove the following:

Theorem 20. *Let $x_{0,i}$, $i=1,2,\ldots,k$, be k positive numbers, independent over $\mathbb{Z}$. Suppose $k\geq 3$. The set $\times_{i=1}^{k} a_i\mathbb{N}_0$ is quasi-expandable for monotonic functions.*

Proof. Let $f:\mathbb{R}\to\mathbb{R}$ be a monotonic $\times_{i=1}^{k} X_i$-additive function, where $X_i = a_i\mathbb{N}_0$ (and $\mathbb{N}_0=\mathbb{N}\cup\{0\}$). Proposition 19 yields that $\alpha=\lim_{x\to\infty}(f(x)/x)$ exists and is finite, where x is of the form $\sum_{i=1}^{k} n_i a_i$ $(n_i\in\mathbb{N}_0)$. We then use $g(x)=f(x)-\alpha x$. We may suppose that f is nondecreasing, so that, for all $x\neq y$ in $\mathbb{R}$,

$$\frac{g(x)-g(y)}{x-y}\geq -\alpha. \tag{39}$$

It is clear that $g(0)=f(0)=0$. Now, let z be of the form

$$z=\sum_{i=2}^{k} a_i t_i, \quad \text{where } t_i\in\mathbb{Z} \quad \text{for } i=2,\ldots,k,$$

and fix $x_1\in X_1$. We define, for a strictly positive integer j,

$$y_i^{(j)} = j a_i t_i \quad \text{if } t_i\geq 0 \quad \text{or} \quad y_i^{(j)} = -(j-1)a_i t_i \quad \text{if } t_i<0$$

and

$$z_i^{(j)} = (j-1)a_it_i \quad \text{if } t_i \geq 0 \quad \text{or} \quad z_i^{(j)} = -ja_it_i \quad \text{if } t_i < 0.$$

We notice that $y_i^{(j)} - z_i^{(j)} = a_it_i$. Applying (39) with $x = x_1 + \sum_{i=2}^k y_i^{(j)}$ and $y = \sum_{i=2}^k x_i^{(j)}$, we get

$$\frac{g(x_1) + \sum_{i=2}^{k} (g(y_i^{(j)}) - g(z_i^{(j)}))}{x_1 + z} \geq -\alpha. \tag{40}$$

But

$$\sum_{j=1}^{n} (g(y_i^{(j)}) - g(z_i^{(j)})) = \begin{cases} g(y_i^{(n)}) & \text{if } t_i \geq 0 \\ g(z_i^{(n)}) & \text{if } t_i < 0. \end{cases}$$

We summarize, with $s_i^{(n)} = y_i^{(n)}$ for $t_i \geq 0$ and $s_i^{(n)} = z_i^{(n)}$ for $t_i < 0$:

$$\sum_{j=1}^{n} (g(y_i^{(j)}) - g(z_i^{(j)})) = g(s_i^{(n)}).$$

We use (40) with $j = 1, 2, \ldots, n$, sum those equations and divide by n,

$$\frac{g(x_1) + \frac{1}{n}\sum_{i=2}^{k} g(s_i^{(n)})}{x_1 + z} \geq -\alpha.$$

But, as a consequence of the definition of the function g, we get $\lim_{x\to\infty}(g(x)/x) = 0$ when $x = \sum_{i=1}^k X_i$. Therefore we also have $\lim_{n\to\infty}(1/n)g(s_i^{(n)}) = 0$. We thus obtain

$$\frac{g(x_1)}{x_1 + z} \geq -\alpha. \tag{41}$$

We shall now use the fact that $k \geq 3$. The set of all possible z is then dense in $\mathbb{R}$. Therefore (41) is valid only if $g(x_1) = 0$. But x_1 was arbitrarily chosen in X_1, so $g(X_1) = 0$. From the symmetry in the role of the a_i's we deduce that $g(\sum_{i=1}^k X_i) = 0$. In other words, for every $x \in \sum_{i=1}^k X_i$, we get

$$f(x) = \alpha x.$$

This proves the quasi-expandability of $\times_{i=1}^k X_i$ for monotonic functions and concludes the proof of Theorem 20.

Exercises and further results

(For Exercises 1–4, see Dhombres 1979, pp. 3.4–3.10.)

1*. Let F be a group with at least one element of order greater than 2. Let G be a group. Prove or disprove that $Z = G \times Y$ is addundant (for $f: G \to F$) if, and only if, the subgroup generated by Y is G.

2*. Generalize Proposition 8 in the following way. Let F and G be two groups. Let Y be a nonempty subset of G. Express, in terms of the subgroup H of G, generated by Y, and of liftings relative to H, the general solution $f: G \to F$ of the Cauchy conditional equation

$$f(x+y)=f(x)+f(y) \quad (x\in G, y\in Y).$$

3. Let G be a locally compact connected topological group and F a group. Suppose that Y is a subset of G of positive Haar measure. Prove or disprove that $Z = G\times Y$ is addundant (for $f:G\to F$).
4. Given an example of a set $Y\subset\mathbb{R}$ of Lebesgue measure zero such that $\mathbb{R}\times Y$ is addundant (with respect to all $f:\mathbb{R}\to\mathbb{R}$).
5. (Erdös 1946.) Give a direct proof of Proposition 12 without using Proposition 8.
6. (Kátai 1967; Máté 1967.) Prove the following result. A function $f:\mathbb{N}\to\mathbb{R}$ such that $\varliminf_{n\to\infty}(f(n+1)-f(n))\geqslant 0$ is a solution of the functional equation
$$f(mn)=f(m)+f(n) \quad \text{for all relative prime } m,n$$
if, and only if, there exists a real constant α such that $f(n)=\alpha\log n$ for all $n\geqslant 1$.
7. Prove or disprove the following statement. Let F be a divisible abelian group and G an abelian group. Suppose that X is a subset of G containing a subgroup H and Y is a subset of G, generating H as a group. Let $f:G\to F$ be a Z-additive function where $Z=X\times Y$. There exists an additive $g:G\to F$ such that $g=f$ on H.
8. (Aczél–Daróczy 1975, pp. 171–2.) Determine the general continuous and strictly monotonic $\psi:[1,\infty[\to\mathbb{R}$ for which there exist $A,B:\mathbb{N}\to\mathbb{R}$ such that equation (24) is satisfied for all $x\in[1,\infty[$, $l\in\mathbb{N}$.

(For Exercises 9–10, see Forti 1983.)

9*. Prove or disprove the following. If a continuous function $h:\mathbb{R}\to\mathbb{R}$ satisfies $h(0)=0$ and $xh(x)>0$ for all $x\neq 0$, then $C_h=\{(x,h(x))\mid x\in\mathbb{R}\}$ is addundant for all functions f which are continuous on a neighbourhood of 0 and for which $f'(0)$ exists.

10*. Suppose the $\mathscr{X}$ maps any function $f:\mathbb{R}\to\mathbb{R}$ into its graph $\mathscr{X}(f)=\{(x,f(x))|, x\in\mathbb{R}\}$. Prove or disprove that, for all functions f which are continuous in a neighbourhood of 0 and for which $f'(0)\notin\{-1,0\}$, $\mathscr{X}$ is additively expandable.

(For Exercises 11–14, see Parsloe 1982.)

11. Let $f:]0,1]\to\mathbb{R}_+$. Are the following two properties equivalent? (under the supposition that the denominators are not 0):

(a) $$\lim_{\substack{x+y\to 0\\ x>0,y>0}}\left|\frac{f(x)+f(y)}{f(x+y)}-1\right|=0,$$

(b) $$\lim_{x\to 0+}\sup_{0<s\leqslant 1}\left|\frac{f(sx)}{f(x)}-s\right|=0.$$

12*. Same question as in the previous exercise, but with
$$\lim_{\substack{x+y\to 0\\ x>0,y>0}}\frac{1}{F(x+y)}\left|\frac{f(x)+f(y)}{f(x+y)}-1\right|=0$$
and
$$\lim_{x\to 0+}\sup_{0<s\leqslant 1}\frac{1}{F(x)}\left|\frac{f(sx)}{f(x)}-s\right|=0$$
in place of (a) or (b), respectively, where $F:]0,1]\to\mathbb{R}_+$ is nondecreasing. In this case we will say that f is *F-locally additive at* $0+$.

13*. A function $f:\bar{\mathbb{R}}_+ \to \mathbb{R}$ is uniformly *ϕ-slowly varying* if, for each bounded interval I of $\bar{\mathbb{R}}_+$,

$$\lim_{x\to\infty} \sup [\phi(x)|f(x+\alpha)-f(x)|]=0,$$

where $\phi:\bar{\mathbb{R}}_+ \to \mathbb{R}_+$ is nondecreasing. Prove or disprove that $f:]0,1]\to\mathbb{R}_+$ is F-locally additive at $0+$ (cf. Exercise 12) if, and only if, $f(x)=x\exp g(\log 1/x)$ $(x\in]0,1])$ where g is ϕ-slowly varying and $\phi(x)=F(e^{-x})^{-1}$.

14*. Suppose $\lim_{x\to 0+}\sum_{n=0}^{\infty}[F(xe^{-n})]/[F(x)]<\infty$. Prove or disprove that $f:]0,1]\to\mathbb{R}_+$ is F-locally additive at $0+$ (cf. Exercise 12) if, and only if, for some $\lambda>0$,

$$\lim_{\substack{x\to 0\\ x>0}} \frac{1}{F(x)}\left|\frac{f(x)}{x}-\lambda\right|=0.$$

15. (Kuczma 1963.) Let g, h be two continuous functions satisfying the conditions of Theorem 1. Prove the following result. Suppose $f:\mathbb{R}\to\mathbb{R}$ is differentiable at 0 and $f(0)>0$. If f satisfies

$$f(g(x)+h(x))=f(g(x))f(h(x))$$

for all x in $\mathbb{R}$, then there exists an $a\in\mathbb{R}$ such that

$$f(x)=e^{ax}.$$

16. Let $k:\bar{\mathbb{R}}_+\to\bar{\mathbb{R}}_+$ be a continuous function such that $0<k(x)<x$ for $x\in\mathbb{R}_+$ and $k(2x)=2k(x)$ for $x\in\bar{\mathbb{R}}_+$. Prove or disprove that $f:\mathbb{R}\to\mathbb{R}$ is a solution of

$$f(x)=f(k(x))+f(x-k(x))\quad (x\in\mathbb{R})$$

if and only if f is additive.

(For Exercises 17–19, cf., e.g., Bachman 1964, pp. 17–22.)

17*. Two valuations (cf. Section 10.6) f_1 and f_2 on a field K are *equivalent* if there exists a positive ρ such that

$$f_1(x)=f_2(x)^{\rho}\quad \text{for all } x\in K.$$

Determine, up to equivalence, all valuations on $\mathbb{Q}$.

18. Let K be a field and A a subring of K. Let F be an algebraically closed field and $f:A\to F$ a nontrivial homomorphism. Prove that there exists a $\phi:K\to F\cup\{\infty\}$ with the following properties:
 (i) the restriction of ϕ to A coincides with f,
 (ii) $\phi^{-1}(F)$ is a subring U of K,
 (iii) the restriction of ϕ to U is a nontrivial homomorphism,
 (iv) if $\phi(a)=\infty$, then $\phi(a^{-1})=0$.

19*. Let F be a field and $f_1, f_2, \ldots, f_n$ nontrivial and not equivalent valuations. Let λ_i $(i=1,\ldots,n)$ be elements of F. Prove that, for every $\varepsilon>0$, there exists a $\lambda\in F$ such that

$$f_i(\lambda-\lambda_i)<\varepsilon\quad (i=1,\ldots,n).$$

20. Show that every divisible abelian group G has a Hamel basis, i.e., there exists

a subset H of G such that every x in G can be written in a unique way as

$$x = \sum_{i=1}^{n} s_i x_i,$$

with $s_i \in \mathbb{Q} \setminus \{0\}$ and the x_i are distinct elements of H $(i = 1, \dots, n)$.

21*. (Piccard 1939, Mehdi 1964.) Let $S \subset \mathbb{R}$ have the Baire property but not be meagre (for definitions see Section 2 just before Theorem 10). In analogy to Theorem 2.6, prove that $S + S$ contains an interval of positive length. Then, in analogy to Theorem 2.8, prove that $f(x) = cx$ (c an arbitrary real constant) gives all additive functions which are bounded from one side on a nonmeagre subset of $\mathbb{R}$ with the Baire property.

22. Does Theorem 5 remain true if the class of odd functions is replaced by that of continuous functions (differentiable at 0)?

(For Exercises 23 and 24, cf. Bachman 1964, pp. 16–17.)

23*. Let K be a field. Prove that two nontrivial valuations f_1 and f_2 are equivalent (see Exercise 17) if, and only if, $f_2(x) < 1$ whenever $f_1(x) < 1$.

24. Let K be a field and f a nontrivial valuation on K. A null-sequence $\{x_n\}$ in (K, f) is a sequence of points $x_n \in K$ such that $\lim_{n\to\infty} f(x_n) = 0$. Prove that two nontrivial valuations in K are equivalent (see Exercise 17) if, and only if, they have the same null sequences.

25*. (Daróczy–Kátai 1985.) Let G be a compact abelian group and let $A(G)$ be the set of all $f: \mathbb{N} \to G$ such that

$$f(mn) = f(m) + f(n) \quad \text{for all } m, n \in \mathbb{N}.$$

Let $K(G)$ be the set of all $f \in A(G)$ such that

$$\lim_{n\to\infty} (f(n+1) - f(n)) = 0.$$

Let $D(G)$ be the set of all $f \in A(G)$ such that

$$g' = \lim_{k\to\infty} f(n_k + 1)$$

exists if $g = \lim_{k\to\infty} f(n_k)$ exists for all $n_1 < n_2 < \cdots < n_k < n_{k+1} < \cdots$. Is $K(G) = D(G)$?

17

Mean values, mediality and self-distributivity

One encounters several means in mathematics and its applications. Two well known examples are the geometric mean $(xy)^{1/2}$ and the arithmetic mean $(x+y)/2$. Both are special cases of the so called quasiarithmetic mean which we have already introduced in Chapter 15. Our purpose in this chapter is to characterize these means by some of their common properties and for this purpose we use a functional equations (mediality, self-distributivity). We do the same for the more general quasilinear means (no symmetry supposed, cf. (16.20)).

Here we write M as an operation $\circ$, that is, we write $x \circ y$ for $M(x, y)$, so

$$x \circ y = k^{-1}\left(\frac{k(x)+k(y)}{2}\right) \quad (x, y \in I), \tag{1}$$

where I is an (open, closed, half-closed, finite or infinite) real interval of positive length and $k: I \to \mathbb{R}$ a continuous and strictly monotonic function. The characterization will be carried out with aid of the mediality (or bisymmetry) condition (1.12) which we now write as

$$(x \circ y) \circ (z \circ w) = (x \circ z) \circ (y \circ w) \quad (x, y, z, w \in I). \tag{2}$$

It is clear from (2) that we have to suppose that the range of $\circ$ is a subset of I. Obviously, (1) satisfies (2). There are many works on this identity both in abstract algebra and in the theory of mean values, a few of the latter being Aczél (1948*b*, 1966*c* pp. 278–87); Fuchs (1950); Sigmon (1968, 1970).

Theorem 1. *There exists a continuous and strictly monotonic function* $k: I \to \mathbb{R}$ *with which* (1) *holds if, and only if,* $\circ: I^2 \to I$ *is continuous and strictly increasing in both variables, satisfies* (2) *and*

$$x \circ x = x \quad \textit{for all } x \in I \tag{3}$$

(*idempotence*), *and is also commutative* (*symmetric*), *that is*,

$$x \circ y = y \circ x. \tag{4}$$

Proof. The 'only if' part is obvious. In the proof of the 'if' part, we restrict outselves first to finite closed intervals $I = [a, b]$, $a < b$. We will construct the inverse

$$f = k^{-1} \tag{5}$$

of k. We can choose the domain of f to be exactly $[0, 1]$, that is, $f:[0,1] \to [a, b]$, as shall be seen. If k is to satisfy (1), then f has to satisfy

$$f\left(\frac{u+v}{2}\right) = f(u) \circ f(v) \quad \text{for all } u, v \in [0, 1]. \tag{6}$$

We define

$$f(0) = a, \quad f(1) = b, \tag{7}$$

$$f(\tfrac{1}{2}) = a \circ b, \quad f(\tfrac{1}{4}) = a \circ (a \circ b), \quad f(\tfrac{3}{4}) = (a \circ b) \circ b, \quad \text{etc.}, \tag{8}$$

more exactly, we define f on the dyadic fractions $m/2^n$ $(0 \leqslant m \leqslant 2^n)$ (cf. the proof of Theorem 8.1) by (7) and by recursion with respect to n:

$$f\left(\frac{2k+1}{2^{n+1}}\right) = f\left(\frac{k+1}{2^n}\right) \circ f\left(\frac{k}{2^n}\right) \quad (0 \leqslant 2k+1 \leqslant 2^{n+1}).$$

We need only a definition for odd m. However, we also notice, using (3), that

$$f\left(\frac{2k}{2^{n+1}}\right) = f\left(\frac{k}{2^n}\right) = f\left(\frac{k}{2^n}\right) \circ f\left(\frac{k}{2^n}\right).$$

(This last equation will be convenient later, because we will have the chance of choosing larger powers of 2 in the denominators if this makes the calculation simpler.) So we have

$$f\left(\frac{2k+\delta}{2^{n+1}}\right) = f\left(\frac{k+\delta}{2^n}\right) \circ f\left(\frac{k}{2^n}\right) \quad \text{where } \delta \text{ is either 0 or 1.} \tag{9}$$

We first prove by induction on n that (6) is satisfied for such dyadic u, v. It is true for $n = 0$ (and $n = 1$) by (7) and (8). Suppose that (6) is already satisfied for dyadic fractions with denominator 2^n. Let

$$u = \frac{2k+\delta}{2^{n+1}}, \quad v = \frac{2l+\varepsilon}{2^{n+1}} \quad \text{where } \delta \text{ and } \varepsilon \text{ are either 0 or 1}$$

(by enlarging both numerator and denominator, one can always get a common denominator in u and v). Then, by (9), (4), (2) and by the induction hypothesis,

$$\begin{aligned} f(u) \circ f(v) &= f\left(\frac{2k+\delta}{2^{n+1}}\right) \circ f\left(\frac{2l+\varepsilon}{2^{n+1}}\right) \\ &= \left[f\left(\frac{k+\delta}{2^n}\right) \circ f\left(\frac{k}{2^n}\right)\right] \circ \left[f\left(\frac{l}{2^n}\right) \circ f\left(\frac{l+\varepsilon}{2^n}\right)\right] \end{aligned}$$

$$= \left[f\left(\frac{k+\delta}{2^n}\right) \circ f\left(\frac{l}{2^n}\right) \right] \circ \left[f\left(\frac{k}{2^n}\right) \circ f\left(\frac{l+\varepsilon}{2^n}\right) \right]$$
$$= f\left(\frac{k+l+\delta}{2^{n+1}}\right) \circ f\left(\frac{k+l+\varepsilon}{2^{n+1}}\right).$$

Now $k+l+\delta$ and $k+l+\varepsilon$ differ either by 0 or by 1 unit, so, by (9) and, if necessary, (4),

$$f\left(\frac{k+l+\delta}{2^{n+1}}\right) \circ f\left(\frac{k+l+\varepsilon}{2^{n+1}}\right) = f\left(\frac{2k+2l+\delta+\varepsilon}{2^{n+2}}\right) = f\left(\frac{u+v}{2}\right).$$

This proves (6) for dyadic u, v. Also f, so defined, will be strictly *increasing* on the set of dyadic fractions in $[0, 1]$ because $\circ$ is strictly increasing in both variables and $f(0) < f(1)$ by (7). Indeed, if f is increasing for dyadic fractions with denominators 2^n, then, from this, from (3), and from (9),

$$f\left(\frac{2k}{2^{n+1}}\right) = f\left(\frac{k}{2^n}\right) = f\left(\frac{k}{2^n}\right) \circ f\left(\frac{k}{2^n}\right) < f\left(\frac{k}{2^n}\right) \circ f\left(\frac{k+1}{2^n}\right) = f\left(\frac{2k+1}{2^{n+1}}\right)$$

and

$$f\left(\frac{2k+1}{2^{n+1}}\right) = f\left(\frac{k+1}{2^n}\right) \circ f\left(\frac{k}{2^n}\right) < f\left(\frac{k+1}{2^n}\right) \circ f\left(\frac{k+1}{2^n}\right)$$
$$= f\left(\frac{k+1}{2^n}\right) = f\left(\frac{2k+2}{2^{n+1}}\right).$$

Now we extend the definition of f to all real $u \in [0, 1]$. Since $f(0) = a$ and $f(1) = b$ are already defined, we restrict ourselves to $u \in]0, 1[$. Let $\{d_n\}$ be an increasing and $\{D_n\}$ a decreasing sequence of dyadic fractions (different from u), tending to u, such that d_n and D_n are the fractions nearest to u (but different from u) with denominators 2^n. Then

$$d_1 \leqslant \cdots \leqslant d_{n-1} \leqslant d_n \leqslant d_{n+1} \leqslant \cdots < u < \cdots \leqslant D_{n+1} \leqslant D_n \leqslant D_{n-1} \leqslant \cdots \leqslant D_1, \tag{10}$$

The function f, being increasing on the set of dyadic fractions, satisfies

$$f(d_1) \leqslant \cdots \leqslant f(d_{n-1}) \leqslant f(d_n) \leqslant f(d_{n+1}) \leqslant \cdots$$
$$\leqslant f(D_{n+1}) \leqslant f(D_n) \leqslant f(D_{n-1}) \leqslant \cdots \leqslant f(D_1).$$

The sequences $\{f(d_n)\}$ and $\{f(D_n)\}$ being increasing, bounded from above, or decreasing, bounded from below, respectively, both have limits, say

$$\lambda = \lim_{n \to \infty} f(d_n) \quad \text{and} \quad L = \lim_{n \to \infty} f(D_n). \tag{11}$$

Since f is increasing for dyadic fractions, therefore on sequences of dyadic fractions, tending to u from below or above, different from $\{d_n\}$, $\{D_n\}$, the limits of f also exist and are λ and L respectively. If $\lambda \neq L$ then, by the above definition (11),

$$\lambda < L. \tag{12}$$

Since $\circ$ is strictly increasing and satisfies (3),

$$\lambda = \lambda \circ \lambda < \lambda \circ L < L \circ L = L. \tag{13}$$

There exist subsequences $\{d_{n_k}\}$ and $\{D_{n_k}\}$ such that

$$\frac{d_{n_k} + D_{n_k}}{2} > u \quad \text{for all } k = 1, 2, \ldots.$$

Thus

$$f(d_{n_k}) \circ f(D_{n_k}) = f\left(\frac{d_{n_k} + D_{n_k}}{2}\right) \geqslant L$$

and, if k tends to ∞,

$$\lambda \circ L \geqslant L,$$

in contradiction to (13). So (12) cannot hold and

$$\lambda = L.$$

We define, for any u in $[0, 1]$,

$$\bar{f}(u) = \lambda = L.$$

If u is dyadic, then we clearly get $\lambda \leqslant f(u) \leqslant L$ and so $f(u) = \bar{f}(u)$. Thus $\bar{f}$ extends f to $[0, 1]$. We will denote $\bar{f}$ simply by f.

This function f is strictly increasing on $[0, 1]$. Indeed, take arbitrary $u < v$. Let D be an upper dyadic approximation of u and d a lower dyadic approximation of v such that $D < d$. Then

$$f(u) \leqslant f(D) < f(d) \leqslant f(v).$$

As a monotonic function f can have, at most, gap discontinuities. If it had one, then $\lambda < L$ (cf. (11), (12)) would hold there. But we have just proved $\lambda = L$. The continuity at 0 and 1 is equally obvious. So f is continuous on $[0, 1]$ and, since it satisfies (6) for dyadic u, v, by continuity, (6) is satisfied for all $u, v \in [0, 1]$.

Since f is continuous and strictly monotonic and (7) holds, it has a unique inverse $k = f^{-1}: [a, b] \to [0, 1]$ for which, by (6), we have (1). This proves the Theorem 1 for closed intervals $I = [a, b]$.

If I is open (possibly infinite) on one side or both, take a (not necessarily strictly) decreasing sequence $\{\alpha_n\}$ and an increasing $\{\beta_n\}$ in I, $\alpha_n < \beta_n$, $(n = 1, 2, \ldots)$, tending to the two extremities of I. By what we have proved already, there exist continuous (and strictly monotonic) functions $k_n: [\alpha_n, \beta_n] \to \mathbb{R}$ such that

$$x \circ y = k_n^{-1}\left[\frac{k_n(x) + k_n(y)}{2}\right] \quad \text{on } [\alpha_n, \beta_n]. \tag{14}$$

As we have seen in Corollary 15.5, k_n is determined only up to a nonzero multiplicative and an additive constant (of course, then the range will not necessarily be $[0, 1]$ anymore and k_n may be strictly decreasing, but this makes no difference – that is why we have required only strict monotonicity of k in Theorem 1). Since, by definition, $\alpha_{n-1}, \beta_{n-1} \in [\alpha_n, \beta_n]$, we can choose each k_n (beginning with k_2) so that

$$k_n(\alpha_{n-1}) = k_{n-1}(\alpha_{n-1}) \quad \text{and} \quad k_n(\beta_{n-1}) = k_{n-1}(\beta_{n-1}).$$

But then (see Theorem 15.4),

$$k_n(x) = k_{n-1}(x) \quad \text{for all } x \in [\alpha_{n-1}, \beta_{n-1}] \tag{15}$$

(k_{n-1} is not defined elsewhere). So we have a sequence of functions which are extensions of each other on a sequence of extending intervals. As their limit, we get a function $k: I \to \mathbb{R}$ and, from the formulae (14) and (15), we have

$$x \circ y = k^{-1}\left(\frac{k(x) + k(y)}{2}\right) \quad \text{for all } x, y \in I.$$

This concludes the proof of Theorem 1.

Going back to the notation M, we have (almost) proved the following:

Corollary 2. *The general solution $M: I^2 \to I$ of the system*

$$M(x, x) = x, M(x, y) = M(y, x), M[M(x, y), M(z, w)] = M[M(x, z), M(y, w)]$$
$$(x, y, z, w \in I)$$

of functional equations in the class of continuous functions, strictly monotonic in each variable, is given by

$$M(x, y) = k^{-1}\left(\frac{k(x) + k(y)}{2}\right) \quad (x, y \in I),$$

where $k: I \to \mathbb{R}$ is an arbitrary continuous and strictly monotonic function.

Proof. We only have to show that M is strictly *increasing* in both variables. First, there cannot exist $y_1 < y_2$, say, such that $x \mapsto M(x, y_1)$ is strictly increasing while $x \mapsto M(x, y_2)$ is strictly decreasing. Indeed, otherwise, for any $x < x'$, we would have $M(x, y_1) - M(x', y_1) < 0$, while $M(x, y_2) - M(x', y_2) > 0$, and so, by the continuity of M, there would exist a $y \in]y_1, y_2[$ such that $M(x, y) - M(x', y) = 0$, contrary to the strict monotonicity of $x \mapsto M(x, y)$. Also M, being symmetric, cannot increase in one variable and decrease in the other. Finally, M cannot be strictly decreasing in both variables, because then, for $x < x'$, $x = M(x, x) > M(x, x') > M(x', x') = x'$.

This contradiction concludes the proof of Corollary 2, in view of Theorem 1.

In Proposition 15.9 we have already characterized a nonsymmetric (arithmetic) mean value. We now use our above conditions without the symmetry (4) to characterize the following generalizations of quasi-arithmetic means (1):

$$x \square y = k^{-1}[(1-q)k(x) + qk(y)], \tag{16}$$

which we call *quasilinear means*, cf. (16.20). (Again, k is continuous and strictly monotonic on our fundamental interval I. The constant q is different from 0 and 1. We denote the operation here by $\square$, because we want to reserve $\circ$ in this section for symmetric means to which we intend to reduce $\square$.) Cf. Aczél 1948*b*; Fuchs 1950.

So we suppose that the operation $\square$ is strictly monotonic in both variables (this time it is not necessarily increasing), continuous, and satisfies

$$(x \square y) \square (u \square v) = (x \square u) \square (y \square v) \quad (x, y, u, v \in I) \tag{17}$$

and

$$z \square z = z \quad (z \in I). \tag{18}$$

We note that these conditions imply [with $x=y$ or $u=v$ in (17), respectively] the self-distributivity equations

$$x \square (u \square v) = (x \square u) \square (x \square v) \quad \text{and} \quad (x \square y) \square u = (x \square u) \square (y \square u) \quad (x, y, u, v \in I). \tag{19}$$

Of course, (16) has all these properties.

As indicated above, we define an operation $\circ$ from the given operation $\square$. This operation $\circ$ will satisfy *all* conditions of Theorem 1 (including *symmetry*). Take $a \in I$ arbitrarily. We define

$$z = x \circ y$$

as the solution of

$$(a \square z) \square (z \square a) = (a \square x) \square (y \square a). \tag{20}$$

Thus

$$[a \square (x \circ y)] \square [(x \circ y) \square a] = (a \square x) \square (y \square a). \tag{21}$$

Equation (20) has indeed exactly one solution z, increasing with x and y, since both the left hand side and the right hand side are continuous and strictly monotonic in z and strictly monotonic in the same direction in x and y. Indeed, they are all strictly increasing if $\square$ increases with both factors and all strictly decreasing if $\square$ increases with one factor and decreases with the other. As before, decrease of $\square$ in both factors is excluded by (18), while continuity excludes that, say, $x \mapsto x \square y_1$ be decreasing while $x \mapsto x \square y_2$ be increasing but all $x \mapsto x \square y$ still be strictly monotonic. Also the range of both sides of (20) is the same: $(a \square I) \square (I \square a)$. This last fact, the continuity and the strict monotonicity of $z \mapsto (a \square z) \square (z \square a)$, assures the existence of a unique solution z of (20), while the strict monotonicity, in the same sense, of both sides of (20) assures that z increases with both x and y. Also the continuity and, by (21) and (17), the symmetry (4) of $\circ$, thus defined, are clear and, putting

$y = x$ into (21), we see that $\circ$ satisfies (3). So, in order to be able to apply Theorem 1, we have only to show that $\circ$ is medial.

For this we need the following important lemma (Aczél 1948*b*, Fuchs 1950):

Lemma 3. *Under the above suppositions we have*

$$(x \circ y) \square (u \circ v) = (x \square u) \circ (y \square v) \quad \textit{for all } x, y, u, v \in I. \tag{22}$$

Proof. We apply (19), (17), (21), (17), (19) and (21) in this order:

$$\begin{aligned}
&\{a \square [(x \circ y) \square (u \circ v)]\} \square \{[(x \circ y) \square (u \circ v)] \square a\} \\
&\quad = \{[a \square (x \circ y)] \square [a \square (u \circ v)]\} \square \{[(x \circ y) \square a] \square [(u \circ v) \square a]\} \\
&\quad = \{[a \square (x \circ y)] \square [(x \circ y) \square a]\} \square \{[a \square (u \circ v)] \square [(u \circ v) \square a]\} \\
&\quad = [(a \square x) \square (y \square a)] \square [(a \square u) \square (v \square a)] \\
&\quad = [(a \square x) \square (a \square u)] \square [(y \square a) \square (v \square a)] = [a \square (x \square u)] \square [(y \square v) \square a] \\
&\quad = \{a \square [(x \square u) \circ (y \square v)]\} \square \{[(x \square u) \circ (y \square v)] \square a\}.
\end{aligned}$$

The strict monotonicity of $z \mapsto (a \square z) \square (z \square a)$ gives (22) and the lemma is proved.

We are ready now to prove the mediality of $\circ$. We make use of (21), (3), (4) and of (22) twice:

$$\begin{aligned}
&\{a \square [(x \circ y) \circ (u \circ v)]\} \square \{[(x \circ y) \circ (u \circ v)] \square a\} = [a \square (x \circ y)] \square [(u \circ v) \square a] \\
&\quad = [(a \circ a) \square (y \circ x)] \square [(u \circ v) \square (a \circ a)] = [(a \square y) \circ (a \square x)] \square [(u \square a) \circ (v \square a)] \\
&\quad = [(a \square y) \square (u \square a)] \circ [(a \square x) \square (v \square a)].
\end{aligned}$$

By the mediality (17) of $\square$, however, y and u may be interchanged on the far right of this series of equalities, so also on the far left. This and the strict monotonicity of $z \mapsto (a \square z) \square (z \square a)$ yield the required mediality of $\circ$. Thus, by Theorem 1, there exists a continuous, increasing k such that

$$x \circ y = k^{-1}\left(\frac{k(x) + k(y)}{2}\right) \quad (x, y \in I).$$

Substitute this into (22) with

$$X := k(x), \quad Y := k(y), \quad U := k(u), \quad V := k(v), \quad G(X, U) := k[k^{-1}(X) \square k^{-1}(U)] \tag{23}$$

and get

$$G\left(\frac{X+Y}{2}, \frac{U+V}{2}\right) = \frac{G(X, U) + G(Y, V)}{2}.$$

By the continuity of $\square$ and k, the new variables X, Y, U, V range over an interval $J = k(I)$ and G is continuous on J^2. Thus, by Proposition 15.3, G is linear, that is, there exist constants p, q, r such that

$$G(X, U) = pX + qU + r$$

or, with (23),

$$x \square u = k^{-1}[pk(x) + qk(u) + r]. \tag{24}$$

Now, (18) yields $p + q = 1$, $r = 0$ and the strict monotonicity of $\square$ excludes $p = 0$ or $q = 0$. So we have proved the following:

Theorem 4. *The general continuous, strictly monotonic solutions* $\square$ *of* (17) *and* (18) *are given by*

$$x \square y = k^{-1}[(1-q)k(x) + qk(y)] \quad (x, y \in I),$$

where $k: I \to \mathbb{R}$ *is an arbitrary continuous and strictly monotonic function and* q *is an arbitrary constant, different from* 0 *and* 1.

One can determine the general continuous and strictly monotonic solutions of (17) alone (see Aczél 1948*b*, 1966*c*, pp. 287–92), they happen to be of the form (24) (cf. Exercise 2). Fuchs (1950) has generalized all this to operations on arbitrary ordered sets.

Another characterization of the quasilinear mean values (16) can be given by means of the equations

$$x \square (y \square z) = (x \square y) \square (x \square z) \tag{25}$$

and

$$(x \square y) \square z = (x \square z) \square (y \square z) \tag{26}$$

of *self-distributivity*, mentioned above as (19). We will give the general continuous operation $\square$, strictly increasing in both variables (factors) (supposing strict monotonicity would be sufficient instead of strict increase of $\square$ in both variables, cf. Remark after Corollary 6 below), satisfying (25) and (26). We do this by showing that $\square$ satisfies (18) and the mediality (17). The first part of the following proof is due to Hosszú 1959*b*, for the second see Aczél 1964*a*. The deduction of (18) is simple. Put $x = y = z$ into (25) and get, by the strict monotonicity of $\square$,

$$z \square z = z, \tag{27}$$

that is, (18).

In order to establish mediality, suppose, for instance, that v lies between t and u and define $f(v)$ by

$$(s \square t) \square (u \square v) = (s \square u) \square [t \square f(v)]. \tag{28}$$

This is a valid definition. Indeed, by the strict monotonicity of $\square$ *at most* one such $f(v)$ exists. If $v = t$, then by (26) and (27),

$$(s \square t) \square (u \square t) = (s \square u) \square t = (s \square u) \square (t \square t)$$

and, comparing this with (28), we see that

$$f(t) = t. \tag{29}$$

Similarly, if $v = u$, we have

$$(s \square t) \square (u \square u) = (s \square t) \square u = (s \square u) \square (t \square u)$$

and

$$f(u) = u. \tag{30}$$

By the continuity of $\square$, (28) has now *at least* one solution $f(v)$ (as long as $v \in [t, u]$; in the other cases (28) has to be slightly altered) and thus f is indeed uniquely defined.

We want to obtain $f(v) = v$ on $[t, u]$. First we prove that

$$f(x \square y) = f(x) \square f(y) \tag{31}$$

(on $[t, u]^2$). Indeed, by (28) and (25),

$$(s \square u) \square [t \square f(x \square y)] = (s \square t) \square [u \square (x \square y)] = (s \square t) \square [(u \square x) \square (u \square y)]$$
$$= [(s \square t) \square (u \square x)] \square [(s \square t) \square (u \square y)]$$
$$= \{(s \square u) \square [t \square f(x)]\} \square \{(s \square u) \square [t \square f(y)]\}$$
$$= (s \square u) \square \{[t \square f(x)] \square [t \square f(y)]\} = (s \square u) \square \{t \square [f(x) \square f(y)]\}.$$

Thus the strict monotonicity of $\square$ gives (31). We already have (29) and (30). By (27) and since $\square$ is strictly increasing, $\square$ is also internal

$$x = x \square x < x \square y < y \square y = y \quad \text{whenever } x < y.$$

So, for $F(x, y) = x \square y$, $H(u, v, x, y) = u \square v$, all conditions of Theorem 15.4 are satisfied. Of course, (31) is also satisfied by the identity function ($f(v) \equiv v$). Thus, indeed,

$$f(v) = v$$

(for all $v \in [t, u]$ and similarly for all other subintervals of our fundamental interval I). Compared to (28) we have that

$$(s \square t) \square (u \square v) = (s \square u) \square (t \square v),$$

that is, $\square$ *is medial.* Thus we have proved the following (cf. Theorem 4):

Theorem 5. *Every continuous operation $\square$ on an interval I, which is strictly monotonic (increasing) and which satisfies* (25) *and* (26) *on I^3, satisfies also* (18) *and* (17) *on I or I^4 respectively.*

Corollary 6. *The general continuous strictly increasing solutions of* (25) *and* (26) *are given by*

$$x \square y = k^{-1}[(1 - q)k(x) + qk(y)], \tag{16}$$

where k is an arbitrary continuous strictly monotonic function and $q \in]0, 1[$ an arbitrary constant.

Remark. If $y \mapsto x \square y$ is strictly decreasing, q would be negative. If $x \mapsto x \square y$ is strictly decreasing, $q > 1$ would hold. By more sophisticated algebraic and topological means Strambach (1975) has established that, under some further assumptions, the result of this corollary remains essentially true without supposing (25) (cf. Exercise 18.14).

Strict monotonicity in the above results can be replaced by cancellativity ($x \mapsto x \Box y$ and $y \mapsto x \Box y$ are injections, cf. Exercise 5).

Notice that (6) and

$$k(x \circ y) = \frac{k(x) + k(y)}{2}, \quad k(x \Box y) = (1 - q)k(x) + qk(y) \quad (q \in]0, 1[), \tag{32}$$

which follow from (1) and (16), respectively, are of the form (15.21) and the conditions of Theorem 15.4 are satisfied. But here our main concern was not uniqueness but existence. However, Theorems 1, 4 and Corollaries 2, 6 give *necessary and sufficient conditions for the existence of continuous and strictly increasing solutions of the functional equations* (32).

Exercises and further results

1. Let I be an interval of $\mathbb{R}$ and $M(x, y)$ a quasi-arithmetic mean on I. Determine the general continuous solution $f: \mathbb{R} \to \mathbb{R}$ of

$$f(x + y) + f(M(x, y)) = f(x + M(x, y)) + f(y) \quad (x, y \in I).$$

(For Exercises 2–5, see Aczél 1966c, pp. 287–99.)

2*. Show that every (in each variable) continuous and strictly monotonic solution $\Box: I^2 \to I$ (I is an open, closed or half-closed, finite or infinite proper real interval) of (17) alone is of the form

$$x \Box y = k^{-1}[ak(x) + bk(y) + c] \quad (x, y \in I), \tag{33}$$

where a, b, c are constants, $ab \neq 0$, and $k: I \to \mathbb{R}$ is continuous and strictly monotonic.

3. Given the operation $\Box$ in (33), state to what extent the constants a, b, c and the function k in (33) are determined.

4. Let α, β be the endpoints of I (possibly infinite). What relations among α, β, a, b, c have to hold in (33), in order that $I \Box I \subseteq I$ be valid? ($I \Box I = \{x \Box y \mid x \in I, y \in I\}$.)

5. Show that, in Exercise 2 and in Theorems 4, 5 and Corollaries 2, 6 the requirement of strict monotonicity of the operations can be replaced by their cancellativity ($\Box$ is *cancellative* if $z_1 \Box t = z_2 \Box t$ or $t \Box z_1 = t \Box z_2$ for one t implies $z_1 = z_2$).

(For Exercises 6–11, see Aczél 1966c, pp. 245–69.)

6. Let I be a proper real interval and $(I, \circ)$ a continuous group. Show that there exists a continuous and strictly monotonic $g: I \to \mathbb{R}$ such that

$$x \circ y = g^{-1}[g(x) + g(y)] \quad (x, y \in I). \tag{34}$$

What restrictions are imposed on the nature of I (open, closed, half-closed) by this result?

7*. How is the result of the previous exercise modified, if $(I, \circ)$ is a continuous cancellative semigroup? (cf. Exercise 5 for the definition of 'cancellative'.)

8. Take $\circ: I^2 \to I$, where $I = [e, b[$, $e \in \mathbb{R}$, $b \in \mathbb{R}$ and $e < b$. Let $c \in]e, b[$. Suppose that $e \circ x = x$ and $(x \circ y) \circ z = x \circ (y \circ z)$ holds if $x, y, z, x \circ y, y \circ z, (x \circ y) \circ z$ are all in $[e, c[$ (conditional associativity equation). Determine on $[e, c[^2$ all operations $\circ$ satisfying these equations which are also continuous and cancellative (cf. Exercise 5).

9*. Let I be an (open, closed, half-closed, finite or infinite) proper real interval, $F: \mathbb{R} \times I \to I$, $x \mapsto F(t, x)$ continuous for all $t \in \mathbb{R}$, $t \mapsto F(t, x_0)$ continuous for at least one $x_0 \in I$, and $t \mapsto F(t, x)$ nonconstant for all $x \in I$. Show that, in this class, the general solution of the functional equation

$$F[s, F(t, x)] = F(s + t, x) \quad (s, t \in \mathbb{R}; x \in I) \tag{35}$$

is given by

$$F(t, x) = h^{-1}[t + h(x)] \quad (t \in \mathbb{R}, x \in I), \tag{36}$$

where $h: I \to \mathbb{R}$ is a continuous and strictly monotonic function. Show that I has to be open.

10*. How is the result of the previous exercise modified if (35) is supposed only for $s, t \in]c, \infty[$ $(c \geqslant 0)$ but F is supposed to be strictly increasing in both variables and continuous in x.

11. Given $\circ$ and F, state to what extent are the functions g or h in (34) and (36), respectively, determined.

12*. Try to generalize the results of Exercises 1–11 from $\mathbb{R}$ (and I) to more general (ordered) topological spaces.

18

Generalized mediality. Connection to webs and nomograms

This chapter is a natural sequel to Chapter 17 and establishes connections to webs and nomograms. We generalize the mediality equation which we used in the previous chapter to characterize quasi-arithmetic means, to an equation with six unknown functions (operations).

We can generalize the mediality equation (17.2), in the same manner as the Pexider equation (4.25) generalized the Cauchy equation (2.1), by allowing all unknown functions – operations this time – to be different. In order not to have to invent five new operation signs similar to $\circ$, we denote the six operations simply by $1, 2, 3, 4, 5, 6$ and write the new equation of *generalized mediality* as

$$(x1y)2(z3w) = (x4z)5(y6w). \tag{1}$$

We can do this in good conscience because, for all practical purposes, this will be the only mathematical use for (real) numbers in this chapter (other than as indices).

The only exceptions are the following remarks about *nomography*, which will lead us to some simple facts in the *theory of webs* which, in their turn, will help us solve (1) under rather general conditions. *Nomograms* are visual calculation aids to represent and evaluate multiplace functions (functions of several, at least two, variables) in the plane(s). We will restrict ourselves here to functions of two variables (two-place functions) and the nomograms we deal with are mainly (modifications of) the *contour line representation* of such functions. We get these in the simplest way by drawing, in the coordinate plane, curves through those points to which the same value for the function is assigned. For instance, Figure 15 shows this kind of nomogram of the function associating $(x(1-y))^{1/2}$ to positive real values of x and to $y < 1$. With the aid of this diagram, we can determine, for any given pair of numbers

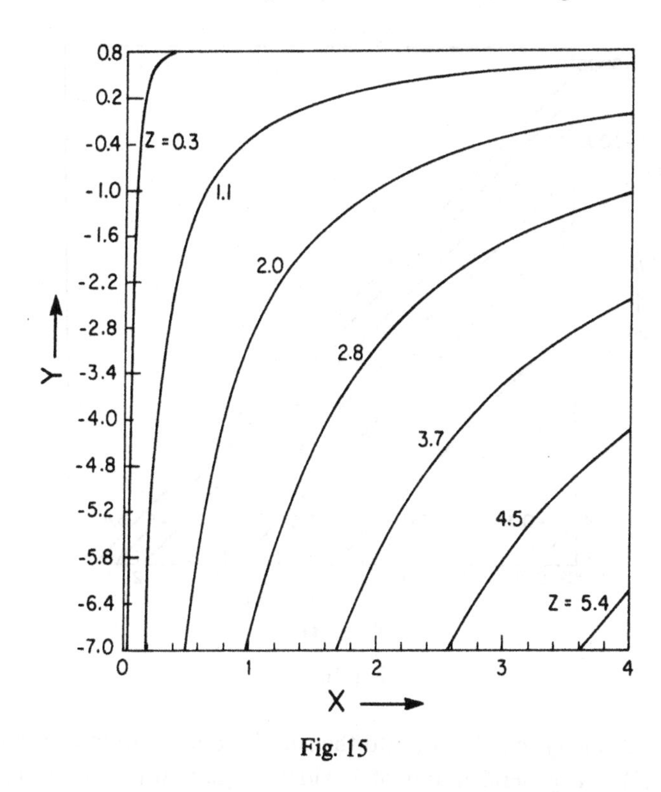

Fig. 15

$x > 0$, $y < 1$, the value of $z = (x(1-y))^{1/2}$ exactly or approximately by interpolation between the curves drawn on the diagram. Similarly, we can determine y, given x and z and also x, given y and z.

Here some difficulty lies partly in the great amount of work required to draw all these curves and partly in the fact that interpolating, even approximately, is more difficult between such curves than between straight lines. Therefore, replacing curves by straight lines (called anamorphosis) is of great importance. This can be done sometimes by replacing the linear scales on the axes by nonequidistant scales. For instance, $z = (x(1-y))^{1/2}$ can also be written as $2\log z = \log x + \log(1-y)$. If we introduce the new variables $x' = \log x$, $y' = \log(1-y)$, $z' = 2\log z$ (that is, we use logarithmic scales), then the relation between x, y and z will be transformed into $z' = x' + y'$ and Figure 15 into Figure 16, consisting of parallel (but not equidistant) straight lines intersecting the horizontal axis in a 135° angle (there is also an advantage in not having the straight lines too steep or too flat).

In general, the problem of *parallel anamorphosis* consists in finding, for a given two-place function F, mappings f, g, h such that

$$h[F(x,y)] = f(x) + g(y), \tag{2}$$

if they exist, which means that the contour line representation of F goes over

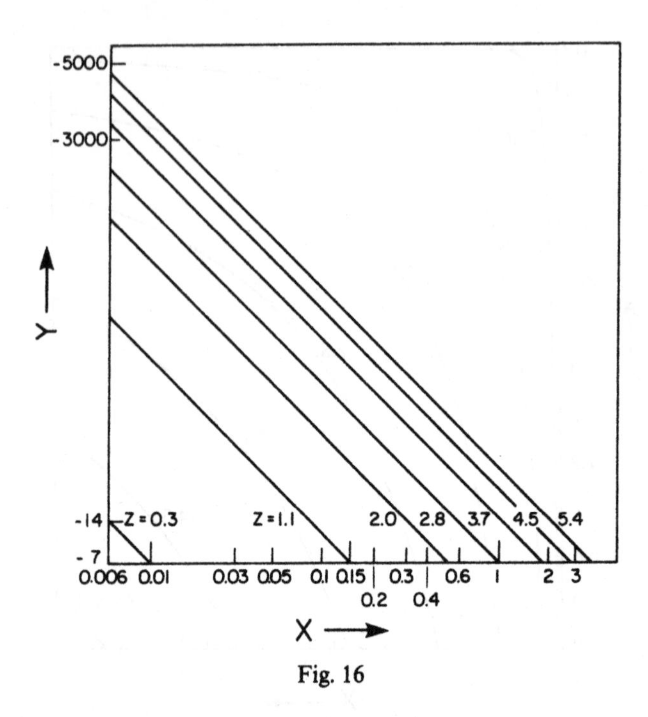

Fig. 16

into a *parallel nomogram*, that is, into the simple configuration of Figure 16. Of course, (2) is a generalization of Pexider's equation (4.25). Notice that both in the original representation (Figure 15) and in the result of the parallel anamorphosis (Figure 16), there is more than one family of lines: the coordinate lines, $x = \text{const.}$, $y = \text{const.}$ (or $x' = \text{const.}$, $y' = \text{const.}$), also form two bundles of parallel straight lines each. There are values of x, y and z assigned to each line of the first, second or third family, respectively, and by checking the point of intersection of a given x-line and y-line, the number attached to the z-line going through that point is the corresponding z-value. (Similarly, x can be determined from y and z, and y from x and z.)

Somewhat more general are the *straight nomograms* where all three families consist of (not necessarily parallel) straight lines (they result from *anamorphosis*).

We also mention the other important kind of nomograms, the *alignment charts*. In projective duality, points correspond to straight lines, points lying on the same straight line to straight lines intersecting in one point, and vice versa. Remember that the three sets of straight lines in a straight nomogram represent the variables x, y, z and the intersection of an x-, y-, and z-line in the same point corresponded to the functional relation $z = F(x, y)$ among the respective x, y, and z values. Therefore, in alignment charts, the duals of straight nomographs, sets of points (curves) correspond to the x, y and z values and, in particular, such x, y, and z values are connected by $z = F(x, y)$, if, and only if, the respective x-, y-, and z-points are on the same straight line (Figure 17).

The duals of straight nomograms consisting of three bundles of straight lines, each going through one point, finite or infinite, are the alignment charts with three straight scales (Figure 18). If those three points lie on one straight line – the infinite line of the plane, if all three bundles are parallel as in parallel

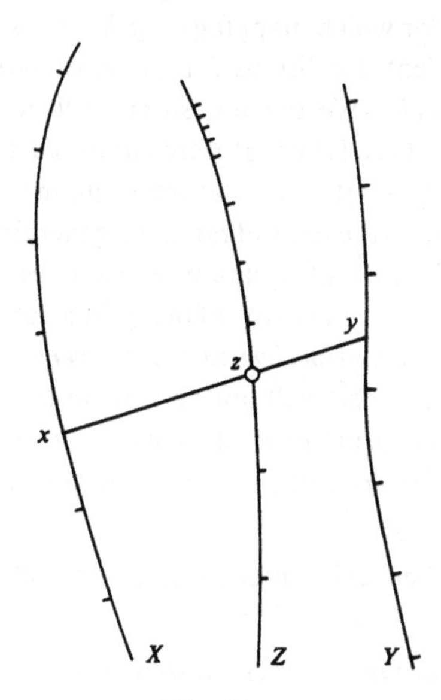

Fig. 17

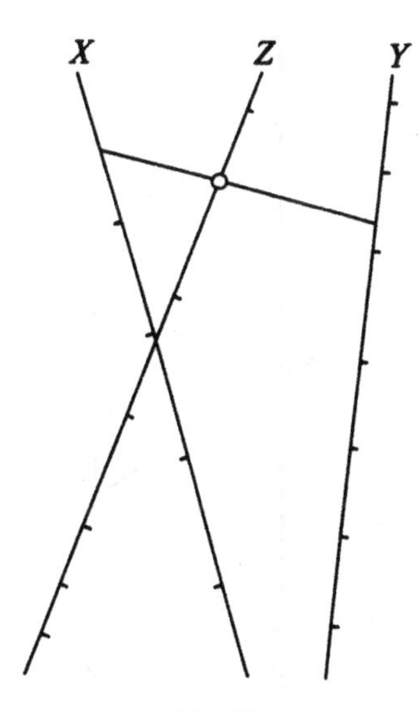

Fig. 18

nomograms – then the three straight lines of the alignment chart go through the same point. Again this can be in infinity and then we get *alignment charts with three parallel straight scales* (Figure 19). If ϕ, ψ, χ are the scale functions and p, q the distances of the scales, we then have

$$(p+q)\chi(z) = p\phi(x) + q\psi(y),$$

which, of course, is equivalent to (2). This again emphasizes the importance of the class of functions for which mappings f, g, h exist, so that (2) is satisfied.

Necessary and sufficient conditions for the existence of such maps are given in the *theory of webs.* We did not specify till now the domains and range of F or domains where (1) or (2) were supposed to be valid.

In our example $z = (x(1-y))^{1/2}$, the natural domain is the set $\{(x, y) | x > 0, y < 1\}$ and $\mathbb{R}_+ = \{z | z > 0\}$ is the natural range. In general we deal with slightly generalized groupoids. In view of equation (2), we have *three sets* X, Y, Z. A *generalized groupoid* $(X, Y, Z; F)$ consists of these three sets and of an *operation* (function, mapping) F, *attaching an element* $z \in Z$ *to every pair* (x, y) $(x \in X, y \in Y)$. Since multiplication of numbers will not be used in the rest of this chapter, for simplicity we will, as is usual in algebra, write xy (later $x1y$, etc.) instead of $F(x, y)$ and $(X, Y, Z; \cdot)$ instead of $(X, Y, Z; F)$. A generalized groupoid is a *generalized quasigroup* if

for each $x \in X, y \in Y, z \in Z$ there exist unique $x' \in X, y' \in Y$ such that $x'y = z, xy' = z$. (3)

Groupoids and *quasigroups* are generalized groupoids resp. generalized

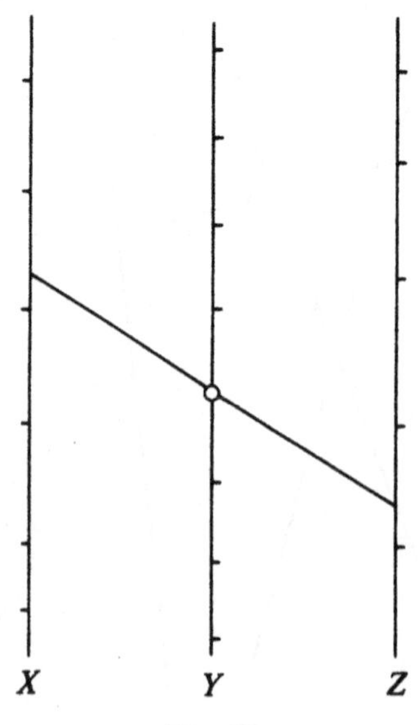

Fig. 19

quasigroups with $X = Y = Z$. A *loop* is a quasigroup with a *neutral element* e such that

$$ze = ez = z \quad \text{for all } z \in Z. \tag{4}$$

The usual algebraic concept of isomorphy is replaced by *isotopy*. Two generalized groupoids (or quasigroups) $(X, Y, Z; 1)$ and $(X', Y', Z'; 2)$ are *isotopic* if there exist bijections $f: X \to X'$, $g: Y \to Y'$, $h: Z \to Z'$ such that

$$h(x1y) = f(x)2g(y) \quad \text{for all } x \in X, \quad y \in Y. \tag{5}$$

The triple (f, g, h) is the *isotopism*. The relation between $(X, Y, Z; 1)$ and $(X', Y', Z'; 2)$, established by (5), is called an *isotopy*. In this sense, if f, g, h in (2) are injections into $\mathbb{R}$ (one-to-one), then (2) *means that* $(X, Y, Z; F)$ *is isotopic to* $(R_1, R_2, R_3; +)$ where the $R_i (i = 1, 2, 3)$ are subsets of the set $\mathbb{R}$ and $+$ is the common addition of reals. As mentioned before, we will give, with the aid of the theory of webs, conditions for equations generalizing (2), where $(R_1, R_2, R_3; +)$ will be replaced by an arbitrary abelian group.

A *web* belonging to a generalized groupoid may be defined as consisting of all *points* (x, y) $(x \in X, y \in Y)$ and of three families of lines. An X-, Y- or Z-line is defined by

$$X_a = \{(x, y) \mid x = a, y \in Y\}, \quad Y_b = \{(x, y) \mid x \in X, y = b\},$$
$$\text{and} \quad Z_c = \{(x, y) \mid xy = c, x \in X, y \in Y\}$$

respectively. We get all X-, (Y-, Z-) lines by letting $a, (b, c)$ go through the whole set X, (Y, Z). A point (x, y) *lies on* X_a (or Y_b or Z_c), if $(x, y) \in X_a$ (or $\in Y_b$ or $\in Z_c$). Every point $(a, b) \in X \times Y$ lies on exactly one X-line (X_a), exactly one Y-line (Y_b) and exactly one Z-line (Z_{ab}). The lines X_a and Z_c *meet* in the point (x, y) if $(x, y) \in X_a$ and $(x, y) \in Z_c$ (similarly for Y_b). Since no point lies on more than one line of the same family, two lines belonging to the same family *never* meet. An X-line (X_a) and a Y-line (Y_b) meet in *exactly one* point $[(a, b)]$. If (and only if) the web belongs to a *generalized quasigroup*, then an X- (or Y-) and a Z-line meet in *exactly one* point.

We say that a web (or the underlying generalized groupoid) satisfies the *Thomsen condition* (T for short) if

$$x_1 y_2 = x_2 y_1 \quad \text{and} \quad x_1 y_3 = x_3 y_1 \quad \text{together imply}$$
$$x_2 y_3 = x_3 y_2 \quad (x_1, x_2, x_3 \in X, y_1, y_2, y_3 \in Y). \tag{6}$$

The meaning of T on webs can be seen on Figure 20 (which represents, of course, the real case). *T is invariant under isotopy.* This T is the necessary and sufficient condition which we have promised. (Thomsen 1929; because of easier accessability we quote Aczél 1965*a* here and in what follows, in particular in the Exercises and Further Results at the end of the chapter, instead of the primary sources.)

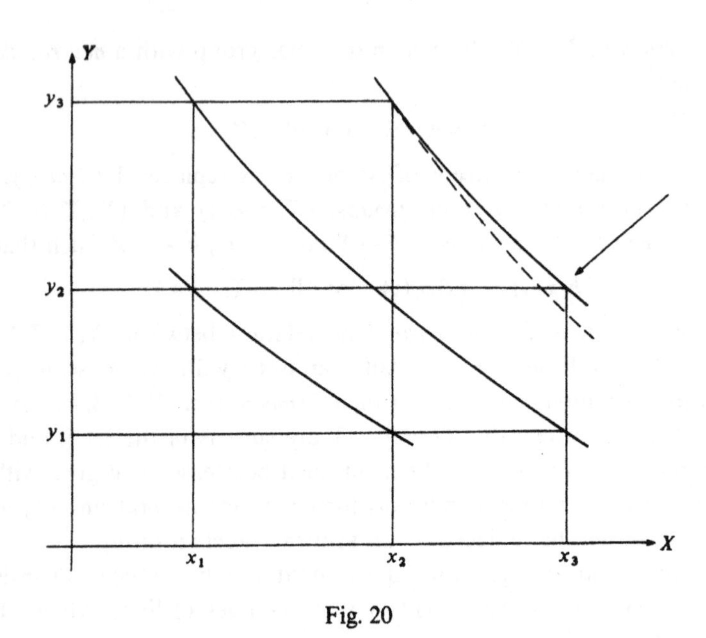

Fig. 20

Lemma 1. *A generalized quasigroup* $(X, Y, Z; \cdot)$ *is isotopic to an abelian group if, and only if, it satisfies the Thomsen condition* (6).

Proof. The 'only if' part can be easily checked.

In order to prove the 'if' part, we first construct, on Z, an isotopic loop which thus also satisfies T and then prove that it is an abelian group. Let $u \in X$, $v \in Y$ be arbitrary but fixed for the time being. Since $(X, Y, Z; .)$ is a generalized quasigroup, an operation $\circ$ is completely and unambiguously defined on Z by

$$(xv)\circ(uy) = xy \quad (x \in X, y \in Y). \tag{7}$$

Indeed, given $s \in Z$, $t \in Z$, $u \in X$, $v \in Y$ there exist unique $x \in X$, $y \in Y$ such that

$$s = xv, \quad t = uy \tag{8}$$

and thus, for all $s, t \in Z$, $s \circ t$ is uniquely defined by (8) and (7). (The need for this new operation is the reason why we use generalized quasigroups rather than generalized groupoids. The solvability conditions can, however, be weakened. See, for instance, the references at the end of this chapter.) Comparison of (5) and (7) shows that $(X, Y, Z; .)$ *is isotopic to* $(Z, Z, Z; \circ)$[$(Z, \circ)$ for short], since f, g, h defined by $f(x) = xv$, $g(y) = uy$ and $h(z) = z$ are bijections, as we have just seen. Also $(Z, \circ)$ is a quasigroup because, evidently, under isotopy a generalized quasigroup remains a generalized quasigroup (a quasigroup in this case). Finally, putting $x = u$ or $y = v$ into (7) shows that $e = uv$ is a neutral element, as defined by (4). This concludes the proof that $(Z,\circ)$ *is a loop isotopic to* $(X,Y,Z;.)$. As a matter of fact, we have found *several loops isotopic to* $(X,Y,Z;\circ)$, *depending upon the choice of* $u\in X$, $v\in Y$.

We prove now that, if T holds, $(Z,\circ)$ is an abelian group (cf. Aczél 1981*c* for this proof). As we have seen, if T is satisfied on $(X,Y,Z;.)$ it holds also on the isotopic $(Z,\circ)$. Choose in (6), $y_1=e$, $y_2=s$, $y_3=t$, $x_1=r$, $x_2=r\circ s$, $x_3=r\circ t$. Then

$$x_1\circ y_2=r\circ s=x_2\circ y_1,\quad x_1\circ y_3=r\circ t=x_3\circ y_1,$$

that is, the hypotheses of T are fulfilled, thus the conclusion

$$x_2\circ y_3=x_3\circ y_2,$$

that is,

$$(r\circ s)\circ t=(r\circ t)\circ s \tag{9}$$

holds for all $r,s,t\in Z$. Put $r=e$ into (9) and get

$$s\circ t=t\circ s,$$

that is, $(Z,\circ)$ is *commutative*. This changes (9) into

$$t\circ(r\circ s)=(r\circ s)\circ t=(r\circ t)\circ s=(t\circ r)\circ s,$$

that is, $(Z,\circ)$ is also *associative*. Since $(Z,\circ)$ is a quasigroup (even a loop) we have proved that $(Z,\circ)$ is an abelian group. Since it is isotopic to $(X,Y,Z;.)$, Lemma 1 is proved.

We now return to the generalized mediality equation (1):

$$(x1y)2(z3w)=(x4z)5(y6w)\quad\text{for all } x\in X,\quad y\in Y,\quad z\in Z,\quad w\in W, \tag{10}$$

with the unknown functions (operations) $1:X\times Y\to S$, $2:S\times T\to G$, $3:Z\times W\to T$, $4:X\times Z\to U$, $5:U\times V\to G$, $6:Y\times W\to V$. Often, for the sake of brevity, we will write just the operation instead of the generalized groupoid (quasigroup), that is, 1 for $(X,Y,S;1)$, etc. We prove the following (Aczél–Belousov–Hosszú 1960; Aczél 1965*a*; Taylor 1972*b*, 1973*a*, 1978).

Theorem 2. *The generalized quasigroups* 1, 2, 3, 4, 5, 6 *satisfy the equation* (10) *if, and only if, there exists an operation* $\circ$ *such that* $(G,\circ)$ *is an abelian group and there exist bijections* $f:X\to G$, $g:Y\to G$, $h:Z\to G$, $k:W\to G$, $m:S\to G$, $n:T\to G$, $p:U\to G$, $q:V\to G$ *such that, for all* $x\in X$, $y\in Y$, $z\in Z$, $w\in W$, $s\in S$, $t\in T$, $u\in U$, $v\in V$

$$s2t=m(s)\circ n(t), \tag{11}$$

$$x1y=m^{-1}[f(x)\circ g(y)], \tag{12}$$

$$z3w=n^{-1}[h(z)\circ k(w)], \tag{13}$$

$$u5v=p(u)\circ q(v), \tag{14}$$

$$x4z=p^{-1}[f(x)\circ h(z)], \tag{15}$$

$$y6w=q^{-1}[g(y)\circ k(w)]. \tag{16}$$

Proof. We prove first that the *Thomsen condition* (6) *holds for* 1. So we suppose

$$x_1 1 y_2 = x_2 1 y_1 \tag{17}$$

and

$$x_1 1 y_3 = x_3 1 y_1 \tag{18}$$

($x_1, x_2, x_3 \in X$, $y_1, y_2, y_3 \in Y$). In view of (17), we get from (10)

$$(x_2 4z)5(y_1 6w) = (x_2 1 y_1)2(z3w) = (x_1 1 y_2)2(z3w) = (x_1 4z)5(y_2 6w). \tag{19}$$

Since $(X, Z, U; 4)$ and $(Y, W, V; 6)$ are generalized quasigroups, there exist $z_1, z_2 \in Z$, $w_1, w_2 \in W$ such that, for given x_1, x_3, u, y_1, y_3, v,

$$x_1 4z_1 = x_3 4z_2 = u \quad \text{and} \quad y_1 6w_1 = y_3 6w_2 = v. \tag{20}$$

We put $z = z_1$, $w = w_1$ into (19) and get, with (20),

$$(x_2 4z_1)5(y_3 6w_2) = (x_3 4z_2)5(y_2 6w_1),$$

or, with (10),

$$(x_2 1 y_3)2(z_1 3w_2) = (x_3 1 y_2)2(z_2 3w_1). \tag{21}$$

However, from (20),

$$(x_1 4z_1)5(y_3 6w_2) = (x_3 4z_2)5(y_1 6w_1)$$

and, again by (10),

$$(x_1 1 y_3)2(z_1 3w_2) = (x_3 1 y_1)2(z_2 3w_1).$$

By (18), the expression in the first brackets on both sides are equal. But 2 is a quasigroup (therefore 2 is cancellative, cf. Exercise 17.5), so

$$z_1 3w_2 = z_2 3w_1$$

and now (21) gives, for the same reason,

$$x_2 1 y_3 = x_3 1 y_2,$$

so the Thomsen condition is indeed satisfied for 1.

Thus, by Lemma 1, $(X, Y, S; 1)$ is isotopic to an abelian group $(G, \circ)$ which proves (12).

Next we prove that *all generalized quasigroups* 1, 2, 3, 4, 5, 6 *are isotopic* (cf. (5)).

We fix $z = a$, $w = b$ in (10) in order to get

$$(x1y)2(a3b) = (x4a)5(y6b).$$

Since the mappings defined by $f(x) = x4a$, $g(y) = y6b$ and $h(s) = s2(a3b)$ ($x \in X$, $y \in Y$, $s \in S$) are bijections, 1 and 5 are isotopic. Similar proofs show that 3 and 5 are isotopic and so are 2 and 4 and also 2 and 6. So now 1, 3, 5 are pairwise isotopic and so are 2, 4, 6. In order to establish the isotopy of all six generalized quasigroups, we have only to show that an even and an odd numbered, say 1 and 4, are isotopic. For this we choose constants $c \in T$, $d \in V$. For each $y \in Y$, there exists a unique $w = \phi(y)$ such that $y6w = d$ and, for this $w \in W$, there exists a unique $z = \psi(w) \in Z$ such that $z3w = c$. Putting these z

and w into (10) we get

$$(x1y)2c = (x4\psi[\phi(y)])5d,$$

which shows that 1 and 4, and thus all six generalized quasigroups are isotopic. By taking the connection (10) among them into consideration, one gets, after replacing $(G, \circ)$ by an isomorphic abelian group, eventually (11)–(16), which completes the proof of Theorem 2.

The conditions in this theorem may be weakened (in particular the restriction that all generalized groupoids be generalized quasigroups) and other, similar functional equations containing several unknown multiplace functions can be solved in such generality. See, for instance, Aczél–Belousov–Hosszú 1960; Hosszú 1963*d*, Aczél 1966*c*, pp. 310–19; Taylor 1972*b*, 1973*a*, 1978.

Exercises and further results

(For Exercises 1–11, see, e.g., Aczél 1965*a*.)

1. We say that a web satisfies the *Reidemeister condition*, if

$$R{:}(x_1y_2 = x_2y_1 \ \& \ x_1y_4 = x_2y_3 \ \& \ x_3y_2 = x_4y_1) \Rightarrow x_3y_4 = x_4y_3.$$

Show that, for (generalized) quasigroups, every web satisfying the Thomsen condition T ((6)) also satisfies R: $T \Rightarrow R$ (draw diagrams).

2. Show that a generalized quasigroup is isotopic to a group if, and only if, it satisfies R, as defined in Exercise 1 (draw a diagram).
3. Use the results of Exercise 1 and 2 to give another proof of Lemma 1.
4. The *Bol conditions* B_1, B_2, B_3 are defined by

$$B_1{:}(x_1y_2 = x_2y_1 \ \& \ x_1y_4 = x_2y_3 \ \& \ x_2y_2 = x_4y_1) \Rightarrow x_2y_4 = x_4y_3,$$
$$B_2{:}(x_1y_2 = x_2y_1 \ \& \ x_1y_4 = x_2y_2 \ \& \ x_3y_2 = x_4y_1) \Rightarrow x_3y_4 = x_4y_2,$$
$$B_3{:}(x_1y_2 = x_2y_1 \ \& \ x_1y_4 = x_2y_3 = x_3y_2 = x_4y_1) \Rightarrow x_3y_4 = x_4y_3.$$

Show that, for (generalized) quasigroups, $(B_1 \ \& \ B_2) \Rightarrow B_3$ (draw diagrams).

5. Show that a generalized quasigroup satisfies B_1 or B_2 if, and only if, it is isotopic to a loop $(L, \circ)$ satisfying

$$y \circ [z \circ (y \circ x)] = [y \circ (z \circ y)] \circ x, \tag{22}$$

or

$$[(x \circ y) \circ z] \circ y = x \circ [(y \circ z) \circ y], \tag{23}$$

respectively, for all $x, y, z \in L$ (draw diagrams). *Remark*: No equation is known to correspond exactly to B_3.

6*. A loop $(L, \circ)$ is *Moufang* if it satisfies

$$[(x \circ y) \circ z] \circ y = x \circ [y \circ (z \circ y)] \quad \text{for all } x, y, z \in L.$$

Prove that a loop is Moufang if, and only if, both (22) and (23) are satisfied. Combine this result with Exercises 4 and 5 and draw conclusions.

7. The *hexagonal condition* is

$$H{:}(x_1y_2 = x_2y_1 \ \& \ x_1y_3 = x_2y_2 = x_3y_1) \Rightarrow x_2y_3 = x_3y_2.$$

Show that a quasigroup satisfies H if, and only if, in *every* loop isotopic to it $(x\circ x)\circ x = x\circ(x\circ x)$ is satisfied (draw a diagram).

8. In loops we define the left inverse x^{-1} by $x^{-1}\circ x = e$ (e is the neutral element). Show that H (see Exercise 7) is satisfied in a quasigroup if, and only if, in every isotopic loop the left inverses are also right inverses (that is, $x\circ x^{-1} = e$).

9*. In loops we define $x^0 = e$, $x^{n+1} = x\circ x^n$ $(n \geqslant 0)$, $x^{-n} = (x^{-1})^n$ $(n \geqslant 1$, cf. Exercise 8). Prove that H is satisfied in a quasigroup iff in every isotopic loop $x^{m+n} = x^m \circ x^n$ $(m, n \in \mathbb{N})$ (draw diagrams).

10*. Let I be a proper real interval forming a loop under the continuous operation $\circ$, which satisfies, for the powers x^m $(m \in \mathbb{N})$ as defined in Exercise 9, $x^{m+n} = x^m \circ x^n$ $(m, n \in \mathbb{N})$. Show that $(I, \circ)$ is o-isomorphic to $(\mathbb{R}, +)$, that is, there exists a continuous strictly monotonic function $f: I \to \mathbb{R}$, such that

$$x\circ y = f^{-1}[f(x) + f(y)] \quad (x, y \in I).$$

Prove that, as a consequence, for continuous (generalized) quasigroups on I the closure conditions T ((6)), R (Exercise 1), B_1, B_2, B_3 (Exercise 4) and H (Exercise 7) are equivalent.

11. Show that the Thomsen condition T ((6)) is satisfied in a (generalized) quasigroup iff *every* loop isotopic to it is commutative $(x\circ y = y\circ x)$. Compare the result to Lemma 1.

12. Show that a (generalized) quasigroup satisfies B_1 or B_2 if, and only if, in *every* loop isotopic to it $(x\circ x)\circ y = x\circ(x\circ y)$ or $(x\circ y)\circ y = x\circ(y\circ y)$ holds, respectively.

13*. (Aczél 1965*a*, 1966*c*, pp. 311–13; Taylor 1972*b*, 1978.) Determine the general solutions 1, 2, 3, 4 of the following generalized associativity equation on quasigroups: $(x1y)2z = x3(y4z)$. Can you weaken the supposition that the underlying sets should be quasigroups with respect to all four operations?

14*. (Strambach 1975.) Let I be a real interval forming a quasigroup under the continuous operation $\Box$. Determine all such $(I, \Box)$ which satisfy $(x\Box y)\Box u = (x\Box u)\Box(y\Box u)$ (cf. (17.26)).

19

Further composite equations. An application to averaging theory

In the preceding chapters, we sometimes encountered functional equations where the unknown function acted on a combination of the variable and the function itself, that is, composite equations. This chapter investigates various such composite equations and some applications.

The first and second sections give a rather general method for finding subgroups of a given continuous group of transformations by solving various composite functional equations. General solutions are determined without any regularity assumptions. In the third section, we solve, among others, the typical composite equation

$$f(x + f(y)) = f(y + f(x)),$$

when x and y belong to an abelian group. This functional equation arose from an averaging problem in turbulent fluid motions. It gave birth to several families of linear operators satisfying additional equations. Thus we have, among many others, the Reynolds operator, the averaging operator, the interpolating operator, the extension operator, the multiplicatively symmetric operator, derivation operators, etc. Using some functional analysis (the Riesz representation theorem for example), in the last three sections we give representations for these operators, and their properties when they are bounded and act on spaces of continuous functions over a compact space (with respect to the uniform norm).

We elaborate a bit on the above points, mainly on the subject of the first two sections:

It is always important to determine all subgroups of a given continuous group (or semigroup) of transformations. The classical Lie method for determining one parameter subgroups supposes strong regularity. Methods using functional equations can bring some insight under weaker assumptions (Gołąb–Schinzel 1959; Aczél–Gołąb 1960, pp. 24–86, 1970; Gołąb–Krzeszowiak 1970; and Dhombres 1984*b*, among others). In order to be concrete, we will work in the following setting. Let G and F be two nonempty sets and $*$ be a binary operation defined on the cartesian product $\mathscr{G} = G \times F$. We will suppose that $(\mathscr{G}, *)$ is a semigroup (or a group). We look for subsemigroups H of $(\mathscr{G}, *)$.

We say that a subset H of $\mathscr{G}$ depends *faithfully* on a set E when there exists a mapping $g: E \to H$, written as $g(u) = (\alpha(u), \beta(u))$, such that

(i) g is onto ($g(E) = H$),

and either

(ii) $\beta(E) = F$, $\beta(u) = \beta(u')$ implies $\alpha(u) = \alpha(u')$,

or

(iii) $\alpha(E) = G$, $\alpha(u) = \alpha(u')$ implies $\beta(u) = \beta(u')$.

If E, F and G are topological spaces, $\mathscr{G}$ is equipped with the product topology, and the operation $*$ is continuous, then we say that H depends faithfully and continuously on E if g is also continuous and, in case (ii), β admits locally a continuous lifting (that is, for every x in F, there exists an open neighbourhood U of x and a continuous $\gamma: F \to E$ such that $\beta \circ \gamma$ is the identity function on U) or, in case (iii), α admits locally a continuous lifting.

In other words, when H depends faithfully and continuously on E, we may consider E as a convenient set of parameters for H. We will say for short that H has a *faithful continuous parametrization.* Our purpose is to determine all *subsemigroups* H of $(\mathscr{G}, *)$ which have such a parametrization, without too strong regularity suppositions. For any pair u, v in E, we have a w in E such that

$$(\alpha(w), \beta(w)) = (\alpha(u), \beta(u)) * (\alpha(v), \beta(v)). \tag{1}$$

Let us take case (ii), for example. The equation $f(\beta(u)) = \alpha(u)$ defines unambiguously a function $f: F \to G$ (because $\beta(E) = F$ and because $\beta(u) = \beta(u')$ implies $\alpha(u) = \alpha(u')$). Since $g: E \to H$ and thus also $\alpha: E \to G$ are continuous, and since $\beta: E \to F$ is a surjection which admits locally a continuous lifting, this function f is continuous. Using $x = \beta(u)$, $y = \beta(v)$ and $z = \beta(w)$, we deduce from (1) that

$$(f(z), z) = (f(x), x) * (f(y), y). \tag{2}$$

With (2), both z and $f(z)$ are determined by the right hand side so that (2) provides a functional equation for f. Our programme is then to solve this equation so that, using the definition of f, we get the description of H for which we were looking. The advantage of this method is that it shows how relatively unimportant the set E of parameters is. Naturally, in case (iii), we get another functional equation and we proceed in the same way. In this chapter we will deal with such functional equations and with similar composite equations.

19.1 One-parameter subgroups of affine groups

The functional equation (2) depends on the binary operation $*$ and reduces to classical functional equations when $*$ is a sum or product. We give two examples.

We first suppose that both F and G are the additive group $(\mathbb{R}, +)$ and $(\mathscr{G}, *)$ is $G \oplus F$, that is, $(\mathbb{R}^2, +)$. Then the functional equation (2) yields the Cauchy equation

$$f(x+y) = f(x) + f(y) \quad (x, y \in \mathbb{R}). \tag{3}$$

If we suppose that F is the additive group $(\mathbb{R}, +)$, G is the multiplicative group $(\mathbb{R}\backslash\{0\}, \cdot)$ and $(\mathscr{G}, *)$ is $G \oplus F$, then (2) yields Cauchy's exponential equation

$$f(x+y) = f(x)f(y) \quad (x, y \in \mathbb{R}). \tag{4}$$

Let us now take more sophisticated binary operations $*$. We begin with

$$(\alpha, \beta) * (\alpha', \beta') = (\alpha\alpha', \alpha\beta' + \beta), \tag{5}$$

where $(G, \cdot)$ is a multiplicative semigroup acting over an additive group $(F, +)$. The semigroup $(\mathscr{G}, *)$ can be considered to consist of all proper affine transformations from F into F $(x \mapsto \alpha x + \beta, \beta \in F, \alpha \in G\backslash\{0\})$. Equation (2) yields

$$f(x + f(x)y) = f(x)f(y). \tag{6}$$

First considered by Aczél 1957*a* and Gołąb–Schinzel 1959 (cf. Aczél–Gołąb 1970), equation (6) is a typical composite functional equation in the sense that it says something about the action of f on an expression containing the values of f; here that expression is $x + f(x)y$.

A good technique for such composite functional equations is to look for the subset S of periods of f. For $f: F \to G$, the subset S is defined as the set of all y in F such that

$$f(x+y) = f(x) \quad \text{for all } x \text{ in } F. \tag{7}$$

As F is a group, it is easy to check that S is a subgroup of F. (If y_1, y_2 are in S, then $f(x + y_1 + y_2) = f(x + y_1) = f(x)$. Therefore $y_1 + y_2$ also belongs to S. Also, (7) can be rewritten as $f(x - y) = f(x)$. Thus $-y$ belongs to S. So, of course, does 0.)

But (6) yields more properties of S. Namely, $x_2 - x_1 \in S$ if $f(x_1) = f(x_2)$ and if $f(x_1)^{-1}$ exists in $(G, \cdot)$. Given an arbitrary x in F, there is then a y in F such that $x = x_1 + f(x_1)y$. Then, using (6) and $f(x_1) = f(x_2)$ repeatedly, we get

$$\begin{aligned} f(x) &= f(x_1 + f(x_1)y) \\ &= f(x_1)f(y) \\ &= f(x_2)f(y) \\ &= f(x_2 + f(x_2)y) \end{aligned}$$

$$= f(x_2 + f(x_1)y)$$
$$= f(x + x_2 - x_1)$$

proving that $x_2 - x_1 \in S$.

We prove now our first result concerning the functional equation (6):

Proposition 1. *A continuous $f:\mathbb{R}\to\mathbb{R}$ is a solution of* (6) *if, and only if, f has one of the mutually exclusive forms:*

$$f(x) = ax + 1 \qquad (a \in \mathbb{R}), \tag{8}$$

$$f(x) = 0, \tag{9}$$

$$f(x) = \sup(ax + 1, 0) \quad (a \in \mathbb{R}^* = \mathbb{R}\setminus\{0\}). \tag{10}$$

Proof. By continuity, S is a closed subset of $\mathbb{R}$ and therefore a closed subgroup. It is known (cf. Exercise 1) that a closed subgroup under addition is either $\mathbb{R}$ or $\alpha\mathbb{Z}$ for some $\alpha \in \mathbb{R}$.

If $S = \mathbb{R}$, then f is a constant function, either $f(x) = 1$, which is of the form (8) with $a = 0$, or $f(x) = 0$, that is, (9). From now on, we exclude case (9).

Suppose now $S = \alpha\mathbb{Z}$. From (6), $f(0) = 1$, if $f(x) \not\equiv 0$. By continuity, there exists a $y_0 \neq 0$ such that $f(y_0) \neq 0$. Suppose also that $f(x) \neq 0$. By (6),

$$0 \neq f(x)f(y_0) = f(x + f(x)y_0) = f(y_0 + f(y_0)x).$$

Then $z = x - y_0 + f(x)y_0 - f(y_0)x$ belongs to S. If $z \neq 0$, then $\alpha \neq 0$, and f is periodic and continuous, and therefore bounded in absolute value and reaching its bound at a point x_0. As $f(x_0) \neq 0$, in every neighbourhood of x_0 there exist at least two distinct points x_1, x_2 $(x_1 < x_0 < x_2)$ for which $f(x_1) = f(x_2) \neq 0$. So $x_1 - x_2 \in \alpha\mathbb{Z}$. But $\alpha \neq 0$ is impossible because, by a convenient choice of the neighbourhood of x_0, the points $x_1 \neq x_2$ can be chosen in it so that $|x_1 - x_2| < \alpha$. Thus we must have $\alpha = 0$ and $z = 0$.

So, when $f(x) \neq 0$, we have obtained $f(x) = ax + 1$ where $a = (f(y_0) - 1)/y_0$. From this, by the assumed continuity of f, we deduce forms (8), (10) and

$$f(x) = \inf(ax + 1, 0) \quad (x \in \mathbb{R}), \tag{11}$$

where a is a real constant, different from 0.

Conversely, (8), (9) and (10) are continuous solutions of (6). However, (11) is not a solution of (6) because, if $f(x) \neq 0$, then we would have $f(x) < 0$, which is a contradiction to $f(x)^2 = f(x + xf(x))$ which follows by putting $y = x$ into (6). This completes the proof of Proposition 1.

We may now return to the problem of finding subgroups of the group of all proper affine transformations from $\mathbb{R}$ onto $\mathbb{R}$.

Proposition 2. *The only subsemigroups of the group of all proper linear transformations from $\mathbb{R}$ onto $\mathbb{R}$, which have a faithful continuous parametri-*

zation, are the group of all translations

$$x \mapsto x + \beta; \quad (\beta \in \mathbb{R})$$

and the groups H_a $(a \in \mathbb{R})$

$$x \mapsto \alpha(x + a) - a \quad (\alpha \in \mathbb{R}^* = \mathbb{R} \backslash \{0\}).$$

Proof. In the present situation $G = \mathbb{R}^*$, $F = \mathbb{R}$ and $g(u) = (\alpha(u), \beta(u))$, we first deal with the case (ii) (see the beginning of this chapter) of a faithful parametrization and look for the continuous solutions $f: \mathbb{R} \to \mathbb{R}^* = \mathbb{R} \backslash \{0\}$ of the functional equation (6). So f should be nowhere 0 and Proposition 1 yields $f(x) \equiv 1$ as the only possibility. Since the parametrization is faithful, the only possible subsemigroup is in fact the group of all translations.

Now we take case (iii) and thus we look for the (continuous) solutions $f: \mathbb{R}^* \to \mathbb{R}$ of the functional equation:

$$f(xy) = xf(y) + f(x) \quad (x, y \in \mathbb{R}^*). \tag{12}$$

From the symmetry of the left hand side of (12) we deduce that

$$(x - 1)f(y) = (y - 1)f(x).$$

And so, for some $a \in \mathbb{R}$,

$$f(x) = a(x - 1). \tag{13}$$

(We did not use continuity in this part of the proof.) Conversely, for any $a \in \mathbb{R}$, (13) is a (continuous) solution of (12).

Returning to the determination of subsemigroups, for each value a in $\mathbb{R}$ we get H_a, the family of all transformations $x \mapsto \alpha(x + a) - a$ where α runs through $\mathbb{R}^* = \mathbb{R} \backslash \{0\}$. It turns out that H_a is a subgroup. This concludes the proof of Proposition 2.

By solving (6) and (12) in a more general situation, we obtain a more general result:

Proposition 3. *Let F be a real topological linear space and let* $\mathscr{G}$ *be the group of all proper affine transformations from F onto F. The only subsemigroups of* $\mathscr{G}$ *which have faithful continuous parametrizations are the group of all translations*

$$x \mapsto x + \beta \quad (\beta \in F)$$

and the groups H_a $(a \in F)$ *defined by*

$$x \mapsto \alpha(x + a) - a \quad (\alpha \in \mathbb{R}^*).$$

Proof. Here $G = \mathbb{R}^*$ and F is a real topological linear space. $\mathscr{G}$ is the group of those affine transformations from F onto F, which are of the form

$$x \mapsto \alpha x + \beta \quad (\alpha \in \mathbb{R}^*, \beta \in F).$$

We will call these the *proper* affine transformations from F onto F. Equation (12) for $f:\mathbb{R}^* \to F$ has again the solution (13), where this time a runs through F. This yields the family H_a, a subgroup of $\mathscr{G}$. Equation (6) for $f:F \to \mathbb{R}^* = \mathbb{R}\backslash\{0\}$ needs more careful study. For the sake of generality, we shall determine all continuous solutions $f:F \to \mathbb{R}$ of (6). We will need the following:

Proposition 4. *Let F be a real topological linear space. The continuous solutions $f:F \to \mathbb{R}$ of (6) are of one of the following mutually exclusive forms:*

$$f(x) = \langle x, x^* \rangle + 1, \quad \textit{where } x^* \in F^*, \tag{14}$$

(F^* is the dual topological space of F),

$$f(x) = 0, \tag{15}$$

and

$$f(x) = \sup(\langle x, x^* \rangle + 1, 0), \quad \textit{where } x^* \in F^*\backslash\{0\}. \tag{16}$$

(Note that here F^* is the dual of F and *not* $F\backslash\{0\}$.)

Proof. With $y = 0$ in (6), we deduce that either $f(0) = 1$ or $f(x) \equiv 0$. The latter is (15). From now on, we shall suppose that $f(0) = 1$.

We first restrict f to one dimensional linear subspaces of F. We define, for $x_0 \in F\backslash\{0\}$, a function $g:\mathbb{R} \to \mathbb{R}$ by

$$g(\lambda) = f(\lambda x_0). \tag{17}$$

From (6) we deduce that g also satisfies (6) because $g(\lambda + g(\lambda)\mu) = f(\lambda x_0 + f(\lambda x_0)\mu x_0) = f(\lambda x_0)f(\mu x_0) = g(\lambda)g(\mu)$. Thus g has one of the forms (8) or (10) ((9) being excluded by our assumption $g(0) = f(0) = 1$).

A second step is to prove that the set I of all $x \in F$, for which $f(x) = 1$, is a closed hyperplane of F. From our hypothesis, $f(0) = 1$, and so I is not empty. Moreover, if x and y are in I, from (6) we see that $x + y$ also belongs to I. If $x_0 \in I$, using g as defined in (17), we get $g(0) = g(1) = 1$. This excludes the form (10) for g (else $1 = g(1) = \sup(a + 1, 0)$, so $a = 0$, contrary to the supposition $a \neq 0$) and gives $a = 0$ in (8), that is, $g(\lambda) \equiv 1$. In other words, $x_0 \in I$ for all $\lambda \in \mathbb{R}$. We thus have proved that I is a linear subspace of F. It is also closed due to the continuity of f. Suppose now that x_1 and x_2 are two vectors generating a plane in F, which have only 0 in common with I. Using (17), we define g_1 and g_2 (with x_0 replaced by x_1 and x_2 respectively). Because of forms (8) or (10), valid for g_1, g_2, there exist real numbers λ_1 and λ_2 for which $g_1(\lambda_1) < 1$ and $g_2(\lambda_2) > 1$ (the cases where $g_1 \equiv 1$ or $g_2 \equiv 1$ are impossible since we suppose the existence of x_1 and x_2 in the complement of I). Now consider the segment $[\lambda_1 x_1, \lambda_2 x_2]$, that is, the set of all $z_\lambda = \lambda(\lambda_1 x_1) + (1 - \lambda)(\lambda_2 x_2)$ where $0 \leqslant \lambda \leqslant 1$. Its image by f is a bounded closed interval in $\mathbb{R}$ with one extremity to the left of 1 and the other to the right of 1. Therefore the image contains 1. Thus there exists a point z_λ in the segment for which $f(z_\lambda) = 1$. In other words, a nonzero linear combination of x_1 and x_2 is in I, which contradicts our definition of these vectors x_1 and x_2. Therefore, either $I = F$, yielding (14) with $x^* = 0$, or I is a proper closed hyperplane of F. In this last case, there exists $y^* \in F^* \backslash \{0\}$ such that I is the set of all x in F for which $\langle x, y^* \rangle = 0$. Let x_0 be in $F \backslash I$. For any x in F, there exists a real λ and a $y \in I$ such that $x = y + \lambda x_0$. We compute $f(x)$, using $f(y) = 1$:

$$\begin{aligned} f(x) = f(y + \lambda x_0) &= f(y + \lambda x_0 f(y)) \\ &= f(y) f(\lambda x_0) \\ &= g(\lambda) \\ &= g\left(\frac{\langle x, y^* \rangle}{\langle x_0, y^* \rangle}\right). \end{aligned}$$

For $a \in \mathbb{R} \backslash \{0\}$, we define $x^* = a/(\langle x_0, y^* \rangle) y^*$, an element of $F^* \backslash \{0\}$. According to form (8) or (10) of the function g, we deduce that $f(x) = \langle x, x^* \rangle + 1$ for all x in F (that is, (14) with $x^* \neq 0$) or $f(x) = \sup(\langle x, x^* \rangle + 1, 0)$, that is, (16).

Conversely, a simple computation shows that $f: F \to \mathbb{R}$, defined by (14), (15) or (16), is a continuous solution of (6). This concludes the proof of Proposition 4.

End of proof of Proposition 3. Returning to the determination of subsemigroups of $\mathscr{G}$, we exclude all solutions of (6) which take the value zero somewhere. Then only $f(x) \equiv 1$ remains. This concludes the proof of Proposition 3.

The general solution of (6) for $f: F \to K$ is also known without any regularity condition when F is a linear space over a commutative field K. Two results of more theoretical than practical use were found the same year: a theorem characterizing the solutions of (6) (Javor 1968) and an explicit construction of all solutions (Wołodźko 1968). We begin with the first result:

Theorem 5. *Let $f: F \to K$ where F is a linear space over a commutative field K. A not identically zero function f is a solution of* (6) *if, and only if, there exists an additive subgroup L of F, a multiplicative subgroup Λ of $K \backslash \{0\}$ and a function $l: \Lambda \to F$ such that*

(a) *For all $\lambda \in \Lambda$, $x \in L$, also $\lambda x \in L$ and conversely, if $y \in L$, $\lambda \in \Lambda$, then there exists an $x \in L$ satisfying $\lambda x = y$ (i.e. $\lambda L = L$ for every $\lambda \in \Lambda$),*
(b) *$l(\lambda) \in L$ if, and only if, $\lambda = 1$,*
(c) *$l(\lambda_1 \lambda_2) - l(\lambda_1) - \lambda_1 l(\lambda_2) \in L$ for all λ_1, λ_2 in Λ,*
(d) *$f(x) = \lambda$ if $x \in l(\lambda) + L$ and $f(x) = 0$ if $x \notin \bigcup_{\lambda \in \Lambda} (l(\lambda) + L)$.*

Proof. We suppose that $f:F\to K$ is not identically zero. We define L as the subset of all x in F such that $f(x)=1$, $\ker f$ as the kernel of f, M as the nonempty subset $F\backslash\ker f$ and Λ as the image of M under f ($\Lambda=f(M)$).

As before, our first step is to characterize the group of periods of f.

First step: *L is the set of all periods of f*. As f is not identically zero, in the same way as in Proposition 4, we get $f(0)=1$, proving that L is not empty.

Let $x\in L$. For any y in F, we get

$$f(x+y)=f(x+f(x)y)=f(x)f(y)=f(y).$$

Therefore L is a subset of the group of periods of f. Conversely, let x be a period of f, that is, $f(x+y)=f(y)$ for all y in F. Let $y=0$. We deduce that $f(x)=1$ and so $x\in L$. As a byproduct we get that $(L,+)$ is a subgroup of $(F,+)$.

We define a binary operation $*$ on M by $x*y=x+f(x)y$.

Second step: *$(M,*)$ is a group*. We can now write (6) in the form

$$f(x*y)=f(x)f(y). \tag{17}$$

From (17) it is clear that $x\in M$ and $y\in M$ imply $x*y\in M$. We have already seen that 0 belongs to L, therefore to M. Moreover, 0 is a neutral element for $*$ ($0*y=y=y*0$).

The law $*$ is associative, as can be seen from the following computation:

$$\begin{aligned}(x*y)*z&=x+f(x)y+f(x+f(x)y)z\\&=x+f(x)y+f(x)f(y)z\\&=x+f(x)(y+f(y)z)\\&=x+f(x)(y*z)\\&=x*(y*z).\end{aligned}$$

Finally, $x'=-f(x)^{-1}x$ is the inverse element of x with respect to $*$ if $x\in M$. In fact, we easily get $x*x'=0$. From (17), we deduce that $1=f(x)f(x')$ and so $f(x')=f(x)^{-1}$. So, indeed, $x'*x=0$.

Third step: *$(\Lambda,\cdot)$ is a multiplicative subgroup of $K\backslash\{0\}$ and $(L,*)$ is a normal subgroup of $(M,*)$*. This follows from the fact that f is a homomorphism from the group $(M,*)$ into the multiplicative group $K\backslash\{0\}$, the image of which is Λ. The kernel of this homomorphism is L. As K is a commutative field, $(L,*)$ is a normal subgroup of $(M,*)$.

Fourth step: *$\lambda L=L$ holds for all $\lambda\in\Lambda$*. Since $(L,*)$ is a normal subgroup of $(M,*)$, we get for all x in M

$$L*x=x*L.$$

But $f(L)=1$. Therefore $L*x=L+x$. But we have also $x*L=x+f(x)L$. So

$$L+x=x+f(x)L.$$

This yields $L=f(x)L$. So $L=\lambda L$ for all $\lambda\in\Lambda$. This last equality is exactly the meaning of property (a) in Theorem 5.

We now proceed to the construction of $l:\Lambda\to F$. As the subgroup Λ is isomorphic, under the mapping f, to the quotient group M/L (quotient relative to the operation $*$), we may find a mapping $l:\Lambda\to M$, for which

$$f(l(\lambda))=\lambda, \quad \text{for each } \lambda \text{ in } \Lambda, \tag{18}$$

by using a lifting from M/L into M (relative to L) and by identifying Λ and M/L. When this is done, property (b) of Theorem 5 becomes obvious.

For any x in M, a coset for L relative to x (that is, a set $L*x(=x*L)$) is a subset $f^{-1}(f(x))$ and it also coincides with the subset $L+x$. Therefore, using (17), we deduce that $l(\lambda_1\lambda_2)$ belongs to $l(\lambda_1)*l(\lambda_2)+L$. We write

$$\begin{aligned} l(\lambda_1\lambda_2)-l(\lambda_1)*l(\lambda_2)&=l(\lambda_1\lambda_2)-l(\lambda_1)-f(l(\lambda_1))l(\lambda_2)\\ &=l(\lambda_1\lambda_2)-l(\lambda_1)-\lambda_1 l(\lambda_2). \end{aligned}$$

This proves property (c) of Theorem 5.

Now, if $x\in l(\lambda)+L$, that is, if x is in the coset for L relative to $l(\lambda)$, then $f(x)=\lambda$. But, if $x\notin M=\bigcup_{\lambda\in\Lambda}(l(\lambda)+L)$, or equivalently, if $x\in\ker f$, then $f(x)=0$. This proves the property (d) of Theorem 5.

Conversely, let properties (a), (b), and (c) be satisfied for some Λ, l and let L be as given in Theorem 5. A function $f:F\to K$ is unambiguously defined by (d) if we can show that the two subsets $l(\lambda_1)+L$ and $l(\lambda_2)+L$ have no element in common for $\lambda_1\neq\lambda_2$ $(\lambda_1,\lambda_2\in\Lambda)$. Suppose $l(\lambda_1)=l(\lambda_2) \bmod L$, that is,

$$l(\lambda_1)=l(\lambda_2)+y, \quad y\in L.$$

There exists $\lambda_3\in\Lambda$ such that $\lambda_2=\lambda_1\lambda_3$, because Λ is a subgroup of $K\backslash\{0\}$. Using (c), we obtain that $l(\lambda_2)=l(\lambda_1)+\lambda_1 l(\lambda_3) \bmod L$. Thus $\lambda_1 l(\lambda_3)=0 \bmod L$. Using (a), we deduce $l(\lambda_3)=0 \bmod L$. But (b) yields $\lambda_3=1$ and so $\lambda_1=\lambda_2$ which was to be proved.

Using the definition of f from (d), we notice that L is contained in the subgroup of periods for f. In fact, if $x\in l(\lambda)+L$, it is clear that $f(x+L)=f(x)$. And, if $x\notin\bigcup_{\lambda\in\Lambda}(l(\lambda)+L)$, then $(x+L)\cap(\bigcup_{\lambda\in\Lambda}(l(\lambda)+L))=\varnothing$. Therefore $f(x+L)=f(x)$.

Let us now compute $f(x+f(x)y)$ and $f(x)f(y)$. Several cases have to be investigated:

If $x\notin\bigcup_{\lambda\in\Lambda}(l(\lambda)+L)$, then $f(x)=0$. Therefore $f(x+f(x)y)=f(x)=0$. But $f(x)f(y)=0$. Thus (6) is satisfied, independently of the choice of y.

If $x\in l(\lambda_1)+L$, we shall make a further distinction between two subcases. If $y=l(\lambda_2)+L$, then, for some z_1,z_2 in L, we may write $x+f(x)y$ as

$$l(\lambda_1)+z_1+f(x)(l(\lambda_2)+z_2)=l(\lambda_1)+z_1+\lambda_1(l(\lambda_2)+z_2).$$

Using (b) and (c), this last expression is equal to $l(\lambda_1\lambda_2) \bmod L$. Therefore

$f(x+f(x)y)=f(l(\lambda_1\lambda_2))=\lambda_1\lambda_2$. But $f(x)f(y)=f(l(\lambda_1))f(l(\lambda_2))=\lambda_1\lambda_2$. Thus (6) is satisfied in this case too.

If $y\notin\bigcup_{\lambda\in\Lambda}(l(\lambda)+L)$, then $f(y)=0$. To end the proof of Theorem 5, we just have to show that $f(x+f(x)y)=0$ if $f(x)f(y)=0$. Suppose the contrary. That would imply the existence of a $\lambda\in\Lambda$ such that

$$x+f(x)y\in l(\lambda)+L.$$

Let $\lambda_2=\lambda/\lambda_1$. We compute that $f(x)y=l(\lambda)-l(\lambda_1)(\operatorname{mod} L)$. Using (c), we deduce that $f(x)y=\lambda_1 l(\lambda_2)\operatorname{mod} L$. But $f(x)=\lambda_1$. Thus

$$\lambda_1(y-l(\lambda_2))=0\operatorname{mod} L.$$

Because of (a), this yields

$$y-l(\lambda_2)=0\operatorname{mod} L$$

or $f(y)=\lambda_2$, which contradicts our assumption $f(y)=0$. This concludes the proof of Theorem 5.

The impractical aspect of Theorem 5 is that, in order to get a solution of (6), we have to explicitly construct a function l satisfying (c), whereas the choice of a convenient l for f is somewhat arbitrary (any lifting from M/L to M relative to L will do). The following construction, which we will call construction W (Wołodźko), alleviates this difficulty:

Construction W. We start from any multiplicative subgroup Λ' of $K\backslash\{0\}$ and we arbitrarily define $l:\Lambda'\to F$. Then we consider any subset L' of F containing all possible values $l(\lambda_1\lambda_2)-l(\lambda_1)-\lambda_1 l(\lambda_2)$ for λ_1,λ_2 in Λ'. We construct L'' as the subset $\Lambda' L'$ of F and L as the additive subgroup of F generated by L''. We finally choose a subgroup Λ of Λ' in such a way that $l(\lambda)\in L$ if and only if $\lambda=1$. Such a choice is always possible because we may verify that $l(1)\in L$. Therefore we have at least the choice to take $\Lambda=\{1\}$.

Let us check that Λ, l and L, as obtained through the construction W, satisfy properties (a), (b) and (c) of Theorem 5. Properties (b) and (c) follow from the construction. It is not difficult to verify that $\Lambda L\subset L$. Let $y\in L$ and $\lambda\in\Lambda$. Then $\lambda^{-1}\in\Lambda$ and so $x=\lambda^{-1}y\in L$. Therefore $y=\lambda x\in\lambda L$. This proves that $\lambda L=L$ for every $\lambda\in\Lambda$ and so $\Lambda L=L$. Thus (a) holds too. Therefore, from any construction W and defining $f:F\to K$ by (d), we get a solution f of (6).

Conversely, any solution of (6) can be obtained from the definition (d) of Theorem 5 by a construction W. It is enough to start from $\Lambda'=\Lambda$, where Λ is given by Theorem 5, then use l and L as given by that theorem.

We have obtained the following:

Proposition 6. *All solutions $f:F\to K$ of (6) can be obtained from some construction W using definition (d) of Theorem 5.*

Proposition 6 gives the required tool in order to find any subsemigroup with a faithful parametrization in the family of all affine transformations on a linear space F over a commutative field K. If we restrict ourselves to proper affine transformations ($x \mapsto \alpha x + \beta, \alpha \in K \backslash \{0\}, \beta \in F$), we get the following theorem:

Theorem 7. *Let F be a linear space over a commutative field K. Let $\mathscr{G}$ be the group of all proper affine transformations from F onto F. The only subsemigroups of $\mathscr{G}$ which admit a faithful parametrization are the group of all translations*

$$x \mapsto x + \beta \quad (\beta \in F),$$

and the groups H_a,

$$x \mapsto \alpha(x + a) - a \quad (\alpha \in K \backslash \{0\}),$$

where $a \in F$.

The difference between Theorem 7 and Proposition 3 is that here we do not make any continuity assumption concerning the parametrization. With the functional equation (12) for $f: K \backslash \{0\} \to F$, the solution is (13) with $a \in F$. With the functional equation (6), we look for a solution $f: F \to K \backslash \{0\}$. In other words, using the notations of the proof of Theorem 5, we must have $M = F$. We shall prove that in this case $\Lambda = \{1\}$ and so $F = M = L$, that is, $f \equiv 1$.

For contradiction, we suppose that $\lambda_1 \in \Lambda$ and $\lambda_1 \neq 1$. There exists an x in F such that $f(x) = \lambda_1$. As $M = F$, there exists λ_2 in Λ such that $(1 - \lambda_1)^{-1} x \in l(\lambda_2) + L$. We compute that $x = l(\lambda_2) - \lambda_1 l(\lambda_2) \bmod L$ and that $x = l(\lambda_1) \bmod L$. Therefore $l(\lambda_1) + \lambda_1 l(\lambda_2) = l(\lambda_2) \bmod L$. But (c), in Theorem 5, yields $l(\lambda_1 \lambda_2) = l(\lambda_2) \bmod L$. Thus $\lambda_1 \lambda_2 = \lambda_2$, which is impossible since $\lambda_1 \neq 1$. This concludes the proof of Theorem 7.

The analogue of Theorem 7, when we replace $\mathscr{G}$ by other groups of transformations, for example, by $GL(F)$, is still an open question.

19.2 Another example of determining one-parameter subgroups

Let us take a second example with a different binary operation $*$. We start from $F = G$ and suppose that F is a ring. For $*$ we use the following:

$$(\alpha, \beta) * (\alpha', \beta') = (\lambda \alpha \alpha', \alpha \beta' + \beta \alpha'),$$

where λ is a fixed parameter in F.

When λ is a unit for F, we may consider $(\mathscr{G}, *)$ as the multiplicative semigroup of all matrices $\begin{pmatrix} \alpha & 0 \\ \beta & \alpha \end{pmatrix}$ where $\alpha \in F$ and $\beta \in F$ (this is the Clifford group over F if we restrict α to $F \backslash \{0\}$).

In order to find all subsemigroups of $(\mathscr{G}, *)$ which have a faithful continuous parametrization when F is a topological ring, we are confronted with the problem of finding the solutions of the following two functional equations:

$$f(\lambda xy) = xf(y) + f(x)y \quad (x, y \in F) \tag{21}$$

and

$$f(xf(y) + f(x)y) = \lambda f(x)f(y) \quad (x, y \in F). \tag{22}$$

When $\lambda = 1$, then (21) reduces to the classic functional equation of derivation (cf. Exercise 2.11). Equation (22) is a composite functional equation and appears rather difficult to solve in a general setting. However, in the case where $F = \mathbb{R}$ and under a continuity assumption for f, the solution of (22) has been found (Brillouët 1982).

Theorem 8. *Suppose $\lambda > 0$ and let $f: \mathbb{R} \to \mathbb{R}$ be a continuous function. This function is a solution of* (22) *if, and only if, f is of one of the following forms:*

$$f = 0, \tag{23}$$

$$f = \frac{1}{\lambda} \tag{24}$$

and, only in the case $\lambda = 2$, of one of the additional forms:

$$f(x) = Ax \tag{25}$$

and

$$f(x) = \sup(Ax, 0), \tag{26}$$

where $A \in \mathbb{R} \backslash \{0\}$ is a constant.

Proof. The function (23) is clearly a solution of (22) and from now on we may exclude it. Therefore we suppose the existence of an x_0 in $\mathbb{R}$ such that $f(x_0) = \gamma \neq 0$.

We define $g: \mathbb{R} \to \mathbb{R}$ by

$$g(y) = \gamma y + x_0 f(y). \tag{27}$$

A simple computation from (22) yields

$$g(g(y)) = (\lambda + 1)\gamma g(y) - \lambda \gamma^2 y. \tag{28}$$

When $\gamma = 1$, (28) is called the equation of linear iteration of order 2, the general continuous solution of which is known even in a more general setting than $\mathbb{R}$ (Dhombres 1977*b*). We may adapt the construction to the present case. Because of (27), g is continuous. As $\lambda \gamma^2 \neq 0$, (28) yields that $y_1 \neq y_2$ implies $g(y_1) \neq g(y_2)$. Therefore g is strictly monotonic. Moreover, (28) prevents the limit of g, when y goes to ∞ (or to $-\infty$), from being finite. Thus g is a bijection from $\mathbb{R}$ onto $\mathbb{R}$.

By induction, we get the following formulas for the iterates g^n of g, defined by $g^0(y) = y$ and $g^n(y) = g^{n-1}(g(y))$ $(n = 1, 2, \ldots)$,

$$g^n(y) = \frac{1-\lambda^n}{1-\lambda}\gamma^{n-1}g(y) - \lambda\frac{1-\lambda^{n-1}}{1-\lambda}\gamma^n y \quad \text{for } \lambda \neq 1 \tag{29}$$

and

$$g^n(y) = n\gamma^{n-1}g(y) - (n-1)\gamma^n y \quad \text{for } \lambda = 1. \tag{30}$$

In (29) and (30), n is a nonnegative integer. However, as g is a bijection, it is not difficult to prove that (29) and (30) both remain valid for negative integers, if we define g^{-1} as the inverse function of g and $g^{-k}(x) = (g^{-1})^k(x)$ $(k = 2, 3, \ldots)$.

We now want to investigate various possibilities according to the value of γ.

First case. $f(x) \neq 1$ *for all* $x \in \mathbb{R}^*$. Our aim is to prove that in this case f is constant and equal to $1/\lambda$. From (22), we have either $f(0) = 0$ or $f(0) = 1/\lambda$.

(a) *Suppose* $f(0) = 0$. As f is continuous and never reaches 1, we get $f(x) < 1$ for all x in $\mathbb{R}$. Let $x = y$ in (22). We deduce that

$$f(2xf(x)) = \lambda f(x)^2 \quad (x \in \mathbb{R}). \tag{31}$$

Therefore we get $|f(x)| < \lambda^{-1/2}$ for all x in $\mathbb{R}$. Using (31) for the induction, we get

$$|f(x)| < \lambda^{-(2^{-1}+2^{-2}+\cdots+2^{-n})}.$$

If $\lambda > 1$, letting n go to infinity we conclude that $|f(x)| \leqslant 1/\lambda$. This inequality (even the strict inequality) is obviously true for $0 < \lambda \leqslant 1$. In other words $|\gamma| \leqslant 1/\lambda$. However, since f is continuous and equal to zero at 0, we may always suppose $|\gamma| < 1/\lambda$.

We then choose μ so that $\lambda|\gamma| < \mu < 1$ and define $h: \mathbb{R} \to \mathbb{R}$ by $h(y) = x_0 f(y)$. This function h is bounded in absolute value; let M be the least upper bound of h. Notice that $x_0 \neq 0$ $(\gamma \neq 0)$ and therefore $M > 0$. By the continuity of h, there exists a $y_0 \in \mathbb{R}$ for which

$$|h(y_0)| > \mu M.$$

As g is a bijection, there exists *a* z_0 such that $g(z_0) = y_0$. Therefore

$$\begin{aligned} h(y_0) = h(g(z_0)) &= h(\gamma z_0 + h(z_0)) \\ &= \lambda\gamma h(z_0). \end{aligned}$$

Thus

$$\mu M < |h(y_0)| = \lambda|\gamma h(z_0)|.$$

This inequality yields $|h(z_0)| > M$, which is a contradiction. Therefore, the subcase $f(0) = 0$ is impossible.

(b) *Suppose* $f(0) = 1/\lambda$. We shall now have to investigate three different cases: $0 < \lambda < 1$, $\lambda = 1$ and $\lambda > 1$.

If $0 < \lambda < 1$, as $f(0) = 1/\lambda$ and as f never reaches 1, we obtain that $f(x) > 1$ for all x in $\mathbb{R}$. In a similar way as in case (a), we compute that $f(x) \geqslant 1/\lambda$ for all x in $\mathbb{R}$. We make use of (29) for negative integers n. Assume for contradiction that $f(x) \not\equiv 1/\lambda$. We may suppose, without loss of generality, that $f(x_0) = \gamma > 1/\lambda$. As $\gamma > 1$ and $\lambda\gamma > 1$, we have

$$\lim_{n \to -\infty} g^n(y) = 0 \quad (y \in \mathbb{R}).$$

But $g^{n+1}(y) = g(g^n(y))$, so that $g(0) = 0$ and, consequently, $f(0) = 0$ (notice that $x_0 \neq 0$). This contradicts our hypothesis about f. Therefore f is a constant, and is equal to $1/\lambda$ (case (24)).

Let $\lambda = 1$. If there exists an x_0 for which $\gamma = f(x_0) > 1$, we must have $x_0 \neq 0$. In the same way as previously, we deduce that $g(0) = 0$. But $g(0) = x_0 \neq 0$, a contradiction. If there exists an x_0 for which $\gamma = f(x_0) < 1$, we have $x_0 \neq 0$ again. Using (30) for large positive integers n, we deduce in a similar way that $g(0) = 0$. This contradicts $g(0) = x_0 \neq 0$. So the only possibility is $f \equiv 1$.

If $\lambda > 1$, we get $f(x) < 1$ for all x in $\mathbb{R}$. As previously, we deduce that $f(x) \leqslant 1/\lambda$. We prove that $f(x) \equiv 1/\lambda$. If not, there exists an $x_0 \neq 0$ such that $\gamma = f(x_0) < 1/\lambda < 1$. Using (29) for large positive integers n, we obtain that $g(0) = 0$ which contradicts $g(0) = x_0 \neq 0$. This contradiction proves (24).

Second case. There exists an $x_0 \in \mathbb{R} \backslash \{0\}$ such that $f(x_0) = 1$. Equation (28) becomes

$$g(g(y)) = (\lambda + 1)g(y) - \lambda y \quad (y \in \mathbb{R}). \tag{32}$$

We make use of the following Proposition 9 (Dhombres 1977*b*. 1979, pp. 6.7–6.19), the proof of which is postponed to the end of the proof of Theorem 8. To state Proposition 9, we make use of the following terminology:

A function $g: \mathbb{R} \to \mathbb{R}$ is of *type* (λ, a, b), where $\lambda \in \mathbb{R}$ and $-\infty \leqslant a < b \leqslant \infty$, if

$$g(y) = \begin{cases} \lambda y + (1-\lambda)a & y < a \\ y & a \leqslant y \leqslant b \\ \lambda y + (1-\lambda)b & y > b. \end{cases}$$

(We adopt such natural conventions as, for example, when $b = \infty$, then there is no $y > \infty$ case.)

A function $g: \mathbb{R} \to \mathbb{R}$ is of *type* (λ, δ) if $g(y) = \lambda y + \delta$ for all y in $\mathbb{R}$.

Proposition 9. *A continuous* $g: \mathbb{R} \to \mathbb{R}$ *satisfies* (32), *for* $\lambda > 0$, *if and only if* g *is of one of the following types:*

$$\lambda \neq 1: \quad \textit{type } (\lambda, a, b) \textit{ or type } (\lambda, \delta),$$
$$\lambda = 1: \quad \textit{type } (1, \delta).$$

When Proposition 9 has been proved, we may *end the proof of Theorem* 8 in the following way:

If $\lambda = 1$, then $g(y) = y + \delta$ for some $\delta \in \mathbb{R}$. From (27) with $\gamma = 1$, $f(y) = \delta/x_0$. But $f(x_0) = 1$. Thus $f \equiv 1$, that is, we have the case (24) with $\lambda = 1$.

If $\lambda \neq 1$, we have two types to investigate for g. Type (λ, δ) yields $f(x) = [(\lambda - 1)x + \delta]/x_0$ and $1 = f(x_0) = \lambda - 1 + (\delta/x_0)$. We write

$$f(x) = (\lambda - 1)\frac{x}{x_0} + (2 - \lambda). \tag{33}$$

Put (33) into both sides of the functional equation (31). We get a polynomial of degree two in x, which is identically zero. All coefficients must be zero, which yields

$$\frac{(2-\lambda)(\lambda-1)^2}{x_0^2} = 0, \quad \frac{-2(2-\lambda)(\lambda-1)^2}{x_0} = 0 \quad \text{and} \quad (2-\lambda)(\lambda-1)^2 = 0.$$

As $\lambda \neq 1$, the only possibility is $\lambda = 2$. In this case (33) gives $f(x) = x/x_0$, that is (25). Conversely, (25) provides a solution of (22).

If g is of type (λ, a, b) then f is of the following form:

$$f(x) = \begin{cases} \dfrac{\lambda - 1}{x_0}(x - a) & \text{if} \quad x \leqslant a \\ 0 & \text{if } a \leqslant x \leqslant b \\ \dfrac{\lambda - 1}{x_0}(x - b) & \text{if} \quad x \geqslant b. \end{cases}$$

Let $x = y \in [a, b]$. As $f(y) = f(x) = 0$, we deduce from (22) that $f(0) = \lambda f(x) f(y) = 0$. Therefore $0 \in [a, b]$. We distinguish two cases:

(i) $(\lambda - 1)/x_0 > 0$. We shall prove that $a = -\infty$, $b = 0$ and $\lambda = 2$. As $f(x_0) = 1$ and as $f(x) \leqslant 0$ for $x \leqslant b$, we see that $x_0 > b$. Thus $(\lambda - 1)(x_0 - b)/x_0 = 1$ which gives $\lambda \geqslant 2$ and $b < \infty$. Let $y = b$ and $x \geqslant b$ in (22). We get $f(bf(x)) = \lambda f(b) f(x) = 0$. But, as $(\lambda - 1)/x_0 > 0$, if $b > 0$, we can find an x large enough so that $bf(x) > b$. Then $f(bf(x)) \neq 0$, which contradicts $f(bf(x)) = 0$. Therefore $b = 0$. Consequently $(\lambda - 1)(x_0 - 0)/x_0 = 1$ and so $\lambda = 2$.

Suppose now $a > -\infty$. With $y = a$ and $x < a$ in (22), we deduce that $f(af(x)) = \lambda f(x) f(a) = 0$, $a \leqslant 0$ (as $b = 0$). If $a < 0$, then

$$af(x) = a\frac{\lambda - 1}{x_0}(x - a) > 0.$$

Thus

$$f(af(x)) = \frac{\lambda - 1}{x_0}\left(a\frac{\lambda - 1}{x_0}(x - a) - b\right).$$

We can find an $x < a$ for which $f(af(x)) \neq 0$, contradicting $f(af(x)) = 0$. Thus $a = -\infty$. For f we deduce the form (26) with $A > 0$. If $a = 0$, we get the form (25) again.

(ii) $(\lambda - 1)/x_0 < 0$. In a similar way, we get $\lambda = 2$, $b = \infty$ and $a = 0$. Moreover we deduce the form (26) with $A < 0$ for f. This concludes the proof of Theorem 8, except that we have to prove Proposition 9.

Proof of Proposition 9. As (32) is a special case of (28), a continuous solution of (32) is a bijection (except if it is identically zero).

If $\lambda = 1$, let $f(x) = g(x) - x$. Then we deduce for f the following equation

$$f(x + f(x)) = f(x). \tag{34}$$

This is called *Euler's equation* (Euler 1764). The analogue of (30), in the present situation, is

$$g^n(x) = x + nf(x) \quad (n \in \mathbb{Z}). \tag{35}$$

If $f \equiv 0$, then $g(x) = x$ for all x in $\mathbb{R}$, that is, type $(1, 0)$. If $f(x_0) \neq 0$ for some x_0 in $\mathbb{R}$, then every point x in $\mathbb{R}$ belongs to some interval with endpoints $x_0 + nf(x_0) = g^n(x_0)$ and $x_0 + n'f(x_0) = g^{n'}(x_0)$ for some integers n and n'. Using the monotony of g^n, we deduce that, for all $m, g^m(x)/m$ belongs to an interval with extremities $(x_0/m) + ((n + m)/m)f(x_0)$, $(x_0/m) + (n' + m/m)f(x_0)$. Letting m go to infinity, we get immediately that $\lim_{m\to\infty} g^m(x)/m = f(x_0)$ and, from (35), $\lim_{m\to\infty} g^m(x)/m = f(x)$. Therefore f is constant so that g is of type $(1, \delta)$ for some δ in $\mathbb{R}$. In particular, we have proved that *all continuous solutions of Euler's functional equation* (34) *are constant.*

If $\lambda \neq 1$ $(0 < \lambda < \infty)$, we define $h: \mathbb{R} \to \mathbb{R}$ by

$$h(x) = \frac{g(x) - \lambda x}{1 - \lambda}.$$

Equation (32) yields for all x in $\mathbb{R}$

$$h(g(x)) = h(x). \tag{36}$$

By induction, we get

$$g^n(x) = x + (\lambda^n - 1)(x - h(x)) \quad (n \in \mathbb{Z}). \tag{37}$$

Let us first suppose that $0 < \lambda < 1$. Letting n grow to infinity, we deduce from (37)

$$\lim_{n\to\infty} g^n(x) = h(x) \quad (x \in \mathbb{R}).$$

The function h is continuous, like g. Therefore we get from (36)

$$h^2(x) = h(x) \tag{38}$$

Let $I = \{x \mid x \in \mathbb{R}, h(x) = x\}$. This subset is not empty (because of (38)) and is closed since h is continuous. But I is the range of h (see (38)). Therefore I is a connected subset of $\mathbb{R}$ (a topological retract of $\mathbb{R}$). If I is a singleton,

$$I = \{a\},$$

then $h(x) = a$ for all x and so $g(x) = \lambda x + \delta$ with $\delta = (1 - \lambda)a$. This gives type (λ, δ) for g.

If I is not a singleton, then it contains a proper interval. By the definition of I, we have $h(x) = x$ for all x in I and so $g(x) = x$. In particular, g is an increasing function on I. But this implies that g is an increasing function on $\mathbb{R}$, as g is a continuous bijection from $\mathbb{R}$ onto $\mathbb{R}$. The same result is true for all g^n ($n \in \mathbb{N}$). From

$$\lim_{n\to\infty} g^n(x) = h(x)$$

we deduce that h is nondecreasing on $\mathbb{R}$. But h takes all its values in $\mathbb{R}$. Therefore we get the following three cases.

$I = [a, b]$, where we suppose $-\infty < a < b < \infty$. We know now that h is nondecreasing on $\mathbb{R}$, $h(x) = x$ for all $x \in I$, and $h(x) \in I$ for all $x \in \mathbb{R}$. So we see that $h(x) = b$ for $x \geqslant b$ and $h(x) = a$ for $x \leqslant a$. Thus $g(x) = \lambda x + (1-\lambda)b$ for $x \geqslant b$, $g(x) = x$ for $x \in [a, b]$ and $g(x) = \lambda x + (1-\lambda)a$ for $x \leqslant a$. This is type (λ, a, b).

$I =]-\infty, b]$, where we suppose $b < \infty$. We easily deduce that $g(x) = \lambda x + (1-\lambda)b$ for $x \geqslant b$ and $g(x) = x$ otherwise. This is type $(\lambda, -\infty, b)$.

$I = [a, \infty[$, where we suppose $a > -\infty$. In the same way, we get type (λ, a, ∞).

In order to complete the proof of Proposition 9, we now have to investigate the case $\lambda > 1$. But g is a bijection and (36) can be written as $h(G(x)) = h(x)$, where $G(x) = g^{-1}(x)$, the inverse function of g. Then (37) for negative n yields, with $\Lambda = \lambda^{-1}$,

$$G^n(x) = x + (\Lambda^n - 1)(x - h(x)). \tag{39}$$

But $0 < \Lambda < 1$ and we obtain that $\lim_{n\to\infty} G^n(x) = h(x)$ and we have (38) as well. In other words, we are back, with $0 < \Lambda < 1$, to the previous case and its conclusions. This concludes the proof of Proposition 9.

For the solution of (32) in the case $\lambda \leqslant 0$, not covered by Proposition 9, see Exercise 11 and Dhombres 1979, pp. 6.17–6.19.

Theorem 8 provides us with what is needed for finding subgroups of the Clifford group of $\mathbb{R}$, that is, for the group of all matrices $\begin{pmatrix} \alpha & 0 \\ \beta & \alpha \end{pmatrix}$ where $\alpha \in \mathbb{R}^* = \mathbb{R}\backslash\{0\}$, $\beta \in \mathbb{R}$.

Theorem 10. *Let $\mathscr{G}$ be the Clifford group of $\mathbb{R}$. The only subsemigroups of $\mathscr{G}$ which have a faithful continuous parametrization are the group of all matrices $\begin{pmatrix} 1 & 0 \\ \beta & 1 \end{pmatrix}$ with $\beta \in \mathbb{R}$, and the groups H_a ($a \in \mathbb{R}$) of all matrices of the form*

$$\begin{pmatrix} \alpha & 0 \\ a\alpha \log|\alpha| & \alpha \end{pmatrix} \quad \textit{with } \alpha \in \mathbb{R}^*.$$

Proof. The problem reduces to finding the general continuous solutions of (21) and of (22) when $\lambda = 1$. More exactly, in the case of (21), we look for those continuous $f:\mathbb{R}^* = \mathbb{R}\backslash\{0\} \to \mathbb{R}$, which are solutions of (21) with $\lambda = 1$. Dividing both sides of (21) by xy ($\neq 0$), we see that g defined by $g(x) = f(x)/x$ satisfies

$$g(xy) = g(x) + g(y) \quad (x, y \in \mathbb{R}^*).$$

As g is continuous on $\mathbb{R}^*$, Theorem 3.3 gives

$$g(x) = a \log|x| \quad (x \in \mathbb{R}^*).$$

Thus

$$f(x) = ax \log|x| \quad (x \in \mathbb{R}^*). \tag{40}$$

From (40), we obtain the family H_a. One can easily verify that H_a is in fact a subgroup of $\mathscr{G}$.

In the case of (22), we look for the continuous solutions $f:\mathbb{R} \to \mathbb{R}^*$. Theorem 8 gives $f(x) \equiv 1$, that is, the family of all matrices $\left(\begin{smallmatrix} 1 & 0 \\ \beta & 1 \end{smallmatrix}\right)$ ($\beta \in \mathbb{R}$). One can see immediately that this is a subgroup. This concludes the proof of Theorem 10.

19.3 Two more composite equations

In order to solve (22), we have used (28) or (32). However, our proof for Proposition 9 strongly depends on the fact that we work in $\mathbb{R}$. We would like to solve composite functional equations on more general structures, for example abelian groups. In this section we make essential use of group theory (mainly cosets and liftings).

To introduce such composite equations, let us make a simple change in (32) by defining $f(y) = g(y) - y$. Suppose that g satisfies (32). Then f satisfies

$$f(y + f(y)) = \lambda f(y). \tag{41}$$

From Proposition 9 we know all continuous solutions of (41) for $f:\mathbb{R} \to \mathbb{R}$ and $\lambda > 0$ (cf. Exercise 11 for the case $\lambda \leqslant 0$). Take $\lambda = 2$ and change (41) by introducing a second variable:

$$f(x + f(y)) = f(x) + f(y). \tag{42}$$

Any solution $f:G \to G$ of (42), where $(G, +)$ is an abelian group, is a solution of (41) with $\lambda = 2$ (but the converse is not true).

In order to state the following proposition, we need some notation. Let $(G, +)$ be an abelian group and H be a subgroup of G. We denote by π_H the canonical epimorphism from G onto G/H and by h a lifting relative to H, that is, a mapping $h:G/H \to G$ such that $\pi_H \circ h$ is the identity mapping on G/H.

Proposition 11. *Let $(G, +)$ be an abelian group. A function $f:G \to G$ satisfies (42) if, and only if, there exists a subgroup H of G and a lifting h relative to H such that*

$$f(x) = x - h(\pi_H(x)) \quad (x \in G). \tag{43}$$

Proof. Suppose that $f:G\to G$ is of the form (43) for H and h as described. Then we notice that $\pi_H(f(x))=\pi_H(x)-\pi_H(h(\pi_H(x)))=0$, so that $f(x+f(y))=x-h(\pi_H(x))+f(y)=f(x)+f(y)$, which is (42).

Conversely, let $f:G\to G$ be a solution of (42) and define H as the image of G under f ($H=f(G)$). From (42) we obtain with x replaced by $x-f(y)$, that

$$\begin{aligned}f(x)&=f(x-f(y)+f(y))\\&=f(x-f(y))+f(y).\end{aligned}$$

Therefore $f(x-f(y))=f(x)-f(y)$. In other words, H is a subgroup of G. Define $g(x)=x-f(x)$. From (42) we get that

$$g(x+f(y))=g(x). \tag{44}$$

Equation (44) means that all elements of H are periods of g. Moreover $\pi_H(g(x))=\pi_H(x)$ as $\pi_H(f(x))=0$. We then see that $g(x)$ belongs to the same coset of x and depends only on the coset of x. Therefore there exists a lifting h for H and $g(x)=h(\pi_H(x))$. So we obtain (43) for f, as asserted.

Proposition 11 can be generalized to the nonabelian case (see Exercise 18) and in the commutative version for more general structures than groups (see Exercise 19). When $(G,+)$ is commutative, an interesting aspect of (42) is its symmetry

$$f(x+f(y))=f(f(x)+y). \tag{45}$$

This equation (45) is interesting as a typical composite functional equation. It can be solved completely when $(G,+)$ is an abelian group (Dhombres 1972*b* and 1977*a*).

Theorem 12. *Let $(G,+)$ be an abelian group. A function $f:G\to G$ satisfies (45) for all x,y in G if, and only if, there exist two subgroups H and I of G, with $H\subseteq I$ and two liftings h and i:*

$$h:G/I\to G/H \quad \text{and} \quad i:I/H\to I,$$

such that

$$f(x)=i[\pi_H(x)-h(\pi_I(x))] \quad \text{for all } x\in G. \tag{46}$$

Proof. We have identified $(G/H)/(I/H)$ with G/I as usual in group theory and use the notations introduced before stating Proposition 11. We have then, by definition, $\pi_H\circ i\circ\pi_H(x)=\pi_H(x)$ for all x in I.

Let $f:G\to G$ be defined by (46). We notice that

$$\pi_H(x)-h(\pi_I(x))\in I/H \quad \text{for all } x\in G,$$

and that $f(G)\subseteq I$. In particular, we get

$$\pi_H(f(x))=\pi_H(x)-h(\pi_I(x)) \quad \text{for all } x\in G,$$
$$\pi_I(f(x)+y)=\pi_I(y) \quad \text{and} \quad \pi_I(x+f(y))=\pi_I(x) \quad \text{for all } x,y\in G.$$

Therefore

$$\begin{aligned} f(x+f(y)) &= i[\pi_H(x+f(y)) - h(\pi_I(x))] \\ &= i[\pi_H(x) - h(\pi_I(x)) + \pi_H(y) - h(\pi_I(y))] \\ &= i[\pi_H(y+f(x)) - h(\pi_I(y))] \\ &= f(y+f(x)). \end{aligned}$$

Thus f satisfies (45).

Conversely, let $f:G\to G$ be a solution of (45). As already mentioned, for such functional equations we look for the subgroup of periods of f. We begin with the following computation:

$$\begin{aligned} f(x+f(y+f(z))) &= f(x+f(f(y)+z)) \\ &= f(f(x)+f(y)+z) \\ &= f(f(y)+f(x)+z) \\ &= f(y+f(f(x)+z)) \\ &= f(y+f(x+f(z))) \\ &= f(f(y)+x+f(z)). \end{aligned}$$

Thus

$$f(x+f(y+f(z))) = f(x+f(y)+f(z)) \quad \text{for all } x,y,z\in G. \tag{47}$$

Let H be the subgroup of G generated by all the elements of the form $f(y+f(z))-f(y)-f(z)$ where y,z are arbitrary elements in G. Because of (47), all elements of H are periods of f. Therefore we may define a mapping $g:G/H\to G/H$ by

$$g(\pi_H(x)) = \pi_H(f(x)).$$

We can easily obtain a functional equation for g:

$$\begin{aligned} g[\pi_H(x)+g(\pi_H(y))] &= \pi_H[f(x+f(y))] = \pi_H[f(x)+f(y)] \\ &= g(\pi_H(x)) + g(\pi_H(y)). \end{aligned}$$

In other words, for all x,y in G/H, we get

$$g(x+g(y)) = g(x)+g(y).$$

This is the functional equation (42) for $g:G/H\to G/H$, whose solutions we know from Proposition 11. There exists a subgroup H' of G/H such that

$$g(x) = x - h'(\pi_{H'}(x)) \quad \text{for all } x \text{ in } G/H,$$

where $\pi_{H'}$ is the canonical epimorphism from G/H onto $(G/H)/H'$ and $h':(G/H)/H'\to G/H$ a lifting relative to H'. There exists a subgroup I of G such that $H\subset I$ and H' can be identified with I/H and we identify $(G/H)/(I/H)$ with G/I. We denote by h the lifting $h:G/I\to G/H$ relative to I/H. We have obtained

$$\pi_H(f(x)) = \pi_H(x) - h(\pi_I(x)).$$

Finally, as $\pi_H(x) - h(\pi_I(x)) \in I/H$, with a lifting $i: I/H \to I$ relative to H, we get

$$f(x) = i[\pi_H(x) - h(\pi_I(x))] \quad \text{for all } x \in G.$$

This concludes the proof of Theorem 12.

Theorem 12 has been generalized for nonabelian groups but with additional assumptions concerning f. It remains an open problem to find the general solution there as well as in the case of an abelian quasigroup. There is another way of stating Theorem 12.

Theorem 13. *Let $f: G \to G$ be a function on an abelian group $(G, +)$. This function satisfies* (45) *if and only if $f = f_2 \circ f_1$ where $f_1: G \to G$ and $f_2: I \to I$ are two functions such that $I = f_1(G)$ is a subsemigroup of G and f_1 satisfies* (42) *while f_2 satisfies*

$$f_2(x + f_2(y)) = f_2(x + y) \quad (x, y \in I). \tag{48}$$

Proof. Suppose $f = f_2 \circ f_1$ with the properties stated in Theorem 13 for f_1 and f_2. Then $I = f_1(G)$ is a subsemigroup of G. Take a $z \in G$ such that $f_1(z) = f_2 \circ f_1(y)$. We get

$$\begin{aligned} f(x + f(y)) &= f_2 \circ f_1[x + f_1(z)] \\ &= f_2(f_1(x) + f_1(z)) = f_2(f_1(x) + f_2 \circ f_1(y)) \\ &= f_2(f_1(x) + f_1(y)). \end{aligned}$$

By symmetry,

$$f(x + f(y)) = f(y + f(x)).$$

Now let f satisfy (45) and define, using the notations of Theorem 12, $f_1: G \to G$ by

$$f_1(x) = x - \tilde{h}(h(\pi_I(x))),$$

where $\tilde{h}: G/H \to G$ is any lifting relative to H. We know that $f_1(x) \in I$ for all x in G. A simple computation shows that f_1 satisfies (42) and that I is in fact the range of f_1 (take $x \in I$; for example, $f_1(x) = x - \tilde{h}(h(0))$). Define $f_2: I \to I$, using the lifting i introduced in Theorem 12, by

$$f_2(x) = i(\pi_H(x)).$$

The function f_2 satisfies (48), because $\pi_H \circ i \circ \pi_H(x) = \pi_H(x)$. Then we can compute $f_2 \circ f_1(x)$:

$$f_2 \circ f_1(x) = i[\pi_H(x) - \pi_H(\tilde{h}(h(\pi_I(x))))] = i[\pi_H(x) - h(\pi_I(x))] = f(x),$$

as asserted.

Equations (42) and (48) are particular cases of equation (45). We have found the general solution of (42) in a group (Proposition 11). We leave it as an exercise to find the general solution of (48) (Exercise 16).

Corollary 14. *A function $f:\mathbb{R}\to\mathbb{R}$ is a nonconstant continuous solution of* (45) *if, and only if, there exists a real constant a and*

$$f(x)=x+a \quad (x\in\mathbb{R}).$$

Proof. We can see easily that for any a in $\mathbb{R}$, $f(x)=x+a$ is a continuous solution of (45). —Conversely, let f be a continuous solution of (45). By continuity, the subgroup of periods for f is $\alpha\mathbb{Z}$ for some $\alpha\in\mathbb{R}$.

But, for the subgroup H defined in Theorem 12, $H\subset\alpha\mathbb{Z}$ has to hold. Therefore $f(y+f(z))-f(y)-f(z)$ has its values in $\alpha\mathbb{Z}$. By continuity, we get for some b in $\alpha\mathbb{Z}$ that

$$f(y+f(z))=f(y)+f(z)+b.$$

Using $f_1(y)=f(y)+b$, and $f(y+b)=f(y)$, we see that f_1 satisfies (42). Using Proposition 11, we see that the range of f_1 is a subgroup H_1 of $\mathbb{R}$. By the continuity of f_1, and as f_1 is not a constant function since f is not constant, we get $H_1=\mathbb{R}$. Thus $h_1(\pi_{H_1}(x))$ is a constant, say b_1. By Proposition 11, $f_1(x)=x-b$, and so $f(x)=x+b-b_1=x+a$, which ends the proof of Corollary 14.

19.4 Reynolds and averaging operators

The origin of the functional equations (42), (45) and (48) is in averaging theory applied to turbulent fluid motion. The Navier–Stokes equation which describes the motion of a newtonian fluid is the nonlinear partial differential equation

$$\rho\frac{\partial\mathbf{V}}{\partial t}=\rho\mathbf{f}-\operatorname{grad}p+\nu\rho\Delta\mathbf{V}-\rho\operatorname{div}(\mathbf{V}\otimes\mathbf{V}), \tag{49}$$

where $\mathbf{V}(t,M)$ is the velocity at a point $M=(x_1,x_2,x_3)$ and at time t, the components of which are V_1,V_2 and V_3. In (49), $\Delta\mathbf{V}$ is a vector whose components are $\Delta V_1,\Delta V_2$ and ΔV_3 where Δ is the Laplacian, p is the (scalar) pressure, $\mathbf{f}$ an external force and $\operatorname{div}(\mathbf{V}\otimes\mathbf{V})$ is a nonlinear term, which denotes a vector whose components are equal to

$$\sum_{j=1}^{3}\frac{\partial}{\partial x_i}(V_iV_j)\quad\text{for } i=1,2,3.$$

For incompressible fluids, we must add the condition

$$\operatorname{div}\mathbf{V}=0. \tag{50}$$

At the end of the nineteenth century O. Reynolds looked for a linear operator P acting on $\mathbf{V}$ and on $\mathbf{p}=(p,p,p)$ in such a way that $P(\mathbf{V})$ satisfies a Navier–Stokes equation with a supplementary term, the turbulent one, considered as added to the external force:

$$\rho\frac{\partial P(\mathbf{V})}{\partial t} = P(\mathbf{f} - \operatorname{div}[(\mathbf{V} - P(\mathbf{V}))\otimes(\mathbf{V} - P(\mathbf{V}))]) + \nu\rho\Delta P(\mathbf{V})$$
$$- \rho \operatorname{div}(P(\mathbf{V})\otimes P(\mathbf{V})) - \operatorname{grad}(P(p)). \tag{51}$$

In order to deduce (51) from (50) under some regularity assumptions for the linear operator P, it is enough to suppose that

$$P(f(x)^2) = (Pf(x))^2 + P((f(x) - P(x))^2).$$

Replacing f by $f+g$, and using the linearity of P, we obtain the so-called *Reynolds identity*

$$P(f\cdot Pg + g\cdot Pf) = Pf\cdot Pg + P(Pf\cdot Pg). \tag{52}$$

Let $\mathscr{A}$ be a commutative algebra of functions. A linear operator $P:\mathscr{A}\to\mathscr{A}$ satisfying (52) for all f,g in $\mathscr{A}$, is called a *Reynolds operator*.

An *averaging operator* $P:\mathscr{A}\to\mathscr{A}$, on the algebra $\mathscr{A}$, is a linear operator such that for all f,g in $\mathscr{A}$:

$$P(f\cdot Pg) = Pf\cdot Pg. \tag{53}$$

There are close relations between Reynolds operators and averaging operators (Kampé de Fériet 1957; Rota 1964; Miller 1966*b*, 1968, Fox–Miller 1968; Gamlen–Miller 1968; Dhombres 1971*b*).

In what follows we often use somewhat deeper results of functional analysis concerning representations of Banach algebras. The reader who is not knowledgeable on this subject, but does not want to skip these parts, may consult Dunford–Schwartz 1963, pp. 859–84. We have to specialize our algebra $\mathscr{A}$. In fact, from the point of view of partial differential equations the algebras of the type L^∞, the algebra of real Lebesgue bounded functions on $\mathbb{R}^n$ is interesting. But a classical result of functional analysis states that L^∞, as a Banach algebra, is isomorphic to $C_{\mathbb{R}}(X)$, the algebra of real valued continuous functions over a compact Hausdorff space X. This space X has very special properties (hyperstonean space). In particular, X is totally disconnected. This explains the interest in the following result:

Theorem 15. *Let P be a continuous Reynolds operator over $C_{\mathbb{R}}(X)$, the Banach algebra of all real valued continuous functions on a compact space X. Suppose X is a totally disconnected compact Hausdorff space. Then P is an idempotent averaging operator.*

Here we suppose that P is continuous when $C_{\mathbb{R}}(X)$ is equipped with the uniform norm $\|f\| = \sup_{x\in X}|f(x)|$. We shall *prove* Theorem 15 in three steps.

The first step is combinatorial. It consists of proving that the following formula holds for all $n \geqslant 1$ and all f in $C_R(X)$:

$$nP(f(Pf)^{n-1}) = (Pf)^n + (n-1)P((Pf)^n). \tag{54}$$

For $n=1$, this equality is trivial. For $n=2$, it follows from the definition (52) of Reynolds operators. We now proceed by induction. We suppose the formula true for n. Then

$$P(f(Pf)^n) = nP(f \cdot P(f(Pf)^{n-1})) - (n-1)P(f \cdot P[(Pf)^n]).$$

Using the Reynolds identity twice, we get

$$\begin{aligned}(n+1)P(f(Pf)^n) = n[&Pf \cdot P(f(Pf)^{n-1}) + P(Pf \cdot P(f(Pf)^{n-1}))] \\ &- (n-1)(Pf \cdot P[(Pf)^n] + P(Pf \cdot P[(Pf)^n]) - P[(Pf)^{n+1}]).\end{aligned}$$

Replacing twice $nP(f(Pf)^{n-1})$ by the value given in the induction formula, we get

$$\begin{aligned}(n+1)P(f(Pf)^n) = (Pf)^{n+1} &+ (n-1)P[(Pf)^n] \cdot Pf + P[(Pf)^{n+1}] \\ &+ (n-1)P(Pf \cdot P[(Pf)^n]) - (n-1)P[(Pf)^n] \cdot Pf \\ &+ (n-1)P[(Pf)^{n+1}] - (n-1)P(P[(Pf)^n] \cdot Pf),\end{aligned}$$

and so

$$(n+1)P(f(Pf)^n) = (Pf)^{n+1} + nP[(Pf)^{n+1}],$$

which ends the proof of (54).

The second step is a topological one. We prove that the space $C_R(X)$ is the closed linear hull of its idempotents if X is a compact totally disconnected Hausdorff space. An idempotent of $C_R(X)$ is the characteristic function of a 'clopen' subset of X (both open and closed in X). Because the family of clopen subsets of X is stable under finite intersections, we get that the subset of idempotents of $C_R(X)$ is stable under multiplication. The linear hull of this subset of idempotents of $C_R(X)$ is then a subalgebra B of $C_R(X)$. This algebra separates points of X when X is totally disconnected (and compact Hausdorff) because in such a space the intersection of all clopen sets containing an x is reduced to the singleton $\{x\}$. The Stone–Weierstrass theorem states that the algebra B is dense in $C_R(X)$.

The third and final step is to compute the action of P on some idempotent of $C_R(X)$. Let Pf be an idempotent in the image $P(C_R(X))$. Using (54) we get

$$P(f \cdot Pf) = \frac{1}{n} Pf + \left(1 - \frac{1}{n}\right) P(Pf).$$

Letting n grow to infinity, we obtain $P(Pf) = P(f \cdot Pf)$ and so

$$P(Pf) = Pf. \tag{55}$$

We therefore get P as the identity operator over the idempotents in the image C of $C_R(X)$ under P $(C = P(C_R(X)))$.

Because of the functional equation (52), written in the form

$$P(f \cdot Pg + g \cdot Pf - Pf \cdot Pg) = Pf \cdot Pg,$$

we see that C is a subalgebra of $C_R(X)$. Using once more the Stone–Weierstrass theorem, we see that the closure $\bar{C}$ of C in $C_R(X)$ is isomorphic to a $C_R(Y)$ where Y is a compact totally disconnected Hausdorff space. Step 2 can be applied with (55), proving that P is the identity on $\bar{C}$. In other words, P is an idempotent operator. We replace f by Pf in the Reynolds identity (52) and use $P(Pf) = Pf$ in order to get

$$P(Pf \cdot Pg) + P(g \cdot Pf) = Pf \cdot Pg + P(Pf \cdot Pg).$$

Thus $P(g \cdot Pf) = Pf \cdot Pg$, which is the functional equation of an averaging operator. This concludes the proof of Theorem 15.

The structure of all continuous averaging linear operators over $C_R(X)$ is known if X is a compact Hausdorff space (Birkhoff 1950, Kelley 1958). If $P: C_R(X) \to C_R(X)$, we define $\mathscr{P}(x)$ as the subset of all y in X such that $Pf(x) = Pf(y)$ for all f in $C_R(X)$. This $\mathscr{P}(x)$ is a closed subset of X.

Proposition 16. *A continuous linear operator $P: C_R(X) \to C_R(X)$, where X is a compact Hausdorff space, is an averaging operator if, and only if, for any f in $C_R(X)$, the value of Pf at any point x depends only upon the values of f on the subset $\mathscr{P}(x)$ of x.*

Proof. If $Pf(x)$ depends only upon the values of f on the closed subset $\mathscr{P}(x)$ then, as Pg is a constant function on $\mathscr{P}(x)$, we deduce from the linearity of P that $P(f \cdot Pg)(x) = Pf(x)Pg(x)$, that is, we have (53).

Conversely, suppose (55) is satisfied for each x in X. There exists a real Radon measure μ_x on X such that

$$Pf(x) = \int_X f(y)\, d\mu_x(y).$$

Equation (53) can therefore be written as

$$0 = \int_X f(y)[Pg(y) - Pg(x)]\, d\mu_x(y).$$

As f is arbitrary in $C_R(X)$, we deduce that $Pg(y) = Pg(x)$ on the support of the measure μ_x. (Therefore the support of μ_x is a subset of $\mathscr{P}(x)$ and $Pf(x) = \int_{\mathscr{P}(x)} f(y)\, d\mu_x(y)$.) In other words, $Pf(x)$ only depends on the values of f on the equivalence class $\mathscr{P}(x)$ of x. This concludes the proof of Proposition 16.

In the last decades an increasing interest has been manifested in linear operators satisfying additional functional equations ('algebraic identities').

We have just given two such examples, the Reynolds and the averaging operators.

Here we will not develop the very close relation between averaging operators and the operations of conditional expectation as defined in probability theory on spaces $L^p(\Omega, \mathscr{F}, \mu)$.

Another class of linear operators, the Baxter operators, have been defined on commutative algebras of functions by aid of a functional equation, quite similar to (52),

$$P(f\cdot Pg + g\cdot Pf) = Pf\cdot Pg + P(f\cdot g).$$

Their use originates in queuing theory (see, e.g., Kingman 1966), and they are interesting for their combinatorial and other properties which are rather different from those of the Reynolds operators. They have been studied in depth, for instance by Atkinson (1963); Miller (1966*a*, 1969); Rota (1969*a*, *b*); Dhombres (1971*b*); and Nguyen Huu Bong (1976). We refer the reader to these works. Cf. also Exercises 31, 32 and 33. Linear operators satisfying other (but related) equations will be the object of the next two sections.

19.5 Interpolating and extension operators

We will now turn, in connection with Theorem 13, to operators satisfying the functional equation (48) in the form (56) below. (In the rest of this chapter we assume familiarity with properties of Radon measures on compact Hausdorff spaces.)

A linear operator $P: C_{\mathbf{R}}(X) \to C_{\mathbf{R}}(X)$ is called an *interpolating operator* if, for all f, g in $C_{\mathbf{R}}(X)$,

$$P(f\cdot Pg) = P(fg). \tag{56}$$

With any linear operator $P: C_{\mathbf{R}}(X) \to C_{\mathbf{R}}(X)$ we define its interpolator (int P) as the subset of all x in X such that $Pf(x) = f(x)$ for all f in $C_{\mathbf{R}}(X)$. It is often empty, but never in the case of an interpolating operator.

Proposition 17. *Let $P: C_{\mathbf{R}}(X) \to C_{\mathbf{R}}(X)$ be a continuous linear operator. Then P is an interpolating operator if, and only if,* int P, *the interpolator of P, is not empty and $Pf = 0$ for every $f \in C_{\mathbf{R}}(X)$ which is zero on* int P.

Proof. If P is an interpolating operator, let int P be its interpolator. For each x in X, there exists a Radon measure μ_x on X so that $Pf(x) = \int_X f(y)\, d\mu_x(y)$. Then (56) yields

$$\int_X f(y)(g(y) - Pg(y))\, d\mu_x(y) = 0.$$

Therefore $g(y) = Pg(y)$ for every y belonging to the support of μ_x. In other words, the interpolator of P is not empty and contains, for every x in X, the support of μ_x. Therefore, if $f(y)=0$ for all $y \in \operatorname{int} P$, we get $\int_X f(y)\,d\mu_x(y)=0$ for all x in X, and so $Pf(x)=0$, that is, $Pf=0$.

Conversely, let P be a bounded linear operator on $C_R(X)$ with a nonempty interpolator, such that $Pf=0$ when f is equal to zero on $\operatorname{int} P$. Let x be a point in X and μ_x be the Radon measure on X for which

$$Pf(x)=\int_X f(y)\,d\mu_x(y) \quad (f \in C_R(X)).$$

Clearly, $\operatorname{int} P$ is a closed subset of X. If the support of μ_x were not a subset of $\operatorname{int} P$, then there would exist an $f \in C_R(X)$ which is zero on $\operatorname{int} P$ and for which $Pf(x)=\int_X f(y)d\mu_x(y) \neq 0$. This contradicts our hypothesis. Therefore the support of μ_x is a subset of $\operatorname{int} P$ for all $x \in X$. Thus

$$P(f\cdot Pg)(x)=\int_X f(y)Pg(y)\,d\mu_x(y)=\int_{\operatorname{int} P} f(y)Pg(y)\,d\mu_x(y)$$

$$=\int_{\operatorname{int} P} f(y)g(y)\,d\mu_x(y)=\int_X f(y)g(y)\,d\mu_x(y)=P(fg)(x),$$

as asserted. Interpolating operators are directly connected to *linear extension operators*.

Let Y be a closed subset of X. A bounded linear operator $E\colon C_R(Y)\to C_R(X)$ is a linear extension operator relative to Y if $Ef(y)=f(y)$ for all y in Y and all f in $C_R(Y)$.

Proposition 18. *Let Y be a closed subspace of a compact Hausdorff space X. There exists a linear extension operator relative to Y if, and only if, there exists a linear interpolating operator $P\colon C_R(X)\to C_R(X)$ having Y as its interpolator.*

Proof. Suppose E is a linear extension operator relative to Y. Then, for any x in X, we get a Radon measure μ_x supported by Y and

$$Ef(x)=\int_Y f(t)\,d\mu_x(t).$$

Define $P\colon C_R(X)\to C_R(X)$ by

$$Pf(x)=\int_X f(t)\,d\mu_x(t).$$

The interpolator of P is easily shown to be Y. Moreover, by Proposition 17, P is an interpolating operator.

Conversely, let P be an interpolating operator, with $Y=\operatorname{int} P$. Then, for any f in $C_R(Y)$, we define $Ef=P(\tilde{f})$ where $\tilde{f}$ is any continuous extension of f to all of X (Tietze's extension theorem). By Proposition 17, we notice that E is a bounded linear operator and does not depend upon the chosen extension of f. Moreover, E is a linear extension operator. This concludes the proof of Proposition 18.

An interesting problem, still unsolved in the general case, is to characterize those compact spaces for which every closed subspace of X is an interpolator for some interpolating operator on $C_{\mathbf{R}}(X)$ (cf. Dhombres 1971*b*).

There is a functional equation satisfied both by averaging operators and by interpolating operators. We will say that $P:C_{\mathbf{R}}(X)\to C_{\mathbf{R}}(X)$ is a *multiplicatively symmetric operator* if, for all f,g in $C_{\mathbf{R}}(X)$,

$$P(f\cdot Pg)=P(Pf\cdot g), \tag{57}$$

(cf. (45)).

There are many such operators as can be seen from the following theorem which we state without proof (cf. Dhombres 1968, where it is proved in the algebra $C(X)$ of complex valued continuous functions over X).

Theorem 19. *Let $P:C_{\mathbf{R}}(X)\to C_{\mathbf{R}}(X)$ be a linear, continuous and idempotent operator of norm* 1. *Suppose that the range of P contains a strictly positive function. Then, for all f,g in $C_{\mathbf{R}}(X)$, the operator P satisfies* (57).

In the case where $\|P\|=1$ and $P(\mathbf{1})=\mathbf{1}$, and for $P:C_{\mathbf{R}}(X)\to C_{\mathbf{R}}(X)$, multiplicatively symmetric operators coincide with the so-called *exaves*, as introduced by Pełczyński 1968, pp. 15–27. We introduce some notation.

Let A,B be compact spaces and $\pi:A\to B$ be a continuous function. Let $[\pi]:C_{\mathbf{R}}(B)\to C_{\mathbf{R}}(A)$ denote the linear operator of composition defined by

$$[\pi](f)=f\circ\pi.$$

A linear *exave* P for $\pi:A\to B$ is a linear operator $P:C_{\mathbf{R}}(A)\to C_{\mathbf{R}}(B)$, such that $[\pi]=[\pi]\circ P\circ[\pi]$. Here we will not prove the equivalence of linear exaves with multiplicatively symmetric operators (cf. Dhombres 1979, pp. 7.28–7.34) but just sketch some results.

Let P be a bounded multiplicatively symmetric operator on $C_{\mathbf{R}}(X)$. We define A to be the set of all x in X such that, for all f,g in the algebra $C_{\mathbf{R}}(X)$, we get

$$P(f\cdot Pg)(x)=Pf(x)Pg(x).$$

This closed set A is called the *averaging set* of the operator P. When P is markovian (that is, $\|P\|=1$ and $P(\mathbf{1})=\mathbf{1}$), then it can be proved that A is not empty. We define B to be a quotient space of X, namely the set of all subsets $\mathscr{P}(x)$, with the quotient topology ($y\in\mathscr{P}(x)$ if and only if $Pf(y)=Pf(x)$ for all $f\in C_{\mathbf{R}}(X)$).

Then $B=X/\mathscr{P}$ and we will denote by π the canonical quotient mapping $X\to X/\mathscr{P}$. By $\tilde{\pi}$ we denote its restriction to A. We finally define $\tilde{P}:C_{\mathbf{R}}(A)\to$

$C_R(B)$. Let $\tilde{g}$ be an extension of g ($\in C_R(A)$). By definition,

$$(\tilde{P}g)(\pi(x)) = (P\tilde{g})(x).$$

One can show that $\tilde{P}$ does not depend upon the choice of the extension of g. Moreover, it can be proved that $\tilde{P}$ is a markovian linear exave for $\tilde{\pi}$.

Conversely, we associate a multiplicatively symmetric operator with a markovian exave. Let A, B be two compact Hausdorff spaces and $\tilde{\pi}: A \to B$ be a continuous mapping. Denote by $\tilde{\pi}(A)$ the image of A under $\tilde{\pi}$. Let $\pi: X \to B$ be any compact fibre bundle over B, extending the given fibre bundle $\tilde{\pi}: A \to \pi(A)$. By $R: C_R(X) \to C_R(A)$, we will denote the restriction operator on $C_R(X)$, associated with A:

$$Rf(a) = f(a) \quad \text{for all } f \text{ in } C_R(X),\ a \text{ in } A.$$

If we define $P: C_R(X) \to C_R(X)$ by

$$Pf(x) = \tilde{P}(Rf)(\pi(x)),$$

we can prove that P is a markovian multiplicatively symmetric operator.

Let Y be a closed subspace of a compact Hausdorff space X and let $\sim$ be a closed equivalence relation on X. We denote by $\tilde{C}_R(X)$, identified with $C_R(X/\sim)$, the subalgebra of $C_R(X)$ containing all functions which are constant on each class of equivalence. (We define $\tilde{C}_R(Y)$ similarly, by restricting $\sim$ to Y). *A linear q-extension operator for Y and $\sim$* is a bounded linear operator $Q: \tilde{C}_R(Y) \to C_R(X)$, such that $Qf(y) = f(y)$ for all f in $\tilde{C}_R(Y)$ and y in Y.

With these definitions, multiplicatively symmetric operators can be investigated and the following theorem proved (see Dhombres 1976*b*):

Theorem 20. *Let $P: C_R(X) \to C_R(X)$ be a markovian operator. Then P is multiplicatively symmetric if, and only if,*

$$P = Q \circ S \circ R,$$

where R is the restriction operator associated with some closed subset Y of X, S is an averaging markovian operator on $C_R(Y)$ for which $S(C_R(Y)) = R(P[C_R(X)])$, Q is a markovian linear q-extension operator for Y and for the equivalence relation $\mathcal{P}$.

19.6 Derivation operators

Reynolds operators are closely related to derivation operators. Formally, suppose P is invertible and is a Reynolds operator. We define the linear operator D by

$$Df = P^{-1}(f) - f,$$

where P^{-1} is the inverse of the operator P. Then D satisfies

$$D(f \cdot g) = Df \cdot g + f \cdot Dg. \tag{58}$$

This functional equation (58) is the defining equation of a *derivation operator* (cf. (21) and Exercise 2.11).

Let $C(T)$ be the algebra of all complex valued continuous functions defined over $\mathbb{R}$ and of period 2π. We say that $P:C(T)\to C(T)$ is *stationary* if $P(T_h(f))=T_h(Pf)$ for all h in $\mathbb{R}$ and all f in $C(T)$ where

$$T_h(f)(x)=f(x+h),$$

T_h being the translation operator.

It is easy to prove that the only continuous linear and stationary derivation operator $D:C(T)\to C(T)$ is the zero operator.

Indeed, consider the functions e_n defined by $e_n:x\mapsto e^{inx}$ where n is any integer. These functions are the only eigenfunctions for all operators T_h. Since the order of application of P and T_h can be reversed, the functions e_n are also eigenfunctions of P, which gives

$$P(e_n)=a_n e_n, \tag{59}$$

where the a_n are complex constants. Then (58) yields

$$a_{n+m}=a_n+a_m \quad (n,m\in\mathbb{Z}).$$

From this simple functional equation (the Cauchy equation on $\mathbb{Z}^2$) we get that, for some λ in C, $a_n=\lambda n$. But P is continuous for the uniform norm, which implies

$$\|P(e_n)\|=|\lambda|\cdot|n|\leqslant\|P\|.$$

The only possibility is $\lambda=0$ and thus $P\equiv 0$.

A classical theorem in analysis states that the result holds also without the stationarity supposition.

Theorem 21. *Let X be a compact Hausdorff space and let $C(X)$ be the Banach algebra of all complex valued continuous functions over X. The only linear operator $D:C(X)\to C(X)$, which is a derivation, is the zero operator.*

Proof. By the linearity, it is enough to prove that $Df=0$ for any real valued f in $C(X)$. Suppose first that f has a square root which is an element of $C_{\mathbb{R}}(X)$, that is, $f(x)=g(x)^2$. Equation (58) yields $Df=2gD(g)$. Therefore, if g is zero at some point x_0 in X, then $D(g^2)(x_0)=0$. Clearly, any real valued f in $C(X)$, zero at x_0, can be written as a difference of squares, that is, $f(x)=g(x)^2-h(x)^2$ with $0=g(x_0)=h(x_0)$. As a result, for any real valued f in $C(X)$, zero at x_0, $Df(x_0)=0$. If we let $\mathbf{1}$ be the function everywhere equal to 1 and x_0 be an arbitrary element of X, we deduce that $D(f-f(x_0)\mathbf{1})(x_0)=0$. But (58) yields $D(\mathbf{1})\equiv 0$. Therefore we obtain, for any x_0 in X,

$$Df(x_0)=0,$$

yielding $D\equiv 0$, as asserted, because x_0 was chosen arbitrarily.

Theorem 21 can be generalized to any commutative complex Banach algebra, that is an algebra A which is a complete normed space satisfying the inequality $\| fg \| \leqslant \| f \| \cdot \| g \|$ for all f, g in A. The idea of the generalization is the following (Singer–Wermer 1955). Let λ be any complex number and suppose first that $D: A \to A$ is a bounded linear derivation operator. We define $e^{\lambda D}$ as a bounded linear operator:

$$e^{\lambda D} = \sum_{n=0}^{\infty} \frac{\lambda^n D^n}{n!} \quad (\text{where } D^\circ = I).$$

Let $\chi: A \to C$ be a multiplicative linear form on A. We consider

$$Q(\lambda, f) = \chi(e^{\lambda D}(f)),$$

a multiplicative linear functional. We calculate $Q(\lambda, fg)$:

$$\begin{aligned} Q(\lambda, fg) &= \sum_{n=0}^{\infty} \frac{\lambda^n}{n!} \chi(D^n(fg)) \\ &= \sum_{n=0}^{\infty} \lambda^n \sum_{i+j=1} \frac{\chi(D^i(f))}{i!} \frac{\chi(D^j(g))}{j!}, \end{aligned}$$

where i, and j, are nonnegative integers. Using the absolute convergence of the series (Cauchy's product), we get

$$Q(\lambda, fg) = \left(\sum_{n=0}^{\infty} \frac{\lambda^n}{n!} \chi(D^n(f)) \right) \left(\sum_{n=0}^{\infty} \frac{\lambda^n}{n!} \chi(D^n(g)) \right).$$

Therefore, for all f, g in A

$$Q(\lambda, fg) = Q(\lambda, f) Q(\lambda, g). \tag{61}$$

It turns out that every nonzero multiplicative linear functional on A is bounded and even of norm 1:

$$|Q(\lambda, f)| \leqslant \| f \|. \tag{62}$$

But the mapping $\lambda \mapsto Q(\lambda, f)$, for any given f in A, is an entire function. The majorization (62), combined with Liouville's theorem, implies that $Q(\lambda, f) = Q(0, f) = 1$ for all $f \in C(X)$. As a consequence, $\chi(D^n f) = 0$ for $n \geqslant 1$. Finally, $\chi(Df) = 0$, for any multiplicative linear form on A and for any f in A. In a commutative complex Banach algebra, the set of all f in A such that $\chi(f) = 0$ for every multiplicative linear form χ on A is the *radical* of A. A *semisimple algebra* is an algebra where the radical consists of the zero element alone. It is an easy exercise to show that the Banach algebra $C(X)$ is semisimple. We have thus proved the following:

Theorem 22. *The only bounded linear derivation on a semisimple complex Banach algebra is the zero operator.*

Notice that, in Theorem 21, no regularity was assumed for the derivation operator, in contrast to Theorem 22. So Theorem 22 does not contain Theorem 21 as a particular case. However, it can be proved that Theorem 22 remains true without the regularity (boundedness) assumption for the operator of derivation D. This theorem has been considerably extended.

Exercises and further results

1. Prove that S is a closed proper subgroup of the topological additive group $\mathbb{R}$ if, and only if, there exists a real number α such that $S = \alpha\mathbb{Z}$.
2. (E.g. Hewitt–Ross 1963, p. 587.) Consider the set $H = \{(x, \alpha x) | x \in \mathbb{R}\}$, where $\alpha \in \mathbb{R}\backslash\{0\}$ is a constant. Then $(H, +)$ is a closed subgroup of the topological additive group on $\mathbb{R}^2$. We denote by ϕ the canonical mapping from $\mathbb{R}^2$ onto the topological quotient group $\mathbb{R}^2/\mathbb{Z}^2$. Prove or disprove that $\phi(H)$ is a closed subgroup of $\mathbb{R}^2/\mathbb{Z}^2$ if, and only if, α is a rational number and that $\phi(H)$ is a proper dense subgroup of $\mathbb{R}^2/\mathbb{Z}^2$ if, and only if, α is irrational.
3. Prove that a characteristic function of a nonempty subset of $\mathbb{R}$ is a solution of the Gołąb–Schinzel equation (6)
$$f(x + f(x)y) = f(x)f(y) \quad (x, y \in \mathbb{R})$$
if and only if it is the characteristic function of a subgroup of $(\mathbb{R}, +)$. Does this result remain true if we replace $\mathbb{R}$ by an arbitrary ring?
4. (Cf. Dhombres 1982*c*.) Solve the Pexider analogue of equation (6), that is, determine all $f, g, h: F \to \mathbb{R}$ which satisfy
$$f(x + g(x)y) = g(x)h(y) \quad (x, y \in F),$$
where F is a real linear space.
5. (Lawrence 1981*a*.) Let D be a division ring. Prove that all solutions $f: D\backslash\{0\} \to D$ of the generalization
$$f(xy) = xf(y) + f(x) \quad (x, y \in D\backslash\{0\})$$
of (12) are of the form $f(x) = (x - 1)a$ $(x \in D)$ where $a \in D$ is an arbitrary constant.
6. Prove Proposition 1 by using Theorem 5.
7. (Cf. Popa 1965.) Prove or disprove that a Lebesgue measurable $f: \mathbb{R} \to \mathbb{R}$ satisfying
$$f(x + f(x)y) = f(x)f(y) \quad (x, y \in \mathbb{R})$$
is either continuous or almost everywhere zero.

(For Exercises 8–9, see Wołodźko 1968.)

8. Let $f: \mathbb{R} \to \mathbb{R}$ be such that 0 is not a cluster point of the subgroup of periods of f. Suppose that $f(\mathbb{R}) \neq \{-1, 0, 1\}$. Prove or disprove that
$$f(x + f(x)y) = f(x)f(y) \quad \text{for all } x, y \in \mathbb{R}$$
if, and only if, there exists a multiplicative subgroup G of $\mathbb{R}\backslash\{0\}$ and a constant $a \neq 0$ such that
$$f(x) = \begin{cases} 1 + ax & \text{if } (1 + ax) \in G \\ 0 & \text{if } (1 + ax) \notin G. \end{cases}$$
9. Find all continuous functions $f: \mathbb{C} \to \mathbb{C}$ satisfying
$$f(x + f(x)y) = f(x)f(y) \quad \text{for all } x, y \in \mathbb{C}.$$

10*. (Brillouët–Dhombres 1986.) Suppose that $\mathbf{F}(\mathbf{x})$ is a 2×2 diagonal matrix for all $\mathbf{x}$ in $\mathbb{R}^2$. Find all continuous solutions $\mathbf{F}$ of $\mathbf{F}(\mathbf{x}+\mathbf{F}(\mathbf{x})\mathbf{y}) = \mathbf{F}(\mathbf{x})\mathbf{F}(\mathbf{y})$ for all $(\mathbf{x}, \mathbf{y} \in \mathbb{R}^2)$ where the products $\mathbf{F}(\mathbf{x})\mathbf{y}$ and $\mathbf{F}(\mathbf{x})\mathbf{F}(\mathbf{y})$ are products of a matrix and a vector or of two matrices, respectively.

(For Exercises 11–13, see Dhombres 1979, pp. 6.10–6.19.)

11. Find all continuous solutions $g:\mathbb{R} \to \mathbb{R}$ of the functional equation

$$g(g(x)) = (\lambda + 1)g(x) - \lambda x$$

in the case $\lambda \leqslant 0$.

12*. Let $0 < \lambda < 1$ and G be a real linear Hausdorff space. A *regular iterative convex decomposition* of G is a splitting of G into closed subsets S_x $(x \in I)$ such that $S_x \cap S_{x'} = \varnothing$ if $x \neq x'$, I is a closed and nonempty subset of G, $S_x \cap I = \{x\}$, $\lambda y + (1-\lambda)x \in S_x$ for $y \in S_x$, and the mapping $h:G \to I$, defined by $h(y) = x$ for $y \in S_x$ is continuous. Prove or disprove that $g:G \to G$ satisfies the functional equation

$$g(g(x)) = (\lambda + 1)g(x) - \lambda x \quad (x \in G)$$

if, and only if, there exists a regular iterative convex decomposition of G such that $h(x) = [g(x) - \lambda x]/(1-\lambda)$.

13. What happens to the solutions of the equation in the previous exercise when $\lambda > 1$ or $\lambda \leqslant 0$?

14*. Give examples in $\mathbb{R}^n$ of regular iterative convex decompositions as defined in Exercise 12.

15. (Dhombres 1979, p. 6.40.) Prove or disprove that all nonconstant continuous solutions of the functional equation

$$f(xf(y)) = f(yf(x)) \quad (x, y \in \mathbb{R})$$

are of one of the forms

(i) $f(x) = ax$,
(ii) $f(x) = \sup(bx, cx)$,
(iii) $f(x) = \inf(dx, -dx)$,

where a, b, c, d are real constants such that $a \neq 0$, $b \leqslant 0 \leqslant c$, $b < c, d > 0$, but otherwise arbitrary.

16. (Dhombres 1977*a*.) Let $(G, *)$ be a group. Find all solutions $f:G \to G$ of the functional equation

$$f(x * f(y)) = f(x * y) \quad (x, y \in G).$$

17. (Dhombres 1984*c*.) Find all $f:\mathbb{R} \to \mathbb{R}$ such that $f(x)f(y) - f(xy)$ depends only upon $x + y$ (cf. Exercises 4.5, 4.24).

18. (Dhombres 1977*a*.) Let $(G, *)$ be a group. Prove or disprove the following. A function $f:G \to G$ satisfies the functional equation

$$f(x * f(y)) = f(x) * f(y) \quad (x, y \in G)$$

if, and only if, there exists a subgroup H (not necessarily normal) and a lifting $h:(G/H)_l \to G$ such that

$$f(x) = (h(\pi_H(x)))^{-1} * x \quad (x \in G),$$

where $G/H)_l$ is the set of all left cosets $x*H$ and π_H the canonical mapping from G onto $(G/H)_l$.

(For Exercises 19–22, see Dhombres 1979, pp. 6.41–6.50.)

19*. Let $(G,*)$ be a cancellative abelian monoid. (A monoid is a semigroup with a neutral element. For the definition of cancellativity, see Exercise No. 17.5.) Prove or disprove that $f:G\to G$ satisfies

$$f(x*f(y))=f(x)*f(y) \quad \text{for all } x,y\in G$$

if, and only if, there exists a group $(\bar{G},*)$ and a subgroup $(H,*)$ of $(\bar{G},*)$ such that $G\subset\bar{G}$ and

$$f(x)=h(\pi_H(x))^{-1}*x \quad (x\in G),$$

where h is a lifting relative to H.

20*. Let $(G,*)$ be a cancellative abelian monoid. Find the general operations $\Delta:G^2\to G$ satisfying

$$(x*y)\Delta z=x*(y\Delta z) \quad \text{for all } x,y,z\in G.$$

21. Let $(G,*)$ be a simple abelian group. Prove or disprove that the only associative operations $\Delta:G^2\to G$ satisfying $(x*y)\Delta z=x*(y\Delta z)$ are $x\Delta y=x*y*a$ for some $a\in G$ or $x\Delta y=x$.

22*. Prove the following result. Let $(G,+)$ be an abelian group and Δ an operation $\Delta:G^2\to G$ such that $e\Delta e=e$. Suppose that $((x+y)\Delta z)-(y\Delta z)$ depends only upon x. Then the operation Δ is associative if, and only if, there exists a subgroup I of G, which is a factor of G, such that $G=G/I\oplus I$, there exists a subgroup H of I, a lifting $h:I/H\to I$ with $h(e)=e$ and a mapping $i:G\to G/I$ for which $i(e)=e$ and $i(y+z)=z$ for all $y\in I$, $z\in i(I)$, such that

$$x\Delta y=\pi_I(x+y)-h(\pi_H(y))+i(y) \quad \text{for all } x,y\in G.$$

23. (Cf. Dhombres 1976c.) Prove or disprove the following. Let Δ be an associative binary operation on the real axis. Suppose $((x+y)\Delta z)-(y\Delta z)$ depends only upon x. Suppose also that both $x\mapsto x\Delta y$ and $y\mapsto x\Delta y$ are continuous for all $x,y\in\mathbb{R}$. Suppose finally that there exist x_0, y_0 such that $x\mapsto x_0\Delta x$ and $x\mapsto x\Delta y_0$ are not constant. Then there exists a real constant a such that $x\Delta y=x+y+a$ for all $x,y\in\mathbb{R}$.

24. (Cf. Dhombres 1979, pp. 6.50–6.55.) Let $(G,+)$ be an abelian group. Find all $f:G\to G$, with $f(0)=0$, for which the operation $\Box$ defined by $x\Box y=f(x+f(y))$ is associative.

25. (Dhombres 1979, pp. 6.2–6.6.) Prove or disprove that $f:\bar{\mathbb{R}}_+\to\bar{\mathbb{R}}_+$ is a continuous solution of $f(xf(x))=f(x)^2$ $(x\in\bar{\mathbb{R}}_+)$ if, and only if, f has one of the following forms

(a) $f(x)=1 \quad (x\in\bar{\mathbb{R}}_+)$,

(b) $f(x)=\mu x \quad (x\in\bar{\mathbb{R}}_+)$,

(c) $f(x)=\begin{cases}\dfrac{x}{\lambda} & \text{for } 0\leqslant x<\lambda\\ 1 & \text{for } x\geqslant\lambda,\end{cases}$

where $\mu,\lambda,\lambda_1,\lambda_2$ are arbitrary real constants satisfying $\mu\geqslant 0$, $\lambda>0$, $0<\lambda_1<\lambda_2$.

(For Exercises 26–8, see Dhombres 1979, pp. 7.4–7.9 and 7.16–7.19.)

26. The algebra of all complex valued continuous functions on $\mathbb{R}$ with period 2π being denoted by $C(T)$, prove or disprove that a nonzero continuous stationary linear operator $P:C(T)\to C(T)$ is a Reynolds operator if, and only if, there exists an integer k and a complex number $s\notin k\mathbb{Z}$ or $s=\infty$ such that

$$P = R_{s,k}\circ P_k,$$

where

$$P_0 f = \frac{1}{2\pi}\int_{-\pi}^{\pi} f(t)dt,$$

$$P_k f(x) = \frac{f(x)+f\left(x+\frac{2\pi}{k}\right)+\cdots+f\left(x+\frac{k-1}{k}2\pi\right)}{k} \quad (k\neq 0),$$

$$R_{s,0}f(x) = f(x),$$

$$R_{s,k}f(x) = \frac{s}{2\sin(\pi s/k)}\int_{-\pi/k}^{\pi/k} e^{-its} f\left(x-t-\frac{\pi}{k}\right)dt,$$

and

$$R_\infty f(x) = R_{s,0}f(x) = f(x) \quad (R_{s,0}=R_\infty \text{ is the identity operator}).$$

27. Characterize those continuous linear operators on $P:C_R(T)\to C_R(T)$ which are multiplicatively symmetric (see (57)) and stationary.

28*. Characterize those continuous linear operators $P:C_R(T)\to C_R(T)$ which are stationary and which satisfy $P(\mathbf{1})=\mathbf{1}$ and

$$P(fPg+gPf) = \alpha P(fg)+(1-\alpha)Pf\cdot Pg + P(Pf\cdot Pg) \quad \text{for all } f,g\in C_R(T).$$

29*. (Brillouët–Dhombres 1986.) Find all continuous solutions $f:E\to\mathbb{R}$ of

$$f(xf(y)+yf(x)) = 2f(x)f(y) \quad (x,y\in E),$$

where E is a real topological vector space.

30. Let $\phi:[0,1]\to\mathbb{R}$ satisfy $|\phi(x)|\leqslant|x|$ for all x in $[0,1]$. Find all differentiable $f:[0,1]\to\mathbb{R}$, with $f(0)=0$, which satisfy the functional differential equation

$$f'(x) = f(\phi(x))$$

for all $x\in[0,1]$.

31*. (Dhombres 1971*b*.) Let $\mathbb{T}$ be the compact space $\mathbb{R}/\mathbb{Z}$. Let $1<p<\infty$. Find all bounded linear operators $P:L^p(\mathbb{T})\to L^p(\mathbb{T})$ such that
   (i) P is a Baxter operator,
   (ii) P commutes with translations (that is, $P(\tau_h f)=\tau_h(Pf)$ for all h in $\mathbb{T}$, where $\tau_h f(x) = f(x-h)$).

(For Exercises 32–33 see Miller 1966*a*.)

32*. Let A be a Banach algebra with an identity e and $L(A)$ be the Banach algebra of all bounded linear operators on A into A. For $P\in L(A)$, we define res P, the resolvent of P, as the set of all complex λ such that $(\lambda I - P)$ has an inverse in $L(A)$. Similarly we define res a, for $a\in A$ as the set of all $\lambda\in\mathbb{C}$ such that $(\lambda e - a)^{-1}\in A$ (cf. Exercise 11.31). Let $P\in L(A)$ be such that

$$P(x\cdot Py + Px\cdot y) = Px\cdot Py - P(\theta xy) \quad (x,y\in A),$$

where θ is a fixed element of A. Prove that

$$(\lambda I - P)^{-1}x = \frac{x}{\lambda} + \frac{1}{\lambda} f_\lambda \cdot P(f^{-1}(\lambda e - \theta)^{-1}x)$$

where

$$f_\lambda = (P - \lambda I)^{-1}e,\ \lambda \in (\text{Res}\, P) \cap (\text{Res}\, \theta) \text{ and } f_\lambda^{-1} \in A.$$

33*. Let B be the algebra of all polynomials (in x) with complex coefficients. Let $P: B \to B$ be a linear operator such that $P(1) = x$. Show that P is a Baxter operator if, and only if,

$$P(x^n) = \frac{1}{n+1} B_{n+1}(1+x) \qquad (n \geqslant 1),$$

where B_n is the nth Bernoulli polynomial, defined by

$$z\frac{e^{zx}-1}{e^z-1} = \sum_{n=1}^{\infty} B_n(x)\frac{z^n}{n!}.$$

34. (Volkmann–Weigel 1984.) Find all continuous functions $f: \mathbb{R} \to \mathbb{R}$ satisfying

$$f(xf(y) + yf(x) - xy) = f(x)f(y).$$

35. (Benz 1983.) Show that all $f, g: \mathbb{R}^2 \to \mathbb{R}$ satisfying

$$f\left(x+z, y+\frac{1}{z}\right) - f(x, y) - z^2\left(g\left(x+z, y+\frac{1}{z}\right) - g(x, y)\right) = 0 \quad \text{for all } x, y \in \mathbb{R},\ z \in \mathbb{R}^*$$

are given by

$$f(x, y) = d(x) + \alpha x + \beta,$$
$$g(x, y) = d(y) - \alpha y + \gamma,$$

where $\alpha, \beta, \gamma \in \mathbb{R}$ and $d: \mathbb{R} \to \mathbb{R}$ is a derivation (cf. Exercise 2.11).

20

Homogeneity and some generalizations. Application to economics

The homogeneity of a function can be expressed by a functional equation. In this chapter, we introduce more general functional equations for which we give the general solutions and apply them in mathematical economics.

At the end of Chapter 15, in (15.47), we encountered the *homogeneity equation*:

$$f(x_1u, x_2u, \ldots, x_nu) = u^c f(x_1, x_2, \ldots, x_n) \quad (x_1, x_2, \ldots, x_n, u \in R_+). \tag{1}$$

(There we had $c = 1$ and $n = 2$; also the domain is here slightly different.) We deal now with two generalizations of (1). The first is a functional equation with two unknown functions, one n-place, the other one-place:

$$f(x_1u, x_2u, \ldots, x_nu) = g(u) f(x_1, x_2, \ldots, x_n) \quad (x_1, x_2, \ldots, x_n, u \in R_+). \tag{2}$$

This is the equation of *generalized homogeneity*. By repeated use of (2) we get

$$\begin{aligned} g(uv) f(x_1, x_2, \ldots, x_n) &= f(x_1uv, x_2uv, \ldots, x_nuv) \\ &= g(v) f(x_1u, x_2u, \ldots, x_nu) = g(u)g(v) f(x_1, \ldots, x_n). \end{aligned} \tag{3}$$

With the exception of the trivial case,

$$f(x_1, x_2, \ldots, x_n) = 0 \quad (\text{for all } x_1, x_2, \ldots, x_n) \text{ and } g \text{ arbitrary}, \tag{4}$$

it follows from that

$$g(uv) = g(u)g(v) \quad \text{for all } u, v \in R_+, \tag{5}$$

which is Cauchy's power equation (3.23). Again with the exception of the case $g(u) = 0$ for all $u \in \mathbb{R}_+$, which, by (2), implies $f(x_1, x_2, \ldots, x_n) = 0$ for all $x_1, x_2, \ldots, x_n \in \mathbb{R}_+$ and thus falls under (4), we have, from (5) with $u = 1$ or $u = 1/v$,

$$g(1) = 1 \quad \text{and} \quad g\left(\frac{1}{v}\right) = \frac{1}{g(v)} \quad (v \in \mathbb{R}_+),$$

respectively. So, if we substitute $u = 1/x_1$ into (2), we get, with the notation

$h(t_2,t_3,\ldots,t_n)=f(1,t_2,t_3,\ldots,t_n)$,

$$f(x_1,x_2,\ldots,x_n)=g(x_1)h\left(\frac{x_2}{x_1},\frac{x_3}{x_1},\ldots,\frac{x_n}{x_1}\right) \quad \text{for all } x_1,x_2,\ldots,x_n\in\mathbb{R}_+. \quad (6)$$

It is easy to check that both (4) and (6), with arbitrary h and with g satisfying (5), satisfy (2). So we have the following (cf. also Proposition 3.6).

Theorem 1. *The general solutions* $f:\mathbb{R}_+\to\mathbb{R}$, $g:\mathbb{R}_+\to\mathbb{R}$ *of* (2) *are given by* (4) *and by* (6), $h:\mathbb{R}_+^{n-1}\to\mathbb{R}$ *being an arbitrary function, while* g *is an arbitrary solution of* (5). *If, in particular,* g *is continuous at a point or bounded on a set of positive measure, then the general solutions in these classes of admissible functions (but with no restrictions on* f*) are given by* (4) *(with* g *an arbitrary function in that class) and by*

$$\left.\begin{aligned} &g(u)=u^c \quad (u\in\mathbb{R}_+;\ c \text{ an arbitrary constant}) \text{ and} \\ &f(x_1,x_2,\ldots,x_n)=x_1^c h\left(\frac{x_2}{x_1},\frac{x_3}{x_1},\ldots,\frac{x_n}{x_1}\right) \quad (x_1,\ldots,x_n\in\mathbb{R}_+;\ h \text{ arbitrary}).\end{aligned}\right\} \quad (7)$$

In the latter case (2) *reduces to* (1).

Even easier to treat, by the substitution $u=x_1^{-1/c_1}$, is the following generalization of (1):

Proposition 2. *The general solution* $f:\mathbb{R}_+^n\to\mathbb{R}$ *of the functional equation*

$$f(x_1u^{c_1},x_2u^{c_2},\ldots,x_nu^{c_n})=u^cf(x_1,x_2,\ldots,x_n) \quad (x_1,x_2,\ldots,x_n,u\in\mathbb{R}_+) \quad (8)$$

[where, say, $c_1\neq 0$*, but otherwise* $c,c_1,c_2,\ldots,c_n$ *are arbitrary (fixed) real constants] is given by*

$$f(x_1,x_2,\ldots,x_n)=x_1^{c/c_1}h(x_2x_1^{-c_2/c_1},x_3x_1^{-c_3/c_1},\ldots,x_nx_1^{-c_n/c_1}) \quad (x_1,x_2,\ldots,x_n\in\mathbb{R}_+), \quad (9)$$

with arbitrary $h:\mathbb{R}_+^{n-1}\to\mathbb{R}$.

(Of course, all solutions are trivial if $c\neq 0$ but $c_1=c_2=\cdots=c_n=0$.)

For references see, e.g., Wellstein 1910; Halphen 1911; Favre 1917; Lions 1958; Aczél 1966c, pp. 231–2, 304–10.

Such equations have applications, for instance, in mathematical economics.

Thus, Eichhorn (1972) defines *economic effectiveness* as a real function f of the time t, the capital x, the yield y (which may be negative) and the profitability z (which may be 0) of a production process which has the following properties for all $y\in\mathbb{R}$, $t,u,x\in\mathbb{R}_+$, $z\in\bar{\mathbb{R}}_+$:

$$f(t,x,y,z) \gtreqless 0 \text{ according to } y \gtreqless 0 \quad (10)$$

(the effectiveness is positive, zero or negative as is the yield),

$$f(ut, x, uy, z) = f(t, ux, uy, z) = f(t, x, y, z) \tag{11}$$

(the effectiveness does not change if both the time and the yield or both the capital and the yield increase or decrease by a factor of u),

$$f(t, x, uy, z) = uf(t, x, y, z) \tag{12}$$

(the effectiveness increases (decreases) by a factor of u if, with the same capital and profitability, during the same time, the yield increases (decreases) by a factor of u) and

$$f(t, x, y, uz) = g(u)f(t, x, y, z), \tag{13}$$

where $g: \mathbb{R}_+ \to \mathbb{R}$ is in increasing function. The last property describes the increase of effectiveness with increasing profitability.

Theorem 3. *The general solution of the system* (10)–(13) $(t, u, x \in \mathbb{R}_+, y \in \mathbb{R}, z \in \bar{\mathbb{R}}_+)$ *for increasing* g *is given by*

$$f(t, x, y, z) = a_{\operatorname{sign} y} \frac{y}{tx} z^c, \tag{14}$$

where a_0, a_1, a_{-1} *and* c *are arbitrary positive constants.*

Proof. Functions defined by (14) evidently satisfy all conditions (10), (11), (12), and (13). Conversely, apply both equalities in (11) (with $u = t$ or $u = x$, respectively), then (12) (with $u = |y|/tx$):

$$\begin{aligned} f(t, x, y, x) &= f\left(t \cdot 1, x, t \cdot \frac{y}{t}, z\right) = f\left(1, x, \frac{y}{t}, z\right) = f\left(1, x \cdot 1, x \cdot \frac{y}{tx}, z\right) \\ &= f\left(1, 1, \frac{y}{tx}, z\right) = f\left(1, 1, \frac{|y|}{tx} \operatorname{sign} y, z\right) \\ &= \frac{|y|}{tx} f(1, 1, \operatorname{sign} y, z) \end{aligned} \tag{15}$$

(this remains true, by (10), also for $y = 0$ as does (14), so we may ignore $y = 0$ in what follows). On the other hand, by (13), we have

$$g(uv)f(1, 1, s, z) = f(1, 1, s, uvz) = g(u)g(v)f(1, 1, s, z) \quad (s = \pm 1, u > 0, v > 0)$$

from which, by (10), we get

$$g(uv) = g(u)g(v) \quad \text{for all } u > 0, \quad v > 0.$$

For increasing functions g, Proposition 3.6 now gives $g(u) = u^c$ $(u > 0)$ with $c > 0$ and so (13) yields

$$f(1, 1, s, z \cdot 1) = z^c f(1, 1, s, 1) = s a_s z^c \quad \left(\text{with } a_s = \frac{1}{s} f(1, 1, s, 1), s = \pm 1\right)$$

for $z > 0$. Putting $t = x = 1$, $y = s$, $z = 0$ into (13) with $g(u) = u^c$ $(c > 0)$ gives

$f(1,1,s,0)=0$. All this, together with (15), proves (14), since, by (10), $a_s>0$ and (14) remains true also for $y=0$. This concludes the proof of Theorem 2.

(Eichhorn (1972) supposed a somewhat weaker condition instead of (13) and accordingly got a slightly more complicated result.)

There are many more applications of homogeneous functions and their generalizations to mathematical economics (cf. Eichhorn 1978). One concerns production functions. The following is a modified and somewhat more general result than can be found for instance in Eichhorn 1978, pp. 69–97, 184–227. The nonnegative variables $x_1, x_2, \ldots, x_n$ are called 'input quantities' or *input variables* and $F(x_1, x_2, \ldots, x_n)$ the *maximum output* (in quantity or value, say) obtainable from them through a production process; $F: \bar{\mathbb{R}}_+^n \to \bar{\mathbb{R}}_+$ is called a *production function*. Note that $F(x_1, x_2, \ldots, x_n)$ is supposed to be *nonnegative*. We will use the following somewhat weaker versions of the strict monotonicity which is usually supposed regarding production functions. First, F is increasing and nonconstant for small values of its variables:

There exist positive constants $d_1, d_2, \ldots, d_n$ such that

$$\left.\begin{array}{l} F(x_1,\ldots,x_{j-1},x_j,x_{j+1},\ldots,x_n) \leqslant F(x_1,\ldots,x_{j-1},x'_j,x_{j+1},\ldots,x_n) \\ \textit{whenever } 0<x_k<d_k\ (k=1,2,\ldots,j-1,j+1,\ldots,n),\ 0<x_j<x'_j<d_j; \\ \textit{furthermore} < \textit{holds for at least one pair } x_j<x'_j\ (j=1,2,\ldots,n). \end{array}\right\} \quad (16)$$

For the purpose of the Theorem 5 below, it is sufficient to suppose that

$$\left.\begin{array}{l} \textit{for every } j\in\{1,2,\ldots,n\} \textit{ there exist constants } c_j, d_j \textit{ and positive} \\ \textit{constants } b_{j,1}, b_{j,2},\ldots,b_{j,j-1},b_{j,j+1},\ldots,b_{j,n} \textit{ such that } 0\leqslant c_j<d_j \textit{ and} \\ F(b_{j,1},\ldots,b_{j,j-1},x_j,b_{j,j+1},\ldots,b_{j,n}) \leqslant F(b_{j,1},\ldots,b_{j,j-1},x'_j,b_{j,j+1},\ldots,b_{j,n}) \\ \textit{whenever } c_j<x_j<x'_j<d_j; \textit{furthermore} < \textit{holds for at least one such pair} \\ x_j<x'_j. \end{array}\right\} \quad (17)$$

This version is just slightly stronger than the requirement that the production function be *nonconstant in all variables*.

The remaining suppositions are conditions of homogeneity and generalized homogeneity, one being the homogeneity of degree 1 in all variables:

$$F(tx_1, tx_2, \ldots, tx_n) = tF(x_1, x_2, \ldots, x_n) \quad \text{for all } x_j \geqslant 0\ (j=1,2,\ldots,n), \quad t>0. \tag{18}$$

The last condition is one of partial generalized homogeneity:

$$\left.\begin{array}{l} \textit{There exist functions } g_j: \mathbb{R}_+ \to \mathbb{R}\ (j=1,2,\ldots,n) \textit{ such that} \\ F(tx_1,\ldots,tx_{j-1},x_j,tx_{j+1},\ldots,tx_n) = g_j(t)F(x_1,\ldots,x_{j-1},x_j,x_{j+1},\ldots,x_n) \\ \quad \textit{for all } t>0,\ x_k\geqslant 0\ (k=1,2,\ldots,n; j=1,2,\ldots,n). \end{array}\right\} \quad (19)$$

The condition (18) seems to be rather restrictive. But, in the opinion of many economists, *if a production function is not homogeneous (of degree 1) then not all possible inputs have been taken into consideration.* Mathematicians may agree, in a sense, because the following holds:

Proposition 4. *Let $f:\bar{\mathbb{R}}_+^{n-1}\to\mathbb{R}$ be an arbitrary function and let $c>0$ be an arbitrarily given constant. There exists exactly one function $F:\bar{\mathbb{R}}_+^{n-1}\times\mathbb{R}_+\to\mathbb{R}$ which satisfies*

$$F(x_1,x_2,\dots,x_{n-1},c)=f(x_1,x_2,\dots,x_{n-1}) \text{ on } \bar{\mathbb{R}}_+^{n-1} \tag{20}$$

and is homogeneous of degree 1

$$F(tx_1,tx_2,\dots,tx_{n-1},tx_n)=tF(x_1,x_2,\dots,x_{n-1},x_n)$$
$$\text{for all } x_1,x_2,\dots,x_{n-1}\in\bar{\mathbb{R}}_+, x_n>0, t>0. \tag{21}$$

The definition of F can be extended so that (21) *remains true for $x_n=0$ too.*

Indeed, (21) and (20) imply

$$F(x_1,x_2,\dots,x_{n-1},x_n)=\frac{x_n}{c}F\left(\frac{x_1}{x_n}c,\frac{x_2}{x_n}c,\dots,\frac{x_{n-1}}{x_n}c,c\right)$$
$$=\frac{x_n}{c}f\left(\frac{x_1}{x_n}c,\frac{x_2}{x_n}c,\dots,\frac{x_{n-1}}{x_n}c\right),$$

and, conversely, the function F so defined, together with

$$F(x_1,x_2,\dots,x_{n-1},0):=0, \tag{22}$$

if needed, always satisfies (20) and (21), as asserted.

There is no compelling reason in choosing (22); $(x_1,x_2,\dots,x_{n-1})\mapsto F(x_1,x_2,\dots,x_{n-1},0)$ could be an arbitrary homogeneous function of $n-1$ variables and of degree 1 (so the *uniqueness of F does not extend to $x_n=0$*), but the definition (22) will be convenient later. Proposition 4 may be less (or more) surprising, if we notice that similar statements are true regarding homogeneous functions of any degree and for generalized homogeneous functions. These may be considered as extension propositions.

If (at least) one of the input variables remains unchanged (maybe because it has not been considered), then we cannot hope for homogeneity of degree 1 anymore. But condition (19) postulates that there should be at least some kind of generalized homogeneity in those cases too. We will prove the following:

Theorem 5. *A function $F:\bar{\mathbb{R}}_+^n\to\bar{\mathbb{R}}_+$ satisfies* (17), (18) *and* (19) *if, and only if,* *there exist constants $b, a_1, a_2, \ldots, a_n$ such that*

$$0 < a_j < 1, \quad \sum_{j=1}^{n} a_j = 1, \quad b > 0, \tag{23}$$

$$g_j(t) = t^{1-a_j} \quad (t \in \mathbb{R}_+), \tag{24}$$

and

$$F(x_1, x_2, \ldots, x_n) = b x_1^{a_1} x_2^{a_2} \cdots x_n^{a_n} \quad \text{on } \bar{\mathbb{R}}_+^n. \tag{25}$$

These functions F are called the *Cobb-Douglas production functions.* In previous characterizations (see Eichhorn 1978, pp. 185–9) the result (24) was a supposition.

Proof. Just as (5) has followed from (2), we get from (19)

$$g_j(tu) = g_j(t) g_j(u) \quad \text{for all } t, u \in \mathbb{R}_+, \quad j = 1, 2, \ldots, n, \tag{26}$$

since, by (17), F is not identically 0. Also, by (17) and (19), every g_j is different from 0 at at least one point and therefore everywhere on $\mathbb{R}_+$. Being nonnegative by (26), g_j is thus everywhere positive ($j = 1, 2, \ldots, n$), as in Chapter 3. Using (18) and (19) (repeatedly) we get

$$\begin{aligned}
&F(x_1, x_2, x_3, \ldots, x_n) \\
&= x_1 x_2 x_3 \cdots x_n F\left(\frac{1}{x_2 x_3 x_4 \cdots x_n}, \frac{1}{x_1 x_3 x_4 \cdots x_n}, \frac{1}{x_1 x_2 x_4 \cdots x_n}, \ldots, \frac{1}{x_1 x_2 x_3 \cdots x_{n-1}}\right) \\
&= x_1 g_1\left(\frac{1}{x_1}\right) x_2 x_3 \cdots x_n \\
&\quad \cdot F\left(\frac{1}{x_2 x_3 x_4 \cdots x_n}, \frac{1}{x_3 x_4 x_5 \cdots x_n}, \frac{1}{x_2 x_4 x_5 \cdots x_n}, \ldots, \frac{1}{x_2 x_3 x_4 \cdots x_{n-1}}\right) \\
&= x_1 g_1\left(\frac{1}{x_1}\right) x_2 g_2\left(\frac{1}{x_2}\right) x_3 \cdots x_n \\
&\quad \cdot F\left(\frac{1}{x_3 x_4 \cdots x_n}, \frac{1}{x_3 x_4 \cdots x_n}, \frac{1}{x_4 x_5 \cdots x_n}, \ldots, \frac{1}{x_3 x_4 \cdots x_{n-1}}\right) = \cdots \\
&= x_1 g_1\left(\frac{1}{x_1}\right) x_2 g_2\left(\frac{1}{x_2}\right) \cdots x_n g_n\left(\frac{1}{x_n}\right) F(1, 1, 1, \ldots, 1) \text{ on } \mathbb{R}_+^n.
\end{aligned} \tag{27}$$

From (17) and since $F(1, 1, 1, \ldots, 1) \geq 0$, $x_k g_k(1/x_k) > 0$ ($k = 1, 2, \ldots, n$), we see that $x_j \mapsto x_j g_j(1/x_j)$ is increasing on $]c_j, d_j[$ and therefore

$$x_j g_j\left(\frac{1}{x_j}\right) \leq d_j g_j\left(\frac{1}{d_j}\right), \quad g_j(t) \leq \frac{d_j}{c_j} g_j\left(\frac{1}{d_j}\right) \quad \text{on} \quad]1/d_j, 1/c_j[\quad (j = 1, 2, \ldots, n).$$

Thus g_j is bounded from above on a proper interval and therefore (Proposition 3.6) $g_j(t) = t^{1-a_j}$ ($t \in \mathbb{R}_+$) for some constant $a_j > 0$ (since $x_j \to x_j g_j(1/x_j)$ is increasing). So we have (24), furthermore (27) goes over into (25) on $\mathbb{R}_+^n$, that is, into

$$F(x_1,x_2,\ldots,x_n)=bx_1^{a_1}x_2^{a_2}\cdots x_n^{a_n}\quad (x_1>0,x_2>0,\ldots,x_n>0). \tag{28}$$

By (17) and since F is nonnegative, $b,a_1,a_2,\ldots,a_n$ are all positive. Putting (28) into (18) we also get $\sum_{j=1}^n a_j=1$, thus $a_j<1$ $(j=1,2,\ldots,n)$. So (23) is established.

If at least one of the x_k, say x_j is 0 then, from (18) and (19)

$$\begin{aligned}tF(x_1,\ldots,x_{j-1},0,x_{j+1},\ldots,x_n)&=F(tx_1,\ldots,tx_{j-1},0,tx_{j+1},\ldots,tx_n)\\&=g_j(t)F(x_1,\ldots,x_{j-1},0,x_{j+1},\ldots,x_n).\end{aligned}$$

Since $g_j(t)=t$ is excluded by $g_j(t)=t^{1-a_j}$ with $0<a_j<1$, we must have

$$F(x_1,\ldots,x_{j-1},0,x_{j+1},\ldots,x_n)=0\quad (j=1,2,\ldots,n),$$

(cf. (22)). So (25) holds also if some x_j are 0. Conversely, the functions given by (24) and (25) with (23) satisfy the conditions (19), (18), (17) (and (16)), which concludes the proof of Theorem 5.

In mathematical economics several other requirements regarding production functions have also been stated. We will see now that some of them are *always* satisfied by Cobb–Douglas production functions, that is, they are consequences of (16) (or (17)), (18) and (19). Other 'requirements' can *not* be satisfied by them.

The *law of (eventually) diminishing marginal returns* states that

$$\left.\begin{aligned}&x_j\mapsto F(x_1,\ldots,x_j,\ldots,x_n)\quad (j=1,2,\ldots,n)\\ \textit{and}\quad&\\&t\mapsto F(tx_1,\ldots,tx_k,x_{k+1},\ldots,x_n)\quad (k=1,2,\ldots,n)\end{aligned}\right\}\tag{29}$$

are *concave*, at least *from a point* (x_{j_0} or t_0) *on*. This is clearly satisfied by (25) with (23). But it *does not follow from just two of* (19), (18), (17) *or* (16). Indeed, (25) with $b>0$, $\sum_{j=1}^n a_j=1$ but at least one $a_k<0$ satisfies (18), (19) (with $g_j(t)=t^{a_j}, j=1,2,\ldots,n$) but not (16), (17) and $x_k\mapsto F(x_1,\ldots,x_k,\ldots,x_n)$ is convex, not concave. Similarly, (25) with $b,a_1,a_2,\ldots,a_n\in\mathbb{R}_+$, but $\sum_{j=1}^n a_j>1$ and one $a_k>1$ satisfies (16) (and thus (17)), (19) (with $g_j(t)=t^{1-a_j}$ again), but not (18) and $x_k\mapsto F(x_1,\ldots,x_k,\ldots,x_n)$ is convex. Finally, $F(x_1,x_2,\ldots,x_n)=(x_1^2+x_2^2+\cdots+x_n^2)^{1/2}$ satisfies (16) (and thus (17)), (18), but not (19) (with any g_j) and every $x_j\mapsto F(x_1,\ldots,x_j,\ldots,x_n)$ is convex.

The *law of (eventually) diminishing average returns* requires that

$$x_j\mapsto F(x_1,\ldots,x_j,\ldots,x_n)/x_j\quad (j=1,2,\ldots,n)$$

and

$$t\mapsto F(tx_1,\ldots,tx_k,x_{k+1},\ldots,x_n)/t\quad (k=1,2,\ldots,n)$$

are *decreasing for large enough* x_j or t. (In both laws $x_1,\ldots,x_k$ may be replaced by other subsets of $\{x_1,x_2,\ldots,x_n\}$.) The law of diminishing average returns is also true for (25) with (23), but it follows already from (16) and (18), without

(19). Indeed,

$$\frac{1}{x_j}F(x_1,\ldots,x_{j-1},x_j,x_{j+1},\ldots,x_n)=F\left(\frac{x_1}{x_j},\ldots,\frac{x_{j-1}}{x_j},1,\frac{x_{j+1}}{x_j},\ldots,\frac{x_n}{x_j}\right),$$

$$\frac{1}{t}F(tx_1,\ldots,tx_k,x_{k+1},\ldots,x_n)=F\left(x_1,\ldots,x_k,\frac{x_{k+1}}{t},\ldots,\frac{x_n}{t}\right),$$

and, by (16), F is increasing for small variables. The last law would also hold if F were homogeneous of degree smaller than 1 but larger than 0.

Another 'law', that of 'initially increasing and eventually decreasing marginal returns', which postulates that there exist $x_{10},\ldots,x_{n0},t_0$ such that the functions (29) be increasing and convex for $x_j \leqslant x_{j_0}$ and $t \leqslant t_0$ but concave for $x_j > x_{j_0}, t > t_0$ *contradicts* (17), (18) and (19), because it is not satisfied by (25) (with (23)).

Exercises and further results

(For Exercises 1–4, see Eichhorn 1978, pp. 141–2, 152–5, 174–9, 210.)

1. A *price level* depending upon the prices of n commodities is a function $P:\mathbb{R}_+^n \to \mathbb{R}_+$ for which there exist $c_j \in \mathbb{R}_+$ $(j=1,2,\ldots,n)$ such that

$$P(\lambda p_1,\ldots,\lambda p_{j-1},p_j,\lambda p_{j+1},\ldots,\lambda p_n)=\lambda^{c_j}P(p_1,\ldots,p_n) \quad \text{for all } p_1,\ldots,p_n,\lambda\in\mathbb{R}_+ \text{ and all } j=1,\ldots,n.$$

Determine all such functions P.

2. A *price index* describing the change of prices of n commodities from a base year $((p_1^0,\ldots,p_n^0)=\mathbf{p}^0)$ to the current year $((p_1,\ldots,p_n)=\mathbf{p})$ is a function $I:\mathbb{R}_+^{2n}\to\mathbb{R}_+$. Some of the suppositions usually made about I are $(\mathbf{p}_0\in\mathbb{R}_+^n, \mathbf{p}\in\mathbb{R}_+^n, \lambda\in\mathbb{R}_+)$:

$$I(\mathbf{p}^0,\lambda\mathbf{p})=\lambda I(\mathbf{p}^0,\mathbf{p}), \tag{30}$$

$$I(\lambda\mathbf{p}^0,\lambda\mathbf{p})=I(\mathbf{p}^0,\mathbf{p}), \tag{31}$$

$$I(\lambda\mathbf{p}^0,\mathbf{p})=\frac{1}{\lambda}I(\mathbf{p}^0,\mathbf{p}), \tag{32}$$

$$I(\mathbf{p}^0,\lambda\mathbf{p}^0)=\lambda, \tag{33}$$

in particular,

$$I(\mathbf{p}^0,\mathbf{p}^0)=1. \tag{34}$$

(Try to give motivations for these suppositions.) Show that (30) and (34) imply (33), that (30) and (31) imply (32), further that (30) and (32) imply (31), and that (31) and (32) imply (30). Determine all functions I satisfying (30) and (31).

3. Also in economics, in the theory of multisectoral growth, functions $\mathbf{H}:\mathbb{R}_+^n\to\mathbb{R}_+^n$ are applied for which there exist functions $\phi:\mathbb{R}_+^n\to\mathbb{R}_+$, nondecreasing in each variable on an n-dimensional interval, and satisfying

$$\mathbf{H}(\lambda_1x_1,\ldots,\lambda_nx_n)=\phi(\lambda_1,\ldots,\lambda_n)\mathbf{H}(x_1,\ldots,x_n) \quad \text{for all } \lambda_1,\ldots,\lambda_n,x_1,\ldots,x_n\in\mathbb{R}_+.$$

Determine all such functions $\mathbf{H}$ and ϕ.

4. In addition to the economic effectiveness, which we have determined in Theorem 3, also a technical effectiveness $T:\mathbb{R}_+^{m+2}\to\mathbb{R}$ is defined in economics, depending upon the cost s, the time t, and the output $\mathbf{u}=(u_1,\dots,u_m)$ of a production process. Some of the suppositions usually made are the existence of increasing functions $\psi,\chi,\rho,\sigma,\tau:\mathbb{R}_+\to\mathbb{R}$ and of a decreasing $\phi:\mathbb{R}_+\to\mathbb{R}$ such that

$$T(\lambda s,\lambda t,\lambda\mathbf{u})=T(s,t,\mathbf{u}),\tag{35}$$

$$T(\lambda s,\lambda t,\mathbf{u})=\phi(\lambda)T(s,t,\mathbf{u}),\tag{36}$$

$$T(s,\lambda t,\lambda\mathbf{u})=\psi(\lambda)T(s,t,\mathbf{u}),\tag{37}$$

$$T(\lambda s,t,\lambda\mathbf{u})=\chi(\lambda)T(s,t,\mathbf{u}),\tag{38}$$

$$T(\lambda s,t,\mathbf{u})=\rho(\lambda)T(s,t,\mathbf{u}),\tag{39}$$

$$T(s,\lambda t,\mathbf{u})=\sigma(\lambda)T(s,t,\mathbf{u}),\tag{40}$$

$$T(s,t,\lambda\mathbf{u})=\tau(\lambda)T(s,t,\mathbf{u}),\tag{41}$$

$(\lambda,s,t\in\mathbb{R}_+,\mathbf{u}\in\mathbb{R}_+^m)$. Determine those of these conditions which imply all others. Determine all functions $T,\phi,\psi,\chi,\rho,\sigma,\tau$ satisfying these conditions.

5. (Aczél 1976*b*.) Show that, in the previous exercise, the validity of (36), (37), and (38), with decreasing $\phi:\mathbb{R}_+\to\mathbb{R}$ and increasing $\psi,\chi:\mathbb{R}_+\to\mathbb{R}$ implies that (41) holds with an increasing $\tau:\mathbb{R}\to\mathbb{R}$.

(For Exercises 6–8, see Aczél 1969*a*, pp. 57–62; and Eichhorn 1978, pp. 23–4, 94–7.)

6. A generalization of (2) (for $n=2$) is

$$f(xu,yu)=g(x,y,u)f(x,y)\quad(x,y,u\in\mathbb{R}_+).\tag{42}$$

But this is meaningless, since for every $f:\mathbb{R}_+^2\to\mathbb{R}_+$ there exists a $g:\mathbb{R}_+^3\to\mathbb{R}_+$ such that this equation (42) is satisfied, namely

$$g(x,y,u)=\frac{f(xu,yu)}{f(x,y)}.\tag{43}$$

However, the same argument that has led from (2) to (5), leads from (42) to

$$g(x,y,tu)=g(xu,yu,t)g(x,y,u)\quad(x,y,t,u\in\mathbb{R}_+).\tag{44}$$

Show that the general solution $g:\mathbb{R}_+^3\to\mathbb{R}_+$ of (44) is of the form (43) where $f:\mathbb{R}_+^2\to\mathbb{R}_+$ is an *arbitrary* function (this statement is independent of (42)).

7. The special case

$$f(xu,yu)=h\left(\frac{x}{y},u\right)f(x,y)\quad(x,y,u\in\mathbb{R}_+)$$

of (42) does make sense. Determine its general solutions $f:\mathbb{R}_+^2\to\mathbb{R}_+$, $h:\mathbb{R}_+^2\to[a,\infty[$ $(a>0)$.

8. Same question as in Exercise 7 but with the additional suppositions that $u\mapsto h(z,u)$ is increasing for a $z>0$ and $f(x,y)\geqslant f(s,t)$ if $x\geqslant s$ and $y\geqslant t$.

9. (Eichhorn 1978, pp. 90–3.) Let $f:\bar{\mathbb{R}}_+^n\to\mathbb{R}$ be homogeneous of degree $c\in]0,1]$:

$$f(x_1u,\dots,x_nu)=u^cf(x_1,\dots,x_n)\quad(x_1,\dots,x_n,u\in\bar{\mathbb{R}}_+),\tag{45}$$

monotonic in the sense that $f(y_1,\dots,y_n)\geqslant f(x_1,\dots,x_n)$ if $y_j\geqslant x_j$ $(j=1,\dots,n)$, continuous at hyperplanes i.e.

$$\lim_{x_j \to 0} f(x_1,\ldots,x_{j-1},x_j,x_{j+1},\ldots,x_n) = f(x_1,\ldots,x_{j-1},0,x_{j+1},\ldots,x_n)$$

and quasi-concave, that is, the set $L(u) = \{(x_1,\ldots,x_n) | f(x_1,\ldots,x_n) \geq u\}$ is convex. Show that f is concave on $\bar{\mathbb{R}}^n_+$.

(For Exercises 10–12, see e.g. Aczél 1966c, pp. 324–5, 377–8.)

10. Show that (45) (for arbitrary $c > 0$) implies, if f is (partially) differentiable, Euler's partial differential equation

$$\sum_{k=1}^{n} x_k \frac{\partial f}{\partial x_k} = cf.$$

Conversely, find the general differentiable solution of this partial differential equation.

11. Determine the general solutions $\mathbf{h}: \mathbb{R}^{2n} \to \mathbb{R}^n$, differentiable at the origin, of $\mathbf{h}(\lambda \mathbf{x}, \lambda \mathbf{y}) = \lambda \mathbf{h}(\mathbf{x}, \mathbf{y})$ $(\lambda \in \mathbb{R}, \mathbf{x} \in \mathbb{R}^n, \mathbf{y} \in \mathbb{R}^n)$.
12. Suppose, in addition to the conditions of Exercise 11, that, whenever $\mathbf{x}$ and $\mathbf{y}$ are rotated simultaneously in $\mathbb{R}^n$, the vector $\mathbf{h}(\mathbf{x}, \mathbf{y})$ undergoes the same rotation. Prove that there exist scalar constants $a, b \in \mathbb{R}$ such that $\mathbf{h}(\mathbf{x}, \mathbf{y}) = a\mathbf{x} + b\mathbf{y}$ for all $x, y \in \mathbb{R}^n$.
13. (Eichhorn-Gehrig 1982.) This question arises in the theory of measures of inequality (say, of incomes) in mathematical economics. Depending on whether we suppose the inequality to remain unchanged when all incomes are multiplied by the same factor or when the same amount is added to each income, we get the functional equations

$$I(tu_1,\ldots,tu_n) = I(u_1,\ldots,u_n) \quad (u_1,\ldots,u_n,t > 0),$$

or

$$I(x_1 + s,\ldots,x_n + s) = I(x_1,\ldots,x_n) \quad (x_1,\ldots,x_n,s > 0).$$

Do there exist functions $I: \mathbb{R}^n_+ \to \mathbb{R}_+$ satisfying *both* of these equations? Determine all such functions. Determine all continuous solutions.

21

Historical notes

As the size of the bibliography (which contains only a fraction of the existing literature on functional equations) indicates, it would be hopeless to try to present in one chapter of this book an even approximately complete history of functional equations (in several variables). So we restrict ourselves to a few notes (cf. also Aczél 1966*c*, pp. 5–12; Dhombres 1986) on the beginnings of this part of mathematics, some milestones in its development and a sketchy panorama of its present aspirations and applications.

21.1 Definition of linear and quadratic functions by functional equations in the Middle Ages and application of an implied characterization by Galileo

The emergence of functional equations was necessarily connected to the development of the notion of function, but we cannot, of course, go into the details of that history here. Because of the absence of any notion of function it would be very contrived to interpret passages of Euclid or Archimedes as even disguised formulations of functional equations. One should also differentiate between *stating*, for *given* functions, properties which amount to functional equations satisfied by them, and *determining all* functions with such properties, that is, *solving* these functional equations.

An important historical role of functional equations has been the *definition* of functions by functional equations (or their paraphrases). But often it was not shown (though it was implied) that these functions are the *only* solutions of these equations. An early example is that of Oresme (1347, 1352) who used a variant of coordinates and who determined 'uniformly deformed' (affine) 'qualities' by what amounts to the functional equation

$$\frac{f(x_1)-f(x_2)}{f(x_2)-f(x_3)}=\frac{x_1-x_2}{x_2-x_3} \quad \text{for all } x_1,x_2,x_3 \text{ with } x_1>x_2>x_3. \tag{1}$$

The relevant passages in Oresme 1352 say:

"*Qualitas vero uniformiter difformis est cuius omnium trium punctorum proportio distantiae inter primum et secundum ad distantiam inter secundum et tertium est sicut proportio excessus primi supra secundum ad excessum secundi supra tertium in intensione.... Omnis autem qualitas se habens alio modo a predicto dicitur difformiter difformis et potest describi negative, scilicet qualitas que non est in omnibus partibus subiecti equaliter intensa nec omnium trium punctorum ipsius proportio excessus primi supra secundum ad excessum secundi supra tertium est sicut proportio distantiarum eorum.*"

('A uniformly deformed quality is one for which, given any three points, the ratio of the distance between the first and second point to that between the second and third equals the ratio of the excess in intensity of the first over the second to that of the second over the third....Every quality which behaves in any way other than those previously described is said to be deformedly deformed. It can be described negatively as a quality which is neither everywhere equally intense nor for which, given any three points, the ratio of the excess of the first over the second to the excess of the second over the third equals the ratio of their distances'.)

The 'previously described qualities' were the 'uniform' (constant) ones and those just defined as 'uniformly deformed' (affine).

If in (1) we choose x_2 half-way between x_1 and x_3, we get from (1) the 'Jensen equation' (15.19)

$$f\left(\frac{x_1+x_3}{2}\right)=\frac{f(x_1)+f(x_3)}{2}. \tag{2}$$

This argument is related to what became known as the 'Merton theorem', so called because the authors (Swineshead, Heytesbury, Dumbleton), to whom this result is usually attributed, worked in Merton College in Oxford (founded in 1264). This result was known to Oresme and often repeated in later textbooks, thereby showing its pedagogical importance.

The Merton theorem states that the distance covered by an object moving on a straight line with 'uniformly deformed' speed equals that covered by an object whose speed is uniform (constant) and equal to the arithmetic mean of the initial and final speeds of the first object. In modern notation, if $x(t)$ denotes the distance covered by the first object and $v(t)$ its speed, this translates into the functional equation (for two functions)

$$x(t_2)-x(t_1)=\frac{v(t_1)+v(t_2)}{2}(t_2-t_1).$$

Oresme (1352) used it in the form

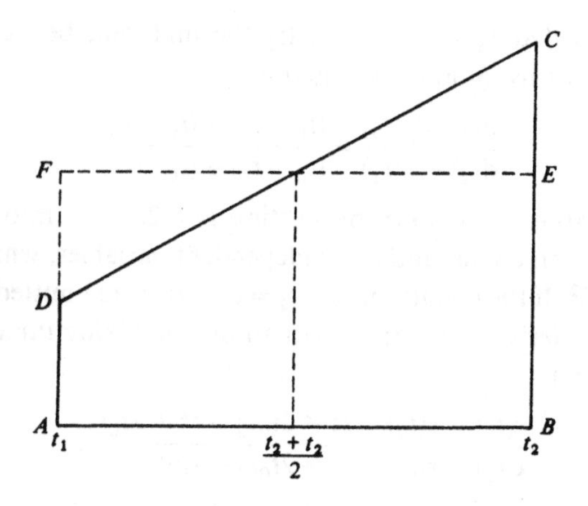

Fig. 21

$$x(t_2) - x(t_1) = v\left(\frac{t_1 + t_2}{2}\right)(t_2 - t_1) \tag{3}$$

(in modern notation) thus implying

$$v\left(\frac{t_1 + t_2}{2}\right) = \frac{v(t_1) + v(t_2)}{2}. \tag{4}$$

In modern language, the Merton theorem connects a simple fact of integral calculus with the functional equation (4) characterizing affine functions. Oresme (1352) gave a geometric justification of this result, using the equality of the areas of a trapezoid and a rectangle (Figure 21) to prove (3).

Oresme went further than the Merton theorem since he stated (3), not only for motions but, more abstractly, for his general 'qualities' (he mentions the amounts of heat and even of 'whiteness' as examples) and, as we will see, even 'solved' a related functional equation. In Oresme 1352 the equation (3) is stated explicitly as "*Omnis qualitas, si fuerit uniformiter difformis, ipsa est tanta quanta foret qualitas eiusdem subiecti vel equalis uniformis secundum gradum puncti medii eiusdem subiecti....*" ('Every quality, if it is uniformly deformed, is as much as would be the quality of the same or equal subject that is uniform according to the degree of the middle point of the same subject....')

In modern notation (4) would lead to $v(t) = at + b$ and this would transform (3) into

$$x(t_2) - x(t_1) = \left(a\frac{t_1 + t_2}{2} + b\right)(t_2 - t_1). \tag{5}$$

Earlier, Oresme (1347) actually chose a 'final' t_0 for which $v(t_0) = 0$ (thus $b = -at_0$ in (5)) and then designated three points (of which only two are independent) so that $t_3 - t_2 = t_2 - t_1$ (t_2 the midpoint between t_1 and t_3). This trick transforms (5) into the equation

$$\frac{x(t_3) - x(t_2)}{x(t_2) - x(t_1)} = \frac{(t_0 - t_2) + (t_0 - t_3)}{(t_0 - t_1) + (t_0 - t_2)} \tag{6}$$

between two ratios. Of course, by putting $t_3 = 2t_2 - t_1$ into (6), we get a functional equation containing two independent variables, which is satisfied by $x(t) = (t_0 - t)^2$. If the condition $t_3 - t_2 = t_2 - t_1$ were omitted then, instead of (6), we would derive from (5), the equation (containing three independent variables t_1, t_2, t_3)

$$\frac{x(t_3) - x(t_2)}{x(t_2) - x(t_1)} = \frac{t_3^2 - 2t_0t_3 - t_2^2 + 2t_0t_2}{t_2^2 - 2t_0t_2 - t_1^2 + 2t_0t_1},$$

which shows even clearer that x can be a polynomial of second degree. But this, of course, is far beyond the mathematical notions, notations, and techniques of that age.

However, Oresme (1347) recognized the *quadratic* nature of the process by dealing with (6) in the arithmetically restricted case where $t_0 - t_1$, $t_0 - t_2, t_0 - t_3$ are consecutive multiples of a segment, so that the ratio of the changes equals the ratio of consecutive odd numbers:

$$\frac{x(t_0 - (n+1)t) - x(t_0 - nt)}{x(t_0 - nt) - x(t_0 - (n-1)t)} = \frac{2n+1}{2n-1}. \tag{7}$$

In Oresme's (1347) words: "...*subiecto taliter (per aliqueas partes equales) diviso, et vocetur semper pars remissior prima, proportio partialium qualitatum... est sicud series imparium numerorum*" ('...with a subject so divided (into equal parts) and with the farthest part called the first, the ratio of partial qualities...is as the series of odd numbers'). Oresme drew Figure 22 as illustrated. In the same work (Oresme 1347) he reached the conclusion "...

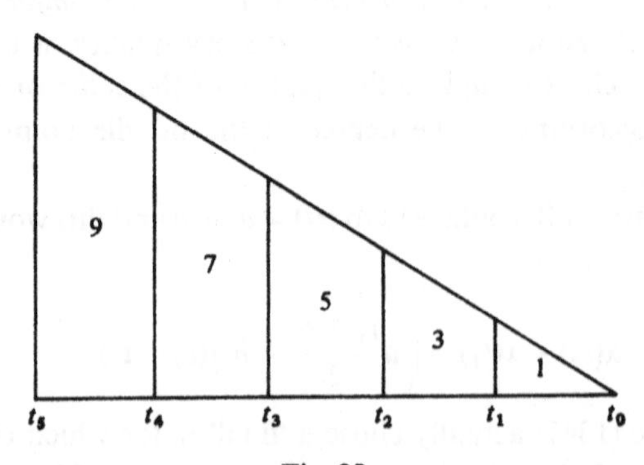

Fig. 22

quod omnis subiecti uniformiter difformis ad non gradum proportio qualitatis totius ad qualitatem partis terminatae ad non gradum est sicud proportio totius subiecti ad istam partem, proportio duplicata" ('...that, in the case of every subject, uniformly deformed till the degree zero, the ratio of the whole quality to the quality of a part, terminated at zero degree, is as the square of the ratio of the whole subject to that part'). In modern notation (t_0 is still the value for which $v(t_0)=0$),

$$\frac{x(t)}{x(t')}=\left(\frac{t_0-t}{t_0-t'}\right)^2$$

or $x(t)=c(t_0-t)^2$ (c happens to be the half of a in (5)). He has also verified that, conversely, the quadratic functions so defined satisfy equation (7).

Almost three centuries later, Galileo (Galilei 1638) applied the variant

$$\frac{x((n+1)t)-x(nt)}{x(nt)-x((n-1)t)}=\frac{2n+1}{2n-1} \tag{8}$$

of exactly this functional equation (7) and its solution to the fall of bodies. So he may have been the first to apply a functional equation to physics. Note that the classification in the Middle Ages, mentioned above, seems to have been purely theoretical, while Galileo examined actual motions, such as the fall of bodies. He used the modification (8) of (7) and, accordingly, Figure 23 (essentially the mirror image of Figure 22), since he took $t_0=0$ as the time when the motion started with 0 speed. His description of (8) is remarkably similar to that of Oresme: "...*si fuerint quotiunque tempora aequalia consequenter sumpta... quibus conficiantur spatia... ipsa spatia erunt inter se ut numeri impari....*" ('...if there are several equal times taken consecutively... in which spaces... are traversed, then these spaces will be to each other as the odd numbers....'). And so is his conclusion: "*Si aliquid mobile motu uniformiter accelerata descendat ex quiete, spatia, quibuscunque temporibus ab ipso peracta, sunt inter se in duplicata ratione eorundem temporum*

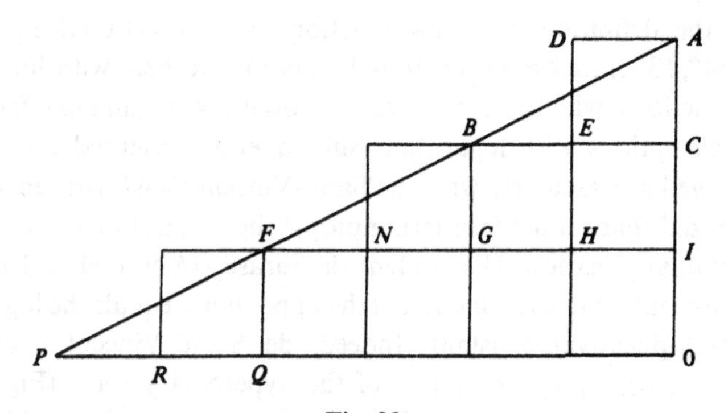

Fig. 23

nempe ut eorundem temporum quadrata". ('If an object starts with uniformly accelerated motion from immobility, the distances covered by it have as ratio the square of the ratio of the respective times, that is the ratio of the squares of the times'.) Galileo was so clearly aware of the fact that the functional equation (8) characterizes the quadratic law (although he did not prove it in both directions), that it was (8) which he verified in his experiments for the fall of bodies in order to show that the process is indeed quadratic.

He also gave another proof of the Merton theorem via Cavalieri's 'method of indivisibles'. And, as is well known, it was he who announced that 'the laws of nature are written in mathematical language'.

21.2 The functional equations of the logarithm and of the exponential function

Oresme, and even Euclid, knew and used the following properties of powers (exponential functions):

$$\frac{a^m}{a^n} = a^{m-n} \quad \text{and} \quad a^{m+n} = a^m a^n$$

(for positive integer exponents), or certain special cases (Oresme, even for some rational exponents which he had introduced). Stifel (1544) extended the definition and these equations to arbitrary (not only positive) integers.

But it was Bürgi (1620, written earlier) who used them, and Briggs (1624; cf. also Kepler 1624) who used the corresponding property of logarithms

$$\log xy = \log x + \log y$$

to *construct* logarithms (which were created by them and by Napier) using a kind of continuity and thus solving the functional equation

$$g(xy) = g(x) + g(y), \tag{9}$$

in a sense, by *constructing the solution.* We may consider this as the first example of the definition of a new function by its functional equations. (Oresme (1347, 1352) had used functional equations to deal with linear and quadratic functions which, in one way or other, were already familiar.) Technical calculations with logarithms subsequently obscured this aspect. But it reemerged spectacularly when de Saint-Vincent (1647, written around 1625) recognized that (in modern terminology) the integral of $1/x$ satisfies a similar functional equation. His student de Sarasa (1649) deduced from it that this definite integral (as a function of the upper limit) equals the logarithm up to a (multiplicative) constant. Indeed, de Saint Vincent chose the coordinates $(x_1, y_1), (x_2, y_2), (x_3, y_3), \ldots$ of the hyperbola $y = 1/x$ (Figure 24) so that the areas under the hyperbola between them be equal. By Archimedes'

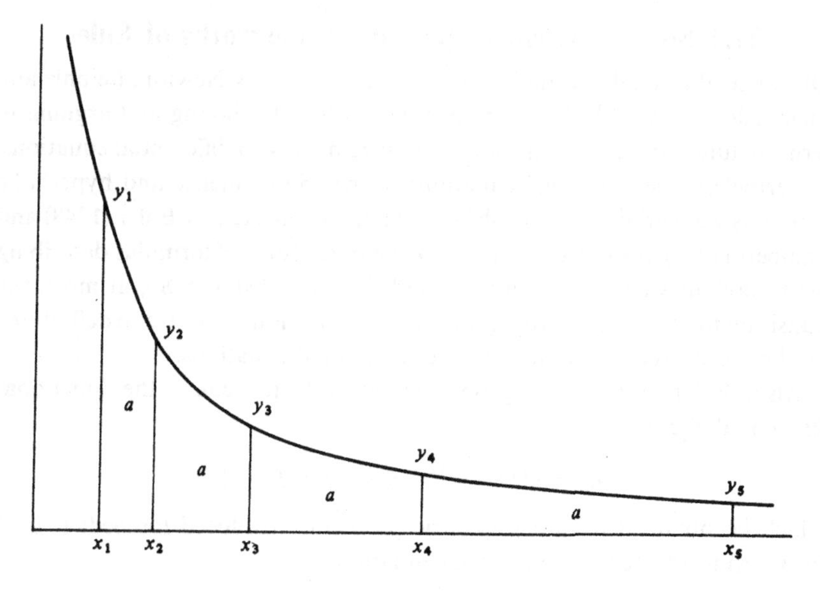

Fig. 24

exhaustion method he established

$$\frac{y_1}{y_2} = \frac{y_2}{y_3} = \frac{y_3}{y_4} \cdots,$$

that is

$$\frac{x_2}{x_1} = \frac{x_3}{x_2} = \frac{x_4}{x_3} \cdots.$$

The first equation of this chain gives $x_2^2 = x_1 x_3$, while the condition on the areas yields

$$g(x_2) - g(x_1) = g(x_3) - g(x_2),$$

where $g(x)$ denotes the area under the hyperbola between $(1, 1)$ and $(x, 1/x)$. So we get $2g(x_2) = g(x_1) + g(x_3)$, that is,

$$2g((x_1 x_3)^{1/2}) = g(x_1) + g(x_3).$$

Since $g(1) = 0$, we have $2g(x^{1/2}) = g(x)$ and so

$$g(x_1 x_3) = g(x_1) + g(x_3),$$

that is, the equation (9). Then de Sarasa stated that, *since the logarithm has the same property, therefore* $g(x)$ *is the logarithm up to a multiplicative constant* (which need not be 1 because of the different bases of the logarithms used). In de Sarasa (1649)'s words: "*unde hae superficies supplere possunt locum logarithmorum*" ('therefore these areas may be used in place of logarithms').

21.3 Some functional equations in the works of Euler

Of course, the definite formulation of calculus given by Newton, Leibniz and their followers, mainly Euler, immensely furthered thinking and arguing in terms of functions. But the applications were mostly to differential equations. Nevertheless, the functional equations of the trigonometric and hyperbolic functions were gradually assembled and finally collected by Euler (1748) and Lambert (1768), respectively, but of course in the form of formulas describing the behaviour when arguments are added, subtracted, etc. So, at most, the transition from one such property to the other without recourse to definitions can be considered as manipulation of functional equations.

Also, Euler (1764) used geometric methods to reduce the functional differential equation

$$g[x+g(x)g'(x)]^2 = g(x)^2[1+g'(x)^2]$$

(which he obtained from a geometric problem introduced for pedagogical purposes) to the genuine functional equation

$$f[x+f(x)] = f(x), \tag{10}$$

by defining $f(x) = g'(x)g(x)$. This is a functional equation in a single variable, but we mention it here because it is equation (19.34), one of the few equations in a single variable treated in some detail in this book. Euler 'solved' equation (10) using geometric methods, involving infinitesimals, which he then also applied to similar equations. He did not recognize that all continuous solutions of (10) are constant and so he could hardly have noticed that the solutions he had found were discontinuous (in the contemporary as well as in the modern sense).

On the other hand, Euler also solved functional equations in several variables. He (Euler 1768) actually mentioned parenthetically as a self-evident fact that "...*functio homogenea nullius dimensionis, quae ergo posito* $y = ux$ *abit in functionem ipsius* u" ('a homogeneous function of the 0th order...for $y = ux$ changes into a function of u'), which is the case $n = 2$, $g(u) \equiv 1$ of Theorem 20.1. As to the more general case $g(u) = u^c$, Euler (1755) gave an example of the reduction of a functional equation to a partial differential equation, since he derived the 'Euler equation' (cf. Exercise 20.10)

$$\sum_{k=1}^{n} x_k \frac{\partial f}{\partial x_k} = cf \tag{11}$$

from

$$f(x_1u, x_2u, \ldots, x_nu) = u^c f(x_1, \ldots, x_n),$$

and also showed (Euler 1770) that the general (differentiable) solution of (11) (for positive variables) is given by

$$f(x_1, x_2, \ldots, x_n) = x_1^c \Phi\left(\frac{x_2}{x_1}, \ldots, \frac{x_n}{x_1}\right).$$

21.4 Functional equations arising from physics

While the functional equations mentioned in the previous section were connected to geometry and differential equations, a more important development came from physics with the advent of what we will call the functional method, a variant of (relative) axiomatization.

Much more explicitly than the arguments of Oresme, Galileo, and de Sarasa, quoted in the first two sections of this chapter, suggest, the situation was now realized to be as follows. The laws (properties) of a physical process are known by observation and by 'common sense guided by reason' (Descartes), and eventually an explicit description (function) is found or conjectured. Of course, it must have these properties, but this is not enough, because other functions may also have the same properties. So it is important to show that *only* the given function satisfies these 'laws', and this often leads to functional equations. It is also an advantage to characterize these functions by as few (and weak) equations and other conditions as possible, because then less and easier checking has to be done to recognize the function.

A typical example, which has served as the point of departure for this book, is that of the composition of forces or, in more modern language, the determination of the resultant of two vectors. Leading mathematicians had worked on the motivation of this construction from the end of the 17th century throughout the 18th, 19th and even 20th centuries. D. Bernoulli (1726) sharply rejected an experimental 'proof' which Newton gave in 1687 through the composition of motions: "*Nihil in illa demonstratione ut falsam rejicio, sed quaedam ut obscuram, quaedam ut non necessario veram*" ('I do not say that this proof is false, but that it is obscure and not necessarily true'). Instead, he proposed a proof using geometry and analysis.

D'Alembert (1769) returned to the problem and used the functional equation

$$\phi(x+y) + \phi(x-y) = 2\phi(x). \tag{12}$$

(cf. Exercise 2.2), which he solved through a rather lengthy procedure of differentiation, first with respect to x,

$$\phi'(x+y) + \phi'(x-y) = 2\phi'(x),$$

then with respect to y,

$$\phi'(x+y) - \phi'(x-y) = 0,$$

thus obtaining

$$\phi'(x+y)=\phi'(x).$$

He still did not recognize that this means that ϕ' is constant, but could get the solution only after differentiating again both with respect to x and to y:

$$\phi''(x+y)=\phi''(x) \quad \text{and} \quad \phi''(x+y)=0, \text{ thus } \phi''(x)=0.$$

In the same work (d'Alembert 1769), the variant (cf. Exercise 14.7)

$$f(x+y)+f(x-y)=f(x)f(y) \tag{13}$$

of the functional equation (1.6)

$$g(x+y)+g(x-y)=2g(x)g(y), \tag{14}$$

$(g(x)=\frac{1}{2}f(x))$, now called d'Alembert's equation, is used for the same purpose. He found

$$f(x)=c^{xA/2}+c^{-xA/2}$$

as its solution (actually as its general, twice differentiable, not identically zero solution, though, as with (12), he did not spell out the suppositions). Since it was obvious from the composition of forces that the first zero of f is at $\pi/2$, this yielded

$$f(x)=2\cos x,$$

and that is what he needed to obtain the parallelogram rule. An improvement is due to Poisson (1804, 1805).

D'Alembert, Monge, Lagrange, Laplace and others applied (implicitly or explicitly) several other functional equations to the same problem, including equations in a single variable, such as

$$f(x)^2=2+f(2x)$$

(obtained by putting into (13) first $y=0$ with $f(x)\not\equiv 0$ then $y=x$) and

$$f(x)^2+f(a-x)^2=1.$$

Darboux (1875) used the Cauchy equation (1.2)

$$f(x+y)=f(x)+f(y)$$

for the same purpose. Schimmack (1903, 1909*b*) and Picard (1928, pp. 6–17) consolidated and improved the d'Alembert–Poisson–Darboux argument.

Actually, d'Alembert had previously (d'Alembert 1747*a*, *b*) dealt with a related functional equation (containing three unknown functions)

$$f(x+y)-f(x-y)=g(x)h(y)$$

in connection with the problem of the vibrating string, by reduction to differential equations.

D'Alembert's equation (14) was solved by Cauchy (1821) under the supposition of continuity. We will return to this later, but mention here another proof of the same result, which is of even wider significance. Andrade (1900) assumed continuity, but integrability would have been sufficient. Also, his method works for other equations (cf. Sections 5.1, 5.4 and Chapters 13, 14). He integrated (14) with respect to y and got, denoting the antiderivative of g by G,

$$G(x+y)-G(x-y)=2g(x)[G(y)-G(0)],$$

from which he first drew the conclusion that g is differentiable, then that it is twice differentiable. Now two differentiations of (14) and separation of the variables give the differential equation

$$g''(x)=Cg(x).$$

21.5 The binomial theorem and Cauchy's equations

After its successes in physics, the functional method was applied to mathematics itself.

An important example is Newton's binomial theorem,

$$(1+x)^\alpha = 1+\alpha x+\frac{\alpha(\alpha-1)}{2!}x^2+\cdots+\frac{\alpha(\alpha-1)\cdots(\alpha-n+1)}{n!}x^n+\cdots, \quad (15)$$

where α is any complex number and x a complex number whose absolute value is less than one. Because of our familiarity with the proof using Taylor's theorem, we have forgotten how important (even crucial) a result (15) was for analysis of the 18th and 19th centuries. For integer α it goes back at least to Pascal. For noninteger α, especially for $\alpha=\pm\frac{1}{2}$, this formula was essential for the local analysis of most curves in which 18th- and 19th-century mathematicians were interested. That is what Newton emphasized and why (15) carries his name.

The problem was to obtain (15) for nonrational values of α and even for complex values. Previously a proof for $\alpha=\frac{1}{2}$ was formally extended to all real and complex values of α. Euler tried many times, using derivatives, to get a correct and convincing proof, but there was a vicious circle, as (15) was used at the very beginning of the calculus to compute the derivatives of power functions (e.g. Euler 1755). Later Euler (1774) proved (15) – but only for rational exponents – by finding the solutions $f:\mathbb{Q}\to\mathbb{R}$ of the functional equation

$$f(\alpha+\beta)=f(\alpha)f(\beta) \quad (\alpha,\beta\in\mathbb{Q}).$$

Then, in one of his influential textbooks, Lacroix (1797), whose ambition was to present an encyclopedic and rigorous treatment of all contemporary analysis and its applications, picked up the problem of proving (15) – by using functional equations. He was quite modern: "*Toute quantité dont la valeur dépend d'une ou plusieurs autres quantités est dite fonction de ces dernières, soit qu'on sache ou qu'on ignore par quelles opérations il faut passer pour remonter de celles-ci à la première.*" ('Every quantity, whose value depends upon one or more quantities, is called a function of the latter, whether one knows or one does not know by what operations the first quantity is obtained from the others.')

However, like Euler, Lacroix supposed that the series expansion of $(1+x)^\alpha$ exists. He showed that the coefficients can be determined recursively from the first and, denoting the first coefficient by $f(\alpha)$ "the letter f being the abbreviation of the word function", verified that f satisfies what later came to be called the Cauchy equation

$$f(\alpha+\beta)=f(\alpha)+f(\beta) \tag{16}$$

"*équation qui exprime la propriété caractéristique de la fonction représentée par f*" ('an equation which describes the characteristic property of the function denoted by f'). Since, from the definition, $f(1)=1$ and thus from (16) $f(m+1)=f(m)+1$, Lacroix obtained $f(m)=m$ for positive integer m, then $f(m/n)=m/n$ $(m,n\in\mathbb{N})$ and, finally, $f(-q)=-f(q)=-q$, so that he has $f(q)=q$ for all $q\in\mathbb{Q}$. He promises a proof that $f(\alpha)=\alpha$ also holds for irrational and even complex α, but did not get this directly. Instead, he first turned to the series expansions of the exponential and logarithmic functions, this time using the exponential and logarithmic equations $h(x+y)=h(x)h(y)$ and

$$g(xy)=g(x)+g(y), \tag{9}$$

which yielded the coefficients in the power series of a^x and thus also of e^x. He then got the power series of $\log(1+x)$ from the logarithmic functional equation (9) and then established from $\log(1+x)^\alpha=\alpha\log(1+x)$ that

$$f(\alpha)=\alpha$$

for all (real and, intuitively, also for complex) α, thus avoiding Euler's vicious circle. No limit or continuity argument was used and, on the other hand, he did not worry about proving convergence, though he occasionally worked on accelerating it numerically.—Lacroix used d'Alembert's equation (13) similarly to obtain the power series of the cosine function (and later also of the sine).

Cauchy (1821; written earlier) went further. In this monograph, he not only tried to be as rigorous as the mathematics of his time permitted, but he organized his material in a linear way, let us say along the *euclidean pattern.* Recall that one of the achievements of the *Elements* was the idea of putting things one after another in a specific logical order. Some few basic concepts were introduced by Cauchy first (functions, limits, continuity, etc.)

in the vaguely defined setting of measurable quantities, which we call nowadays real numbers (which were not clearly defined until Dedekind and Cantor defined them around 1870). From there on, with a rare sense of economy, Cauchy tried to obtain as much as he could without introducing the derivative.

What really appears in this work, when considered in its entirety, is an elegant and systematic construction of a large bulk of the mathematical knowledge, based on the technique of convergence, which is fundamental to (15). *It is* exactly *in this construction that functional equations play an important role.* And this aspect deserves to be explained. Let us quickly recall Cauchy's method leading to the proof of (15). We define

$$f(\alpha) = 1 + \alpha x + \frac{\alpha(\alpha-1)}{2!}x^2 + \cdots + \frac{\alpha(\alpha-1)\cdots(\alpha-n+1)}{n!}x^n + \cdots. \tag{17}$$

A whole chapter of Cauchy's book was devoted to convergence criteria for such power series, both in the real and in the complex case. With their aid, convergence was established for all complex α if $|x| < 1$. Then Cauchy used a theorem asserting that continuity is invariant under limit processes in order to get the continuity of f. In this generality, Cauchy's theorem is false because he did not suppose uniform convergence. Abel (1826*b*), in a beautiful paper on the binomial theorem, noticed Cauchy's mistake and gave, with the aid of the corresponding functional equations for multiplace functions (cf. Section 5.1), what can be considered as the first rigorous proof of Newton's binomial theorem (15) for complex exponents. He even examined the convergence of the series in (15) at $x = 1$ and $x = -1$. But, surprisingly, it is Cauchy again, who in 1853 gave the exact definition of uniform convergence for which his theorem on continuous functions becomes true.

Then, using a technique he specially devised for multiplying two power series, and some combinatorics, Cauchy (1821) arrived at the same equation as Euler:

$$f(\alpha + \beta) = f(\alpha)f(\beta).$$

He had prepared his way well, because in an earlier chapter he had already solved this functional equation for real valued continuous functions f. A lengthy but clever proof was also provided for the case of complex valued functions (by fixing one particular root of unity) and this proof is necessary to get (15); it is not just a mathematical curiosity. Finally, $f(\alpha) = A^{\alpha}$ and so (15) was proved. Almost every result in Cauchy's book was used and functional equations played the key role around which technical methods were developed.

Of course, it is common knowledge that Cauchy brought rigor to analysis, in the sense that he initiated a stereotypical style in order to carry out familiar proofs in analysis via a standardized use of limits, of epsilon arguments and of inequalities (even though he made the systematic error regarding uniform properties mentioned above). His proofs are typical for the intermediate value theorem, the fundamental theorem of calculus, the definition of the definite integral of continuous functions, Cauchy's root test for the convergence of positive series, etc. Some of these innovative methods appeared in Cauchy 1821. The introductory statement of the book is famous and we quote just one passage which emphasizes the need for a rigorous treatment of analysis.

"*Quant aux méthodes, j'ai cherché à leur donner toute la rigueur qu'on exige en géométrie, de manière à ne jamais recourir aux raisons tirées de la généralité de l'algèbre. Les raisons de cette espèce, quoique assez communément admises, surtout dans le passage des séries convergentes aux séries divergentes, et de quantités réelles aux expressions imaginaires, ne peuvent être considérées, ce me semble, que comme des inductions propres à faire pressentir quelquefois la vérité, mais qui s'accordent peu avec l'exactitude si vantée des sciences mathématiques.*"

('As to the methods, I tried to give them the rigour required in geometry so that I should never have to take recourse to reasoning brought in from the generalities of algebra. Such arguments, although used rather often, mainly in going over from convergent to divergent series and from real quantities to complex expressions, seem to me just intuitive inferences which let us sometimes guess the correct results but which do not give the certainty needed so much in the mathematical sciences.')

As mentioned before, in Cauchy's book we find the first systematic treatment of some basic functional equations, the now classical Cauchy equations

$$f(x+y)=f(x)+f(y), \tag{16}$$

$$f(x+y)=f(x)f(y), \tag{18}$$

$$f(xy)=f(x)+f(y),$$

$$f(xy)=f(x)f(y), \tag{19}$$

and the d'Alembert equation

$$g(x+y)+g(x-y)=2g(x)g(y). \tag{14}$$

One of Cauchy's reasons for presenting them is, quite clearly, to prove how powerful and practical are some rather abstract concepts introduced by him at the beginning of his book: the concept of a limit quantified by inequalities and the concept of a continuous function. Functional equations serve as proof of his '*savoir-faire*'. In this way, Cauchy easily found the continuous solutions of these equations for numerical functions defined for one 'real' variable. And he really was the first to have done so without stronger regularity assumptions. More important is the fact that he emphasized the need for some regularity assumptions at all.

Here we give the statement of the problem set in Cauchy 1821 and the crucial point at which he uses limits and continuity: "*Déterminer la fonction $\phi(x)$ de manière qu'elle reste continue entre deux limites réelles quelconques de la variable x, et que l'on ait pour toutes les valeurs réelles des variables x et y*

$$\phi(x+y) = \phi(x) + \phi(y).\text{''} \tag{20}$$

('We have to determine the function $\phi(x)$ so that it remains continuous between any two values of the variable x and that one should have, for all real values of the variables x and y, (20).')

Like Lacroix, Cauchy gets $\phi[(m/n)\alpha] = (m/n)\phi(\alpha)$ for positive integer m, n; "*Puis, en supposant que la fraction m/n varie de manière à converger vers un nombre quelconque μ, et passant aux limites, on trouvera*

$$\phi(\mu\alpha) = \mu\phi(\alpha).\text{''} \tag{21}$$

('Then, supposing that the fraction m/n varies in such a manner that it converges to μ and passing to the limit, one gets (21).') From positive real μ Cauchy goes over to $\mu \leqslant 0$ and then chooses $\alpha = 1$. He verifies that $\phi(x) = ax$ satisfies (20) for arbitrary a and is continuous.

Then Cauchy gives the general continuous solution of (18) in the same way, after having established from $f(2x) = f(x)^2$ the nonnegativity of f. For this he had to define (in an appendix), A^x for real exponents, again by tending with rational exponents to the real limit x.

Later, as mentioned earlier, he extended the result to complex valued functions, defining A^x for complex A and x with great care. This led to a proof of Newton's binomial theorem (15) for complex x ($|x| < 1$). He also realized that the convergence of the series must be proved and he did prove it.

We can appreciate Cauchy's achievements in precision if we consider that some of his famous contemporaries worked with functional equations only a few years earlier and did not even reach Lacroix's degree of exactness.

Legendre (1791; this was the first of many editions of his popular geometry monograph) tried to get around the euclidean system and methods of proof

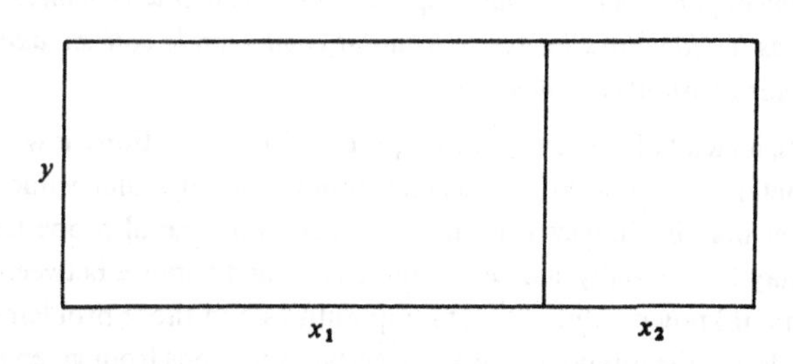

Fig. 25

by (often only tacitly assumed) *ad hoc* axioms and arguments, in which he used functional equations occasionally. In solving them, he sometimes jumped from a result, proved not even for rationals but only for integers, to all real numbers. For instance, in 'proving' the formula for the area $\Phi(x, y)$ of a rectangle with sides x and y, he (tacitly) assumed the additivity of areas and got (Figure 25)

$$\Phi(x_1 + x_2, y) = \Phi(x_1, y) + \Phi(x_2, y).$$

From this $\Phi(kx, y) = k\Phi(x, y)$ follows in the usual way for positive integers k, which Legendre writes as

$$\frac{\Phi(kx, y)}{kx} = \frac{\phi(x, y)}{x},$$

and infers that $\Phi(x, y)/x$ is independent from x. By symmetry he 'proves' similarly that $\Phi(x, y)/xy$ is independent of y and thus constant. So $\Phi(x, y) = cxy$ (c depending on the choice of units).

The above argument amounts, of course, to jumping from the consequence $\phi(kx) = k\phi(x)$ of Cauchy's equation (20) to $\phi(tx) = t\phi(x)$ for all positive real t.

Gauss (1809) gave an ingenious characterization of the normal distribution with the aid of a functional property which reduced to Cauchy's equation (see Exercises 2.1, 3.6, 3.7), but his solution of that equation shows the same gap.

In addition to his mathematical rigour, Cauchy (1821) showed a deep appreciation of the nature of functional equations. The following fundamental statements by Cauchy contain his views on the nature of this subject:

"*Lorsque au lieu de fonctions entières on considère des fonctions quelconques, dont on laisse la forme entièrement arbitraire, on ne peut plus réussir à les déterminer d'après un certain nombre de valeurs particulières quelque grand que soit ce même nombre; mais on y parvient quelquefois dans le cas où l'on suppose connues certaines propriétés générales des fonctions... il y a une grande difference entre les questions où il s'agit de calculer les valeurs inconnues de certaines quantités et des questions dans lesquelles on se propose de découvrir la nature inconnue de certaines fonctions d'après des propriétés données. En effet dans le premier cas, les valeurs des quantités inconnues se trouvent finalement exprimées par le moyen d'autres quantités connues et déterminées, tandis que dans le second cas les fonctions inconnues peuvent comme on le voit ici, admettre dans leur expression des constantes arbitraires...*"

('If one considers, instead of entire functions, quite arbitrary functions, it will no longer be possible to determine these from a certain number of particular values, however large this number may be; however, if one knows certain general properties of these functions, one may occasionally succeed... there is a big difference between problems of calculating the unknown values of certain quantities and those problems in which one attempts to derive the unknown nature of certain functions from given properties. Indeed, in the first case the values of the unknown quantities are expressed by means of other known and specific quantities, whereas in the second case the expression of the unknown functions can, as we see here, admit arbitrary constants.')

21.6 Cauchy equations after Cauchy

In sketching more recent developments in the study of the Cauchy equation and other related functional equations, we must digress, of course, from the chronological order, even more than we have previously done in these historical notes, because more and more equations were investigated simultaneously.

Darboux (1875) applied Cauchy's equation (20) to the problem of the composition of forces, with which we have dealt in Section 4 of this chapter, at the same time as he reduced the regularity supposition from continuity everywhere to *continuity at a point.* He also showed that it can be replaced by the requirement that the solution be *monotonic* or nonnegative for nonnegative variables. Five years later, Darboux (1880) applied the same equation to what he called the fundamental theorem of projective geometry (cf. Theorem 6.3). This time he supposed *boundedness on an interval.* Comprehensive treatments of this and other equations and their applications were given in Pincherle 1906, 1912 (the latter is better).

Further weakening of the regularity supposition for obtaining the solution $\phi(x) = ax$ of Cauchy's equation followed rapidly, the most important being the reduction to measurability (Fréchet 1913; Sierpinski 1920*a*, *b*; Banach 1920) or to majorability by a measurable function (Sierpinski 1924), in the then new theory of Lebesgue measure, until a theorem of Steinhaus (1920) (cf. Theorem 2.6) made it possible to require only *boundedness on a set of positive measure* (Kormes 1926; Ostrowski 1929). Many of the previously known results were proved again in new ways by Kac (1937; cf. the proof of Theorem 5.4 in the present book; clearly this is an ingenious improvement on the method of Andrade 1900, which we have mentioned in Section 4 above), Weil (1940), Alexiewicz–Orlicz (1945), Kestelman (1946), etc. Piccard (1939) proved the topological analogue of Steinhaus' theorem (see Exercise 16.21), thus making it possible (Mehdi 1964) to determine all solutions of Cauchy's equation which are *bounded on a nonmeager set.*

A completely new (and at that time surprising) aspect was introduced by Hamel (1905) when, with aid of the basis of real numbers over the rationals which he constructed *for this purpose* (using the axiom of choice), he showed that Cauchy's equation has *discontinuous* solutions and determined *all* of them. Several remarkable properties of the Hamel bases were the objects of many subsequent works, for instance Burstin 1916, Jones 1942*b*, Erdös 1963, etc.

Even without knowing about Hamel bases, or that the axiom of choice is needed for the construction of such bases, it is clear that no discontinuous solution of Cauchy's equation can be constructed without using the axiom of choice. Indeed, every discontinuous solution is nonmeasurable and no nonmeasurable function can be constructed without the use of the axiom of choice (or one of its equivalents). However, the fact that the graph of any discontinuous solution of the Cauchy equation is everywhere dense in $\mathbb{R}^2$ (Theorem 2.3) can be proved without recourse to the axiom of choice or to Hamel bases. Hamel (1905) himself actually used his basis to obtain this result and he formulated it as a general criterion under which all solutions of the Cauchy equation are of the form $f(x)=Ax$: "*Weiß man daher, daß in einem ganz beliebigen Bereiche der Variabeln x eine Lösung $f(x)$ der in Rede stehenden Funktionalgleichung irgend einem Werte B nicht beliebig nahe kommt, so darf man schließen, daß $f(x)$ stetig ist und die Form Ax hat.*" ('Thus, if one knows that in some arbitrarily chosen domain of the variable x a solution $f(x)$ of this functional equation does not get arbitrarily close to a value B, then one can infer that $f(x)$ is continuous and has the form Ax.') However, the proof does not really depend on Hamel bases and the same proof without them goes back at least to Wilson 1899 (concerning equation (18)) and to San Juan 1946 (concerning (16)).

The solutions of each Cauchy equation establish endomorphisms between algebraic structures over the reals, with addition or multiplication as *single* operations. If, however, the *ring* (or field) of real numbers is considered (with *both* addition and multiplication), no Hamel bases are needed to find all endomorphisms: only the trivial ones (the identity and the identically 0 mappings) exist, that is, $f(x)=x$ and $f(x)=0$ are the only common solutions of (16) and (19) together, without any regularity assumptions. This was shown by Darboux (1880; see Proposition 5.6 in the present book). But this is not so (even if the conjugate is also considered) in the ring (field) of complex numbers as counterexamples constructed by Segre (1947) and Kestelman (1951; written earlier) show (Theorem 5.7). It is interesting to note that Lebesgue (1907*a*) had used a previous (incomplete) construction of such a counterexample (based on the continuum hypothesis) to explain the difference between the axiomatic ('*idéaliste*') and intuitionist ('*empiriste*') points of view.

The set of prime numbers is a countable analogue of a Hamel basis in the case of the logarithmic equation on the domain $\mathbb{N}^2$:

$$f(mn)=f(m)+f(n) \quad (m,n\in\mathbb{N}), \tag{22}$$

which defines the *completely additive number theoretical functions*. So the situation is different from $\mathbb{R}^2$; on $\mathbb{N}^2$ it is easy to determine the *general* solution, without further conditions (see Sections 16.3, 16.5) but the proof that $f(n)=c\log n$ gives the only solutions *under some regularity conditions* such as monotonicity or $\lim_{n\to\infty}[f(n+1)-f(n)]=0$ (Erdös 1946) or

$$\varliminf_{n\to\infty}[f(n+1)-f(n)]\geqslant 0$$

(conjectured by Erdös 1957, proved by Kátai (1967) and Máté (1967)) is somewhat trickier (cf. Theorems 16.15, 16.17 and Exercise 16.6). The latter results were proved even for additive number theoretical functions, that is, for solutions of (22) under the restriction that m and n be *relative prime* numbers. This equation

$$f(mn)=f(m)+f(n),\quad (m,n)=1$$

may be considered as one of the first examples of an essentially *conditional equation* or equation with restricted domain ((14) for $0\leqslant y\leqslant x\leqslant \pi/4$ or (20) for $x\geqslant 0$, $y\geqslant 0$ were more trivial examples).

Another class of conditional Cauchy equations was introduced in the form of alternative equations (e.g. Vincze 1964*b*; Światak 1965) like $f(x+y)^2=(f(x)+f(y))^2$, that is,

$$f(x+y)=f(x)+f(y)\quad \text{if } f(x+y)+f(x)+f(y)\neq 0$$

(cf. Exercises 6.1–6.5).

The related research on *extensions* and quasi-extensions of Cauchy and other equations from subsets to all of $\mathbb{R}^2$ (or more general domains) started with the paper by Daróczy–Losonczi 1967.

21.7 Further equations

Cauchy's exponential equation (18) was used by Bolyai (1832) and Lobachevskiĭ (1837, 1840, 1856), the founders of nonEuclidean geometry, for the derivation of the relationship $s'=se^x$ between the arc lengths of two oricycles with the same centre, whose distance is x. Lobachevskiĭ also derived the formula

$$f(x)=\tan\frac{\Pi(x)}{2}=e^{-x/k}$$

for the angle of parallelism from the functional equation (cf. Exercise 6.7)

$$f(x)^2=f(x+y)f(x-y),$$

which he solved by a method similar to that of Lacroix and Cauchy.

While Babbage's important contributions were mainly to functional equations in a single variable, we quote here a remarkable statement from Babbage 1815: "I am still inclined to think that the evolution of functional equations must be sought by methods peculiarly their own."

Abel made very significant contributions to functional equations. In addition to his work on the binomial series (Abel 1826*b*), which we have already mentioned, he published three other important papers on the subject in the years 1823–27. In the first (Abel 1823) he gave a general method for solving, by differentiation, functional equations containing several unknown functions, a method which, however, needed some clarification (Kiesewetter 1957). In the second (Abel 1826*a*), he solved the system of functional equations

$$F[x, F(y, z)] = F[z, F(x, y)] = F[y, F(z, x)] = F[x, F(z, y)] = F[z, F(y, x)] = F[y, F(x, z)],$$

which is a combination of associativity and commutativity (allegedly this is where the name 'abelian group' comes from). Finally, in the third paper (Abel 1827), the *composite* functional equation

$$\phi(x) + \phi(y) = \psi[xf(y) + yf(x)]$$

(which he writes simply as $\phi x + \phi y = \psi(xfy + yfx)$) is solved for the *three* unknown functions f, ϕ, ψ. All three papers consciously use the method of reducing functional equations to differential equations and contain important basic comments on this method. We quote here from the last-mentioned paper (with correction of an apparent misprint):

"*Durch das Verfahren, welches hier oben die Functionen gab, die der Gleichung* $\phi x + \phi y = \psi(xfy + yfx)$ *genugthun, lassen sich auch die unbekannten Functionen in jeder anderen Gleichung mit zwei unabhängig veränderlichen Größen finden. In der That lassen sich durch wiederholte Differentiationen nach den beiden veränderlichen Größen, so viel Gleichungen finden, als nöthig sind, um beliebige Functionen zu eliminieren, so daß man zu einer Gleichung gelangt, welche nur noch eine dieser Functionen enthält und welche im Allgemeinen eine Differential-Gleichung von irgend einer Ordnung sein wird. Man kann also im Allgemeinen alle die Functionen vermittelst einer einzigen Gleichung finden. Daraus folgt, daß eine solche Gleichung nur selten möglich sein wird. Denn da die Form einer beliebigen Function die in der gegebenen Bedingungs-Gleichung vorkommt, vermöge der Gleichung selbst, von den Formen der andern abhängig sein soll, so ist offenbar, daß man im Allgemeinen keine dieser Functionen als gegeben annehmen kann. So z.B. könnte der obigen Gleichung nicht mehr genug gethan werden, wenn* $f(x)$ *eine andere als die gefundene Gestalt hätte.*"

('By means of the procedure which yielded the functions satisfying the equation $\phi x + \phi y = \psi(xfy + yfx)$, the unknown functions in every other equation with two independent variables can also be found. Indeed, as many equations can be found by repeated differentiations with respect to the two independent variables as are necessary to eliminate arbitrary functions. In this manner, an equation is obtained which contains only one of these functions and which will generally be a differential equation of some order. Thus, it is generally possible to find all the functions by means of a single equation. From this it follows that such an equation can exist only rarely. Indeed, since the form of an arbitrary function appearing in the given conditional equation, by virtue of the equation itself, is to be dependent on the forms of the others, it is obvious that, in general, one cannot assume any of these functions to be given. Thus, for example, the above equation could not be satisfied if $f(x)$ had any other form than that which was found.')

The fifth and thirteenth problems of Hilbert (1900) should also be mentioned. The following remark by Hilbert (1900), concerning the fifth problem, may be repeated here:

"*Überhaupt werden wir auf das weite und nicht uninteressante Feld der Funktionalgleichungen geführt, die bisher meist nur unter Voraussetzung der Differenzierbarkeit der auftretenden Funktionen untersucht worden sind. Insbesondere die von Abel mit so vielem Scharfsinn behandelten Funktionalgleichungen... und andere in der Literatur vorkommenden Gleichungen weisen an sich nichts auf, was zur Forderung der Differenzierbarkeit der auftretenden Funktionen zwingt...*"

('Specifically, we come to the broad and not uninteresting field of functional equations, hitherto largely investigated by assuming differentiability of the occurring functions. Equations treated in the literature, particularly the functional equations treated by Abel with such incisiveness, show no intrinsic characteristics that require the assumption of differentiability of the occurring functions....')

Several functional equations with one or more unknown functions fascinated Schweitzer, who published very little other than abstracts in the *Bulletin of the American Mathematical Society* between 1911 and 1920. However, these abstracts were written in sufficient detail to reveal the author's skill and inventiveness. He was also planning to write a bibliography of functional equations but he seems to have got only to a classification (Schweitzer 1918) of the subject which, however, is interesting in itself.

The equation

$$F[f(x+y), f(x), f(y)] = 0,$$

with rational F, can also be regarded as a functional equation with several unknown functions. It was used by Weierstrass as a starting point in the development of the theory of elliptic functions (cf., for example, Rausenberger 1884; Schwarz 1892). The functional equation (cf. Exercise 15.9)

$$f(x+y) = G[f(x), f(y)],$$

in general (without supposing that G is algebraic) was investigated by Cacciopoli (1928), among others. Of particular interest is the result acheived by Pexider 1903, who solved in an entirely elementary manner the generalizations of Cauchy's functional equations (see Section 4.3)

$$f(x+y) = g(x) + h(y),$$

$$f(x+y) = g(x)h(y), \quad f(xy) = g(x) + h(y), \quad f(xy) = g(x)h(y),$$

for three unknown functions.

Functional equations with several unknown functions were also investigated in the comprehensive article Sincov 1903*a*, which contains many interesting equations.

Wilson (1919) presented a treatment of the connections among many functional equations (and their general solutions; cf. Chapter 13) which are satisfied by the trigonometric and hyperbolic functions, unsurpassed until recently. Most of these equations are special cases of (14.1)

$$f(x+y)+g(x-y)=\sum_{j=1}^{n} f_j(x)g_j(y), \tag{22}$$

which was solved in this generality but under integrability conditions (with some gaps) by Satô (1928) (Kemperman (1957) made the next important contribution to the study of such equations). The solution of the special case $g=0$, that is, of

$$f(x+y)=\sum_{j=1}^{n} f_j(x)g_j(y) \tag{23}$$

under similar conditions goes back farther: Stephanos 1904; Levi–Cività 1913; Stäckel 1913 (for some necessary improvements see Radó 1962). A method for finding the general solutions of (23) without regularity suppositions was constructed by Vincze (1962*a*, *b*, 1963*a*, *b*). A general uniqueness theorem on solutions of another wide class of equations was proved by Haupt (1944).

We also mention Schur 1901, 1928; Perron 1942; and Hosszú 1960, among other works concerning the analogue of Cauchy's power equation for matrices

$$F(\mathbf{XY})=F(\mathbf{X})F(\mathbf{Y})$$

($\mathbf{X}, \mathbf{Y}$ are nth order matrices), and Fréchet 1938 concerning the solution of the conditional matrix equation

$$\mathbf{F}(x,z)=\mathbf{F}(x,y)\mathbf{F}(y,z) \quad (x\leqslant y\leqslant z)$$

(the values of $\mathbf{F}$ are matrices) and its application to probability theory.

It is just one step further to consider

$$f(x+y)=f(x)\circ f(y),$$

with x, y in a semigroup and $f(x)$ a continuous endomorphism on a Banach space ($\circ$ denoting the composition of such endomorphisms) and similar equations for operators. The monograph by Hille 1948 (enlarged as Hille–Phillips 1957) opened the way to the rapidly developing theory of operator semigroups.

Characterizations of arithmetic and quasiarithmetic means, which started with Schiaparelli 1868, resp. Kolmogorov 1930 and Nagumo 1930, gave new impetus to exploring equations for multiplace functions (cf. Chapters 15 and 17 for more recent developments).

21.8 Recent developments

In these historical notes we have avoided, on one hand, the mention of our own results and, on the other, mention of developments in the last two decades. We intend to continue to do so here (although it was not so in other chapters) in the first respect, but, as promised, we want to give at least a sketch of some important directions of recent developments (without mentioning specific authors and works, mainly because of their large numbers).

Cauchy equations continue to fascinate researchers (cf. Chapters 2–4) and there are many more fine points to investigate. There have been several recent successes concerning *generalizations* to more general types of equations such as (22) and, further, with respect to *general solutions*, improving the *regularity* of solutions (cf. Chapter 14), their *existence* and, mainly, *uniqueness* (cf. Chapter 15). Such equations and problems have, as indicated throughout this book, spilled over to *algebra* and *functional analysis* by the consideration of more general and abstract domains and ranges. And, of course, *conditional equations* (cf. Chapters 6, 7, 16) or equations on restricted domains are now the subject of a great volume of research.

We mention three fields which we have just touched on in this book, because they are only marginally related to functional equations in several variables, but where there is also vigorous research activity going on. One concerns functional equations satisfied by and, preferably, characterizing *special functions* (such as elliptic functions, dilogarithms, etc., see, e.g., Exercises 1.8 and 14.11). The other is *iteration* theory (cf. Section 19.2) with its wide ranging connections and promises of applications to dynamical systems, astronomy, automata, combinatorics, etc. While the most important functional equations connected to iteration (those of Abel, Schröder, etc.) are in a single variable, the continuous iterate $f^t(x) = F(t, x)$ is always a solution of the so called translation equation (cf. Exercise 17.9)

$$F[s, F(t, x)] = F(s + t, x),$$

which is an equation in several (three) variables (indeed this and $F(1, x) = f^1(x) = f(x)$ (given) define the continuous iterates and lead to Abel's equation $g[f(x)] = g(x) + 1$). The third field which we only mention is that of *functional inequalities* (cf. Exercises 2.9, 5.4, 5.12, 5.13, 5.14, 5.16, 10.1, 13.12, 13.13), in particular those describing the '*stability*' (or lack of it) of Cauchy's, d'Alembert's and other equations, for instance

$$|g(x + y) - g(x) - g(y)| < \varepsilon \quad \text{or} \quad |f(x + y) - f(x)f(y)| < \varepsilon,$$

(cf. Exercises 11.27, 11.28, 11.29, 11.44), which have applications to characterizations of probability distributions.

Applications of functional equations have a broader range than ever and are as vigorously pursued as ever. We mention among applications within mathematics, for example, some applications to geometry (cf. Section 4.2 and Chapter 6), harmonic analysis (cf. Sections 5.3, 7.2 and 10.5), functional analysis (cf. Chapters 10, 11 and 19), and probability theory (cf. Chapter 12). Outside mathematics or on its borders there are, among others, applications to information theory (cf. Chapter 3, Sections 5.4 and 16.4), physics (cf. Chapter 8 and Section 11.4), economics (cf. Chapter 20) and decision theory (cf. Section 7.3). There are many more applications to these fields than we were able to discuss in the chapters and sections mentioned above in brackets. There are also applications to several other fields of mathematics and the natural, social and behavioral sciences. Moreover, new fields of application continue to open up and new applications are found in the 'old' fields. This, and the progress made in the theory of functional equations as a result of the inner dynamics of the subject, shows the strength and vigour of this theory and its promise for the future.

Notation and symbols

The symbols := or =: mean that the side of the equation, where the colon is, is *defined* by the other side.

As usual, the set of *positive integers* is denoted by $\mathbb{N}$, that of *nonnegative integers* by $\mathbb{N}_0$, that of *integers* by $\mathbb{Z}$, that of *rationals* by $\mathbb{Q}$, that of *real* numbers by $\mathbb{R}$ and that of *complex* numbers by $\mathbb{C}$. Also, $\operatorname{Re} z$ is the *real part*, $\operatorname{Im} z$ the *imaginary part* of z. Furthermore, $\mathbb{R}_+ = \{x|x>0\}$, $\bar{\mathbb{R}}_+ = \{x|x\geqslant 0\}$.

If $S_1, S_2, \ldots, S_n$ are sets, then their *cartesian* product is

$$\mathop{\times}_{k=1}^{n} S_k = S_1 \times S_2 \times \cdots \times S_n = \{(x_1, x_2, \ldots, x_n) | x_k \in S_k \quad (k = 1, 2, \ldots, n)\}.$$

In particular, $S_1^n := \{(x_1, x_2, \ldots, x_n) | x_k \in S_1 \ (k = 1, 2, \ldots, n)\}$. Furthermore,

$$\sum_{k=1}^{n} S_k = S_1 + S_2 + \cdots + S_n = \{x_1 + x_2 + \cdots + x_n | x_k \in S_k \quad (k = 1, 2, \ldots, n)\}$$

is the *sum* of sets (in particular, $S_1 + x = S_1 + \{x\} = \{x_1 + x | x_1 \in S_1\}$), $S_1 - S_2 = \{x_1 - x_2 | x_1 \in S_1, x_2 \in S_2\}$ their *difference*,

$$\prod_{k=1}^{n} S_k = \{x_1 x_2 \cdots x_n | x_k \in S_k \quad (k = 1, 2, \ldots, n)\}$$

their *product* (if the respective operations are defined) and, if $f\colon \times_{k=1}^{n} S_k \to T$ is a function (T a set too), then $f(S_1, S_2, \ldots, S_n) = \{f(x_1, x_2, \ldots, x_n) | x_k \in S_k \ (k = 1, 2, \ldots, n)\}$ ($\subseteq T$) is the *image* of f on these sets. In particular $f(S) = \{f(x) | x \in S\}$. On the other hand, if S is a set and $x \in S$ then

$$S \backslash \{x\}$$

denotes the set of all elements of S *except* x. We use the notation

$$S^* = S \backslash \{0\}$$

except where S^* denotes the dual space of S (which is explicitly stated at those places; P^* denotes the adjoint operator of P). The *characteristic function* of S is denoted by χ_S.

We use g^n as the nth *iterate* of g and so $g^n(x)$ is its value ($g^n(x) = g(g^{n-1}(x), g^1(x) = g(x))$, while $g(x)^n$ denotes the nth power of g.

Vectors in $\mathbb{R}^n$ are often denoted by bold faced letters and so are vector valued functions; $|\mathbf{p}|$ is the *length* of the vector $\mathbf{p}$ and $(\mathbf{p},\mathbf{q})\measuredangle$ the *angle* of $\mathbf{p}$ and $\mathbf{q}$. For *orthogonality* of vectors the symbol $\perp$ is used. *Matrices* are often denoted by capitals or by their entries, as in $[a_{ij}]$. If X is a matrix, then $\det X$ is its determinant.

The *uniform norm* of f is denoted by $\|f\|_\infty$ (sometimes $\|f\|$ for short) while $\langle .,. \rangle$ (or $(.,.),\ldots$) indicates an *inner product*.

A bold faced number (e.g. $\mathbf{0}$ or $\mathbf{1}$) indicates a *constant* function. If not otherwise indicated, log denotes the *natural logarithm*, Li the *dilogarithm*. Furthermore, as usual, a.e. means *almost everywhere*. We write, of course, $A \equiv B \pmod a$ if there exists an $n \in \mathbb{Z}$ such that $A = B + na$. The *Fourier integral* $\int_{-\infty}^{\infty} f(x)\exp(2\pi iax)\,dx$ is denoted by $F(f)$ or by $\hat{f}(a)$. The *convolution* of the functions f and g is denoted by $f * g$. The *kernel* of f, $\ker f$, is the set of zeros of f.

For *ordering* we use the symbol $\prec$. Several symbols, such as $\circ, *, \square$ and also numbers are used to indicate *binary operations*. The *direct product* is denoted by $\oplus$. When talking about *fields*, we do not suppose commutativity (so we usually do not use names like skew fields or sfields either); if commutativity is supposed then the structure is called a *commutative field*.

$\mathscr{A}$: the set of almost periodic functions, p. 148

$A(Z) = \{x \mid \exists y:(x,y)\in Z\} \cup \{y \mid \exists x:(x,y)\in Z\} \cup \{x+y \mid (x,y)\in Z\}$ p. 84

B: spectrum, p. 60

$(\mathscr{B}, *)$: the Bohr group, p. 149

$B_1(X,Y) = \{P: C(X)\to C(Y) \mid P \text{ linear}; \|P\| \leqslant 1\}$, p. 130

$B_1^+(X,Y) = \{P \in B_1(X,Y) \mid \|P\| = 1, P(\mathbf{1}) = \mathbf{1}\}$ ($= B_1^+$ for short), p. 130

$\mathscr{C}F$: the complement of F, p. 81

$C(X)$: Banach space of all continuous functions $f: X \to \mathbb{C}$ equipped with the topology of the uniform norm, p. 130

$C_{\mathbb{R}}(X)$: the subset of $C(X)$ containing only real-valued $f: X \to \mathbb{R}$, p. 130

$C_0(\mathbb{R}) = \{f \mid f:\mathbb{R}\to\mathbb{C}, f \text{ continuous}, \lim_{x\to\pm\infty} f(x) = 0\}$, p. 98

$C_{g,h} = \{(g(x), h(x)) \mid x \in \mathbb{R}\}$, p. 254

C_x: cone of points in $\mathbb{R}^4$ with $x \in \mathbb{R}^4$ as vertex; $C_0 = C$, p. 120

$C_1 = \{(x,y) \mid 0 < y < -x\} \cup \{(x,y) \mid 0 > y > -x\} \cup \{(0,0)\}$, p. 257

δ_{x_0}: Dirac measure at x_0, p. 134

$\Delta_\alpha^1 f(x) = f(x+\alpha) - f(x)$, $\Delta_\alpha^n f(x) = \Delta_\alpha^1(\Delta_\alpha^{n-1} f(x))$ $(n = 2,3,\ldots)$, p. 136

$\mathscr{E}(x_1,x_2) = \{\alpha_1 x_1 + \alpha_2 x_2 \mid \alpha_1, \alpha_2 \in \mathbb{Q}\}$, p. 122

$e_\lambda(x) = e^{i\lambda x}$, p. 61

$G\hat{}$: the dual of the topological group G, p. 110

G/G_0: quotient group, p. 262

$H_a = \{x \mapsto \alpha(x+a) - a \mid \alpha \in \mathbb{R}\backslash\{0\}\}$, p. 313

$H_n(p_1,\ldots,p_n)$: the Shannon entropy, p. 67

$H_n^{(\alpha,D)}(p_1,\ldots,p_n)$: the Rényi entropies, p. 268

$I_x = \{t \mid t \in \mathbb{R}, 0 < |t| < x\}$, $I'_x = \{t \mid t \in I_x, tx > 0\}$, p. 259

$L^\infty(\mathbb{R})$: the set of Lebesgue measurable, essentially bounded functions, p. 96

$M_n(F)$: the groupoid (monoid) of $n \times n$ matrices over F, p. 41

$M(X)$: the set of all bounded Radon measures on X, p. 146

$M_1(X)$: the set of complex valued Radon measures on X of norm $\leqslant 1$, p. 132

$M_1^+(X)$: the set of nonnegative Radon measures on X with total measure 1, p. 132

$\mathscr{P}$: the complex linear space of all trigonometric polynomials, p. 60

π: a canonical epimorphism, p. 262

$\Pi_1(Z) = \{x \mid \exists y:(x, y) \in Z\}$, $\Pi_2(Z) = \{y \mid \exists x:(x, y) \in Z\}$, p. 84

$\mathbb{Q}(E)$: the $\mathbb{Q}$-convex hull of E, p. 124

$Q(f) = \lim\limits_{T \to \infty} \dfrac{1}{2T} \displaystyle\int_{-T}^{T} f(y)dy$, p. 154

$\mathbb{R}_d$: $\mathbb{R}$ equipped with the discrete topology, p. 98

$\operatorname{sign} t = 1$ if $t > 0$, $= -1$ if $t < 0$ and $= 0$ if $t = 0$ (if defined at 0), p. 26

$t(Z) = \{x + y \mid (x, y) \in Z\}$, p. 84

$U_r(a, b) = \{(x, y) \mid |x - a| + |y - b| \leqslant r\}$, p. 90

$\hat{x}(f) = f(x)$, p. 60

ξ: a lifting, p. 262

$\mathscr{Z}$: a mapping associating subsets of $\mathbb{R}^2$ to functions $g: \mathbb{R} \to \mathbb{R}$. The constant mapping $\mathscr{Z}$ is denoted by Z, p. 85

$\mathscr{Z}_1(g) = t^{-1}(\mathscr{C}\operatorname{Ker} g)$, p. 85

$\mathscr{Z}_2(g) = \mathbb{R} \times g(\mathbb{R})$, p. 86.

Hints to selected 'exercises and further results'

Note: We usually do not give hints to exercises which contain easily accessible references or where the solution seems to us quite straightforward or can be done by techniques very similar to those described in the text.

Chapter 1

5. If there exists a z_0 such that $x \mapsto F(x, z_0)$ is nowhere 0, substitute $z = z_0$ into the equation. See also Exercise 6 and reference therein.
7. Show that $f(x/2)^2 = f(x)^2$.
11. Use $\lim_{x \to 0} f(x)$ and $\lim_{x \to \infty} f(x)$ (finite or infinite) and the fact that f has a unique fixed point x_0 ($f(x_0) = x_0$).

Chapter 2

2. Show that $x \mapsto f(x) - f(0)$ is odd, then interchange x and y.
3. If $f(0) \neq 0$ then $f(0) = 1$ may be assumed without loss of generality. Prove then, that $f(x) > 0$ and take logarithms in order to be able to apply the result of Exercise 2.
4. Use the result of Exercise 3.
8. Show that $f \geqslant 0$ on an interval and apply Corollary 5.
10. By using $1/[x(x-1)] = 1/(x-1) - 1/x$ show that $f(x^2) = 2xf(x) - x^2 f(1)$.
12. Use the results of Exercise 11.
18. If ϕ is not constant, show that f has no values strictly between $-2\,a^{1/2}$ and $2\,a^{1/2}$.
21. For the 'if' part, use the Cauchy–Schwarz inequality with $(x, y) = f(x)g(y) + f(y)g(x)$ which is 'almost' an inner product (it does not satisfy $(x, x) = 0 \Leftrightarrow x = 0$).
25. Prove by contradiction, using Theorems 6 and 10.
26. Notice that every point in $[0, 1]$ is midpoint of two points in the Cantor set.
27. Use Theorem 10 (the Hamel construction).

Chapter 3

5. Write $u = x^2$, $v = y^2$ ($u, v \in \bar{R}_+$), $g(u) := f(u^{1/2})$.
6. Use Exercise 5 for $|f|$ and the continuity for $f/|f|$ ($f \neq 0$).
11. If $\lambda \neq 0$, reduce the equation to $g(x+y) = g(x)g(y)$.

Chapter 4

1. Reduce the equation to (2.1).
2. Reduce the equation to (25).
3. Show (in the case of $n = 2$) for elements a and b in the centre of S_1 or S_2, respectively, that $f(a,b)$ is in the centre of $f(S_1,S_2) = \{f(x,y) \mid x \in S_1, y \in S_2\}$ and $-f(a,b)$ commutes with every element of $f(S_1,S_2)$.
6. Yes.
7. Show that $\lim_{\|x\| \to \infty} (\phi(x)/\phi(x+\alpha)) = A(\alpha)$ exists for all $\alpha \in \mathbb{R}^n_+$ and that it satisfies $A(\alpha+\beta) = A(\alpha)A(\beta)$ and $g(\alpha+\beta) = A(\alpha)g(\beta) + g(\alpha)$ for all $\alpha, \beta \in \mathbb{R}^n_+$.
17. Show that $f'(x)$ exists and is equal to $g(x)$ for all $x \in I$. Prove that f is as often differentiable as you need it to be and derive a differential equation for f. See Exercise 18 for another proof.
19. Every linear operator on a finite dimensional normed space is bounded.
21. Prove that m, defined by $m(x) = [f(x)+1]/[f(x)-1]$ when $f(x) \neq 1$, is a homomorphism of the multiplicative group of $\mathbb{R}^* = \mathbb{R}\backslash\{0\}$.
25. Use the Gelfand–Schneider theorem which states that α^β is transcendental whenever α is algebraic ($\alpha \neq 0, 1$), and β is irrational and algebraic.
27. Use Exercise 26.

Chapter 5

5. Show that the image cannot be $\mathbb{R}$ and conclude by using the density of $z\mathbb{Q} + \mathbb{Q}$ in $\mathbb{C}$ for any $z \in \mathbb{C}\backslash\mathbb{R}$.
7. No. (Such an endomorphism would be continuous on the given straight line. Show that this would imply continuity on all of $\mathbb{C}$.)
13. Define $G(x) = g(x+x_0) + f(-x_0)$ and prove that $-G(-h)/h \geqslant [f(x+h) - f(x)]/h \geqslant G(h)$ for $h > 0$.
17. No for $\phi: \mathscr{P} \to \mathbb{C}$. Yes for $\phi: \mathscr{A} \to \mathbb{C}$.
18. Apply Theorem 8.
19. Apply Theorem 8.
21. Apply Theorem 8.
22. Apply Theorem 8.

Chapter 6

5. $f \equiv 0$ or $f(x) = \varepsilon(x)e^{h(x)}$ with $\varepsilon: \mathbb{R} \to \mathbb{R}$ and $\varepsilon(x)^2 = 1$, $h(x+y) = h(x) + h(y)$.
7. If $f(0) \neq 0$, show that $g(x) = f(x)/f(0)$ satisfies $g(u+v) = g(u)g(v)$.
8. $\ker f$ is a bilateral ideal.
10. Examine the kernel of f.

Chapter 7

1. Let $Z = \mathbb{Q}^+ \times \mathbb{Q}^+\sqrt{2}$ where $\mathbb{Q}^+$ is the set of all positive rational numbers, $g(x) = x^2$ if $x \in \Pi_1(Z) \cup \Pi_2(Z)$ and $g(z) = x^2 + y^2$ if $z \in t(Z)$.
2. Use the same set Z as in the solution of Exercise 1, define g by $g(x) = x + 1$ for $x \in \Pi_1(Z) \cup \Pi_2(Z)$ and by $g(z) = z + 2$ for $z \in t(Z)$.
3. Theorem 8 shows that $]1,2[\times]1,2[$ is uniquely quasiexpandable. Show that the

result extends to $[1,2] \times [1,2]$ and that in this case the quasiextension is an extension.

5. $]2,3[^2 \cup]4,6[^2$ in $\mathbb{R}^2$.
7. $f(x) = h(\Pi_H(x)) + g(x - \xi(\Pi_H(x)))$ where H is a subgroup of G, Π_H the canonical epimorphism $G \to G/H$, $g: H \to H$ an additive function, $\xi: G/H \to H$ a lifting, and $h: G/H \to H$ a mapping such that $h(0) = g(\xi(0))$.
8. With the notations of the solution of the previous exercise, $f(x) = \zeta(\Pi_H(x)) + g(x - \zeta(\Pi_H(x)))$, where $\zeta: G/H \to G$ is also a lifting and this time $g(\xi(0)) = \zeta(0)$.
9. Apply Exercise 8.
10. Apply Exercise 7.
11. Similar to the proof of Theorem 4.
13. Cf. Exercise 5.17.

Chapter 8

5. Show that f or $-f$ satisfy the d'Alembert equation.
6. Apply (26) to $2f(x)f(yz^{-1})$ with $x \in G$, $y \in S$, $z \in S$.
7. Apply (26) to $2f(xy^{-1})f(z)$ with $x \in S$, $y \in S$, $z \in G$.
8. Apply (26) to $2f(x)f(uv^{-1}w)$ where $y = uv^{-1}w$ with $u, v, w \in S$ and $x \in G$ and show that $f(xuwv^{-1}) = f(xy)$ and $f(xw^{-1}u^{-1}v) = f(xy^{-1})$ using $f(xyz) = f(xzy)$.
10. In this case (24) goes over into $f(x,y) = \sum_{n=-\infty}^{\infty} \sum_{m=-\infty}^{\infty} e^{inx} e^{imy} \mu(n,m)$, where $\mu(n,m) \geqslant 0$ and $\sum_n \sum_m \mu(n,m) = 1$.
16. Use the nth roots of unity in $\mathbb{C}$.
17. Use Exercise 16 and the fact that G is isomorphic to the product of a finite number of groups $\mathbb{Z}_n$, where n is a power of a prime number.
18. Use the uniform continuity of continuous characters.
20. Use Exercise 18.
23. For $x, y \in S$, define h by $h(xy^{-1}) = 2g(x)g(y) - g(xy)$.

Chapter 9

6. Apply Theorem 2.
8. $f(x) = |g(x)|$, g additive, $g(x_0) = 0$ for an $x_0 \neq 0$, but $g(x) \not\equiv 0$.
10. Apply Theorem 6 to $(-f)$.
13. Use $E + E$ and regularity results from Section 2.1.
15. Take $(B_\alpha)_{\alpha \in \Omega}$, the Borel subsets of $\mathbb{R}^2$, whose projection $P(B_\alpha)$ to the real axis is not countable. For $\alpha \in \Omega$, choose $h_\alpha \in P(B_\alpha)$ so that the h_αs are independent over $\mathbb{Q}$. Define conveniently $f(h_\alpha)$ and extend f to $\mathbb{R}$.

Chapter 10

1. Use Exercise 5.16 to show that $P(f+g) = Pf + Pg$ and $P(fg) = Pf \cdot Pg$.
3. If h were a homeomorphism, then $\mathbb{R}^\wedge$ would be locally compact.
4. Use Theorem 17.
5. Use Theorem 17 and Exercise 4.
7. For a solution F, take a λ such that $F(\phi) = \phi(\lambda)$.
10. Show that $\|f\| = 2^n$.
14. In a finite field every nonzero element is a root of unity.

20. A proof may use Q_μ as in Theorem 19, showing that $Q_\mu(e_\lambda)=0$ if $\lambda\neq 0$ and $Q_\mu(1)=1$; another proof is based on measure theory alone, using Fubini's theorem on $\mathscr{B}$.
23. Generalize the proof of Theorem 7.10.

Chapter 11

8. Use Exercise 2.10.
9. Use $1/[x(x-1)]=1/(x-1)-1/x$ and calculate $f(\lambda^2)$.
25. Separate the odd and even parts of f.
31. Use the fact that every bounded analytic function $F:\mathbb{C}\to\mathscr{A}$ is constant.
45. Prove that linear independence is an orthogonal relation. Apply Exercise 39 to a continuous f on a normed space E. Try to drop the continuity assumption.

Chapter 12

6. Show that the solution $p_0(t)=\begin{cases}0 \text{ for } t>0\\ 1 \text{ for } t=0\end{cases}$ of the 0'th equation of (3) (with $t,u\in\bar{\mathbb{R}}_+$) cannot be imbedded into a system of solutions of (1) and (3) ($k=0,1,2,\ldots,t,u\in\bar{\mathbb{R}}_+$).
11. By use of symmetry or by reduction to the Cauchy equation, determine the first few S_k's then use induction. Cf. also Exercise 12.

Chapter 13

4. Use Exercise 3.
5. Use the fact that $g(z_0)$ and $e-g(2t_0)$ are invertible in B.
6. A nilpotent group can be built up by its ascending central chain
$$\{e\}=Z_0\subset Z_1\subset\cdots\subset Z_n=G,$$
where Z_{i+1}/Z_i is the centre of G/Z_i. Prove the statement by induction on n. For $n=1$, the group G is commutative and one may use the result of Exercise 4.
8. Show that $f(x)f(0)$ satisfies the d'Alembert equation and apply Theorem 16.
9. If $A\neq 0$, then f is odd and $g(x)=-f(x+A)$ satisfies the d'Alembert equation.
12. Show that $f(x)^2=f(-x)^2$.
13. Take $f(x)=e^x$.
14. Either $f(e)=0$ then $f\equiv 0$, or, $f(e)=1$. In the latter case, show that $f(x^{-1})=f(x)^{-1}$, $f(xy)=f(yx)$ and $f(xyx^{-1}y^{-1})=1$. Deduce that $f(x)f(y)=f(y)f(x)$.

Chapter 14

4. Use (14), (15) and Járai 1979, 1986.
9. Cf. Exercise 13.9.
10. Cf. Theorem 13.14.
12. Reduce the equation to (13.53) and (3.14).

Chapter 15

7. If f satisfies the first equation, show that $g(x)=f(x)-f(0)$ is odd and additive.
10. Use Proposition 6.
11. Use Proposition 6.

12. Analogous to Proposition 7.
14. $g(x) = \lambda \exp(h(\log x))$, $x > 0$, $\lambda \in \mathbb{R}$ where h is additive and $g(0) = 0$ or g is constant.
15. Define $h(x) = f(g(f^{-1}(x)))$ and show that h satisfies the Jensen equation. A necessary and sufficient condition is that I is a subset of an n-quadrant, where an n-quadrant is the set of all $x = (x_1, \ldots, x_n)$ in $\mathbb{R}^n$ such that for some $a = (a_1, \ldots, a_n)$

$$x_i \succ_i a_i,$$

where $\succ_i$ means either $>$ or $<$.

16. $F(x) = ax^2 + b$; $a \geqslant 0$, $b \geqslant 0$.
17. $F(x) = a$, $a > 0$.
18. $g(x) = 1/(ax + b)$; $a > 0$, $b > 0$.
20. Apply Theorem 11.9.
22. Determine how F in $u(rx) - u(ry) = F[u(x) - u(y), r]$ depends upon its first variable. Then apply Exercise 21.

Chapter 16

1. If Y generates a subgroup $G_0 \neq G$, use a noncommutative version of Proposition 8.
2. $f(x) = h(\pi(x))g((\xi(\pi(x)))^{-1}x)$, where ξ is a lifting relative to the set of left cosets of the subgroup generated in G by Y.
3. Generalize Theorem 2.6 and apply Proposition 8.
4. The Cantor set is such an example.
7. Apply Proposition 8.
15. Similar to the proof for Theorem 1.
20. Apply Zorn's lemma.
21. Use the proof of Theorem 10.
25. Use the set of accumulation points of $\{f(n)\}$ and show that the function F, defined by $F(g) = g'$, satisfies $2F(g) = F(g + F[F(g)])$.

Chapter 17

1. $f(x) = ax + b$.

Chapter 18

12. Draw a diagram similar to Figure 20, based on techniques similar to those used in the proof of Lemma 1.

Chapter 19

1. Introduce $\alpha = \inf\{x | x > 0; x \in S\}$ and use the euclidean division of real numbers by α.
4. The solution is

$$f(x) = aH(x - \lambda)b,$$
$$g(x) = aH(x - \lambda) \text{ except if } b = 0 \text{ and then } g \text{ is arbitrary},$$
$$h(x) = H(xa)b \text{ except if } a = 0 \text{ and then } h \text{ is arbitrary},$$

where $a, b \in \mathbb{R}$ and $\lambda \in F$ are constants and H satisfies $H(x + yH(x)) = H(x)H(y)$.

11. Prove that g is a bijection and proceed as in the proof of Proposition 9.
12. Define h by $g(x) = x + (\lambda - 1)(x - h(x))$ and show that $h(g(x)) = h(x)$ and $h(h(x)) = h(x)$. Use $I = \{x \mid h(x) = x\}$.
13. Suppose that g is surjective if $\lambda > 1$.
15. Determine the closed subset H so that $f(xH) = f(x)$ for all x in $\mathbb{R}$ and proceed as in the proof of Theorem 12.
19. Define a relation P on G as follows: tPt' when $t' * f(x * t) = t * f(x * t)$ for all x in G. Show that P is a compatible regular equivalence relation on $(G, *)$.
20. Show that $x\Delta y = x * f(y)$, where $f: G \to G$ and $f(x * f(y)) = f(x) * f(y)$.
21. Use Exercise 20.
22. Show that $x\Delta y = g(x) + f(y)$ where g is a homomorphism. Use Proposition 11.
23. Use Exercise 22.
24. The operation $\Box$ is associative if and only if $\Box$ is commutative. Use Theorem 12.
25. Show that $f(x) > 0$ for $x > 0$ if f is not identically zero. Define $F(y) = \log f(e^y)$ and $H(y) = y - F(y)$, $G(y) = y + F(y)$. Prove that $H(G(y)) = H(y)$ and $H(H(y)) = H(y)$.
26. With $P(e_n) = a(n)e_n$, show that $a(n)^{-1} + a(m)^{-1} = a(n+m)^{-1} + 1$ for $a(n)a(m) \neq 0$.
27. $P(e_n) = a(n)e_n$ and $a(n) = a(m)$ if $a(n+m) \neq 0$.
28. $Pf(x) = \alpha \sum_{k=0}^{\infty} (1-\alpha)^k f(x + k\beta)$ for some real β.
30. Show that, for any positive integer n, $|f(x)| \leqslant M|x|^n$ where M is an upper bound of $|f|$ on $[0, 1]$.

Chapter 20

5. Express $T(s, t, \lambda^2\mathbf{u})$ with aid of (38), (37) and (36) (in this order).
13. One possibility is to choose first $s = -(x_1 + \cdots + x_n)/n$, then $u_k = x_k + s\,(k = 1, \ldots, n)$ and $t = (\sum_{k=1}^{n} (x_k + s)^2)^{-1/2}$, except when $x_1 = \cdots = x_n$. Determine I also in the latter, exceptional case. Then, in order to obtain the continuous solutions (if any), make sure that the transition between the two cases is continuous.

BIBLIOGRAPHY

The Bibliography, while rather extensive, is very selective. The selection has not been done by rigid principles. Some works have been selected for their historical interest, many, of course, for the important or intriguing results, others for interesting proofs and many more for interesting and/or unusual applications.

In addition to works directly involving functional equations, the bibliography contains also (primary and) secondary sources on history and references concerning material from other fields which is used in this book. Where available, we have added references to reviews in the *Jahrbuch über die Fortschritte der Mathematik* (= *Fortschr.*, 1868–1943) or *Mathematical Reviews* = *MR*; 1940–; from 1980 on without writing the year, since the volume number consists of the last two digits of the year of publication).

For an almost complete list of works on functional equations see the bibliography in the first author's book, *Lectures on Functional Equations and Their Applications* (Academic Press, New York–London, 1966) supplemented by the series, Works on functional equations in *Aequationes Mathematicae (***1** (1968), 152–91, **3** (1969), 271–312, **7** (1971), 270–319, **13** (1975), 159–96, **18** (1978), 218–56, **21** (1980), 283–317, **23** (1981), 305–39, **24** (1982), 299–316, **26** (1983), 298–313), **29** (1985), 113–21, **30** (1986), 288–99, **33** (1987), 269–95, **36** (1988), 303–18).

c. 1347

Oresme, N. *Questiones super geometriam Euclidis* (Manuscript). Paris. (Ed. H.L.L. Busard. E.J. Brill, Leiden, 1961).

c. 1352

Oresme, N. *Tractatus de configurationibus qualitatum et motuum* (*Manuscript*). Paris. (Ed., transl. and comm. M. Clayett. Univ. of Wisconsin Press, Madison-Milwaukee-London, 1968).

1544

Stifel, M. *Arithmetica integra.* Nürnberg, 1544.

1620

Bürgi, J. *Arithmetische und geometrische Progress Tabulen. Univ. Prague*, 1620.

1624

Briggs, H. *Arithmetica logarithmica.* London, 1624.

Kepler, J. *Chilias logarithmorum ad totidem numeros rotundos.* Marburg, 1624.

1638

Galilei, G. *Discorsi e dimostrazioni matematiche intorno a due nuove scienze.* Leyden, 1638. (*Opere*, Vol. VIII, pp. 209–10. Barbèra, Firenze, 1968).

1647

de Saint-Vincent, G. *Opus geometricum quadraturae circuli et sectionum coni. Problema Austriacum. Plus ultra quadratura circuli.* Antwerpen, 1647.

1649

de Sarasa, A. *Solutio problematis a R.P. Marino Mersenno minimo propositi.* Antwerpen, 1649.

1726

Bernoulli, D. Examen principiorum mechanicae et demonstrationes geometricae de compositione et resolutione virium. *Comm. Acad. Petropol.*, **1**, 126–42.

1747

d'Alembert, J.
(a) Recherches sur la courbe que forme une corde tendue mise en vibration, I. *Hist. Acad. Berlin*, **1747**, 214–19.
(b) Recherches sur la courbe que forme une corde tendue mise en vibration, II. *Hist. Acad. Berlin*, **1747**, 220–49.

1748

Euler, L. *Introductio in analysin infinitorum*. Caput 8. Bousquet, Lausanne (*Opera omnia*. ser. I, Vol. 8, pp. 133–52. Teubner, Leipzig-Berlin, 1922).

1750

d'Alembert, J. Addition au Mémoire sur la courbe que forme une corde tendue mise en vibration. *Hist. Acad. Berlin*, **1750**, 355–60.

1755

Euler, L. *Institutiones calculi differentialis*. Part. Prior. §222, Part. Post. §§70–75. Acad. Imp. Sci., Petropol., 1755 (*Opera omnia*, Ser. 1, Vol. 10, pp. 151, 276–9. Teubner, Leipzig–Berlin, 1913).

1760

de Foncenex, D. Sur les principes fondamentaux de la méchanique. *Misc. Taurinensia*, **2**, 299–322.

1764

Euler, L. De insigne promotione methodi tangentium inversae. *Novi Comm. Acad. Sci. Petropol.*, **10**, 135–55 (*Opera omnia*, Ser. 1, Vol. 27, pp. 365–83. Füssli, Lausanne, 1954).

1768

Euler, L. *Institutiones calculi integralis*, I. §406. Acad. Imp. Sci., Petropol., 1768 (*Opera omnia*, Ser. 1, Vol. 2, pp. 259–60. Teubner, Leipzig-Berlin, 1913).

Lambert, J.H. Observations trigonométriques. *Mém. Acad. Berlin*, **24**, 327–54. (*Opera mathematica*, Vol. 2, pp. 245–69. Füssli, Zürich, 1948).

1769

d'Alembert, J. Mémoire sur les principes de mécanique. *Hist. Acad. Sci. Paris*, **1769**, 278–86.

1770

Euler, L. *Institutiones calculi integralis*, III. §§174–7. Acad. Imp. Sci., Petropol., 1770 (*Opera omnia*, Ser. I. Vol. 13, pp. 116–9. Teubner, Leipzig-Berlin, 1925).

1774

Euler, L. Demonstratio theorematis neutoniani de evolutione potestatum binomii pro casibus quibus exponentes non sunt numeri integri. *Novi Comm. Acad. Sci. Petropol.*, **19**, 103–11 (*Opera Omnia*, Ser. 1, Vol. 15, pp. 207–16. Teubner, Leipzig-Berlin, 1927).

1791

Legendre, A.M. *Éléments de géométrie*. Note II. Didot, Paris, 1791.

1797

Lacroix, S.F. *Traité du calcul différentiel et intégral*. Duprat, Paris, 1797.

1804

Poisson, S.D. Du paraléllogramme des forces. *Corresp. sur l'École Polytechn.* **1**, 356–60.

1805

Poisson, S.D. *Mécanique rationnelle*. Bachelier, Paris, 1805.

1809

Gauss, C.F. *Theoria motus corporum coelestium*. §§175–7. Perthes-Besser, Hamburg, 1809 (*Werke*, Vol. VII, 240–44. Teubner, Leipzig, 1906).

1811

Poisson, S.D. *Traité de mécanique*. Bachelier, Paris, 1811.

1815

Babbage, C. Essay towards the calculus of functions. *Philos. Trans. Roy. Soc. London*, **105**, 389–423.

1821

Cauchy, A.L. *Cours d'analyse de l'École Polytechnique, Vol. I. Analyse algébrique*, Debure, Paris, 1821 (*Oeuvres*, Sér. 2, Vol. 3. Gauthier-Villars, Paris, 1897).

1823

Abel, N.H. Méthode générale pour trouver des fonctions d'une seule quantité variable lorsqu'une propriété de ces fonctions est exprimée par une équation entre deux variables (Norwegian), *Mag. Naturvidenskab.*, **1**, No. 1, 1–10 (*Oeuvres complètes*, Vol. 1, pp. 1–10. Grøndahl & Søn, Christiania, 1881).

1826

Abel, N.H.
(*a*) Untersuchung der Functionen zweier unabhängigen Veränderlichen Grössen x und y, wie $f(x,y)$, welche die Eigenschaft haben, dass $f[z,f(x,y)]$ eine symmetrische Function von x, y und z ist. *J. Reine Angew. Math.*, **1**, 11–5 (*Oeuvres complètes*, Vol. 1, pp. 61–5. Grøndahl & Søn, Christiania, 1881).
(*b*) Untersuchungen über die Reihe $1 + (m/1)x + (m(m-1)/(1.2))x^2 + (m(m-1)(m-2)/(1.2.3))x^3 + \cdots$ *J. Reine Angew. Math.*, **1**, 311–39 (*Oeuvres complètes*, Vol. 1, pp. 219–50. Grøndahl & Søn, Christiania, 1881).

1827

Abel, N.H. Über die Functionen, die der Gleichung $\phi x + \phi y = \psi(xfy + yfx)$ genug thun. *J. Reine Angew. Math.*, **2**, 386–94 (*Oeuvres complètes*, Vol. 1, pp. 389–98. Grøndahl & Søn, Christiania, 1818).

1829

Lobachevskiĭ, N.I. On the foundations of geometry III (Russian) *Kasaner Bote*, **27, 227**–43.

1830

Magnus, L.J. Über die Relationen der Functionen, welche der Gleichung $F_1 y\phi_1 x + F_2 y\phi_2 x + \cdots + F_n y\phi_n x = F_1 x\phi_1 y + F_2 x\phi_2 y + \cdots + F_n x\phi_n y$ genug thun. *J. Reine Angew. Math.*, **5**, 365–73.

1831

Encke, J.F. Über die Begründung der kleinsten Quadrate. *Abh. Berlin Akad. Wiss.*, **1831**, 6–13.

1832

Bolyai, J. *Appendix scientiam spatii absolute veram exhibens: a veritate aut falsitate Axiomatis XI. Euclidei (a priori haud unquam decidenda) independentem: adjecta ad casum falsitatis, quadratura circuli geometrica.* §§23, 24. Marosvásárhely, 1832.

Gauss, C.F.
(*a*) Brief an W. Bolyai 6.3.1832. Manuscript, 1832 (*Werke*, Vol. VIII, pp. 220–4. Teubner, Leipzig, 1900).
(*b*) Zur Astralgeometrie. Manuscript, 1832 (*Werke*, Vol. VIII, pp. 226–8. Teubner, Leipzig, 1900).

1837

Lobachevskiĭ, N.I. New foundations of geometry with a complete theory of parallel lines, IV (Russian). *Kasaner Gelehrte Schriften*, **1837**, No. 1, 3–97.

1840

Lobachevskiĭ, N.I. *Geometrische Untersuchungen zur Theorie der Parallellinien.* §§33, 36. Berlin, 1840. (English transl, G.B. Halsted. Open Court Publ., LaSalle, IL 1914).

1844

Grassmann, H. *Die lineale Ausdehnungslehre.* Wigand, Leipzig, 1844.

1854

Lottner, J.C. Über die Functionen, welche der Gleichung $\phi(x) + \phi(y) = \phi[(fy\cdot Fx + fx\cdot Fy)/\chi(xy)]$ Genüge leisten. *J. Reine Angew. Math.*, **46**, 367–88.

1856

Lobachevskiĭ, N.I. Pangéométrie ou précis de géométrie fondée sur une théorie générale et rigoureuse des parallèles. In *Collection of Learned Articles by the Professors of the Imperial University Kazan Celebrating its 50th Anniversary* (Russian). Vol. 1. Univ. Kazan, Kazan, 1856, pp. 279–340.

1860

Maxwell, J.C. Illustrations of the dynamical theory of gases. *Philos. Mag.*, **19**, 19–32. (*Scientific Papers*, Vol. 1, pp. 377–409. Cambridge University Press, Cambridge, 1890; Dover, New York, 1965).

Stokes, G. On the intensity of light reflected from or transmitted through a pile of plates. *Proc. Roy. Soc. London*, **11**, 545–57 (*Math. and Phys. Papers*, Vol. IV, pp. 145–56. Cambridge University Press, Cambridge, 1904).

1867

Henckel, H. *Theorie der komplexen Zahlensysteme.* Leipzig, 1867.

1868

Schiaparelli, G. On the principle of the arithmetic mean in calculating the results of observations (Italian). *Istit. Lombardo Accad. Sci. Lett. Rend.* (2), **1**, 771–8. *Fortschr.*: **1**, 75 (1868).

1871

Saint Robert, P. De la résolution de certaines équations à trois variables par le moyën d'une règle glissante. *Mem. Accad. Sci. Torino* (2), **25**, 53–72. *Fortschr.*: **3**, 28 (1871).

1872

Glaisher, J.W.L. On the law of facility of errors of observations and on the method of least squares. *Mem. Astronom. Soc. London*, **39**, 75–124. *Fortschr.*: **4**, 92 (1872).

Schlömilch, O. Über die Bestimmung der Wahrscheinlichkeit eines Beobachtungsfehlers. *Z. Math. Phys.*, **17**, 87–8. *Fortschr.*: **4**, 588 (1872).

1873

Stone, E.J. On the most probable result which can be derived from a number of direct determinations of assumed equal values. *Monthly Notices Roy. Astronom. Soc.*, **33**, 570–2. *Fortschr.*: **5**, 122 (1873).

1875

Darboux, G. Sur la composition des forces en statique. *Bull. Sci. Math.* (1), **9**, 281–8. *Fortschr.*: **7**, 550 (1875).

Schiaparelli, G. Sur le principe de la moyenne arithmétique. *Astronom. Nachr.*, **87**, 55–8. *Fortschr.*: **7**, 109 (1875).

1877

Genocchi, A. Sur une mémoire de Daviet de Foncenex et sur les géométries noneuclidiennes. *Mem. Accad. Sci. Torino* (2), **39**, 365–404. *Fortschr.*: **10**, 345 (1878).

Merriman, M. A list of writings relating to the method of least squares with historical and critical notes. *Trans. Connecticut Acad. Arts Sci.* **4**, 151–231. *Fortschr.*: **9**, 154 (1877).

1878

Cayley, A. On a functional equation. *Quart. J. Math.*, **15**, 315–25 (*Collected Math. Papers*, Vol. X, pp. 298–306. Cambridge University Press, Cambridge, 1896). *Fortschr.*: **10**, 290 (1978).

Jensen, J.L.W.V. On the solution of fundamental equations by elementary means (Danish). *Tidsskr. Math.* (4), **2**, 149–55. *Fortschr.*: **10**, 289 (1878).

1880

Darboux, G. Sur le théorème fondamental de la géométrie projective. *Math. Ann.*, **17**, 55–61. *Fortschr.*: **12**, 447 (1880).

1881

Schröder, E. Über eine eigentümliche Bestimmung einer Funktion durch formale Anforderungen. *J. Reine Angew. Math.*, **90**, 189–220. *Fortschr.*: **13**, 325 (1881).

1882

Weierstrass, K. Zur Theorie der Jacobi'schen Functionen von mehreren Veränderlichen. *Sitzungsber. Preuss. Akad. Wiss.* **1882**, 505–9, (*Werke*, Vol. III, pp. 155–9. Mayer & Müller, Berlin, 1903). *Fortschr.*: **14**, 415 (1882).

1884

Koenigs, G. Recherches sur les intégrales de certaines équations fonctionnelles. *Ann. Sci. École Norm. Sup.* (3), Suppl. 3–41. *Fortschr.*: **16**, 376 (1884).

Kronecker, L. Näherungsweise ganzzahlige Auflösung linearer Gleichungen. *Sitzungsber. Preuss. Akad. Wiss.*, **1884**, 1179–93, 1271–99 (*Werke*, Vol. III, pp. 47–100, Teubner, Leipzig, 1899). *Fortschr.*: **16**, 83–4 (1884).

Rausenberger, O. *Functionen einer Variabeln mit einer endlichen Anzahl wesentlicher Discontinuitätspunkte, nebst einer Einleitung in die allgemeine Functionentheorie*. Teubner, Leipzig, 1884. *Fortschr.*: **16**, 334 (1884).

1885

Koenigs, G. Nouvelles recherches sur les équations fonctionnelles. *Ann. Sci. École Norm. Sup.* (3), **2**, 385–404. *Fortschr.*: **17**, 370 (1885).

Phragmén, E. Sur un théorème concernant les fonctions elliptiques. *Acta Math.*, **7**, 33–42. *Fortschr.*: **17**, 425 (1885).

1886

Lecornu, L. Sur le problème de l'anamorphose. *C.R. Acad. Sci. Paris*, **102**, 813–16. *Fortschr.*: **18**, 829 (1886).

Massau, J. *Mémoire sur l'intégration graphique et ses applications.* Vol. III, Chap. III, No. 178. Gauthier-Villars, Paris, 1886. *Fortschr.*: **19**, 285 (1886).

Weierstrass, K. *Zur Determinantentheorie.* (Manuscript, ed. P. Günther). Univ. Berlin, Berlin, 1886 (*Werke*, Vol. III, pp. 271–87. Mayer & Müller, Berlin, 1903). *Fortschr.*: **34**, 23 (1903).

1888

Bertrand, J. Sur la loi de probabilité des erreurs d'observation. *C.R. Acad. Sci. Paris* **106**, 153–6. *Fortschr.*: **20**, 219 (1888).

Tisserand, F. Remarque à l'occasion d'une communication de M. Bertrand. *C.R. Acad. Sci. Paris*, **106**, 231–2. *Fortschr.*: **20**, 220 (1888).

1889

Segre, C.
(*a*) A new field of geometrical research, I (Italian). *Atti Accad. Sci. Torino*, **25**, 276–301. *Fortschr.*: **22**, 609 (1890).
(*b*) A new field of geometrical research, II (Italian). *Atti Accad. Sci. Torino*, **25**, 430–57. *Fortschr.*: **22**, 609 (1890).

1890

Laurent, H. *Traité d'analyse.* Vol. VI, Chap. VI. Gauthier-Villars, Paris, 1890. *Fortschr.*: **22**, 274 (1890).

Segre, C.
(*a*) A new field of geometric research, III (Italian). *Atti Accad. Sci. Torino*, **26**, 35–71. *Fortschr.*: **22**, 609 (1890).
(*b*) A new field of geometric research, IV (Italian). *Atti Accad. Sci. Torino*, **26**, 592–612. *Fortschr.*: **22**, 609 (1890).

1891

Czuber, B. *Theorie der Beobachtungsfehler.* Teubner, Leipzig, 1891. *Fortschr.*: **23**, 235 (1891).

1892

Hermite, C. Sur l'addition des arguments dans les fonctions elliptiques. *Teixeira J. Sci. Math. Astronom.* **11**, 65–66. *Fortschr.*: **24**, 429 (1892).

Schottky, F. Über das Additionstheorem der Cotangente und der Funktion $\zeta(u) = \sigma'(u)/\sigma(u)$. *J. Reine Angew. Math.*, **110**, 324–37. *Fortschr.*: **24**, 427 (1892).

Schwarz, H.A. *Formeln und Lehrsätze zum Gebrauch der elliptischen Functionen.* Springer, Berlin, 1892. Fortschr.: **25**, 757 (1893–94).

1894

Hurwitz, A. Zur Invariantentheorie. *Math. Ann.*, **45**, 381–404 (*Math. Werke*, Vol. II, pp. 508–32. Birkhäuser, Basel, 1933). *Fortschr.*: **25**, 171 (1893–94).

1896

Cantor, M. Funktionalgleichungen mit drei von einander unabhängigen Veränderlichen. *Z. Math. Phys.*, **41**, 161–3. *Fortschr.*: **27**, 312 (1896).

Tisserand, F. *Recueil complémentaire d'exercices sur le calcul infinitésimal.* 2nd ed., Chap. II. Gauthier-Villars, Paris, 1896. *Fortschr.*: **27**, 221 (1896).

1897

Andrade, J. *Leçons de mécanique physique.* Soc. d'Éditions Scient., Paris, 1897. *Fortschr.*: **28**, 613 (1897).

Hayashi, T. On a functional equation treated by Abel (Japanese). *Tokyo Sugaku-Buts.* (1), **8**, 129–34. *Fortschr.*: **31**, 403 (1900).

Jensen, J.L.W.V. On the solution of functional equations with the minimal numbers of suppositions (Danish). *Mat. Tidsskr. B*, **1897**, 25–8. *Fortschr.*: **28**, 346 (1897).

Stäckel, P. Über eine von Abel untersuchte Funktionalgleichung. *Z. Math. Phys.*, **42**, 323–6. *Fortschr.*: **28**, 353 (1897).

1898

Alekseevskiĭ, V.P. On the definition of length in the non-euclidean geometry (Russian). *Soobshch. Kharkovsk. Mat. Obshch.* (2), **6**, 139–53. *Fortschr.*: **29**, 411 (1898).

1899

Hayashi, T. On a functional equation treated by Abel. *Z. Math. Phys.*, **44**, 346–9. *Fortschr.*: **39**, 353 (1899).

Lémeray, E.M.
(*a*) Sur le problème de l'itération. *C.R. Acad. Sci. Paris*, **128**, 278–9. *Fortschr.*: **30**, 351 (1899).
(*b*) Sur les équations functionnelles qui caractérisent les opérations associatives et les opérations distributives. *Bull. Soc. Math. France*, **27**, 130–7. *Fortschr.*: **30**, 351 (1899).

Siacci, F.
(*a*) On the composition of forces in statics and its postulates, I (Italian). *Rend. Accad. Sci. Fis. Mat. Napoli* (3), **5**, 34–9. *Fortschr.*: **30**, 614 (1899).
(*b*) On the composition of forces in statics and its postulates, II (Italian). *Rend.*

Accad. Sci. Fis. Mat. Napoli (3), **5**, 69–78. *Fortschr.*: **30**, 615 (1899).
(*c*) On the composition of forces in statics and its postulates, III (Italian). Rend. Accad. Sci. Fis. Mat. Napoli (3), **5**, 147–51. *Fortschr.*: **30**, 615 (1899).

de la Vallée Poussin, C.J. *Cours d'analyse infinitésimale*. Vol. 1, Introduction §5. Libr. Univ., Louvain; Gauthier-Villars, Paris, 1899. *Fortschr.*: **30**, 267 (1899).

Wilson, E.B. Note on the function satisfying the functional relation $\phi(u)\phi(v) = \phi(u+v)$. *Ann. Math.* (2), **1**, 47–8. *Fortschr.*: **30**, 353 (1899).

Zeiliger, D.N. Über eine Funktionalgleichung (Russian). *Bull. Soc. Phys.-Math. Kazan* (2), **9**, 187–90. *Fortschr.*: **30**, 403 (1899).

Zignano, I. Extension of two problems of Cauchy (Italian). *Atti Accad. Pont. Nuovi Lincei*, **53**, 139–49. *Fortschr.*: **30**, 402 (1899).

1900

Andrade, J. Sur l'équation fonctionnelle de Poisson. *Bull. Soc. Math. France*, **28**, 58–64. *Fortschr.*: **31**, 403 (1900).

Hilbert, D. Mathematische Probleme (Vortrag, gehalten auf dem internationalen Mathematiker-Congress zu Paris, 1900). *Nachr. Ges. Wiss. Göttingen* **1900**, 253–97 (*Gesammelte Abhandl.*, Vol. III, pp. 290–329. Springer, Berlin, 1935). *Fortschr.*: **31**, 68 (1900).

Pexider, J.V. A study on functional equations (Czech). *Časopis Pěst. Mat.*, **29**, 153–95. *Fortschr.*: **31**, 403 (1900).

Severi, F. On the conics which touch or intersect one or more spatial curves (Italian). *Atti Accad. Sci. Torino*, **36**, 74–93. *Fortschr.*: **32**, 553 (1901).

1901

Hölder, O. Die Axiome der Quantität und die Lehre vom Mass. *Ber. Sächs. Gesellsch. Wiss. Leipzig*, **53**, 1–64. *Fortschr.*: **32**, 79 (1901).

Pincherle, S. and Amaldi, U. *The Distributive Operations and their Applications to Analysis* (Italian). Zanichelli, Bologna, 1901. *Fortschr.*: **32**, 75 (1901).

Schur, I. *Über eine Klasse von Matrizen, die sich einer gegebenen Matrix zuordnen lassen*. Dissertation, Berlin, 1901. *Fortschr.*: **32**, 165 (1901).

Sincov, D.M. On the problem of Semikolenov (Russian). *Bull. Soc. Phys.-Math. Kazan* (2), **11**, 13–16. *Fortschr.*: **32**, 391 (1901).

Zeiliger, D.N. On the preceeding paper of D. Sincov 'On the problem of Semikolenov' (Russian). *Bull. Soc. Phys.-Math. Kazan* (2), **11**, 103–13. *Fortschr.*: **32**, 39 (1901).

1903

Hamel, G. Über die Zusammensetzung von Vektoren. *Z. Math. Phys.*, **49**, 362–71. *Fortschr.*: **34**, 621 (1903).

Pexider, J.V. Notiz über Funktionaltheoreme. *Monatsh. Math. Phys.*, **14**, 293–301. *Fortschr.*: **34**, 420 (1903).

Schimmack, R. Über die axiomatische Begründung der Vektoraddition. *Nachr. Ges. Wiss. Göttingen*, **1903**, 317–25. *Fortschr.*: **34**, 621 (1903).

Schur, F. Über die Zusammensetzung von Vektoren. *Z. Math. Phys.*, **49**, 352–61. *Fortschr.*: **34**, 621 (1903).

Sincov, D.M.
(*a*) Notes on the functional calculus (Russian). *Bull. Soc. Phys.-Math. Kazan* (2), **13**, 48–72. *Fortschr.*: **34**, 421 (1903).
(*b*) Über eine Funktionalgleichung. *Arch. Math. Phys.* (3), **6**, 216–17. *Fortschr.*: **34**, 421 (1903).

Volpi, R. Observations for a purely analytical and elementary theory of circular and hyperbolic functions and their relations to the exponential (Italian). *Giorn. Mat. Battaglini*, **41**, 33–46. *Fortschr.*: **34**, 474 (1903).

1904

Schur, I. Über die Darstellung der endlichen Gruppen durch gebrochene lineare Substitutionen. *J. Reine Angew. Math.*, **127**, 20–50. *Fortschr.*: **35**, 155 (1904).

Stephanos, C. Sur une catégorie d'équations fonctionnelles. *Rend. Circ. Mat. Palermo* **18**, 360–2. *Fortschr.*: **35**, 391 (1904).

1905

Hamel, G. Eine Basis aller Zahlen und die unstetigen Lösungen der Funktionalgleichung $f(x+y) = f(x) + f(y)$. *Math. Ann.*, **60**, 459–62. *Fortschr.*: **36**. 446 (1905).

1906

Bernstein, F. Über eine Funktionalgleichung und eine erweiterte Begründung des Gaussschen Fehlergesetzes. *Leipzig. Ber.*, **58**, 228–36. *Fortschr.*: **37**, 265 (1906).

Bonola, R. *The Non-Euclidean Geometry* (Italian). Appendix I. Zanichelli, Bologna, 1906. *Fortschr.*: **37**, 422 (1906).

Jensen, J.L.W.V. Sur les fonctions convexes et les inégalités entre les valeurs moyennes. *Acta Math.* **30**, 175–93. *Fortschr.*: **37**, 422 (1906).

Pincherle, S. Funktionaloperationen und Gleichungen. In *Encyklopädie der mathematischen Wissenschaften mit Einschluss ihrer Anwendungen*, Vol. II. 1.2. Teubner, Leipzig, 1906, pp. 761–81. *Fortschr.*: **37**, 346 (1906).

1907

Lebesgue, H.
(*a*) Sur les transformations ponctuelles transformant les plans en plans, qu'on peut définir par des procédés analytiques. *Atti Accad. Sci. Torino*, **42**, 532–9. *Fortschr.*: **38**, 967 (1907).
(*b*) Contribution à l'étude des correspondances de M. Zermelo. *Bull. Soc. Math. France*, **35** (1907), 202–12. *Fortschr.*: **38**, 96 (1907).

Schiaparelli, G. How one can justify the use of the arithmetic means in calculating the results of observations (Italian). *Istit. Lombardo Accad. Lett. Rend.* (2), **40**, 752–64. *Fortschr.*: **38**, 280 (1907).

1908

Peano, G. *Collection of Mathematical Formulas* (Italian). 5th edn., Fratres Bocca, Torino, 1908. *Fortschr.*: **39**, 84 (1908).

1909

Broggi, U. Sur le principe de la moyenne arithmétique. *Enseign. Math.*, **11**, 14–17. *Fortschr.*: **40**, 286 (1909).

Fréchet, M. Une définition fonctionnelle des polynômes. *Nouv. Ann.* (4), **9**, 145–62. *Fortschr.*: **40**, 389 (1909).

Schimmack, R.
(*a*) Der Satz vom arithmetischen Mittel in axiomatischer Begründung. *Math. Ann.*, **68**, 125–32. *Fortschr.*: **40**, 285 (1909).
(*b*) Axiomatische Untersuchungen über die Vektoraddition. *Abh. Akad. Naturforsch. Halle*, **90**, 5–104. *Fortschr.*: **40**, 148 (1909).

1910

Curtiss, D.R. Relations between the Gramian, the Wronskian and a third determinant connected with the problem of linear independence. *Bull. Amer. Math. Soc.*, **17**, 462–7. *Fortschr.*: **42**, 173 (1911).

van Vleck, E.B. A functional equation for the sine. *Ann. Math.* (2), **11**, 161–5. *Fortschr.*: **41**, 375 (1910).

Wellstein, J. Zwei Funktionalgleichungen. *Arch. Math. Phys.* (3), **16**, 93–100. *Fortschr.*: **41**, 373 (1910).

1911

Carmichael, R.D. A generalization of Cauchy's functional equation. *Amer. Math. Monthly* **18**, 198–203. *Fortschr.*: **42**, 385 (1911).

Halphen, C. Sur les fonctions homogènes. *Rev. Math. Spec.*, **21**, 130–1. *Fortschr.*: **42**, 447 (1911).

1912

Gronwall, T.H. Sur les équations entre trois variables représentables par des nomogrammes à points alignés. *J. Math. Pures Appl.* (6), **8**, 59–102. *Fortschr.*: **43**, 159 (1912).

Pincherle, S. Équations et opérations fonctionnelles. In *Encyclopédie des sciences mathématiques pures et appliquées*, Vol. II_5. 1, II, 26. Gauthier-Villars, Paris; Teubner, Leipzig, 1912.

Schweitzer, A.R. Theorems on functional equations. *Bull. Amer. Math. Soc.*, **19**, 66–70. *Fortschr.*: **43**, 408 (1912).

Soreau, R. Réduction de $F_{123} = 0$ à la forme $f_1 f_3 + f_2 g_3 + h_3 = 0$. *C.R. Acad. Sci. Paris*, **155**, 1065–7. *Fortschr.*: **43**, 160 (1912).

1913

Fréchet, M. On the functional equation $f(x+y) = f(x) + f(y)$ (Esperanto). *Enseign. Math.*, **15**, 390–3. *Fortschr.*: **44**, 399 (1913).

Levi-Cività, T. On the functions which permit an addition formula of the type $f(x+y) = \sum_{i=1}^{n} X_i(x) Y_i(y)$ (Italian). *Atti Accad. Naz. Lincëi Rend.* (5), **22**, Pt 2, 181–3. *Fortschr.*: **44**, 502 (1913).

Stäckel, P. On the functional equation $f(x+y) = \sum_{i=1}^{n} X_i(x) Y_i(y)$ (Italian). *Atti Accad. Naz. Lincei Rend.* (5), **22**, Pt 2, 392–3. *Fortschr.*: **44**, 174 (1913).

Stephanos, C. Sur une propriété caractéristique des déterminants. *Ann. Mat. Pura Appl.* (3), **21**, 233–6. *Fortschr.*: **44**, 174 (1913).

Sutô, O. On some classes of functional equations. *Tôhoku Math. J.*, **3**, 47–62.

van Vleck, E.B. On the functional equation for the sine. Additional note. *Ann. Math.* (2), **13**, 154. *Fortschr.*: **43**, 530 (1912).

1914

Koebe, P. Über diejenigen analytischen Funktionen eines Arguments, welche ein algebraisches Additionstheorem besitzen und die endlich-vieldeutig umkehrbaren Abelschen Integrale. In *Mathematische Abhandlungen H.A. Schwarz zu seinem 50-jährigen Doktorjubiläum gewidmet*. Springer, Berlin, 1914, pp. 192–4. *Fortschr.*: **45**, 699 (1914–15).

Stirling, J. Note on a functional equation employed by Sir George Stokes. *Proc. Roy. Soc. London, Ser. A*, **90**, 237–9. *Fortschr.*: **45**, 516 (1914–15).

Sutô, O.
(*a*) Studies on some functional equations. *Tôhoku Math. J.*, **6**, 1–15. *Fortschr.*: **45**, 1310 (1914–15).
(*b*) Law of arithmetical mean. *Tôhoku Math. J.*, **6**, 79–81. *Fortschr.*: **45**, 1265 (1914–15).
(*c*) Studies on some functional equations. *Tôhoku Math. J.*, **6**, 82–101. *Fortschr.*: **45**, 1310 (1914–15).

1915

Beetle, R.D. On the complete independence of Schimmack's postulates for the arithmetic mean. *Math. Ann.*, **76**, 444–6. *Fortschr.*: **45**, 354 (1914–15).

1916

Burstin, C. Die Spaltung des Kontinuums in c in L-Sinne nichtmessbare Mengen. *Sitzungsber. Akad. Wiss. Wien, Math.-Natur. Kl. Abt. IIa*, **125**, 209–17. *Fortschr.*: **46**, 293–4 (1916–18).

Noether, E. Die Funktionalgleichungen der isomorphen Abbildung. *Math. Ann.*, **77**, 536–45. *Fortschr.*: **46**, 170 (1916–18).

van Vleck, E.B. and Doubler, F.H. A study of certain functional equations for the θ-functions. *Trans. Amer. Math. Soc.*, **17**, 9–49. *Fortschr.*: **46**, 604 (1916–18).

Wilson, W.H. On a certain general class of functional equations. *Bull. Amer. Math. Soc.*, **23**, 392–3. *Fortschr.*: **46**, 711 (1916–18).

1917

Bernstein, S.N. Essay of an axiomatic foundation of probability theory (Russian). *Khark. Zap. Mat. Obshch.* (2), **15**, 209–74. *Fortschr.*: **48**, 596–9 (1921–22).

Favre, A. Sur les fonctions homogènes. *Nouv. Ann.* (4), **17**, 426–8. *Fortschr.*: **46**, 317 (1916–18).

Ostrowski, A. Über einige Losungen der Funktionalgleichung $\phi(x)\cdot\phi(y) = \phi(xy)$. *Acta. Math.*, **41**, 271–84. *Fortschr.*: **46**, 169 (1916–18).

1918

Bell, E.T. A partial isomorph of trigonometry. *Bull. Amer. Math. Soc.*, **25**, 311–21. *Fortschr.*: **47**, 319 (1919–20).

Schweitzer, A.R. On the history of functional equations. *Bull. Amer. Math. Soc.*, **25**, 439.

Wilson, W.H. On a certain general class of functional equations. *Amer. J. Math.*, **40**, 263–82. *Fortschr.*: **46**, 711 (1916–18).

1919

Loewy, A. Axiomatische Begründung der Zinstheorie. *Jber. Deutsch. Math.-Verein.*, **28**, 26–31. *Fortschr.*: **47**, 58 (1919–20).

Lunn, A.C. Some functional equations in the theory of relativity. *Bull. Amer. Math. Soc.*, **26**, 26–34. *Fortschr.*: **47**, 808 (1919–20).

Perron, O. Über Additions- und Subtraktionstheoreme. *Arch. Math. Phys.* (3), **28**, 97–100. *Fortschr.*: **47**, 189 (1919–20).

Wilson, W.H. On certain related functional equations. *Bull. Amer. Math. Soc.*, **26**, 300–312. *Fortschr.*: **47**, 320 (1919–20).

1920

Banach, S. Sur l'équation fonctionnelle $f(x+y) = f(x) + f(y)$. *Fund. Math.*, **1**, 123–4. *Fortschr.*: **47**, 235 (1919–20).

Campbell, N.R. *Physics, the Elements.* Cambridge University Press, 1920, (Reprinted as: *Foundations of Science: the Philosophy of Theory and Experiment*. Dover, New York, 1957.

Mineur, H. Sur les solutions discontinues d'une classe d'équations fonctionnelles. *C.R. Acad. Sci. Paris*, **170**, 793–6. *Fortschr.*: **47**, 235 (1919–20).

Sierpiński, W.
(*a*) Sur l'équation fonctionnelle $f(x+y) = f(x) + f(y)$. *Fund. Math.*, **1**, 116–22. *Fortschr.*: **47**, 235 (1919–20).
(*b*) Sur les fonctions convexes mesurables. *Fund. Math.*, **1**, 125–9. *Fortschr.*: **47**, 235 (1919–20).

Steinhaus, H. Sur les distances des points des ensembles de mesure positive. *Fund. Math.*, **1**, 93–104. *Fortschr.*: **47**, 179 (1919–20).

1921

Minetti, S.
(*a*) On the functional equation $f(x+y) = f(x)\cdot f(y)$, I (Italian). *Atti Accad.*

Minetti, S. (*cont.*)
Naz. Lincei Rend. (5), **31**, 12–15. *Fortschr.*: **48**, 298 (1921–22).
(*b*) On the functional equation $f(x + y) = f(x) \cdot f(y)$, II (Italian). *Atti Accad. Naz. Lincei Rend.* (5), **31**, 202–3. *Fortschr.*: **48**, 298 (1921–2).

Mineur, H. Sur les fonctions qui admettent un théorème d'addition algébrique. *C.R. Acad. Sci. Paris*, **172**, 1461–3. *Fortschr.*: **48**, 392 (1921–22).

1922

Banach, S. and Ruziewicz, S. Sur les solutions d'une équation fonctionnelle de J. Cl. Maxwell. *Bull. Internat. Acad. Polon. Sci. Lett.*, **1922**, 1–8. *Fortschr.*: **48**, 1204 (1921–22).

Mineur, H. Sur les fonctions qui admettent un théorème d'addition algébrique. *Bull. Sci. Math.* (2), **46**, 156–76. *Fortschr.*: **48**, 443 (1921–22).

Myrberg. P. J. *Über Systeme analytischer Funktionen, welche ein Additionstheorem besitzen.* (Preisschriften d. fürstl. Jablonowski. Ges. Leipzig). Teubner, Leipzig, 1922. *Fortschr.*: **48**, 433 (1921–22).

Picard, É.
(*a*) Deux leçons sur certaines équations fonctionnelles et la géométrie noneuclidienne, I. *Bull. Sci. Math.* (2), **46**, 404–16. *Fortschr.*: **48**, 639 (1921–22).
(*b*) Deux leçons sur certaines équations fonctionnelles et la géométrie noneuclidienne, II. *Bull. Sci. Math.* (2), **46**, 425–32. *Fortschr.*: **48**, 639 (1921–22).

1923

Alaci, V.
(*a*) Pseudo-homogeneous functions (Romanian). *Rev. Mat. Timişoara*, **3**, No. 1, 3–4.
(*b*) On pseudo-homogeneous functions (Romanian). *Rev. Mat. Timişoara*, **3**, No. 1, 6–7.

Andreoli, G. On a simple and well-known functional equation (Italian). *Rend Accad. Sci. Fis. Mat. Napoli* (3), **29**, 12–14. *Fortschr.*: **49**, 197 (1923).

Frattini, G.
(*a*) A problem of amplification by isomorphism originating from the theory of relativity, I (Italian). *Atti Accad. Pont. Nuovi Lincei*, **76**, 94–8. *Fortschr.*: **49**, 631 (1923).
(*b*) A problem of amplification by isomorphism originating from the theory of relativity, II (Italian). *Atti Accad. Pont. Nuovi Lincei*, **76**, 146–56. *Fortschr.*: **49**, 631 (1923).

Hille, E. A pythagorean functional equation. *Ann. Math.* (2), **24**, 175–80. *Fortschr.*: **49**, 327 (1923).

Myrberg, P.J. Über das Additionstheorem analytischer Funktionen. *Ann. Acad. Sci. Fenn. Ser. A.*, **19**, No. 17, 1–9. *Fortschr.*: **51**, 293 (1925).

1924

Kaczmarz, S. Sur l'équation fonctionnelle $f(x) + f(x + y) = \phi(y) f[x + (y/2)]$. *Fund. Math.* **6**, 122–9. *Fortschr.*: **50**, 183 (1924).

Ruziewicz, S. Une application de l'équation fonctionnelle $f(x + y) = f(x) + f(y)$ à la décomposition de la droite en ensembles superposables nonmesurables. *Fund. Math.*, **5** (1924), 92–5. *Fortschr.*: **50**, 143 (1924).

Sierpiński, W. Sur une propriété des fonctions de M. Hamel. *Fund. Math.*, **5**, 334–6. *Fortschr.*: **50**, 183 (1924).

1925

Wilson, W.H. Two general functional equations. *Bull. Amer. Math. Soc.*, **31**, 330–4. *Fortschr.*: **51**, 311 (1925).

1926

Caccioppoli, R. On the functional equation $f(x + y) = f(x) + f(y)$ (Italian). *Boll. Un. Mat. Ital.*, **5**, 227–8. *Fortschr.*: **52**, 406 (1926).

Colucci, A. About the functional equation $f(x + y) = f(x) + f(y)$ (Italian). *Giorn. Mat. Battaglini*, **64**, 222–3. *Fortschr.*: **52**, 407 (1926).

Kormes, M. On the functional equation $f(x + y) = f(x) + f(y)$. *Bull. Amer. Math. Soc.*, **32**, 689–93. *Fortschr.*: **52**, 407 (1926).

1927

Ritt, J.F. Real functions with algebraic addition theorems. *Trans. Amer. Math. Soc.* **29**, 361–8. *Fortschr.*: **53**, 245 (1927).

1928

Caccioppoli, R. The functional equation $f(x + y) = F[f(x), f(y)]$ (Italian). *Giorn. Mat. Battaglini* **66**, 69–74. *Fortschr.*: **54**, 440 (1928).

Campbell, N.R. *An Account of the Principles of Measurement and Calculation.* Longmans, Green & Co., London, 1928. *Fortschr.*: **54**, 62, 935 (1928).

Haupt, O. Zur Theorie der

Exponentialfunktion und der Kreisfunktionen. *Sitzungsber. Phys.-Med. Soz. Erlangen*, **60**, 155–60. *Fortschr.*: **54**, 384 (1928).

Hille, E. A class of functional equations. *Ann. Math.* (2), **29**, 215–22. *Fortschr.*: **54**, 330 (1928).

Picard, É *Leçons sur quelques équations fonctionnelles avec des applications à divers problèmes d'analyse et de physique mathématique.* Gauthier-Villars, Paris, 1928. *Fortschr.*: **54**, 426 (1928).

Pólya, G. Über die Funktionalgleichung der Exponentialfunktion im Matrizenkalkul. *Sitzungsber. Preuss. Akad. Wiss.*, **1928**, 96–9. *Fortschr.*: **54**, 111 (1928).

Satô, R. A study of functional equations. *Proc. Phys.-Math. Soc. Japan* (3), **10**, 213–20. *Fortschr.*: **54**, 440 (1928).

Schur, I. Über die stetigen Darstellungen der allgemeinen linearen Gruppe. *Sitzungsber. Preuss. Akad. Wiss.*, **1928**, 100–124. *Fortschr.*: **54**, 148–9 (1928).

1929

Ostrowski, A. Mathematische Miszellen, XIV. Über die Funktionalgleichung der Exponentialfunktion und verwandte Funktionalgleichungen. *Jber. Deutsch. Math.-Verein.*, **38**, 54–62. *Fortschr.*:**55**, 800 (1929).

Thomsen, G. Topologische Fragen der Differentialgeometrie, XII. Schnittpunktsätze in ebenen Geweben. *Abh. Math. Sem. Univ. Hamburg*, **7**, 99–106. *Fortschr.*: **55**, 1014 (1929).

1930

Jessen, B. Über die Verallgemeinerung des arithmetischen Mittels. *Acta Sci. Math. (Szeged)* **5**, 108–16. *Fortschr.*: **57**, 309 (1931).

Kolmogorov, A.N. Sur la notion de la moyenne. *Atti Accad. Naz. Lincei Rend.* (6), **12**, 388–91. *Fortschr.*: **56**, 198 (1930).

Nagumo, M. Über eine Klasse der Mittelwerte. *Japan. J. Math.*, **7**, 71–9. *Fortschr.*: **56**, 198 (1930).

1931

Angheluţa, T. Sur une équation fonctionnelle caractérisant les polynômes. *Bull. Soc. Sci. Cluj*, **6**, 139–45. *Fortschr.*: **58**, 1100 (1932).

de Finetti, B. On the notion of mean value (Italian). *Giorn. Ist. Ital. Attuari*, **2**, 369–96. *Fortschr.*: **57**, 609 (1931).

Montel, P. Sur les fonctions d'une variable réelle qui admettent un théorème d'addition algébrique. *Ann. Sci. École Norm. Sup.* (3), **48**, 65–94. *Fortschr.* **57**, 1397 (1931).

Myrberg, P.J. Sur les systèmes de fonctions qui admettent un théorème d'addition algébrique. *C.R. Acad. Sci. Paris*, **193**, 916–17. *Fortschr.*: **57**, 1398 (1931).

Satô, R. On the solutions of a system of some functional equations. *Japan J. Math.*, **8**, 13–15. *Fortschr.*: **57**, 1434 (1931).

1932

Bohr, H. *Fastperiodische Funktionen.* (Ergebnisse Math. (1), No. 5). Springer, Berlin, 1932. (Reprint 1974). *Fortschr.*: **58**, 264 (1932).

Gołąb, S. Sur les fonctions homogènes, I. Équation d'Euler. *C.R. Soc. Sci. Varsovie*, **25**, 105–10. *Fortschr.*: **59**, 979 (1933).

Mazur, S. and Ulam, S.M. Sur les transformations isométriques d'espaces vectoriels, normés. *C.R. Acad. Sci. Paris*, **194**, 946–8. *Fortschr.*: **58**, 423 (1932).

1933

Favard, J. *Leçons sur les fonctions presque périodiques.* Gauthier-Villars, Paris, 1933. *Fortschr.*: **59**, 996–7 (1933).

1934

Bell, E.T. Exponential polynomials. *Ann. of Math.*, **35**, 258–77. *Fortschr.*: **60**, 295 (1934).

Bohr, H. Again the Kronecker theorem. *J. London Math. Soc.*, **9**, 5–6. *Fortschr.*: **60**, 302 (1934).

Hardy, G.H., Littlewood, J.E., and Pólya, G. *Inequalities.* Cambridge University Press, Cambridge, 1934 (second edn. 1952). *Fortschr.*: **60**, 169 (1934).

Kitagawa, T. On some class of weighted means. *Proc. Phys.-Math. Soc. Japan* (3), **16**, 117–26. *Fortschr.*: **60**, 170 (1934).

Mazur, S. and Orlicz, W.
(*a*) Grundlegende Eigenschaften der polynomischen Operationen, I. *Studia Math.*, **5**, 50–68. *Fortschr.*: **60**, 1074 (1934).
(*b*) Grundlegende Eigenschaften der polynomischen Operationen, II. *Studia Math.*, **5**, 179–89. *Fortschr.*: **60**, 1074 (1934).

Sierpiński, W. Remarques sur les fonctions de plusieurs variables réelles. *Prace Mat.-Fiz.* **41**, 171–5. *Fortschr.*: **60**, 971 (1934).

1935

Aumann, G. Aufbau von Mittelwerten mehrerer Argumente, II. *Math. Ann.* **111**, 713–30. *Fortschr.*: **61**, 200 (1935).

Jordan, P. and von Neumann, J. On inner products in linear, metric spaces. *Ann. Math.* (2), **36**, 719–23. *Fortschr.*: **61**, 435 (1935).

Nakahara, I. Axioms for the weighted means. *Tôhoku Math. J.*, **41**, 424–34. *Fortschr.*: **62**, 200 (1936).

1936

Popoviciu, T. Remarques sur la définition fonctionnelle d'un polynôme d'une variable réelle. *Mathematica (Cluj)*, **12**, 5–12, *Fortschr.*: **62**, 429 (1936).

Sierpiński, W. Sur une fonction universelle de deux variables réelles. *Bull. Internat. Acad. Polon. Sci. Lett. Cl. Sci. Math. Nat. Sér. A.*, **1936**, 8–12. *Fortschr.*: **62**, 232 (1936).

1937

Gołąb, S. Sur une définition axiomatique des nombres conjugués pour les nombres complexes ordinaires. *Opusc. Math. (Cracov.)* **1**, 1–11. *Fortschr.*: **63**, 839 (1937).

Kac, M. Une remarque sur les équations fonctionnelles. *Comment. Math. Helv.*, **9**, 170–171. *Fortschr.*: **63**, 408 (1937).

Lévy, P. *Théorie de l'addition des variables aléatoires.* Gauthier-Villars, Paris, 1937. *Fortschr.*: **63**, 490 (1937).

1938

Doeblin, W. Sur l'équation matricielle $A^{(t+s)} = [A^{(t)}A^{(s)}]$ et ses applications aux probabilités en chaîne. *Bull. Sci. Math.* (2), **62**, 21–32. *Fortschr.*: **64**, 539 (1938).

Fréchet, M. *Recherches théoriques modernes sur le calcul des probabilités, Vol. II: Méthode des fonctions arbitraires. Théorie des événements en chaîne dans le cas d'un nombre fini d'états possibles.* Gauthier-Villars, Paris, 1938. *Fortschr.*: **54**, 536 (1938).

Gołąb, S. Über eine Funktionalgleichung der Theorie der geometrischen Objekte. *Wiadom. Mat.*, **45**, 97–137. *Fortschr.*: **64**, 754 (1938).

1939

Erdös, P. and Wintner, A. Additive arithmetic functions. *Amer. J. Math.*, **61**, 713–21. *MR*: **1**, 40 (1940).

Hostinsky, B. *Equations fonctionnelles relatives aux probabilités continues en chaîne.* (Actualités Sci. Indust., No. 782). Hermann, Paris, 1939. *Fortschr.*: **65**, 1348 (1939).

Krafft, M. Herleitung der trigonometrischen Funktionen aus ihren Funktionalgleichungen. *Deutsch. Math.* **4**, 194–201. *Fortschr.*: **65**, 276 (1939).

Piccard, S. *Sur les ensembles de distances des ensembles de points d'un espace Euclidien.* (Mém. Univ. Neuchâtel, Vol. 13). Université, Neuchâtel, 1939. *MR*: **2**, 129 (1941).

1940

Alt, W. Über die reellen Funktionen einer reellen Veränderlichen welche ein rationales Additionstheorem besitzen. *Deutsch. Math.*, **5**, 1–12. *MR*: **1**, 297 (1940).

Bohnenblust, A. Axiomatic characterization of L_p-spaces. *Duke Math. J.*, **6**, 627–40. *MR*: **2**, 102 (1941).

van der Corput, J.G.

(*a*) Goniometric functions characterized by a functional equation, I (Dutch). *Euclides (Groningen)*, **17**, 55. *Fortschr.*: **66**, 1236–7 (1940).

(*b*) Goniometric functions characterized by a functional equation, II (Dutch). *Euclides (Groningen)*, **17**, 65–75. *Fortschr.*: **66**, 1236–7 (1940).

Lorenzen, P. Ein vereinfachtes Axiomensystem für Gruppen. *J. Reine Angew. Math.*, **182**, 50. *MR*: **2**, 1 (1941).

van der Lyn, G.

(*a*) Sur l équation fonctionnelle $f(x+y) + (f(x-y) = 2f(x)\phi(y)$. *Mathematica (Cluj)*, **16**, 91–6. *MR*: **2**, 134 (1941).

(*b*) Les polynômes abstraits. *Bull. Sci. Math.* (2), **64**, 55–80. *MR*: **1**, 259 (1940).

(*c*) Les polynômes abstraits. (Suite). *Bull. Sci. Math.* (2), **64**, 102–12. *MR*: **2**, 222 (1941).

Weil, A. *L'intégration dans les groupes topologiques et ses applications.* (Actualités Sci. Indust., No. 869), Hermann, Paris, 1940 (second edn. 1965). *MR*: **3**, 198 (1942).

1941

van der Corput, J.G. A remarkable family. *Euclides (Groningen)*, **18**, 50–64. *MR*: **7**, 385 (1946).

Hyers, D.H. On the stability of the linear functional equation. *Proc. Nat. Acad. Sci. USA*, **27**, 222–4. *MR*: **2**, 315 (1941).

1942

Haruki, H. On a certain simultaneous functional equation concerning the elliptic functions. *Proc. Phys.-Math. Soc. Japan* (3), **24**, 450–4. *MR*: **7**, 395 (1946).

Jones, F.B.
(*a*) Connected and disconnected plane sets and the functional equation $f(x) + f(y) = f(x+y)$. *Bull. Amer. Math. Soc.*, **48**, 115–20. *MR*: **3**, 229 (1942).
(*b*) Measure and other properties of a Hamel basis. *Bull. Amer. Math. Soc.*, **48**, 472–81. *MR*: **4**, 4 (1943).

Perron, O. Über eine für die Invariantentheorie wichtige Funktionalgleichung. *Math. Z.* **48**, 136–72. *MR*: **5**, 30 (1944).

1943

Dieudonné, J. Les déterminants sur un corps non commutatif. *Bull. Soc. Math. France*, **71**, 27–45. *MR*: **7**, 3 (1946).

1944

Ficken, F.A. Note on the existence of scalar products in normed linear spaces. *Ann. of Math.*, **45**, 362–6. *MR*: **5**, 270 (1944).

Haupt, O. Über einen Eindeutigkeitssatz für gewisse Funktionalgleichungen. *J. Reine Angew. Math.*, **186**, 58–64. *MR*: **6**, 275 (1945).

Haymann, W.K. A property of the probability integral. *Math. Gaz.*, **28**, 114–15.

Perron, O. Neuer Aufbau der nicht-euklidischen (hyperbolischen) Trigonometrie. *Math. Ann.*, **119**, 247–65. *MR*: **6**, 183 (1945).

Reisch, P. Neue Lösungen der Funktionalgleichung für Matrizen $\phi(X)\phi(Y) = \phi(XY)$. *Math. Z.*, **49**, 411–26. *MR*: **6**, 199 (1945).

Vietoris, L. Zur Kennzeichnung des Sinus und verwandter Funktionen durch Funktionalgleichungen, *J. Reine Angew. Math.*, **186**, 1–15. *MR*: **6**, 271 (1945).

Watson, G.N. On a functional equation. *Math. Gaz.*, **28**, 218–20.

1945

Alexiewicz, A. and Orlicz, W. Remarque sur l'équation fonctionnelle $f(x+y) = f(x) + f(y)$. *Fund. Math.*, **33**, 314–15. *MR*: **8**, 227 (1947).

Hadwiger, H. Über Verteilungsgesetze vom Poissonschen Typus. *Mitt. Verein. Schweiz. Versicherungsmath.*, **45** (1945), 257–77. *MR*: **7**, 215 (1946).

Hyers, D.H. and Ulam, S.M. On approximate isometries. *Bull. Amer. Math. Soc.*, **51**, 288–92. *MR*: **7**, 143 (1946).

Neĭshuler, L.J. On tabulating a class of inexplicit functions of four variables. *Dokl. Akad. Nauk SSSR*, **48**, 461–4. *MR*: **8**, 171 (1947).

1946

Aczél, J. The notion of mean values. *Norske Vid. Selsk. Forh. (Trondheim)*, **19**, 83–6. *MR*: **8**, 504 (1947).

Aczél, J. and Fenyö, I.
(*a*) Über die Theorie der Mittelwerte. *Acta Sci. Math. (Szeged)*, **11**, 239–45. *MR*: **10**, 237 (1949).
(*b*) On fields of forces in which centres of gravity can be defined. *Hungar. Acta Math.*, **1**, 53–60. *MR*: **10**, 159 (1949).

Erdös, P. On the distribution of additive functions. *Ann. of Math.* (2), **47**, 1–20. *MR*: **7**, 416 (1946).

Ghircoiasiu, N. Une équation fonctionnelle caractérisant les coniques. *Mathematica (Cluj)* **22**, 66–8. *MR*: **8**, 84 (1947).

Goła̧b, S. Sur la théorie des objets géométriques. *Ann. Soc. Polon. Math.*, **19**, 7–35. *MR*: **9**, 206 (1948).

Kestelman, H. On the functional equation $f(x+y) = f(x) + f(y)$. *Fund. Math.*, **34**, 144–7. *MR*: **9**, 188 (1948).

Riordan, J. Derivatives of composite functions. *Bull. Amer. Math. Soc.*, **52**, 664–7. *MR*: **8**, 208 (1947).

San Juan, R. An application of Diophantine approximation to the functional equation $f(x_1 + x_2) = f(x_1) + f(x_2)$ (Spanish). *Publ. Inst. Mat. Univ. Litoral*, **6**, 221–4. *MR*: **8**, 27 (1947).

Ulam, J. General means and its applications. *Ann. Soc. Polon. Math.*, **19**, 225–6.

Waraszkiewicz, Z. Sur les fonctions définies par les équations fonctionnelles $f((x+y)/2) = 2f(x)f(y))/(f(x) + f(y))$ *et* $f((x+y)/2) = \lambda f(x) + (1-\lambda)f(y)$, *où* $0 < \lambda < 1$. *Ann. Soc. Polon. Math.*, **19**, 237.

1947

Aczél, J. On mean values and operations defined for two variables. *Norske Vid. Selsk. Forh. (Trondheim)*, **20**, 37–40. *MR*: **9**, 572 (1948).

Bohr, H. *Almost periodic functions.* Chelsea, New York, 1947. (Reprinted 1951). *MR*: **8**, 512 (1947).

Day, M.M. Some characterizations of inner-product spaces. *Trans. Amer. Math. Soc.* **62**, 320–37. *MR*: **9**, 192 (1948).

Dunford, N. and Hille, E. The differentiability and uniqueness of continuous solutions of addition formulas. *Bull. Amer. Math. Soc.*, **53**, 799–805. *MR*: **9**, 95 (1948).

Hoheisel, G. Funktionalgleichung und Differenzierbarkeit bei den trigonometrischen Funktionen. *Math. Ann.*, **120**, 10–11. *MR*: **9**, 274 (1948).

Horváth, J. Sur le rapport entre les systèmes de postulats caractérisant les valeurs moyennes quasi arithmétiques symmétriques. *C.R. Acad. Sci. Paris*, **225**, 1256–7. *MR*: **9**, 337 (1948).

Hyers, D.H. and Ulam, S.M. Approximate isometries in the space of continuous functions. *Ann. of Math.* (2), **48**, 285–9. *MR*: **8**, 588 (1947).

James, R.C. Inner products on normed linear spaces. *Bull. Amer. Math. Soc.*, **53**, 559–66. *MR*: **9**, 42 (1948).

Segre, B. The automorphisms of the complex field and a problem of Corrado Segre (Italian). *Atti Accad. Naz. Lincei Rend.* (8), **3**, 414–20. *MR*: **10**, 231 (1949).

Shannon, C.E.
(*a*) A mathematical theory of communication, I. *Bell System Tech. J.* **27**, 379–423. *MR*:**10**, 133 (1949).
(*b*) A mathematical theory of communication, II. *Bell System Tech. J.* **27**, 623–56. *MR*: **10**, 133 (1949).

Ulam, J. The general mean. In *Sixth Meeting of Polish Mathematicians.* (Dodatek do Roczn. Polsk. Tow. Mat., No. 22). Warszawa, 1950, pp. 57–9.

1948

Aczél, J.
(*a*) Über eine Klasse von Funktionalgleichungen. *Comment. Math. Helv.*, **21**, 247–52. *MR*: **9**, 514 (1948).
(*b*) On mean values. *Bull. Amer. Math. Soc.*, **54**, 392–400. *MR*: **9**, 54 (1948).
(*c*) Un problème de M.L. Fejér sur la construction de Leibniz. *Bull. Sci. Math* (2), **72**, 35–45. *MR*: **10**, 357 (1949).

Gołąb, S. Sur la notion de dérivée covariante. *Colloq. Math.*, **1**, 160.

Hille, E. *Functional Analysis and Semi-groups.* (Colloq. Publ. Amer. Math. Soc., Vol. 31). Amer. Math. Soc., New York, 1948. *MR*: **9**, 594 (1948).

Horváth, J. Note sur un problème de L. Fejér. *Bull. Inst. Politehn. Iasi*, **3**, 164–8. *MR*: **10**, 237 (1949).

Lorch, E.R. On certain implications which characterize Hilbert space. *Ann. of Math.* (2), **49**, 523–32. *MR*: **10**, 129 (1949).

Martis-Biddau, S. On the characterization of some classes of functions (Italian). *Collect. Math.*, **1**, 67–84. *MR*: **11**, 251 (1950).

Mikusiński, J. Sur quelques équations fonctionnelles. *Ann. Soc. Polon. Math.*, **21**, 346.

Montel, P. Sur des équations fonctionnelles caractérisant les polynômes. *C.R. Acad. Sci. Paris*, **226**, 1053–5. *MR*: **9**, 515 (1948).

1949

Aczél, J.
(*a*) Sur les opérations définies pour nombres réels. *Bull. Soc. Math. France*, **76**, 59–64. *MR*: **10**, 685 (1949).
(*b*) Über einparametrige Transformationen *Publ. Math. Debrecen*, **1**, 243–7. *MR*: **12**, 673 (1951).
(*c*) Einige aus Funktionalgleichungen zweier Veränderlichen ableitbare Differentialgleichungen. *Acta Sci. Math. (Szeged)*, **13**, 179–89. *MR*: **13**, 246 (1952).

Aczél, J., Fenyö, I. and Horváth, J. Sur certaines classes de fonctionnelles. *Portugal. Math.*, **8**, 1–11. *MR*: **11**, 584 (1950).

Arrighi, G. On the functional equation $2\phi(x)\phi(y) = \phi(x+y) + \phi(x-y)$ (Italian). *Boll. Un. Mat. Ital.* (3), **4**, 255–7. *MR*: **11**, 598 (1950).

Barnard, G.A. Statistical inference. *J. Roy. Statist. Soc. Ser. B*, **11**, 119–49. *MR*: **11**, 672 (1950).

de Bruijn, N.G. Functions whose differences belong to a given class. *Nieuw. Arch. Wisk.* (2), **23**, 194–218. *MR*: **13**, 332 (1952).

Etherington, I.M.H. Non-associative arithmetics. *Proc. Roy. Soc. Edinburgh Sect. A*, **62**, 442–53. *MR*: **10**, 677 (1949).

Fridman, A.A. On convex functions defined on sets and on the functional equation $f(x+y) = f(x) + f(y)$ (Russian). *Učen. Zap. Kazah. Univ.*, **12**, 21–33.

Gołąb, S. Sur les objets géométriques non différentiels. *Bull. Internat. Acad. Polon. Sci. Lett. Cl. Sci. Math. Nat. Sér. A*, **1949**, 67–72. *MR*: **11**, 690 (1950).

Hua, L.K. On the automorphisms of a sfield. *Proc. Nat. Acad. Sci. USA*, **35**, 386–9. *MR*: **10**, 675 (1949).

Knaster, B. Sur une équivalence pour les fonctions. *Colloq. Math.*, **2**, 1–4. *MR*: **12**, 395 (1951).

Ryll-Nardzewski, C. Sur les moyennes. *Studia Math.*, **11**, 31–7. *MR*: **12**, 12 (1951).

Tamari, D. Caractérisation des semi-groupes à un paramètre. *C.R. Acad. Sci. Paris*, **228**, 1092–4. *MR*: **10**, 508 (1949).
Thielman, H.P.
(*a*) On generalized means. *Proc. Iowa Acad. Sci.*, **56**, 241–7. *MR*: **12**, 12 (1951).
(*b*) On generalized Cauchy functional equations. *Amer. Math. Monthly*, **56**, 452–7. *MR*: **11**, 183 (1950).

1950

Aczél, J. Zur Charakterisierung nomographisch einfach darstellbarer Funktionen durch Differential- und Funktionalgleichungen. *Acta Sci. Math. (Szeged)*, **12A**, 73–86. *MR*: **12**, 541 (1951).
Birkhoff, G. Moyennes des fonctions bornées. In *Algèbre et théorie des nombres*. (Colloq. Internat. CNRS, Vol. 24). CNRS, Paris, 1950, pp. 143–53. *MR*: **13**, 361 (1952).
Fuchs, L. On mean systems. *Acta Math. Acad. Sci. Hungar.*, **1**, 303–20. *MR*: **13**, 922 (1952).
Gołąb, S. Sur les objets géométriques à une composante. *Ann. Soc. Polon. Math.*, **23**, 79–89. *MR*: **12**, 749 (1951).
Good, I.J. *Probability and the Weighing of Evidence*. Appendix III. Charles Griffin, London; Haffner, New York, 1950. *MR*: **12**, 837 (1951).
Halmos, P.R. *Measure Theory*. Van Nostrand, New York, 1950. *MR*: **11**, 504 (1950).
Hewitt, E. and Zuckerman, H.S. A group-theoretic method in approximation theory. *Ann. of Math.* (2), **52**, 557–67. *MR*: **12**, 801 (1951).
Jánossy, L., Rényi, A., and Aczél, J. On composed Poisson-distributions. *Acta Math. Acad. Sci. Hungar.*, **1**, 209–24. *MR*: **13**, 663 (1952).
Lebesgue, H. *Leçons sur les constructions géométriques*. Gauthier-Villars, Paris, 1950. *MR*: **11**, 678 (1950).
Maak, W. *Fastperiodische Funktionen*. (Grundlehren math. Wiss., Vol. 61). Springer, Berlin-Göttingen-Heidelberg, 1950. *MR*: **11**, 101 (1950).
Milkman, J. Note on the functional equations $f(xy) = f(x) + f(y), f(x^n) = nf(x)$. *Proc. Amer. Math. Soc.* **1**, 505–8. *MR*: **10**, 411 (1951).
Sade, A. *Quasigroupes*. Marseille, 1950. *MR*: **13**, 203 (1952).
Thielman, H.P. On a pair of functional equations. *Amer. Math. Monthly*, **57**, 544–7. *MR*: **12**, 680 (1951).

1951

Aczél, J., Kalmár, L., and Mikusiński, J. Sur l'équation de translation. *Studia Math.*, **12**, 112–16. *MR*: **13**, 246 (1952).
Halperin, I. Nonmeasurable sets and the equation $f(x + y) = f(x) + f(y)$. *Proc. Amer. Math. Soc.*, **2**, 221–4. *MR*: **12**, 685 (1951).
Higman, G. and Neumann, B.H. Groups as groupoids with one law. *Publ. Math. Debrecen*, **2**, 215–21. *MR*: **15**, 284 (1954).
Kestelman, H. Automorphisms in the field of complex numbers. *Proc. London Math. Soc.* (2), **53**, 1–12. *MR*: **12**, 812 (1951).
Kuwagaki, A. Sur l'équation fonctionnelle $f(x + y) = R[f(x), f(y)]$. *Mem. Coll. Sci. Univ. Kyoto, Ser. A*, **26**, 139–44. *MR*: **13**, 466 (1952).
Mohr, E. Bemerkung zur Dimensionsanalysis. *Math. Nachr.*, **6**, 145–53. *MR*: **13**, 616 (1952).
Pidek, H. Sur les objets géométriques de classe zéro qui admettent un algèbre. *Ann. Soc. Polon. Math.*, **24**, 111–28. *MR*: **15**, 899 (1954).
Rényi, A. On composed Poisson-distributions, II. *Acta Math. Acad. Sci. Hungar.*, **2**, 83–98. *MR*: **13**, 663 (1952).
Sierpiński, W. *Algèbre des ensembles*. §27. (Monografie Mat., Vol. 23). P.W.N. Warszawa-Wrocław, 1951. *MR*: **13**, 541 (1952).
Thielman, H.P. A note on a functional equation. *Amer. J. Math.*, **73**, 482–4. *MR*: **13**, 43 (1952).

1952

Aczél, J.
(*a*) On composed Poisson-distributions, III. *Acta Math. Acad. Sci. Hungar.*, **3**, 219–24. *MR*: **14**, 770 (1953).
(*b*) Bemerkungen über die Multiplikation von Vektoren und Quarternionen. *Acta Math. Acad. Sci. Hungar.*, **3**, 309–16. *MR*: **15**, 383 (1954).
Bellman, R. A note on scalar functions of matrices. *Amer. Math. Monthly*, **59**, 391. *MR*: **14**, 6 (1953).
de Bruijn, N.G. A difference property for Riemann integrable functions. *Nederl. Akad. Wetensch. Indag. Math.*, **14**, 145–51. *MR*: **13**, 830 (1952).
Kuwagaki, A.
(*a*) Sur les fonctions de deux variables satisfaisant une formule d'addition algébrique. *Mem. Coll. Sci. Univ. Kyoto, Ser. A*, **27**, 139–43. *MR*: **14**, 760 (1953).
(*b*) Sur l'équation fonctionnelle rationnelle

Kuwagaki, A. (*cont.*)
de la fonction inconnue de deux variables. *Mem. Coll. Sci. Univ. Kyoto, Ser. A*, **27**, 145–51. *MR*: **14**, 760 (1953).
(*c*) Sur la fonction analytique de deux variables complexes satisfaisant l'associativité $f\{x, f(y,z)\} = f\{f(x,y), z\}$. *Mem. Coll. Sci. Univ. Kyoto, Ser. A*, **27**, 225–34. *MR*: **15**, 324 (1954).

Szász, P. Neue Herleitung der hyperbolischen Trigonmetrie durch Verwendung der Grenzkugel. *Acta Math. Acad. Sci. Hungar.*, **3**, 327–33. *MR*: **14**, 1007 (1953).

1953

Gáspár, G. Eine neue Definition der Determinanten. *Publ. Math. Debrecen*, **3**, 257–260. *MR*: **17**, 338 (1956).

Hosszú, M.
(*a*) On the functional equation of distributivity. *Acta Math. Acad. Sci. Hungar.* **4**, 159–67. *MR*: **15**, 324 (1954).
(*b*) A generalization of the functional equation of bisymmetry. *Studia Math.* **14**, 100–6. *MR*: **15**, 962 (1954).
(*c*) On the functional equation of autodistributivity. *Publ. Math. Debrecen*, **3**, 83–6. *MR*: **15**, 962 (1954).
(*d*) Some functional equations related with the associative law. *Publ. Math. Debrecen*, **3**, 205–14. *MR*: **17**, 236 (1956).
(*e*) On the functional equation of transitivity. *Acta Sci. Math.* (*Szeged*) **15**, 203–8. *MR*: **16**, 371 (1955).

Khinchin, A.J. The concept of entropy in the theory of probability (Russian). *Uspekhi Mat. Nauk*, **8**, No. 3 (55), 3–20. *MR*: **15**, 238 (1954).

Moser, L. and Lambek, J. On monotone multiplicative functions. *Proc. Amer. Math. Soc.*, **4**, 544–5. *MR*: **15**, 104 (1953).

Redheffer, R.M. A note on the Poisson law. *Math. Mag.*, **26**, 185–8. *MR*: **14**, 1098 (1953).

Schutzenberger, M.P. Une interprétation de certaines solutions de l'équation fonctionnelle $F(x+y) = F(x)F(y)$. *C.R. Acad. Sci. Paris*, **236**, 352–3. *MR*: **14**, 768 (1953).

1954

Gołąb, S. Über den Begriff der kovarianten Ableitung. *Nieuw Arch. Wisk.* (3), **2**, 90–6. *MR*: **16**, 76 (1955).

Maier, W. Inhaltsmessung in R_3 fester Krümmung. Ein Beitrag zur imaginären Geometrie. *Arch. Math.* (*Basel*), **5**, 266–73. *MR*: **16**, 394 (1955).

Pidek, H.
(*a*) Sur un problème de l'algèbre des objets géométriques de classe zéro dans l'espace X_1. *Ann. Polon. Math.*, **1**, 114–26. *MR*: **16**, 173 (1955).
(*b*) Sur un problème de l'algèbre des objets géométriques de classe zéro dans l'espace X_m. *Ann. Polon. Math.*, **1**, 127–34. *MR*: **16**, 174 (1955).

Redheffer, R.M. Novel uses of functional equations. *J. Rational Mech. Anal.*, **3**, 271–9. *MR*: **15**, 763 (1954).

1955

Aczél, J.
(*a*) Remarques algébriques sur la solution donnée par M. Fréchet à l'équation de Kolmogoroff, I. *Publ. Math. Debrecen*, **4**, 33–42. *MR*: **16**, 989 (1955).
(*b*) Über Additions und Subtraktionstheoreme. *Publ. Math. Debrecen*, **4**, 325–33. *MR*: **18**, 488 (1957).
(*c*) Lösung der Vektor-Funktionalgleichung der homogenen und inhomogenen *n*-dimensionalen einparametrigen 'Translation' der erzeugenden Funktion von Kettenreaktionen und des stationären und nicht-stationären Bewegungsintegrals. *Acta. Math. Acad. Sci. Hungar.*, **6**, 131–41. *MR*: **17**, 272 (1956).

Aczél, J. and Varga, O. Bemerkung zur Cayley–Kleinschen Massbestimmung. *Publ. Math. Debrecen*, **4**, 3–15. *MR*: **17**, 177 (1956).

van den Berg, J. Über die Funktionalgleichung $\phi(\alpha x) - \beta\phi(x) = F(x)$, *II. Nieuw. Arch. Wisk.* (3), **3**, 113–23. *MR*: **17**, 375 (1956).

Erdös, P. and Golomb, M. Functions which are symmetric about several points. *Nieuw Arch. Wisk.* (3), **3**, 13–19. *MR*: **16**, 931 (1955).

Fürstenberg, H. The inverse operation in groups. *Proc. Amer. Math. Soc.* **6**, 991–7. *MR*: **17**, 1053 (1956).

Singer, I.M. and Wermer, J. Derivations on commutative normed algebras. *Math. Ann.*, **129**, 260–4. *MR*: **16**, 1125 (1955).

Vaughan, H.E. Characterization of the sine and cosine. *Amer. Math. Monthly*, **62**, 707–13. *MR*: **17**, 631 (1956).

1956

Aczél, J.
(*a*) Beiträge zur Theorie der

geometrischen Objekte, I–II. *Acta Math. Acad. Sci. Hungar.*, **7**, 339–54. *MR*: **19**, 677 (1958).
(*b*) On the introduction of the natural logarithm and exponential function (Hungarian). *Mat. Lapok*, **7**, 101–5.
Aczél, J. and Hosszú, M. On transformations with several parameters and operations in multidimensional spaces. *Acta Math. Acad. Sci. Hungar.*, **7**, 327–38. *MR*: **19**, 41 (1958).
Climescu, A. On the axiomatic definition of determinants (Romanian). *Bul. Inst. Politehn. Iaşi* **2(6)**, Nos. 3–4, 1–7. *MR*: **20**, # 1694 (1959).
Császár, Á. Sur une caratérisation de la repartition normale de probabilités. *Acta Math. Acad. Sci. Hungar.*, **7**, 359–82. *MR*: **19**, 326 (1958).
Faddeev, D.K. On the concept of entropy of a finite probabilistic scheme (Russian). *Uspekhi Mat. Nauk*, **11**, No. 1 (67), 227–31. *MR*: **17**, 1098 (1956).
Fenyö, I. Über eine Lösungsmethode gewisser Funktionalgleichungen. *Acta Math. Acad. Sci. Hungar.*, **7**, 383–96. *MR*: **19**, 152 (1958).
Kolmogorov, A.N. On the representation of continuous functions of several variables by superpositions of continuous functions of a smaller number of variables (Russian). *Dokl. Akad. Nauk. SSSR*, **108**, 179–82. *MR*: **18**, 197 (1957).
Kurepa, S.
(*a*) On some functional equations. *Glasnik Mat.-Fiz. Astronom. Društvo Mat. Fiz. Hrvatske* (2), **11**, 3–5. *MR*: **18**, 217 (1957).
(*b*) Convex functions. *Glasnik Mat.-Fiz. Astronom. Društvo Mat. Fiz. Hrvatske* (2), **11**, 89–94. *MR*: **19**, 408 (1958).
Marcus, S. Sur une généralisation des fonctions de G. Hamel. *Atti Accad. Naz. Lincei Rend.* (8), **20**, 584–9. *MR*: **18**, 794 (1957).
Mycielski, J. and Paszkowski, S. Sur un problème du calcul de probabilité, I. *Studia Math.*, **15**, 188–200. *MR*: **19**, 588 (1958).
Redheffer, R.M. On solutions of Riccati's equation as functions of the initial values. *J. Rational Mech. Anal.*, **5**, 835–48. *MR*: **19**, 558 (1958).
Richter, H. *Wahrscheinlichkeitstheorie.* (Grundlehren math. Wiss., Vol. 86). Springer, Berlin-Göttingen-Heidelberg, 1956. *MR*: **18**, 767 (1957).
Stein, S.K. Foundations of quasigroups. *Proc. Nat. Acad. Sci. U.S.A.*, **42**, 545–6. *MR*: **18**, 111 (1957).
Steinhaus, H. Un problème sur la classification des fonctions continues de plusieurs variables. *Colloq. Math.*, **4**, 264.

1957

Aczél, J.
(*a*) Beiträge zur Theorie der geometrischen Objekte, III–IV. *Acta Math. Acad. Sci. Hungar.*, **8**, 19–52. *MR*: **19**, 677 (1958).
(*b*) Beiträge zur Theorie der geometrischen Objekte, V. *Acta Math. Acad. Sci. Hungar.*, **8**, 53–64. *MR*: **19**, 677 (1958).
Aczél, J. and Egerváry, J. Remarques algébriques sur la solution donnée par M. Fréchet à l'équation de Kolmogoroff, II. *Publ. Math. Debrecen*, **5**, 60–71. *MR*: **19**, 891 (1958).
Aczél, J. and Kiesewetter, H. Über die Reduktion der Stufe bei einer Klasse von Funktionalgleichungen. *Publ. Math. Debrecen*, **5**, 348–63. *MR*: **26**, #6633 (1963).
Arnold, V.I. On the representation of a function of two variables in the form $\chi[\phi(x)+\psi(y)]$ (Russian). *Uspekhi Mat. Nauk (N.S.)*, **12**, No. 2 (74), 119–121. *MR*: **19**, 841 (1958).
Dixmier, J. Quelques propriétés des groupes abéliens localement compacts. *Bull. Sci. Math.* (2), **81**, 38–48. *MR*: **20**, #3926 (1959).
Erdös, P. On the distribution function of additive arithmetical functions and on some related problems. *Rend. Sem. Mat. Fis. Milano*, **27**, 45–49. *MR*: **20**, #7004 (1959).
Gołąb, S. and Pidek, H. Sur l'algèbre des objets géométriques de première classe à une composante. *Ann. Polon. Math.*, **4**, 226–48. *MR*: **20**, #4282 (1959).
Hadwiger, H. *Vorlesungen über Inhalt, Oberfläche und Isoperimetrie* (Grundlehren math. Wiss., Vol. 93). Springer, Berlin-Göttingen-Heidelberg, 1957. *MR*: **21**, #1561 (1960).
Haruki, H. On the functional equation $f(x+y)=F\{f(x),f(y)\}$. *Sci. Rep. Osaka Univ.*, **6**, 11–12. *MR*: **31**, #524 (1966).
Hille, E. and Phillips, R.S. *Functional Analysis and Semi-groups.* (Colloq. Publ. Amer. Math. Soc., No. 31). Amer. Math. Soc., Providence, R.I., 1957. *MR*: **19**, 664 (1958).
Hosszú, M. Functional equations and algebraic methods in the theory of geometric objects. *Publ. Math. Debrecen*, **5**, 294–329. *MR*: **21**, #3518 (1960).
Kampé de Fériet, J. Notion de moyenne dans la théorie de la turbulence. *Rend. Sem.*

Kampé de Fériet (*cont.*)
Mat. Fis. Milano, **27**, 167–207. *MR*: **24**, #A2995 (1962).
Kemperman, J.H.B. A general functional equation. *Trans. Amer. Math. Soc.*, **86**, 28–56. *MR*: **20**, #22 (1959).
Kiesewetter, H. Eine Bemerkung über partielle Differentiationen bei N.H. Abel. *Publ. Math. Debrecen*, **5**, 265–8. *MR*: **22**, #9554 (1961).
Kolmogorov, A.N. On the representation of continuous functions of many variables by superposition of continuous functions of one variable and addition (Russian). *Dokl. Akad. Nauk SSSR*, **114**, 953–6. *MR*: **22**, #2669 (1961).
Robinson, R.M. A curious trigonometric identity. *Amer. Math. Monthly*, **64**, 83–5. *MR*: **18**, 569 (1957).
Sade, A. Quasigroupes obéissant à certaines lois. *Rev. Fac. Sci. Univ. Istanbul Sér. A*, **22**, 151–84. *MR*: **21**, #4987 (1960).
Sorkin, J.I. Rings as sets with one operation subject to a unique identity (Russian). *Uspekhi Mat. Nauk*, **12**, No. 4 (76), 357–62. *MR*: **19**, 1035 (1958).
Stein, S.K. On the foundations of quasigroups. *Trans. Amer. Math. Soc.*, **85**, 228–56. *MR*: **20**; #922 (1959).
Vietoris, L. Vom Grenzwert $\lim_{x\to 0}(\sin x/x)$. *Elem. Math.*, **12**, 8–10.

1958

Aczél, J. Miszellen über Funktionalgleichungen, I. *Math. Nachr.*, **19**, 87–99. *MR*: **21**, #7374 (1960).
Bergmann, A. Über den Zusammenhang von Norm und Spur und ihrer Funktionalgleichungen in gewissen Algebren und dessen Anwendung auf Ringe mit Spurenbedingungen. *Math. Nachr.*, **19**, 237–54. *MR*: **21**, #4165 (1960).
Bruck, R.H. *A Survey of Binary Systems.* (Ergebnisse Math., (2), No. 20). Springer, Berlin-Göttingen-Heidelberg, 1958. *MR*: **20**, #76 (1959).
Dunford, N. and Schwartz, J.T. *Linear Operators. Part I. General theory.* Interscience, New York-London, 1958. *MR*: **22**, #8302 (1960).
Kelley, J.L. Averaging operators on $C_\infty(X)$. *Illinois J. Math.*, **2**, 214–23. *MR*: **21**, #2179 (1960).
Kurepa, S.
(*a*) Semigroups of linear transformations in *n*-dimensional vector space. *Glasnik Mat.-Fiz. Astronom. Društvo Mat. Fiz. Hrvatske* (2), **13**, 3–32. *MR*: **21**, #695 (1960).
(*b*) A cosine functional equation in *n*-dimensional vector space. *Glasnik Mat.-Fiz. Astronom. Društvo Mat. Fiz. Hrvatske* (2), **13**, 169–89. *MR*: **20**, #7156 (1959).
Lions, J.L. Problèmes mixtes abstraits. In *Proc. Intern. Math. Congr. Edinburgh, 1958.* Cambridge University Press, Cambridge, 1960, pp. 389–97. *MR*: **23**, #A3925 (1962).
Moór, A. Über die kovariante Ableitung der Vektoren. *Acta Sci. Math. (Szeged)*, **19**, 237–46. *MR*: **21**, #887 (1960).
Riordan, J. *An Introduction to Combinatorial Analysis.* Wiley, New York, 1958. *MR*: **20**, #3077 (1959).
Tverberg, H. A new derivation of the information function. *Math. Scand.*, **6**, 297–8. *MR*: **21**, #3691 (1960).
Young, G.S. The linear functional equation. *Amer. Math. Monthly*, **65**, 37–8. *MR*: **20**, #4106 (1959).
Zariski, O. and Samuel, P. *Commutative Algebra.* Vol. 1, Chapt. II. Van Nostrand, Princeton, N.J.-Toronto-London-New York, 1958. *MR*: **19**, 833 (1959).

1959

Aczél, J. Beiträge zur Theorie der geometrischen Objekte, VI. *Acta Math. Acad. Sci. Hungar.*, **10**, 1–12. *MR*: **21** #6595 (1960).
Arnold, V.I. On the representation of continuous functions of three variables by superposition of continuous functions of two variables (Russian). *Mat. Sbornik*, **48(90)**, 3–74. *MR*: **22**, #12191 (1961).
Belousov, V.D. Medial systems of quasigroups (Russian). *Uspekhi Mat. Nauk*, **14**, No. 5 (89), 213–14.
Bergmann, A. Ein Axiomensystem für Determinanten. *Arch. Math. (Basel)*, **10**, 243–56. *MR*: **21**, #4168 (1962).
Carroll, F.W. Difference properties for polynomials and exponential polynomials on topological groups. *Notices Amer. Math. Soc.*, **6**, 844.
Ciesielski, C. Some properties of convex functions of higher orders. *Ann. Polon. Math.*, **7**, 1–7. *MR*: **22**, #89 (1961).
Climescu, A. On the definition of trigonometric functions (Romanian). *Bul. Inst. Politehn. Iaşi*, **5(9)**, 35–8. *MR*: **23**, #266 (1962).
Erdös, J. A remark on the paper 'On some functional equations' by S. Kurepa. *Glasnik Mat.-Fiz. Astronom. Drušvto Mat. Fiz. Hrvatske* (2), **14**, 3–5. *MR*: **21**, #7375 (1960).

Gáspár, G. Eine axiomatische Theorie der Dieudonnéschen Determinanten. *Publ. Math. Debrecen*, **6**, 298–302. *MR*: **22**, #4734 (1961).

Gołab, S. Sur l'équation $f(X) \cdot f(Y) = f(X \cdot Y)$ *Ann. Polon. Math.*, **6**, 1–13. *MR*: **21**, #2659 (1960).

Gołąb, S. and Schinzel, A. Sur l'équation fonctionnelle $f[x + y \cdot f(x)] = f(x) \cdot f(y)$. *Publ. Math. Debrecen* **6**, 113–25. *MR*: **21**, #5828 (1960).

Guinand, A.P. Les équations fonctionnelles de la loi d'associativité ternaire. *C.R. Acad. Sci. Paris*, **249**, 23–4. *MR*: **21**, #5831 (1960).

Hosszú, M.
(*a*) A generalization of the functional equation of distributivity. *Acta Sci. Math. (Szeged)*, **20**, 67–80. *MR*: **21**, #434 (1960).
(*b*) Nonsymmetric means. *Publ. Math. Debrecen*, **6**, 1–9. *MR*: **20**, #4109 (1959).
(*c*) A remark on scalar valued multiplicative functions of matrices. *Publ. Math. Debrecen*, **6**, 288–9. *MR*: **22**, #3744 (1961).

Kucharzewski, M. Über die Funktionalgleichung $f(a_k^i) \cdot f(b_k^i) = f(b_\alpha^i a_k^\alpha)$. *Publ. Math. Debrecen* **6**, 181–98. *MR*: **22**, #3742 (1961).

Kuczma, M. Bemerkung zur vorhergehenden Arbeit von M. Kucharzewski. *Publ. Math. Debrecen*, **6**, 199–203. *MR*: **22**, #3743 (1961).

Kurepa, S.
(*a*) Functional equations for invariants of a matrix. *Glasnik Mat.-Fiz. Astronom. Društvo Mat. Fiz. Hrvatske* (2), **14**, 97–113. *MR*: **22**, #5745 (1961).
(*b*) On the quadratic functional. *Acad. Serbe Sci. Publ. Inst. Math.*, **13**, 57–72. *MR*: **24**, #A1040 (1962).

Luce, R.D. On the possible psychophysical laws. *Psychological Rev.*, **66**, 81–95.

Marcus, S. Généralisations, aux fonctions de plusieurs variables, des théorèmes de Alexander Ostrowski et de Masuo Hukuhara concernant les fonctions convexes (J). *J. Math. Soc. Japan*, **11**, 173–6. *MR*: **22**, #12179 (1961).

Moór, A. Über Tensoren, die aus angegebenen geometrischen Objekten gebildet sind. *Publ. Math. Debrecen*, **6**, 15–25. *MR*: **21**, #3877 (1960).

Pfanzagl, J. *Die axiomatischen Grundlagen einer allgemeinen Theorie des Messens.* (Schriftenreihe Stat. Inst. Univ. Wien [N.F.], No. 1). Physica, Würzburg, 1959 *MR*: **21**, #2529 (1960).

Radó, F.
(*a*) Équations fonctionnelles caractérisant les nomogrammes avec trois échelles rectilignes. *Mathematica (Cluj)*, **1(24)**, 143–66. *MR*: **22**, #11238 (1961).
(*b*) Sur quelques équations fonctionnelles avec plusieurs fonctions à deux variables. *Mathematica (Cluj)*, **1(24)**, 321–39. *MR*: **23**, #A1961 (1962).

Stein, S.K.
(*a*) On a construction of Hosszú. *Publ. Math. Debrecen*, **6**, 10–14. *MR*: **21**, #3503 (1960).
(*b*) Left-distributive quasigroups. *Proc. Amer. Math. Soc.*, **10**, 577–8. *MR*: **21**, #7263 (1960).

Straus, E.G. A functional equation proposed by R. Bellman. *Proc. Amer. Math. Soc.* **10**, 860–2. *MR*: **22**, #5831 (1961).

1960

Aczél, J.
(*a*) Verallgemeinerte Addition von Dichten. *Publ. Math. Debrecen*, **7**, 10–15. *MR*: **22**, #7077 (1961).
(*b*) Über die Differenzierbarkeit der integrierbaren Lösungen gewisser Funktionalgleichungen. *Ann. Univ. Sci. Budapest. Eötvös Sect. Math.*, **3–4**, 5–8. *MR*: **27**, #2745 (1964).

Aczél, J., Belousov, V.D., and Hosszú, M. Generalized associativity and bisymmetry on quasigroups. *Acta Math. Acad. Sci. Hungar.*, **11**, 127–36. *MR*: **25**, #4518 (1963).

Aczél, J., Ghermănescu, M., and Hosszú, M. On cyclic equations. *Magyar. Tud. Akad. Mat. Kutató Int. Közl.*, **5**, 215–21. *MR*: **23**, #A2660 (1962).

Aczél, J. and Gołąb, S. *Funktionalgleichungen der Theorie der geometrischen Objekte.* (Monografie Mat., Vol. 39). P.W.N., Warszawa, 1960. *MR*: **24**, #3588 (1962).

Aczél, J., Gołąb, S., Kuczma, M. and Siwek, E. Das Doppelverhältnis als Lösung einer Funktionalgleichung. *Ann. Polon. Math.*, **9**, 183–7. *MR*: **27**, #1731 (1964).

Baxter, G. An analytic problem whose solution follows from a single algebraic identity. *Pacific J. Math.*, **10**, 731–42. *MR*: **22**, #9990 (1960).

Billik, M. and Rota, G.-C. On Reynolds operators in finite-dimensional algebras. *J. Math. Mech.*, **9**, 927–32. *MR*: **23**, #A 2449 (1962).

Chaundy, T.W. and McLeod, J.B. On a functional equation. *Proc. Edinburgh Math. Soc.*, **43**, 7–8. *MR*: **27**, #1732 (1964).

Gantmacher, F.R. *The Theory of Matrices.* (Transld. from Russian by K.A. Hirsch), second edn. Chelsea, New York, 1960. *MR*: **21**, #6372 (1960).

Ghermănescu, M. *Functional Equations* (Romanian). Acad. R.P. Rom., Bucureşti, 1960, *MR*: **23**, #A3931 (1962).

Gołąb, S. Sur les comitants algébriques des tenseurs. *Ann. Polon. Math.*, **9**, 113–18. *MR*: **24**, #A359 (1962).

Good, I.J. Weight of evidence, corroboration, explanatory power, information and the utility of experiments. *J. Roy. Statist. Soc. Ser. B*, **22**, 319–331. *MR*: **22**, #7182 (1961).

Henney, D.R. Quadratic setvalued functions. *Ark. Mat.*, **4**, 377–8. *MR*: **25**, #3940 (1963).

Herman, V. A functional equation which defines rational functions (Romanian). *Gaz. Mat. Fiz. Ser. A*, **12(65)**, 636–9. *MR*: **23**, #A2656 (1962).

Hosszú, M. Remarks on scalar valued multiplicative functions of matrices (Hungarian). *Nehézip. Müsz. Egy. Közl.* **5**, 173–7.

Kucharzewski, M. and Kuczma. M. On linear differential geometric objects with one component. *Tensor* (*N.S.*) **10**, 245–54. *MR*: **24**, #A3590 (1962).

Kurepa, S.

(*a*) A cosine functional equation in Hilbert space. *Canad. J. Math.* **12**, 45–50. *MR*: **22**, #152 (1961).

(*b*) On the functional equation $f(x+y) = f(x)f(y) - g(x)g(y)$. *Glasnik Mat.-Fiz. Astronom. Društvo Mat. Fiz. Hrvatske* (2) **15**, 31–48. *MR*: **25**, #355 (1963).

(*c*) Functional equation $F(x+y)F(x-y) = F^2(x) - F^2(y)$ in *n*-dimensional vector space. *Monatsh. Math.* **64**, 321–9. *MR*: **22**, #8239 (1961).

(*d*) On some functional equations in Banach space. *Studia Math.* **19**, 149–158. *MR*: **22**, #9755 (1961).

Luce, R.D. The theory of selective information and some of its behavioral applications. In *Developments in mathematical psychology, information, learning and tracking.* Glencoe Free Press, Glencoe, IL, 1960, pp. 1–119. *MR*: **22**, #12000 (1961).

Prešić, S.B. Sur l'équation fonctionnelle de translation. *Univ. Beograd. Publ. Elektrotehn. Fak. Ser. Mat. Fiz.* **1960**, No. 44–48, 15–16.

Radó, F. Eine Bedingung für die Regularität der Gewebe. *Mathematica (Cluj)*, **2(25)**, 325–334. *MR*: **23**, #A3771 (1962).

Rådström, H. One parameter semigroups of subsets of a real linear space. *Ark. Mat.*, **4**, 87–97. *MR*: **26**, #3802 (1963).

Rényi, A. On measures of entropy and information. In *Proceedings of the 4th Berkeley Symposium on Mathematical Statistics and Probability, Berkeley, 1960*, Vol. 1. Univ. of Calif. Press, Berkeley, 1961, pp. 547–61. *MR*: **24**, #A2410 (1962).

Rota, G.-C.

(*a*) On the representation of averaging operators. *Rend. Sem. Mat. Univ. Padova*, **30**, 52–64. *MR*: **22**, #2899 (1961).

(*b*) Endomorphismes de Reynolds et théorie ergodique. *C.R. Acad. Sci. Paris*, **250**, 2791–3. *MR*: **22**, #8932 (1961).

(*c*) Représentation des opérateurs de Reynolds. *C.R. Acad. Sci. Paris*, **250**, 2831–3. *MR*: **22**, #8933 (1961).

(*d*) Spectral theory of smoothing operators. *Proc. Nat. Acad. Sci. USA* **46**, 863–8. *MR*: **26**, #2882 (1963).

Ulam, S.M. *A collection of mathematical problems.* Interscience, New York-London, 1960. *MR*: **22**, #10884 (1961).

Vincze, E.

(*a*) Über das Problem der Berechnung der Wirtschaftlichkeit. *Acta Tech. Acad. Sci. Hungar.*, **28**, 33–41.

(*b*) Über die Verallgemeinerung der trigonometrischen und verwandten Funktionalgleichungen. *Ann. Univ. Sci. Budapest. Eötvös Sect. Math.* **3–4**, 389–404. *MR*: **25**, #305 (1963).

1961

Aczél, J.

(*a*) Miszellen über Funktionalgleichungen, II. *Math. Nachr.*, **23**, 39–50. *MR*: **27**, #5056 (1964).

(*b*) Sur une classe d'équations fonctionnelles bilinéaires à plusieurs fonctions inconnues. *Univ. Beograd. Publ. Elektrotehn. Fak. Ser. Mat. Fiz.* **1960**, No. 61–64, 12–20. *MR*: **26**, #508 (1963).

(*c*) Über die Begründung der Additions- und Multiplikationsformeln von bedingten Wahrscheinlichkeiten. *Magyar Tud. Akad. Mat. Kutató Int. Közl.*, **6**, 110–22. *MR*: **32**, #4712 (1966).

(*d*) *Vorlesungen über Funktionalgleichungen und ihre Anwendungen.* Birkhäuser, Basel-Stuttgart, 1961. *MR*: **23**, #A1959 (1962).

(*e*) Über Homomorphismen und Isomorphismen der affinen Gruppen von Körper und von Ringe. *Bul. Inst.*

Politehn. Iaşi **7(11)**, No. 3–4, 7–14. *MR*: **26**, #509 (1963).
Aczél, J., Hosszú, M., and Straus, E.G. Functional equations for products and compositions of functions. *Publ. Math. Debrecen*, **8**, 218–24. *MR*: **26**, #509 (1963).
Cox, R.T. *The algebra of probable inference.* John Hopkins Press, Baltimore, 1961. *MR*: **24**, #A563 (1962).
Daróczy, Z.
(*a*) Notwendige und hinreichende Bedingungen für die Existenz von nichkonstanten Lösungen linearer Funktionalgleichungen. *Acta. Sci. Math. (Szeged)*, **22**, 31–41. *MR*: **24**, #A378 (1962).
(*b*) Elementare Lösung einer mehrere unbekannte Funktionen enthaltenden Funktionalgleichung. *Publ. Math. Debrecen*, **8**, 160–8. *MR*: **27**, #2750 (1964).
Djoković, D.Ž. Sur une équation fonctionnelle cyclique. *Univ. Beograd. Publ. Elektrotehn. Fak. Ser. Mat. Fiz.* **1961**, No. 49–50, 15–16. *MR*: **23**, #A3933 (1962).
Flatto, L. Functions with a mean value property. *J. Math. Mech.* **10**, 11–18. *MR*: **25**, #212 (1963).
Hosszú, M. and Vincze, E. Über die Verallgemeinerungen eines Funktionalgleichungenssystems der Wirtschaftlichkeit. *Magyar Tud. Akad. Mat. Kutató Int. Közl.*, **6**, 313–21. *MR*: **26**, #2759 (1963).
Kiesewetter, H. Struktur linearer Funktionalgleichungen im Zusammenhang mit dem Abelschen Theorem. *J. Reine Angew. Math.* **206**, 113–71. *MR*: **24**, #A936 (1962).
Kuczma, M. On the functional equation $f(x+y)=f(x)+f(y)$. *Fund. Math.* **50**, 387–91. *MR*: **25**, #1378 (1962).
Kurepa, S. On the functional equation $f(x+y)f(x-y)=f^2(x)-f^2(y)$. *Ann. Polon. Math.*, **10**, 1–5. *MR*: **23**, #A3940 (1962).
Mitrinović, D.S. and Djoković, D.Ž. Sur une classe d'équations fonctionnelles cycliques. *C.R. Acad. Sci. Paris*, **252**, 1090–2. *MR*: **24**, #A2160 (1962).
Sakovič, G.N. Solution of a multivariate functional equation (Russian). *Ukrain. Mat. Ž.*, **13**, 173–89. *MR*: **25**, #2337 (1963).
Schmidt, H. Über das Additionstheorem der zyklischen Funktionen. *Math. Z.* **76**, 46–50. *MR*: **24**, #A276 (1962).
Vincze, E. Verallgemeinerung eines Satzes über assoziative Funktionen von mehreren Veränderlichen. *Publ. Math. Debrecen*, **8**, 68–74. *MR*: **24**, #2777 (1962).

1962

Aczél, J.
(*a*) *Ein Blick auf Funktionalgleichungen und ihre Anwendungen.* D.V.W., Berlin, 1962. *MR*: **26**, #6632 (1963).
(*b*) Remarques sur les relations d'équivalence. *Fund. Math.*, **51**, 267–9. *MR*: **26**, #6892 (1963).
Aczél, J., Fladt, K., and Hosszú, M. Lösungen einer mit dem Doppelverhältnis zusammenhängender Funktionalgleichung. *Magyar Tud. Akad. Mat. Kutató Int. Közl.*, **7**, 335–52. *MR*: **27**, #2746 (1964).
Carroll, F.W. Difference properties for continuity and Riemann integrability of locally compact groups. *Trans. Amer. Math. Soc.*, **102**, 284–92. *MR*: **24**, #A3438 (1962).
Choczewski, B. and Kuczma, M. Sur certaines équations fonctionnelles considerées par I. Stamate. *Mathematica (Cluj)*, **4(27)**, 225–33. *MR*: **29**, #5013 (1965).
Daróczy, Z.
(*a*) Über die Funktionalgleichung $\phi[\phi(x)y]=\phi(x)\phi(y)$. *Acta Univ. Debrecen Ser. Fiz. Chem.*, **8**, 125–32.
(*b*) Über die gemeinsame Charakterisierung der zu den nicht vollständigen Verteilungen gehörigen Entropien von Shannon und von Rényi, *Z. Wahrsch. Verw. Gebiete*, **1**, 381–8. *MR*: **32**, #9141 (1966).
Djoković, D.Ž. Sur l'équation fonctionnelle $f(x_3, x_1+x_2)-f(x_1, x_2+x_3)+f(x_2, x_1)-f(x_2, x_3)=0$. *Univ. Beograd. Publ. Elektrotehn. Fak. Ser. Mat. Fiz.*, **1962**, No. 70–76, 7–8. *MR*: **26**, #6631 (1963).
Friedman, A. and Littman, W.
(*a*) Bodies for which harmonic functions satisfy the mean value property. *Trans. Amer. Math. Soc.*, **102**, 147–66. *MR*: **27**, #1611 (1964).
(*b*) Functions satisfying the mean value property. *Trans. Amer. Math. Soc.*, **102**, 167–180. *MR*: **27**, #1612 (1964).
Friedman, D. The functional equation $f(x+y)=g(x)+h(y)$. *Amer. Math. Monthly*, **69**, 769–72.
Garsia, A.M. A note on the mean value property. *Trans. Amer. Math. Soc.* **102**, 181–6. *MR*: **27**, #1613 (1964).
Haruki, H. Another proof of Kaczmarz's theorem. *Sci. Rep. Osaka Univ.*, **11**, 1–2.

Hosszú, M. Note on commutable mappings. *Publ. Math. Debrecen*, **9**, 105–6. *MR*: **26**, #504 (1963).

Ionescu-Tulcea, A. and Ionescu-Tulcea, C. On the lifting property. I. *J. Math. Anal. Appl.*, **3**, 537–46, *MR*: **27**, #257 (1964).

Kurepa, S.
(*a*) A cosine functional equation in Banach algebra. *Acta Sci. Math. (Szeged)*, **23**, 255–67. *MR*: **26**, #2901 (1963).
(*b*) On the functional equation $T_1(t+s)$. $T_2(t-s) = T_3(t)T_4(s)$. *Publ. Inst. Math. (Beograd)*, **2(16)**, 99–108. *MR*: **30**, #507 (1963).

Kuwagaki, A. Sur l'équation fonctionnelle de Cauchy pour les matrices. *J. Math. Soc. Japan*, **14**, 359–66. *MR*: **26**, #507 (1963).

Radó, F. Caractérisation de l'ensemble des intégrales des équations différentielles linéaires homogénes à coefficients constants d'ordre donné. *Mathematica (Cluj)* **4(27)**, 131–43. *MR*: **27**, #2667 (1964).

Rudin, W. *Fourier analysis on groups.* Interscience, New York-London, 1962. *MR*: **27**, #2808 (1964).

Vincze, E.
(*a*) Eine allgemeinere Methode in der Theorie der Funktionalgleichungen, I. *Publ. Math. Debrecen*, **9**, 149–63. *MR*: **26**, #503 (1963).
(*b*) Eine allgemeinere Methode in der Theorie der Funktionalgleichungen, II. *Publ. Math. Debrecen*, **9**, 314–23. *MR*: **27**, #2749 (1964).

Wendel, J.G. Brief proof of a theorem of Baxter. *Math. Scand.*, **11**, 107–8. *MR*: **26**, #4204 (1963).

1963

Aczél, J. Remarks on probable inference. *Ann. Univ. Sci. Budapest. J. Eötvös Sect. Math.*, **6**, 3–11. *MR*: **29**, #6207 (1965).

Aczél, J. and Daróczy, Z.
(*a*) Charakterisierung der Entropien positiver Ordnung und der Shannonschen Entropie. *Acta Math. Acad. Sci. Hungar.*, **14**, 95–121. *MR*: **32**, #9140 (1966).
(*b*) Über verallgemeinerte quasilineare Mittelwerte, die mit Gewichtsfunktionen gebildet sind. *Publ. Math. Debrecen*, **10**, 171–90. *MR*: **29**, #6961 (1965).
(*c*) Sur la caractérisation axiomatique des entropies d'ordre positif, y comprise l'entropie de Shannon. *C.R. Acad. Sci. Paris*, **257**, 1581–4. *MR*: **27**, #5635 (1964).

Aczél, J. and Hosszú, M. On concomitants of mixed tensors. *Ann. Polon. Math.* **13**, 163–71. *MR*: **28**, #545 (1964).

Aczél, J. and Vincze, E. Über eine gemeinsame Verallgemeinerung zweier Funktionalgleichungen von Jensen. *Publ. Math. Debrecen*, **10**, 326–44. *MR*: **29**, #3782 (1965).

Atkinson, F.V. Some aspects of Baxter's functional equation. *J. Math. Anal. Appl.*, **7**, 1–30. *MR*: **27**, #5135 (1964).

Bajraktarević, M. Sur une généralisation des moyennes quasilinéaires. *Publ. Inst. Math. (Beograd)*, **3(17)**, 69–76. *MR*: **30**, #5077 (1965).

Djoković, D.Ž. A theorem on semigroups of linear operators. *Publ. Inst. Math. (Beograd)*, **3(17)**, 129–30. *MR*: **30**, 2089 (1965).

Dunford, N. and Schwartz, J.T. *Linear operators. Part II. Spectral theory. Self-adjoint linear operators in Hilbert space.* Interscience, New York-London, 1963. *MR*: **32**, #6181 (1966).

Eichhorn, W. Lösung einer Klasse von Funktionalgleichungssystemen. *Arch. Math. (Basel)*, **14**, 266–70. *MR*: **27**, #3956 (1964).

Erdös, P. On some properties of Hamel bases. *Colloq. Math.*, **10**, 267–9. *MR*: **28**, #4068 (1964).

Etherington, I.M.H. Note on quasigroups and trees. *Proc. Edinburgh Math. Soc.* (2), **13**, 219–22. *MR*: **28**, #157 (1964).

Flatto, L. Functions with a mean value property, II. *Amer. J. Math.* **85**, 248–70. *MR*: **28**, #1314 (1964).

Flett, T.M. Continuous solutions of the functional equation $f(x+y)+f(x-y) = 2f(x)f(y)$. *Amer. Math. Monthly*, **70**, 392–7. *MR*: **26**, #6635 (1963).

Fuchs, L. *Partially ordered algebraic systems.* Pergamon, Oxford-London-New York-Paris; Addison-Wesley, Reading, Mass.-Palo Alto, Calif.-London, 1963. *MR*: **30**, #2090 (1965).

Haruki, H. On some functional equations, I. *Sci. Rep. Osaka Univ.*, **12**, 1–3.

Hewitt, E. and Ross, K. *Abstract harmonic analysis. Vol. I.* (Grundlehren math. Wiss., Vol. 115). Academic Press, New York; Springer, Berlin-Göttingen-Heidelberg, 1963. *MR*: **28**, #158 (1964).

Hosszú, M.
(*a*) On the explicit form of *n*-group operations. *Publ. Math. Debrecen*, **10**, 88–92. *MR*: **29**, #4816 (1965).
(*b*) On a class of functional equations. *Publ. Inst. Math. (Beograd)*, **3(17)**, 53–5. *MR*: **30** #4080 (1965).

(*c*) On a functional equation treated by S. Kurepa. *Glasnik Mat.-Fiz. Astronom. Društvo Mat. Fiz. Hrvatske*, **18**, 59–60. *MR*: **29**, #2557 (1965).
(*d*) Die Lösung einer verallgemeinerten Bisymmetriegleichung. *Mathematica (Cluj)*, **5(28)**, 39–44. *MR*: **32**, #2763 (1966).
Hosszú, M. and Vincze, E. Über den wahrscheinlichsten Wert. *Acta Math. Acad. Sci. Hungar.*, **14**, 131–6. *MR*: **27**, #6291 (1964).
Kolmogorov, A.N. On the representation of continuous functions of many variables by superposition of continuous functions of one variable and addition. *Amer. Math. Sco. Transl.* (2), **28**, 55–9. *MR*: **27**, #3760 (1964).
Kucharzewski, M. and Kuczma, M. On the functional equation $F(A \cdot B) = F(A) \cdot F(B)$. *Ann. Polon. Math.*, **13**, 1–18. *MR*: **27**, #2747 (1964).
Kuczma, M. A characterization of the exponential and logarithmic functions by functional equations. *Fund. Math.*, **52**, 283–8. *MR*: **27**, #1730 (1964).
Losonczi, L. Über eine multilineare Funktionalgleichung mit mehreren unbekannten Funktionen. *Publ. Inst. Math. (Beograd)*, **3(17)**, 47–52. *MR*: **30**, #2251 (1965).
Phelps, R. R. Extreme positive operators and homomorphisms. *Trans. Amer. Math., Soc.* **108**, 265–74. *MR*: **27**, #6153 (1964).
Sato, T. On the functional equality $|f(x+iy)| = |f(x)||f(iy)|$. *J. College Arts Sci. Chiba Univ.*, **4**, No. 2, 9–10. *MR*: **32**, #2764 (1966).
Schweizer, B. and Sklar, A. Associative functions and abstract semigroups. *Publ. Math. Debrecen*, **10**, 69–81. *MR*: **30**, #1201 (1965).
Vincze, E.
(*a*) Eine allgemeinere Methode in der Theorie der Funktionalgleichungen, III. *Publ. Math. Debrecen*, **10**, 191–200. *MR*: **29**, #3781 (1965).
(*b*) Eine allgemeinere Methode in der Theorie der Funktionalgleichungen, IV. *Publ. Math. Debrecen*, **10**, 283–318. *MR*: **29**, #3781 (1965).

1964

Aczél, J.
(*a*) Ein Eindeutigkeitssatz in der Theorie der Funktionalgleichungen und einige ihrer Anwendungen. *Acta Math. Acad. Sci. Hungar.*, **15**, 355–62. *MR*: **30**, #375 (1965).
(*b*) On a generalization of the functional equations of Pexider. *Publ. Inst. Math. (Beograd)*, **4(18)**, 77–80. *MR*: **30**, #2246 (1965).
(*c*) Zur gemeinsamen Charakterisierung der Entropien α-ter Ordnung und der Shannonschen Entropie bei nicht unbedingt vollständigen Verteilungen. *Z. Wahrsch. Verw. Gebiete*, **3**, 177–83. *MR*: **30**, #969 (1965).
(*d*) Some unsolved problems in the theory of functional equations. *Arch. Math. (Basel)*, **16**, 435–44. *MR*: **30**, #376 (1965).
Andalafte, E.Z. and Blumenthal, L.M. Metric characterization of Banach and Euclidean spaces. *Fund. Math.*, **55**, 23–55. *MR*: **29**, #2625 (1965).
Bachman, G. *Introduction to p-adic numbers and valuation theory*. Academic Press, New York-London, 1964. *MR*: **30**, #90 (1965).
Belousov, V.D. and Hosszú, M. Some problems on ternary quasigroups. *Mat. Vesnik.*, **1(16)**, 319–24. *MR*: **32**, #2503 (1966).
Carroll, F.W. Functions whose differences belong to $L^p[0,1]$. *Indag. Math.*, **26**, 250–5, *MR*: **30**, #1226 (1965).
Daróczy, Z. Über Mittelwerte und Entropien vollständiger Wahrscheinlichkeitsverteilungen. *Acta Math. Acad. Sci. Hungar.*, **15**, 203–10. *MR*: **28**, #3271 (1964).
Djoković, D.Ž. Generalization of a result of Aczél, Ghermănescu and Hosszú. *Magyar Tud. Akad. Mat. Kutató Int. Közl.*, **8**, 51–60. *MR*: **30**, #378 (1965).
Hosszú, M.
(*a*) On local solutions of the generalized functional equation of associativity. *Ann. Univ. Sci. Budapest. Eötvös Sect. Math.*, **7**, 129–32. *MR*: **30**, #4081 (1965).
(*b*) On the Fréchet's functional equation. *Bul. Inst. Politehn. Iaşi*, **10**, 27–8, *MR*: **33**, #2986 (1967).
Howroyd, T.D. The solutions of some functional equations. *Canad. Math. Bull.*, **7**, 279–82. *MR*: **29**, #3779 (1965).
Kucharzewski, M. and Kuczma, M. *Basic concepts of the theory of geometric objects*. (Rozprawy Mat., Vol. 43). P.W.N., Warszawa, 1964, *MR*: **30**, #1467 (1965).
Kuczma, M. A survey of the theory of functional equations. *Univ. Beograd Publ. Elektrotehn. Fak. Ser. Mat. Fiz.*, **1964**, No. 130, 1–64. *MR*: **30**, #5073 (1965).
Kuczma, M. and Zajtz, A. Über die multiplikative Cauchysche Funktional-

Kuczma, M. and Zajtz, A. (*cont.*) gleichung. *Arch. Math.* (*Basel*), **15**, 136–43. *MR*: **29**, #381 (1965).

Kurepa, S.
(*a*) The Cauchy functional equation and scalar product in vector spaces. *Glasnik Mat.-Fiz. Astronom. Društvo Mat. Fiz. Hrvatske* (2), **19**, 23–36. *MR*: **30**, #1331 (1965).
(*b*) On a characterization of the determinant. *Glasnik Mat.-Fiz. Astronom. Društvo Mat. Fiz. Hrvatske* (2), **19**, 189–98. *MR*: **31**, #3436 (1966).

Lee, P.M. On the axioms of information theory. *Ann. Math. Statist.*, **35**, 414–18. *MR*: **28**, #1990 (1964).

Losonczi, L. Bestimmung aller nichtkonstanten Lösungen von linearen Funktionalgleichungen. *Acta Sci. Math.* (*Szeged*), **25**, 250–4. *MR*: **30**, #4084 (1965).

Luce, R.D. A generalization of a theorem of dimensional analysis. *J. Math. Psych.*, **1**, 278–84.

Makai, I. Über Differentialkomitanten erster Ordnung der homogenen linearen geometrischen Objekte in bezug auf einem kontravarianten Vektorfeld. *Publ. Math. Debrecen*, **11**, 191–241. *MR*: **30**, #1468 (1965).

Mehdi, M.R. On convex functions. *J. London Math. Soc.*, **39**, 321–6. *MR*: **28**, #5153 (1964).

Meynieux, R.
(*a*) Sur l'équation fonctionnelle vectorielle $f[x(u), y(v), z(u+v)] = 0$, I. *Ann. Sci. École Norm. Sup.* (3), **81**, 77–106. *MR*: **30**, #5084 (1965).
(*b*) Sur l'équation fonctionnelle vectorielle $f[x(u), y(v), z(u+v)] = 0$, II. *Ann. Sci. École Norm. Sup.* (3), **81**, 107–63. *MR*: **30**, #5084 (1965).

Pisot, C. and Schoenberg, I.J. Arithmetic problems concerning Cauchy's functional equation. *Illinois J. Math.*, **8**, 40–56. *MR*: **30**, #3321 (1965).

Rota, G.-C. Reynolds operators. In *Stochastic processes in mathematical physics and engineering* (*Proc. Sympos Appl. Math., Vol. 16*). Amer. Math. Soc., Providence, R.I., 1964, pp. 70–83. *MR*: **28**, #4349 (1964).

Sakovič, G.N. Functional equations for sums of exponentials (Russian). *Publ. Math. Debrecen*, **11**, 1–10. *MR*: **30**, #2253 (1965).

Vajzović, F. On the functional equation $T_1(t+s)T_2(t-s) = T_3(t)T_4(s)$. *Publ. Inst. Math.* (*Beograd*), **4(18)**, 21–7, *MR*: **30**, #3320 (1965).

Vincze, E.
(*a*) Über eine Verallgemeinerung der Cauchyschen Funktionalgleichung. *Funkcial. Ekvac.*, **6**, 55–62. *MR*: **29**, #6211 (1965).
(*b*) Beitrag zur Theorie der Cauchyschen Funktionalgleichungen. *Arch. Math.* (*Basel*), **15**, 132–5. *MR*: **29**, #390 (1965).

1965

Aczél, J.
(*a*) Quasigroup–nets–nomograms. *Advances in Math.*, **1**, 383–450. *MR*: **33**, #1395 (1967).
(*b*) The Monteiro-Botelho-Teixeira axiom and a "natural" topology in abelian semigroups. *Portugal. Math.*, **24**, 173–7. *MR*: **35**, #4320 (1968).
(*c*) The general solution of two functional equations by reduction to functions additive in two variables and with aid of Hamel-bases. *Glasnik Mat.-Fiz. Astronom. Društvo Mat. Fiz. Hrvatske* (2), **20**, 65–73. *MR*: **33**, #6182 (1967).

Aczél, J. and Erdös, P. The non-existence of a Hamel-basis and the general solution of Cauchy's functional equation for nonnegative numbers. *Publ. Math. Debrecen*, **12**, 259–63. *MR*: **32**, #4022 (1966).

Aczél, J. and Hosszú, M. Further uniqueness theorems for functional equations. *Acta Math. Acad. Sci. Hungar.*, **16**, 51–5. *MR*: **32**, #300 (1966).

Bajraktarević, M. Quelques remarques sur ma note "Sur une généralisation des moyennes quasilinéaires". *Publ. Inst. Math.* (*Beograd*), **5(19)**, 97–8. *MR*: **32**, #6091 (1966).

Bellman, R. Functional equations. In *Handbook of mathematical psychology, Vol. III*. Wiley & Sons, New York-London-Sydney, 1965, pp. 487–513. *MR*: **33**, #7144 (1967).

Belousov, V.D.
(*a*) Systems of quasigroups with general identities (Russian). *Uspekhi Mat. Nauk*, **20**, No. 1 (121), 75–146. *MR*: **30**, #3934 (1965).
(*b*) Nonassociative binary systems (Russian). In *Algebra, topology, geometry, 1965*. Akad. Nauk. SSSR Inst. Nauchn. Tekhn. Inf., Moscow, 1967, pp. 63–81. (English translation in *Progress in Mathematics, Vol. 5, Algebra*, Plenum, New York-London, 1969, pp. 57–76. *MR*: **35**, #5537 (1968).

Blumenthal, R.M., Lindenstrauss, J., and Phelps, R.R. Extreme operators into *C*(*K*). *Pacific J. Math.*, **15**, 747–56. *MR*: **35**, #758 (1968).

Borges, R. and Pfanzagl, J. One-parameter exponential families generated by transformation groups. *Ann. Math. Statist.*, **36**, 261–71. *MR*: **31**, #835 (1966).

Campbell, L.L. A coding theorem and Rényi's entropy. *Inform. and Control*, **8**, 423–9. *MR*: **31**, #4638 (1966).

Carroll, F.W. A difference property for polynomials and exponential polynomials on abelian locally compact groups. *Trans. Amer. Math. Soc.*, **114**, 147–55. *MR*: **30**, #2101 (1965).

Djoković, D.Ž.
(*a*) General solution of a functional equation. *Univ. Beograd. Publ. Elektrotehn. Fak. Ser. Mat. Fiz.*, **1965**, No. 132–142, 55–7. *MR*: **34**, #521 (1967).
(*b*) A special cyclic functional equation. *Univ. Beograd Publ. Elektrotehn. Fak. Ser. Mat. Fiz.*, **1965**, No. 143–155, 45–50. *MR*: **34**, #8022 (1967).

Gołąb, S. Einige grundlegende Fragen der Theorie der Funktionalgleichungen. *Glasnik Mat.-Fiz. Astronom. Društvo Mat. Fiz. Hrvatske* (2), **20**, 58–63. *MR*: 33, #4510 (1967).

Gołąb, S. and Losonczi, L. Über die Funktionalgleichung der Funktion Arccosinus, I. *Publ. Math. Debrecen*, **12**, 159–74. *MR*: **32**, #4416 (1966).

Henney, D.R. *N*-parameter semigroups. *Portugal. Math.*, **24**, 30–7. *MR*: **33**, #4177 (1967).

Hille, E.
(*a*) Topics in classical analysis. In *Lectures on modern mathematics, Vol. III*. Wiley & Sons, New York, 1965, pp. 1–57. *MR*: **31**, #2357 (1966).
(*b*) What is a semigroup? In *Studies in real and complex analysis*. Prentice-Hall, Englewood Cliffs, NJ, 1965, pp. 55–66. *MR*: **32**, #1080 (1966).

Hosszú, M.
(*a*) Some functional equations in connection with a theorem of Dubordieu. *Publ. Math. Debrecen*, **12**, 181–7. *MR*: **33**, #2987 (1967).
(*b*) Über eine Verallgemeinerung der Distributivitätsgleichung. *Acta Sci. Math.* (*Szeged*), **26**, 103–7, *MR*: **31**, #2345 (1966).

Jong, J.K. A class of linear functional equations. *Univ. Beograd. Publ. Elektrotehn. Fak. Ser. Mat. Fiz.*, **1965**, No. 143–15, 39–44. *MR*: **34**, #4742 (1967).

Jurkat, W.B. On Cauchy's functional equation. *Proc. Amer. Math. Soc.*, **16**, 683–86. *MR*: **31**, #3744 (1966).

Kiesewetter, H. Über die arc tan-Funktionalgleichung, ihre mehrdeutigen, stetigen Lösungen und eine nichstetige Gruppe. *Wiss. Z. Friedrich-Schiller-Univ. Jena Math.-Natur. Reihe*, **14**, 417–21, *MR*: **38**, #2468 (1969).

Kotz, S. On the solutions of some "isomoment" functional equations. *Amer. Math. Monthly*, **72**, 1072–5. *MR*: **34**, #6367 (1967).

Kucharzewski, M. and Kuczma, M. *Klassifikation der linearen homogenen geometrischen Objekte von Typus J mit drei Komponenten*. (Rozprawy Mat., Vol. 48). P.W.N., Warszawa, 1965, *MR*: **31**, #3966 (1966).

Kuczma, M.
(*a*) Bemerkungen über die Klassifikation der Funktionalgleichungen. *Prace Mat.*, **9**, 169–83. *MR*: **33**, #456 (1967).
(*b*) Einige Sätze über die Reduktion der Ordnung der Funktionalgleichungen. *Prace Mat.*, **9**, 185–92. *MR*: **33**, #457 (1967).

Kurepa, S.
(*a*) Remarks on the Cauchy functional equation. *Publ. Inst. Math.* (*Beograd*), **5**(**19**), 85–8. *MR*: **32**, #4418 (1966).
(*b*) Note on additive functions. *Glasnik Mat.-Fiz. Astronom. Društvo Mat. Fiz. Hrvatske* (2), **20**, 75–8. *MR*: **33**, #458 (1967).
(*c*) Quadratic and sesquilinear functionals. *Glasnik Mat.-Fiz. Astronom. Društvo Mat. Fiz. Hrvatske* (2), **20**, 79–92. *MR*: **33**, #1610 (1967).
(*d*) On a nonlinear functional equation. *Glasnik Mat.-Fiz. Astronom. Društvo Mat. Fiz. Hrvatske* (2), **20**, 243–9. *MR*: **37**, #4451 (1969).

Ling, C.H. Representation of associative functions. *Publ. Math. Debrecen* **12**, 189–212. *MR*: **32**, #7987 (1966).

Makai, I. Cauchy-Gleichungen für Quaternionenfunktionen. *Publ. Math. Debrecen*, **12**, 235–58. *MR*: **32**, #459 (1967).

Miller, J.B. Möbius transforms of Reynolds operators. *J. Reine Angew. Math.*, **218**, 6–16. *MR*: **31**, #610 (1966).

Moszner, Z. Solution générale de l'équation $F(x, y)F(y, z) = F(x, z)$ pour $x \leqslant y \leqslant z$. *C.R. Acad. Sci. Paris*, **261**, 28. *MR*: **31**, #3749 (1966).

Pfanzagl, J. A general theory of measurement—applications to utility. In *Readings in mathematical psychology, Vol. II*. Wiley & Sons, New York-London-Sydney, 1965, pp. 492–502, *MR*: **33**, #7146 (1967).

Pisot, C. and Schoenberg, I.J. Arithmetic problems concerning Cauchy's functional equation, II. *Illinois J. Math.*, **9**, 129–36. *MR*: **30**, #3323 (1965).

Popa, C.G. Sur l'équation fonctionnelle $f[x + yf(x)] = f(x)f(y)$. *Ann. Polon. Math.*, **17**, 193–198. *MR*: **32**, #4420 (1966).

Rényi, A. On the foundations of information theory. *Rev. Inst. Internat. Statist.*, **33**, 1–14. *MR*: **31**, #5712 (1966).

Światak, H. On the equation $[\phi(x + y)]^2 = [\phi(x)g(y) + \phi(y)g(x)]^2$. *Zeszyty Nauk. Uniw. Jagielloń, Prace Mat.*, **10**, 97–104. *MR*: **34**, #527a (1967).

Vasić, P.M. and Dordević, R.Ž. Sur l'équation fonctionnelle cyclique généralisée. *Univ. Beograd. Publ. Elektrotehn. Fak. Ser. Mat. Fiz.*, **1965**, No. 132–142, 33–8. *MR*: **32**, #4421 (1966).

1966

Aczél, J.
(*a*) On strict monotonicity of continuous solutions of certain types of functional equations. *Canad. Math. Bull.*, **9**, 229–32. *MR*: **33**, #7729 (1967).
(*b*) Funktionskomposition, Iterationsgruppen und Gewebe. *Arch. Math. (Basel)*, **17**, 469–75. *MR*: **34**, #3473 (1967).
(*c*) *Lectures on functional equations and their applications.* (Mathematics in Science and Engineering, Vol. 19). Academic Press, New York-London, 1966. *MR*: **34**, #8020 (1967).

Aczél, J. and Pickert, G. Nichtkommutative monotone Gruppen reeller Zahlen. *Arch. Math. (Basel)*, **17**, 292–7. *MR*: **36**, #3855 (1968).

Bass, J. and Dhombres, J. Moyennes des fonctions. *C.R. Acad. Sci. Paris Sér. A*, **262**, 29–31. *MR*: **33**, #4674 (1967).

Belousov, V.D. Balanced identities in quasigroups (Russian). *Mat. Sb.*, **70(112)**, 55–97. *MR*: **34**, #2757 (1967).

Belousov, V.D. and Ryžkov, V.V. On a method of obtaining closure figures (Russian). *Mat. Issled.*, **1**, No. 2, 140–50. *MR*: **36**, #6526 (1968).

de Bruijn, N.G. On almost additive functions. *Colloq. Math.*, **15**, 59–63. *MR*: **33**, #4221 (1967).

Campbell, L.L. Definition of entropy by means of a coding problem. *Z. Wahrsch. Verw. Gebiete*, **6**, 113–18. *MR*: **34**, #7266 (1967).

Climescu, A.
(*a*) Sur l'équation fonctionnelle de la distributivité unilatérale, I. *Bul. Inst. Politehn. Iaşi*, **12(16)**, No. 1–2, 1–6. *MR*: **36**, #558 (1968).
(*b*) Sur l'équation fonctionnelle de la distributivité unilatérale, II. *Bul. Inst. Politehn. Iaşi*, **12(16)**, No. 3–4, 1–6. *MR*: **36**, #558 (1968).

Daróczy, Z. and Györy, K. Die Cauchysche Funktionalgleichungen über diskrete Mengen. *Publ. Math. Debrecen*, **13**, 249–56. *MR*: **34**, #6364 (1967).

Eichhorn, W. Über die multiplikative Cauchysche Funktionalgleichung für die Hamiltonschen Quaternionen und die Cayleyschen Zahlen. *J. Reine Angew. Math.*, **221**, 2–13. *MR*: **32**, #5698 (1966).

Flatto, L. On polynomials characterized by a certain mean value property. *Proc. Amer. Math. Soc.*, **17**, 598–601. *MR*: **33**, #4217 (1967).

Gleason, A.M. The definition of a quadratic form. *Amer. Math. Monthly*, **73**, 1049–56. *MR*: **34**, #7543 (1967).

Gołąb, S. and Losonczi, L. Über die Funktionalgleichung der Funktion Arccosinus, II. Die alternative Gleichung. *Publ. Math. Debrecen* **13**, 183–5. *MR*: **34**, #6366 (1967).

Haruki, H. On the functional equations $|f(x + iy)| = |f(x) + f(iy)|$ and $|f(x + iy)| = |f(x) - f(iy)|$ and on Ivory's theorem. *Canad. Math. Bull.*, **9**, 473–80. *MR*: **34**, #3145 (1967).

Havel, V. Verallgemeinerte Gewebe, I. *Arch. Math. (Brno)*, **2**, 63–70. *MR*: **34**, #120 (1967).

Hosszú, M. On some functional equtions and their classification. *Mat. Vesnik*, **3 (18)**, 139–45. *MR*: **35**, #1998 (1968).

Kingman, J.F.C. On the algebra of queues. *J. Appl. Probability*, **3**, 285–326. *MR*: **34**, #3678 (1967).

Kuczma, M. and Zajtz, A. On the form of real solutions of the matrix functional equation $\phi(x)\phi(y) = \phi(xy)$ for nonsingular matrices ϕ. *Publ. Math. Debrecen*, **13**, 257–62. *MR*: **34**, #3147 (1967).

Kuwagaki, A. Sur les équations fonctionnelles généralisées de Cauchy et quelques équations qui s'y rattachent. *Publ. Res. Inst. Math. Sci. Ser. A*, **2**, 397–422. *MR*: **35**, #2000 (1968).

Miller, J.B.
(*a*) Some properties of Baxter operators. *Acta Math. Acad. Sci. Hungar.*, **17**, 387–400. *MR*: **34**, #4909 (1967), **35**, 1577 (1968).
(*b*) Averaging and Reynolds operators on Banach algebras, I. Representation by derivations and antiderivations.

J. Math. Anal. Appl., **14**, 527–48. *MR*: **33**, #3104 (1966), **35**, 1577 (1968).

Moór, A. and Pintér, L. Untersuchungen über den Zusammenhang von Differential- und Funktionalgleichungen, I. *Publ. Math. Debrecen*, **13**, 207–23. *MR*: **34**, #6264 (1967).

Soya, M. *Cosine operator functions*. (Rozprawy Mat., Vol. 49). P.W.N., Warszawa, 1966.

Vajzović, F. On a functional which is additive on A-orthogonal pairs. *Glas. Mat. Ser. III*, **1(21)**, 75–81. *MR*: **34**, #8198 (1967).

Vasić, P.M. Sur une équation fonctionnelle de J.K. Jong. *Mat. Vesnik*, **3(18)**, 271–4. *MR*: **34**, #8025 (1967).

1967

Aczél, J. General solution of 'isomoment' functional equations. *Amer. Math. Monthly*, **74**, 1068–71. *MR*: **37**, #4440 (1969).

Aczél, J. and McKiernan, M.A. On the characterization of plane projective and complex Moebius-transformations. *Math. Nachr.*, **33**, 315–37. *MR*: **33**, #5806 (1967).

Aczél, J. and Wallace, A.D. A note on generalizations of transitive systems of transformations. *Colloq. Math.*, **17**, 29–34. *MR*: **36**, #867 (1968).

Alexandrov, A.D. A contribution to chronogeometry. *Canad. J. Math.*, **19**, 1119–28. *MR*: **36**, #2101 (1968).

Belousov, V.D. *Foundations of the theory of quasigroups and loops* (Russian). Nauka, Moscow, 1967. *MR*: **36**, #1569 (1968).

Climescu, A. Sur l'équation fonctionnelle de la distributivité unilatérale, III. *Bul. Inst. Politehn. Iaşi* (N.S.), **13(17)**, No. 1–2, 23–4. *MR*: **38**, #444 (1969).

Corrádi, K.A. and Kátai, I. On multiplicative characters. *Acta Sci. Math.* (*Szeged*), **28**, 71–6. *MR*: **35**, #2838 (1968).

Daróczy, Z.
(*a*) Über eine Charakterisierung der Shannonschen Entropie. *Statistica*, **27**, 189–205. *MR*: **35**, #7727 (1968).
(*b*) Über eine Klasse von Funktionalgleichungen im Hilbert-Raum. *Enseign. Math.* (2), **13**, 99–102. *MR*: **36**, #5555 (1968).

Daróczy, Z. and Losonczi, L. Über die Erweiterung der auf einer Punktmenge additiven Funktionen. *Publ. Math. Debrecen*, **14**, 239–45. *MR*: **39**, #1839 (1970).

Denny, J.L. Sufficient conditions for a family of probabilities to be exponential. *Proc. Nat. Acad. Sci. U.S.A.*, **57**, 1184–7. *MR*: **35**, #2396 (1968).

Dhombres, J. Sur une classe de moyennes. *Ann. Inst. Fourier* (*Grenoble*), **17**, 135–56. *MR*: **36**, #5730 (1968).

Djoković, D.Ž. Triangle functional equation and its generalization. *Univ. Beograd. Publ. Elektrotehn. Fak. Ser. Mat. Fiz.*, **1967**, No. 181–196, 47–52. *MR*: **37**, #1830 (1969).

Fischer, P. and Muszély, G. On some new generalizations of the functional equation of Cauchy. *Canad. Math. Bull.*, **10**, 197–205. *MR*: **35**, #7025 (1968).

Haruki, H. On a well-known improper integral. *Amer. Math. Monthly*, **74**, 846–8.

Havrda, J. and Charvát, F. Quantification method of classification processes. Concept of structural a-entropy. *Kybernetika* (*Prague*), **3**, 30–5. *MR*: **34**, #8875 (1967).

Kátai, I. A remark on additive arithmetical functions. *Ann. Univ. Sci. Budapest. Eötvös Sect. Math.*, **12**, 81–3. *MR*: **41**, #1666 (1971).

Kuczma, M. and Zajtz, A. Quelques remarques sur l'équation fonctionnelle matricielle multiplicative de Cauchy. *Colloq. Math.*, **18**, 159–68. *MR*: **36**, #5554 (1968).

Kurepa, S. *Finite dimensional spaces and applications* (Serbo-Croatian). Tehnička Knjiga, Zagreb, 1967.

Kuwagaki, A. *Theory of functional equations* (Japanese). Asakura, Tokyo, 1967.

Lukacs, E. Analytical methods in probability theory. In *Symposium on Probability Methods in Analysis*. (Lecture Notes in Math., Vol. 31). Springer, Berlin-Heidelberg-New York, 1967, pp. 207–38. *MR*: **36**, #2192 (1968).

Máté, A. A new proof of a theorem of P. Erdös. *Proc. Amer. Math. Soc.*, **18**, 159–62. *MR*: **34**, #4234 (1967).

McKiernan, M.A. On vanishing n-th order differences and Hamel bases. *Ann. Polon. Math.*, **19**, 331–6. *MR*: **36**, #4183. (1968).

Miller, J.B. Homomorphisms, higher derivations and derivations on associative algebras. *Acta Sci. Math.* (*Szeged*), **28**, 221–31. *MR*: **35**, #2927 (1968).

Monroe, I. On the generalized Cauchy equation. *Canad. J. Math.*, **19**, 1314–18. *MR*: **36**, #4188 (1968).

Neumann, H. *Varieties of groups*. (Ergebnisse Math., No. 37). Springer, New York, 1967. *MR*: **35**, #6734 (1968).

Saaty, T.L. *Modern nonlinear equations*. Chapter 3. McGraw-Hill, New York-St.

Saaty, T.L. (*cont.*)
Louis-San Francisco-Toronto-London-Sydney, 1967. *MR*: **36**, #1249 (1968).

Schweizer, B. Probabilistic metric spaces—the first 25 years. *New York Statist.*, **19**, No. 2, 3–6.

Schweizer, B. and Sklar, A. The algebra of multiplace vectorvalued functions. *Bull. Amer. Math. Soc.*, **73**, 510–15. *MR*: **35**, #4138 (1968).

Światak, H. On the functional equations $f_1(x_1 + \ldots + x_n)^2 = [\sum_{(i_1,\ldots,i_n)} f_1(x_{i_1}) \ldots f_n(x_{i_n})]^2$. *Ann. Univ. Sci. Budapest. Eötvös Sect. math.*, **10**, 49–52. *MR*: **39**, #3180 (1970).

Targonski, G. *Seminar on functional operators and equations.* (Lecture Notes in Math., Vol. 33). Springer, Berlin-Heidelberg-New York, 1967. *MR*: **36**, #744 (1968).

Vajzović, F. Über das Funktional *H* mit der Eigenschaft $(x, y) = 0 \Rightarrow H(x + y) + H(x - y) = 2H(x)H(y)$. *Glas. Mat. Ser. III*, **2(22)**, 73–80. *MR*: **36**, #4326 (1968).

Wetzel, J.E. On the functional inequality $f(x + y) \geqslant f(x)f(y)$. *Amer. Math. Monthly*, **74**, 1065–8. *MR*: **37**, #4444 (1969).

Wilansky, A. Additive functions. In *Lectures on calculus.* Holden Day, San Francisco-Cambridge-London-Amsterdam, 1967, pp. 97–124.

1968

Aczél, J. On different characterizations of entropies. In *Probability and information theory* (*Proc. Internat. Sympos., McMaster Univ., 1968*). (Lecture Notes in Math., Vol. 89). Springer, Berlin-Heidelberg-New York, 1969, pp. 1–11. *MR*: **43**, #5991 (1972).

Aczél, J. and Fischer, P. Some generalized 'isomoment' equations and their general solutions. *Amer. Math. Monthly*, **75**, 952–8. *MR*: **38**, #4841 (1969).

Aczél, J., Haruki, H., Mckiernan, M.A., and Sakovič, G.N. General and regular solutions of functional equations characterizing harmonic polynomials. *Aequationes Math.*, **1**, 37–53. *MR*: **43**, #5193 (1972).

Bajraktarević, M. Quelques remarques sur les solutions générales de certaines équations fonctionnelles aux plusieurs inconnues. *Mat. Vesnik*, **5(20)**, 497–503. *MR*: **39**, #1840 (1970).

Baker, J.A. On quadratic functionals along rays. *Glas. Mat. Ser. III*, **3(23)**, 215–29. *MR*: **41**, #669 (1971).

Baker, J.A. and Segal, S.L. On a problem of Kemperman concerning Hamel-functions (**P2851**). *Aequationes Math.*, **2**, 114–5. *MR*: **39**, #664 (1970).

Benz, W. Die 4-Punkt-Invarianten in der projektiven Geraden über einem Schiefkörper. *Ann. Polon. Math.*, **21**, 97–101. *MR*: **38**, #510b (1969).

Climescu, A. Un théorème de transfert pour certaines équations fonctionnelles. *Bul. Inst. Politehn, Iaşi Sect. I*, **14(18)**, No. 1–2, 19–22. *MR*: **38**, #5684 (1969).

Daróczy, Z. Über eine Funktionalgleichung im Hilbertraum., *Publ. Inst. Math.* (*Beograd*), **8(22)**, 121–3. *MR*: **37**, #3234 (1969).

Dhombres, J. Opérateurs semi-multiplicatifs de norme unité. *C.R. Acad. Sci. Paris Sér. A*, **266**, 1046–9. *MR*: **38**, #557 (1969).

Eichhorn, W.
(*a*) Über die multiplikativen Abbildungen endlichdimensionaler Algebren in kommutative Halbgruppen. *J. Reine Angew. Math.*, **231**, 10–46. *MR*: **38**, #197 (1969).
(*b*) Funktionalgleichungen in Vektorräumen, Kompositionsalgebren und Systeme partieller Differentialgleichungen. *Aequationes Math.*, **2**, 287–303. *MR*: **39**, #4240 (1970).

Fischer, P. Sur l'équivalence des équations fonctionnelles $f(x + y) = f(x) + f(y)$ et $f^2(x + y) = [f(x) + f(y)]^2$. *Ann. Fac. Sci. Toulouse Math.*, **1968**, 71–3, *MR*: **40**, #1729 (1970).

Forder, H.G. Groups from one axiom. *Math. Gaz.*, **52**, 263–6. *MR*: **38**, #3325 38. *MR*: **38**, #561 (1969).

Fox, R.A.S. and Miller, J.B. Averaging and Reynolds operators on Banach algebras, III. Spectral properties of Reynolds operators. *J. Math. Anal. Appl.*, **24**, 225–38. *MR*: **38**, #561 (1969).

Gallager, C.G. *Information theory and reliable communication.* Wiley & Sons, New York-London-Sydney-Toronto, 1968.

Gamlen, J.L.B. and Miller, J.B. Averaging and Reynolds operators on Banach algebras, II. Spectral properties of averaging operators. *J. Math. Anal. Appl.*, **23**, 183–97. *MR*: **37**, #3365 (1969).

Good, I.J. Corrigendum: Weight of evidence, corroboration, explanatory power, information and the utility of experiments. *J. Roy. Statist. Soc. Ser. B*, **30**, 203. *MR*: **37**, #4919 (1969).

Haruki, H.
(*a*) On parallelogram functional equations. *Math. Z.*, **104**, 358–63. *MR*: **37**, #1828 (1969).
(*b*) On inequalities generalizing the

Pythagorean functional equation and Jensen's functional equation. *Pacific J. Math.*, **20**, 85–90. *MR*: **37**, #6633 (1969).

Javor, P. On the general solution of the functional equation $f[x + yf(x)] = f(x)f(y)$. *Aequationes Math.*, **1**, 235–238. *MR*: **38**, #3644 (1969).

Jessen, B. The algebra of polyhedra and the Dehn-Sydler theorem. *Math. Scand.*, **22**, 241–56. *MR*: **40**, #4860 (1970).

Jessen, B., Karpf, J., and Thorup, A. Some functional equations in groups and rings. *Math. Scand.*, **22**, 257–65. *MR*: **44**, #1963 (1972).

Johnson, B.E. and Sinclair, A.M. Continuity of derivations and a problem of Kaplansky. *Amer. J. Math.*, **90**, 1067–73. *MR*: **39**, #776 (1970).

Jong, J.K. A generalization of a linear functional equation. *Univ. Beograd. Publ. Elektrohn. Fak. Ser. Mat. Fiz.*, **1968**, No. 232, 13–20. *MR*: **39**, #7316 (1970).

Kannappan, Pl.

(*a*) The functional equation $f(xy) + f(xy^{-1}) = 2f(x)f(y)$ for groups. *Proc. Amer. Math. Soc.*, **19**, 69–74. *MR*: **36**, #3006 (1968).

(*b*) A functional equation for the cosine. *Canad. Math. Bull.*, **11**, 495–8.

Kucharzewski, M.

(*a*) Über eine axiomatische Auszeichnung der Determinanten. *Ann. Polon. Math.*, **20**, 199–202. *MR*: **36**, #3798 (1968).

(*b*) Characterisierung des Flächeninhalts mit Hilfe der Funktionalgleichungen. *Ann. Polon. Math.*, **21**, 59–65. *MR*: **38**, #6470 (1969).

Kuczma, M. *Functional equations in a single variable.* (Monografie Mat., Vol. 46). P.W.N., Warszawa, 1968. *MR*: **37**, #4441 (1969).

Lukacs, E. Non-negative definite solutions of certain differential and functional equations. *Aequationes Math.*, **2**, 137–43. *MR*: **39**, #4539 (1970).

Miller, J.B. A formula for the resolvent of a Reynolds operator. *J. Austral. Math. Soc.*, **8**, 447–56. *MR*: **38**, #540 (1969).

Pearson, K.R.

(*a*) Certain topological semigroups in R_1. *J. Austral. Math. Soc.*, **8**, 171–82. *MR*: 37, #1509 (1969).

(*b*) Embedding semirings in semirings with multiplicative unit. *J. Austral. Math. Soc.*, **8**, 183–91. *MR*: **37**, #1510 (1969).

Pełczyński, A. *Linear extensions, linear averagings, and their applications to linear topological classifications of spaces of continuous functions.* (Dissertationes Math. Rozprawy Mat., Vol. 58). Polish Scientific Publishers, Warszawa, 1968. *MR*: **37**, #3335 (1969).

Pfanzagl, J. *Theory of measurement.* Physica-Verlag, Würzburg-Vienna; Wiley & Sons, New York, 1968. *MR*: **40**, #4180 (1970).

Randolph, J.F. *Basic real and abstract analysis.* Chapter 6. Academic Press, New York-London, 1968. *MR*: **37**, #1211 (1969).

Senechalle, D.A. A characterization of inner product spaces. *Proc. Amer. Math. Soc.*, **19**, 1306–12. *MR*: **39**, #761 (1970).

Sigmon, K. Medial topological groupoids. *Aequationes Math.*, **1**, 217–34. *MR*: **39**, #1586 (1970).

Smital, J. On the functional equation $f(x + y) = f(x) + f(y)$. *Rev. Roumaine Math. Pures Appl.*, **13**, 555–61. *MR*: **38**, #2472 (1969).

Świątak, H. On the functional equation $f(x + y - xy) + f(xy) = f(x) + f(y)$. *Mat. Vesnik*, **5(20)**, 177–82. *MR*: **38**, #4844 (1969).

Vajda, I. Axioms for a-entropy of a generalized probability scheme (Czech). *Kybernetika (Prague)*, **4**, 105–12. *MR*: **38**, #1947 (1969).

Wirsing. E. A characterization of log n as an additive arithmetic function. In *INDAM (Roma 1968–69)*. (Symposia Math., Vol. 4). Academic Press, London, 1970, pp. 45–57. *MR*: **42**, #5932 (1971).

Wołodźko, S. Solution générale de l'équation fonctionnelle $f[x + yf(x)] = f(x)f(y)$. *Aequationes Math.*, **2**, 12–29. *MR*: **38**, #1422 (1969).

1969

Aczél, J.

(*a*) *On applications and theory of functional equations.* (Elemente der Mathematik vom höheren Standpunkt aus, Bd. V). Birkhäuser, Basel; Academic Press, New York, 1969. *MR*: **39**, #1838 (1970).

(*b*) Über Zusammenhänge zwischen Differential- und Funktionalgleichungen. *Jber. Deutsch. Math.-Verein.*, **71**, 55–7. *MR*: **41**, #674 (1971).

Aczél. J. and Benz, W. Kollineationen auf Drei- und Vierecken in der Desarguesschen projektiven Ebene und Äquivalenz der Dreiecksnomogramme und der Dreigewebe von Loops mit der Isotopie-Isomorphie-Eigenschaft. *Aequationes Math.*, **3**, 86–92. *MR*: **40**, #6355 (1970).

Bajraktarević, M. Über die Vergleichbarkeit der mit Gewichtsunktionen gebildeten Mittelwerte. *Studia Sci. Math. Hungar.*, **4**, 3–8. *MR*: **40**, #275 (1970).

Baker, J.A. An analogue of the wave equation and certain related functional equations. *Canad. Math. Bull.*, **12**, 837–46. *MR*: **40**, #7663 (1970).

Carroll, F.W. and Koehl, F.S. Difference properties for Banach-valued functions on compact groups. *Indag. Math.*, **31**, 327–32. *MR*: **40**, #1763 (1970).

Choquet, G. *Lectures on analysis. Vol. II: Representation theory.* Benjamin, New York-Amsterdam, 1969. *MR*: **40**, #3253 (1970).

Daróczy, Z. Über die Funktionalgleichung $f(xy) + f(x + y - xy) = f(x) + f(y)$. *Publ. Math. Debrecen*, **16**, 129–32. *MR*: **41**, #7320 (1971).

Haruki, H. On a 'cube functional equation'. *Aequationes Math.*, **3**, 156–9. *MR*: **43**, #5194 (1972).

Hewitt, E. and Zuckerman, H.S. Remarks on the functional equation $f(x + y) = f(x) + f(y)$. *Math. Mag.*, **42**, 121–3. *MR*: **39**, #7313 (1970).

Hosszú, M.
(*a*) Some remarks on the cosine functional equation. *Publ. Math. Debrecen*, **16**, 93–8. *MR*: **41**, #5820 (1971).
(*b*) A remark on the square norm. *Aequationes Math.*, **2**, 190–3. *MR*: **39**, #7317 (1970).

Howroyd, T.D. Some uniqueness theorems for functional equations. *J. Austral. Math. Soc.*, **9**, 176–9. *MR*: **39**, #7309 (1970).

Johnson, B.E. Continuity of derivations on commutative algebras. *Amer. J. Math.*, **91**, 1–10. *MR*: **39**, #7433 (1970).

Kannappan, Pl. On sine functional equation. *Studia Sci. Math. Hungar.*, **4**, 331–3. *MR*: **40**, #3100 (1970).

Kemperman, J.H.B. On the regularity of generalized convex functions. *Trans. Amer. Math. Soc.*, **135**, 69–93. *MR*: **38**, #1223 (1969).

Koehl, F.S. Difference properties for Banach-valued functions. *Math. Ann.*, **181**, 288–96. *MR*: **39**, #5775 (1970).

Kuczma, M.E. On discontinuous additive functions. *Fund. Math.*, **66**, 383–92. *MR*: **41**, #3676 (1971).

Matras, Y. Sur l'équation fonctionnelle $f[xf(y)] = f(x)f(y)$. *Acad. Roy. Belg. Bull. Cl. Sci.* (5), **55**, 731–51. *MR*: **42**, #2204 (1971).

Miller, J.B. Baxter operators and endomorphisms on Banach algebras. *J. Math. Anal. Appl.*, **25**, 503–20. *MR*: **39**, #815 (1970).

Ostrowski, A. Über eine Funktionalgleichung (Bemerkung zur vorstehenden Mitteilung von J. Aczél). *Jber. Deutsch. Math.-Verein.* **71**, 58–9. *MR*: **41**, #675 (1971).

Radó, F. Darstellung nicht-injektiver Kollineationen eines projektiven Raumes durch verallgemeinerte semilineare Abbildungen. *Math. Z.* **110**, 153–70. *MR*: **40**, #1866 (1970).

Rota, G.-C.
(*a*) Baxter algebras and combinatorial identities, I. *Bull. Amer. Math. Soc.*, **75**, 325–9. *MR*: **39**, #5387 (1970).
(*b*) Baxter algebras and combinatorial identities, II. *Bull. Amer. Math. Soc.*, **75**, 330–4. *MR*: **39**, #5387 (1970).

Rota, G.-C. and Mullin, R. On the foundations of combinatorial theory, III. Theory of binomial enumeration. In *Graph theory and its applications* (*Proc. Advanced Sem., Math. Research Center, Univ. of Wisconsin, Madison, Wis., 1969*). Academic Press, New York, 1970, pp. 167–213. *MR*: **43**, #65 (1972).

Segal, S.L. Remark to Problem 28 (**P28R1**). *Aequationes Math.*, **2**, 111–12.

Wirsing, E. Characterisation of the logarithm as an additive function. In *1969 Number Theory Institute, State Univ. New York, Stony Brook, NY.* (Proc. Symp. Pure Math., Vol. 20). Amer. Math. Soc., Providence, RI, 1971, pp. 375–81. *MR*: **54**, #256 (1977).

1970

Aczél, J. Some applications of functional equations and inequalities to information measures. In *Functional equations and inequalities* (*C.I.M.E., III Ciclo, La Mendola, 1970*). Cremonese, Rome, 1971, pp. 1–20. *MR*: **58**, #23210 (1979).

Aczél, J. and Gołąb, S. Remarks on one-parameter subsemigroups of the affine group and their homo- and isomorphisms. *Aequationes Math.*, **4**, 1–10. *MR*: **41**, #8561 (1971).

Alexandrov, A.D. A certain generalization of the functional equation $f(x + y) = f(x) + f(y)$ (Russian). *Sibirsk. Mat. Ž.*, **11**, 264–278 = *Siberian Math. J.*, **11**, 198–209. *MR*: **42**, #714 (1971).

Baker, J.A. A sine functional equation. *Aequationes Math.*, **4**, 56–62. *MR*: **42**, #715 (1971).

Blanuša, D. The functional equation $f(x + y - xy) + f(xy) = f(x) + f(y)$. *Aequationes Math.*, **5**, 63–7, *MR*: **43**, #5198 (1972).

Cashwell, E.D. and Everett, C.J. Extension of the Riesz representation theorem and solution of the Maxwell-Boltzmann functional equation. *Aequationes Math.*, **5**, 315–18. *MR*: **44**, #1960 (1972).

Cooper, J.L.B. Functional equations for linear transformations. *Proc. London Math. Soc.* (3), **20**, 1–32. *MR*: **40**, #7870 (1970).

Dacić, R. The sine functional equation for groups. *Mat. Vesnik*, **7(22)**, 279–84. *MR*: **43**, #5200 (1972).

Daróczy, Z. Generalized information functions. *Inform. and Control*, **16**, 36–51. *MR*: **42**, #7409 (1971).

Denny, J.L. Cauchy's equation and sufficient statistics on arcwise connected spaces. *Ann. Math. Statist.*, **41**, 401–11. *MR*: **41**, #6346 (1971).

Djoković, D.Ž. On homomorphisms of the general linear group. *Aequationes Math.*, **4**, 99–102. *MR*: **46**, #5460 (1973).

Fattorini, H.O. Uniformly bounded cosine functions in Hilbert space. *Indiana Univ. Math. J.*, **20**, 411–25. *MR*: **42**, #2319 (1971).

Fenyö, I. Über die Funktionalgleichung $f(a_0 + a_1x + a_2y + a_3xy) + g(b_0 + b_1x + b_2y + b_3xy) = h(x) + k(y)$. *Acta Math. Acad. Sci. Hungar.*, **21**, 35–46. *MR*: **58**. #29526 (1979).

Ger, R. and Kuczma, M. On the boundedness and continuity of convex functions and additive functions. *Aequationes Math.*, **4**, 157–62. *MR*: **41**, #8650 (1971).

Gołąb, S. Functional equations in geometry. *Zeszyty Nauk. Uniw. Jagielloń. Prace Mat.* **14**, 13–19.

Gołąb, S. and Krzeszowiak, Z. Les équations fonctionnelles et la théorie des groupes. *Rev. Roumaine Math. Pures Appl.* **15**, 1395–406. *MR*: **43**, #760 (1972).

Hamilton, J. Remarks 31 on the translation equation. *Aequationes Math.*, **4**, 254–5.

Hille, E. Meanvalues and functional equations. In *Functional equations and inequalities* (*C.I.M.E., III. Ciclo, La Mendola, 1970*). Cremonese, Rome, 1971, pp. 153–62. *MR*: **58**, #23210 (1979).

Howroyd, T.D.
(*a*) On the equivalence of a functional equation in *n* variables with *n* functional equations in a single variable. *Aequationes Math.*, **4**, 191–7. *MR*: **41**, #7327 (1971).
(*b*) The uniqueness of bounded or measurable solutions of some functional equations. *J. Austral. Math. Soc.*, **11**, 186–90. *MR*: **41**, #8857 (1971).

Kannappan, Pl. and Kurepa, S. Some relations between additive functions-I. *Aequationes Math.*, **4**, 163–75. *MR*: **41**, #8858 (1971).

Kiesewetter, H. Ein Zusammenhang zwischen assoziativen und zyklisch-bisymmetrischen Verknüpfungen. *Aequationes Math.*, **4**, 83–8. *MR*: **42**, #4912 (1971).

Kuczma, M. Some remarks about additive functions on cones. *Aequationes Math.*, **4**, 303–6. *MR*: **42**, #8121 (1971).

Kurepa, S. Functional equations on vector spaces. In *Functional equations and inequalities* (*C.I.M.E., III Ciclo, La Mendola, 1970*). Cremonese, Rome, 1971, pp. 215–31. *MR*: **58**, #23210 (1979).

Larsen, R. The functional equation $f(t)f(s)g(ts) = g(t)g(s)$. *Aequationes Math.*, **5**, 47–53. *MR*: **44**, #1964 (1972).

McKiernan, M.A. Difference and mean value type functional equations. In *Functional equations and inequalities* (*C.I.M.E., III Ciclo, La Mendola, 1970*). Cremonese, Rome, 1971, pp. 259–86. *MR*: **58**, #23210 (1979).

Miller, J.B.
(*a*) Aczél's uniqueness theorem and cellular internity. *Aequationes Math.*, **5**, 319–25. *MR*: **45**, #2359 (1973).
(*b*) Higher derivations on Banach algebras. *Amer. J. Math.*, **92**, 301–31. *MR*: **44**, #4520 (1972).

Ng, C.T. Uniqueness theorems for a general class of functional equations. *J. Austral. Math. Soc.*, **11**, 362–6. *MR*: **42**, #3457 (1971).

Rätz, J.
(*a*) Zur Definition der Lorentztransformationen. *Math.-Phys. Semesterber.*, **17**, 163–7. *MR*: **42**, #3094 (1971).
(*b*) On isometries of inner product spaces. *SIAM J. Appl. Math.*, **18** (1970), 6–9. *MR*: **41**, #4214 (1971).

Rényi, A. *Probability theory.* (Transld. from Hungarian by L. Vekerdi). (North Holland Series Appl. Math., Mech., Vol. 10), North Holland, Amsterdam-London; American Elsevier, New York, 1970. *MR*: **47**, #4296 (1974).

Sigmon, K.N. Cancellative medial means are arithmetic. *Duke Math. J.*, **37**, 439–45. *MR*: **43**, #407 (1972).

Świątak, H. and Hosszú, M.
(*a*) Remarks on the functional equation $e(x, y)f(xy) = f(x) + f(y)$. *Publ. Techn. Univ. Miskolc.*, **30**, 323–5.
(*b*) Notes on functional equations of polynomial form. *Publ. Math. Debrecen*, **17**, 61–6. *MR*: **47**, #674 (1974).

Venkov, B.A. *Elementary number theory.* (Transld. from Russian by H. Alderson). Walters-Noordhoff, Groningen, 1970. *MR*: **42**, #178 (1971).

Vincze, E. Über ein Funktionalgleichungsproblem von E. Hille. *Arch. Math.* (*Basel*), **21**, 498–501. *MR*: **43**, #2378 (1972).

Ward, L.E., Sr.
(*a*) Solution of elementary problem E2176. *Amer. Math. Monthly*, **77**, 310–11.
(*b*) Editorial note, solution of elementary problem E2176. *Amer. Math. Monthly*, **77**, 767–8.

1971

Aczél, J.
(*a*) Some recent applications of functional equations to semigroups. *Mitt. Math. Sem. Giessen*, **1971**, No. 94, 35–48. *MR*: **47**, #2217 (1974).
(*b*) Some recent applications of functional equations to geometry. *J. Geometry*, **11**, 127–42. *MR*: **48**, #7106 (1974).

Aczél, J., Baker, J.A., Djoković, D.Ž., Kannappan, Pl., and Radó, F. Extensions of certain homomorphisms of subsemigroups to homomorphisms of groups. *Aequationes Math.*, **6**, 263–71. *MR*: **45**, #2004 (1973).

Baker, J.A.
(*a*) Regularity properties of functional equations. *Aequationes Math.*, **6**, 243–8. *MR*: **45**, #2360 (1973).
(*b*) D'Alembert's functional equation in Banach algebras. *Acta Sci. Math.* (*Szeged*), **32**, 225–34. *MR*: **48**, #746 (1974).

Bednarek, A.R. and Wallace, A.D. The functional equation $(xy)(yz) = xz$. *Rev. Roumaine Math. Pures. Appl.*, **16**, 3–6. *MR*: **44**, #353 (1972).

Buche, A.B. On the cosine-sine operator functional equations. *Aequationes Math.* **6**, 231–4. *MR*: **45**, #961 (1973).

Daróczy, Z. On the general solution of the functional equation $f(x + y - xy) + f(xy) = f(x) + f(y)$. *Aequationes Math.*, **6**, 130–2. *MR*: **45**, #2352 (1973).

Dhombres, J.
(*a*) Quelques équations fonctionnelles provenant de la théorie des moyennes. *C.R.. Acad. Sci. Paris. Sér. A*, **273**, 989–91. *MR*: **44**, #4119 (1972).
(*b*) *Sur les opérateurs multiplicativement liés.* (Bull. Soc. Math. France Mém., No. 277). Soc. Math. France, Paris, 1971. *MR*: **58**, #30450 (1979).

Eichhorn, W. Effektivität von Produktionsverfahren. *Operations Research Verfahren*, **12**, 98–115.

Ger, R., On some properties of polynomial functions. *Ann. Polon. Math.*, **25**, 195–203. *MR*: **46**, #4035 (1973).

Haruki, H. On a relation between the "square" functional equation and the "square" mean-value property. *Canad. Math. Bull.*, **14**, 161–5. *MR*: **47**, #7263 (1974).

Horinouchi, S. and Kannappan Pl. On the system of functional equations $f(x + y) = f(x) + f(y)$ and $f(xy) = p(x)f(y) + q(y)f(x)$. *Aequationes Math.*, **6**, 195–201. *MR*: **45**, #7338 (1973).

Hosszú, M. On the functional equation $F(x + y, z) + F(x, y) = F(x, y + z) + F(y, z)$. *Period. Math. Hungar.*, **1**, 213–16. *MR*: **44**, #7176 (1972).

Howroyd, T.D. Some existence theorems for functional equations in many variables and the characterization of weighted quasiarithmetic means. *Aequationes Math.*, **7**, 1–15. *MR*: **45**, #7337 (1973).

Itzkowitz, G.L. Measurability and continuity for a functional equation on a topological group. *Aequationes Math.*, **7**, 194–8. *MR*: **47**, #5477 (1974).

Krantz, D.H., Luce, R.D., Suppes, P., and Tversky, A. *Foundations of measurement, Vol. 1. Additive and polynomial representations.* Academic Press, New York-London, 1971. *MR*: **56**, #17265 (1978).

Kurepa, S. Quadratic functionals conditioned on an algebraic basic set. *Glas. Mat. Ser. III*, **6(26)**, 265–75. *MR*: **46**, #7278 (1973).

Losonczi, L. Über eine neue Klasse von Mittelwerten. *Acta. Sci. Math.* (*Szeged*), **32**, 71–81. *MR*: **47**, #421 (1974).

Maier, W. and Kiesewetter, H. *Funktionalgleichungen mit analytischen Lösungen.* (Studia Mathematica — Mathematische Lehrbücher, Band 20). Vandenhoeck & Ruprecht, Göttingen-Zürich, 1971. *MR*: **48**, #2359 (1974).

Paganoni, L. Existence of solutions for a general class of functional equations (Italian). *Istit. Lombardo Accad. Sci. Lett. Rend. A*, **105** (1971), 891–906. *MR*: **46**, #2289 (1973).

Paganoni Marzegalli, S.
(*a*) Uniqueness theorems for the functional equation $f[F(x, y)] = H[f(x), f(y); x, y]$ in metric spaces (Italian). *Atti Accad. Naz. Lincei Rend. Cl. Sci. Fis. Mat. Natur.* (8), **50**, 438–43. *MR*: **46**, #4044 (1973).
(*b*) Extension of some uniqueness theorems for a general class of functional equations in vectorial topological spaces (Italian). *Istit. Lombardo Accad. Sci. Lett. Rend. A*, **105**, 713–20. *MR*: **46**, #570 (1973).

Rätz, J. Zur Linearität verallgemeinerter

Modulisometrien. *Aequationes Math.*, **6**, 249–55. *MR*: **48**, #4048 (1974).

Reich, S. Solution of elementary problem E2176. *Amer. Math. Monthly*, **78**, 675.

Rota, G.-C. and Smith, D.A. Fluctuation theory and Baxter algebras. In *Convegno di Calcolo delle Probabilità (INDAM Roma, 1971)*. (Symposia Math., Vol. 9). Academic Press, London, 1972, pp. 179–201. *MR*: **49**, #7838 (1975).

Soublin, J.-P.
(*a*) Étude algébrique de la notion de moyenne (chapitre 1). *J. Math. Pures Appl.* (9), **50**, 53–96. *MR*: **45**, #436 (1973).
(*b*) Étude algébrique de la notion de moyenne (suite). *J. Math. Pures Appl.* (9), **50**, 97–192. *MR*: **45**, #437 (1973).
(*c*) Étude algébrique de la notion de moyenne (suite et fin). *J. Math. Pures Appl.* (9), **50**, 193–264. *MR*: **45**, 438 (1973).

Światak, H.
(*a*) A proof of the equivalence of the equation $f(x+y-xy)+f(xy)=f(x)+f(y)$ and Jensen's functional equation. *Aequationes Math.*, **6**, 24–9. *MR*: **45**, #766 (1973).
(*b*) Criteria for the regularity of continuous and locally integrable solutions of a class of linear functional equations. *Aequationes Math.*, **6**, 170–87. *MR*: **45**, #763 (1973).

1972

Aczél, J
(*a*) Sur une équation fonctionnelle liée à la théorie des groupes. *Rev. Roumaine Math. Pures Appl.*, **17**, 819–28. *MR*: **48**, #745 (1974).
(*b*) Remarque 14. *Aequationes Math.*, **8**, 176.

Aczél, J. and Nath, P. Axiomatic characterizations of some measures of divergence in information. *Z. Wahrsch. Verw. Gebiete*, **21**, 215–24. *MR*: **46**, #5043 (1973).

Aczél, J. and Vrănceanu, G. Équations fonctionnelles liées aux groupes linéaires commutatifs. *Colloq. Math.*, **26**, 371–83. *MR*: **49**, #11084 (1975).

Ahsanullah, M. and Rahman, M. A characterization of the exponential distribution. *J. Appl. Probab.*, **9**, 457–61. *MR*: **49**, #8158 (1975).

Belousov, V.D.
(*a*) Balanced identities in algebras of quasigroups. *Aequationes Math.*, **8**, 1–73. *MR*: **46**, #7429 (1973).
(*b*) *n-ary quasigroups* (Russian). Shtiintsa, Kishinev, 1972. *MR*: **50**, #7396 (1975).

Belousov, V.D. and Ryžkov, V.V. The geometry of webs (Russian). In *Algebra, topology, geometry, Vol. 10*, Akad. Nauk SSSR Vsesojuz. Inst. Nauchn. i Tekhn. Informacii, Moscow, 1972, pp. 159–88, 191. *MR*: **50**, #8318 (1975).

Benz, W. The *n*-point-invariants of the projective line and crossratio of *n*-tuples. *Ann. Polon. Math.*, **26**, 53–60. *MR*: **46**, #782 (1973).

Bos, W.
(*a*) Mittelwertstrukturen. *Math. Ann.*, **198**, 317–33. *MR*: **48**, #4765 (1974).
(*b*) Die Einbettung von Mittelwertstrukturen in *Q*-Vektorräume. *Math. Z.*, **128**, 207–16. *MR*: **48**, #4766 (1974).

Campbell, L.L. Characterization of entropy of probability distributions on the real line. *Inform. and Control*, **21**, 329–38. *MR*: **48**, #1803 (1974).

Carroll, F.W. Functions whose differences are integrable. *Aequationes Math.*, **8**, 150–1.

Chein, O. and Robinson, D.A. An "extra" law for characterizing Moufang loops. *Proc. Amer. Math. Soc.*, **33**, 29–32. *MR*: **45**, #2068 (1973).

Dacić, D. The sine functional equation for groups (Serbo-Croatian). *Mat. Vesnik*, **9** (**24**), 49–53. *MR*: **48**, #4564 (1974).

Daróczy, Z. Über eine Klasse von Mittelwerten. *Publ. Math. Debrecen*, **19**, 211–7. *MR*: **48**, #6350 (1974).

Dhombres J.
(*a*) Moyennes de fonctions et opérateurs multiplicativement liés. In *Actes du colloque d'analyse fonctionnelle (Univ. de Bordeaux, Bordeaux, 1971)*. (Bull. Soc. Math. France Mém., No. 31–32). Soc. Math. France, Paris, 1972, pp. 143–9. *MR*: **52**, #1402 (1976).
(*b*) Functional equations on semi-groups arising from the theory of means. *Nanta Math.*, **5**, 48–66. *MR*: **49**, #5625 (1975).
(*c*) Sur les opérateurs multiplicativement liés dans les algèbres de dimension finie. *Ann. Inst. H. Poincaré Sect. B*, **8**, 333–63. *MR*: **50**, #5249 (1975).

Eichhorn, W. Systems of functional equations determining the effectiveness of a production process. In *Mathematical models in economics (Proc. Sympos. Warsaw, 1972 and Conf. on von Neumann Models. Warsaw, 1972)*. North-Holland, Amsterdam, 1974, pp. 433–9. *MR*: **51**, #13503 (1976).

Fishburn, P.C. Alternative axiomatizations of one-way expected utility. *Ann. Math.*

Fishburn, P.C. (*cont.*)
Statist., **43**, 1648–51. *MR*: **50**, #3886 (1975).

Gołąb, S. and Świątak, H. Note on inner product in vector spaces. *Aequationes Math.*, **8**, 74–5. *MR*: **46**, #4180 (1973).

Guilmard, A. Sur l'équation fonctionnelle $f^2(x) - f^2(y) = f(x+y)f(x-y)$. *Rev. Math. Spéciales*, **83**, 161.

Hall, R.L. On single-product functions with rotational symmetry. *Aequationes Math.*, **8**, 281–6. *MR*: **47**, #7255 (1974).

Hille, E. *Methods in classical and functional analysis*. Addison Wesley, Reading, Mass.-London-Don Mills, Ont., 1972. *MR*: **57**, #3802 (1979).

Jasińska, E.J. and Kucharzewski, M. Kleinsche Geometrie und Theorie der geometrischen Objekte. *Colloq. Math.*, **26**, 271–9. *MR*: **49**, #3713 (1975).

Kannappan, Pl. On some identities. *Math. Student*, **40**, 260–4. *MR*: **54**, #7686 (1977).

Kisyński, J.
(*a*) On operator-valued solutions of d'Alembert's functional equation. II. *Studia Math.*, **42**, 43–66. *MR*: **47**, #2428b (1974).
(*b*) On cosine operator functions and one-parameter groups of operators. *Studia Math.*, **44**, 93–105. *MR*: **47**, #890 (1974).

Kucharzewski, M. and Wegrzynowski, S. On scalar concomitants of geometric objects and their transitive domains. *Colloq. Math.*, **26**, 263–9. *MR*: **48**, #12338 (1974).

Kuczma, M. Note on additive functions of several variables. *Uniw. Śląski w Katowicach Prace Nauk.–Prace Mat.*, **2**, 49–51. *MR*: **46**, #7750 (1973).

Lajkó, K. Über die allgemeinen Lösungen der Funktionalgeichung $F(x) + F(y) - F(xy) = H(x + y - xy)$. *Publ. Math. Debrecen*, **19**, 219–23. *MR*: **48**, #11833 (1974).

Lax, P., Burstein, S. and Lax, A. *Calculus with applications and computing.* Vol. 1. Courant Inst., New York, 1972. (Reprinted by Springer, New York-Heidelberg-Berlin, 1976). *MR*: **50**, #7428 (1975).

Levine, M.V. Transforming curves into curves with the same shape. *J. Math. Psych.*, **9**, 1–16. *MR*: **45**, #1618 (1973).

Lin, Y.-F. and McWaters, M.M. On the triviality of the law $(xy)(zy) = yz$. *J. London Math. Soc.* (2), **5**, 276–8. *MR*: **46**, #9228 (1973).

Mauclaire, J.-L. Sur la régularité des fonctions additives. *Enseign. Math.* (2) **18**, 167–74. *MR*: **48**, #3898 (1974).

Moszner, Z.
(*a*) Structure de l'automate plein, réduit et inversible. *Aequationes Math.*, **8**, 155–7. *MR*: **47**, #3585 (1974).
(*b*) Sur le prolongement des objets géométriques transitifs. *Tensor* (*N.S.*), **26**, 239–42. *MR*: **48**, #4938 (1974).

Nordon, J. Sur une équation fonctionnelle. *Rev. Math. Spéciales*, **83**, 89–90.

Paganoni, L.
(*a*) The fixed point method for the class $f[F(x, y)] = H[f(x), f(y); x, y]$ of functional equations (Italian). *Atti. Accad. Naz. Lincei Rend. Cl. Sci. Fis. Mat. Natur.*, **52**, 675–81. *MR*: **49**, #3367 (1975).
(*b*) Uniqueness theorems for a general class of functional equations (Italian). *Boll. Un. Mat. Ital.* (4), **6**, 450–61. *MR*: **47**, #9112 (1974).

Pokropp, F. *Aggregation von Produktionsfunktionen. Klein-Nataf-Aggregation ohne Annahmen über Differenzierbarkeit und Stetigkeit.* (Lecture Notes in Econom. and Math. Systems, Vol. 74). Springer, Berlin-New York, 1972. *MR*: **50**, #1704 (1975).

Sharir, M. Characterization and properties of extremal operators into $C(Y)$. *Israel J. Math.*, **12**, 174–8. *MR*: **47**, #5574 (1974).

Smital, J. On boundedness and discontinuity of additive functions. *Fund. Math.*, **76**, 245–53. *MR*: **47**, #9109 (1974).

Sundaresan, K. Orthogonality and nonlinear functions on Banach spaces. *Proc. Amer. Math. Soc.*, **34**, 187–90. *MR*: **45**, #925 (1973).

Székelyhidi, L. The general representation of an additive function on an open point set (Hungarian). *Magyar Tud. Akad. Mat. Fiz. Oszt. Közl.*, **21**, 503–9, *MR*: **48**, #4561 (1974).

Tan, P. Functional equations in mathematical statistics. *Math. Mag.*, **45**, 179–83.

Taylor, M.A.
(*a*) Relational systems with a Thomsen or Reidemeister cancellation condition. *J. Math. Psych.*, **9**, 456–8. *MR*: **47**, #4665 (1974).
(*b*) The generalized equations of bisymmetry, associativity and transitivity on quasigroups. *Canad. Math. Bull.*, **15**, 119–24. *MR*: **51**, #8317 (1976).

Traynor, T. Decomposition of group-valued additive set functions. *Ann. Inst. Fourier* (*Grenoble*), **22**, 131–40. *MR*: **48**, #11439 (1974).

Ušan, J. Globally associative systems of

ternary quasigroups. *Math. Balkanica*, **2**, 270–87. *MR*: **48**, #4179 (1974).

Zajtz, A.

(*a*) Automorphisms of differential groups. *Colloq. Math.*, **26**, 241–8. *MR*: **49**, #11556 (1976).

(*b*) Klassifikation der linearen homogenen geometrischen Objekte von Typus *J* mit messbarer Transformationsregel. *Ann. Polon. Math.*, **26**, 31–43. *MR*: **50**, #14536 (1975).

Zdun, M.C. On the uniqueness of solutions of the functional equation $\psi(x + f(x)) = \psi(x) + \psi(f(x))$. *Aequationes Math.*, **8**, 229–32. *MR*: **47**, #3869 (1974).

1973

Aczél, J. Zusammenhang zwischen einer Abelschen Kurvenkonstruktion und der Thomsen Figur. *Demonstratio Math.*, **6**, 71–5. *MR*: **50**, #336 (1975).

Baron, K. and Ger, R. On Mikusiński-Pexider functional equation. *Colloq. Math.*, **28**, 307–12. *MR*: **49**, #3365 (1975).

Bellman, R. and Giertz, M. On the analytic formulation of the theory of fuzzy sets. *Inform. Sci.*, **5**, 149–56. *MR*: **54**, #2417 (1977).

Bos, W. Die Charakterisierung einer Mittelwertstruktur durch deren zweistellige Mittelwertfunktion. *Arch. Math.* (*Basel*), **24**, 397–401. *MR*: **49**, #3362 (1975).

Brown, R.B. Sequences of functions of binomial type. *Discrete Math.*, **6**, 313–31. *MR*: **53**, #149 (1977).

Choczewski, B. Asymptotic series expansions of continuous solutions of a linear functional equation. *Bull. Acad. Polon. Sci. Sér. Sci. Math. Astronom. Phys.*, **21**, 925–30. *MR*: **49**, #908 (1975).

Daróczy, Z. and Kiesewetter, H. Eine Funktionalgleichung von Abel und die Grundgleichung der Information. *Period. Math. Hungar.*, **4**, 25–8. *MR*: **48**, #8107 (1974).

Dhombres, J. Hemi-multiplicative operators and related operators in finite-dimensional algebras. *Aequationes Math.*, **9**, 284–95. *MR*: **48**, #2612 (1974).

Drobot, V. Discontinuous solutions of the functional equation $f(x + y) = H(f(x), f(y))$. *Funkcial. Ekvac.*, **16**, 165–8. *MR*: **50**, #2729 (1975).

Eichhorn, W.

(*a*) Zur axiomatischen Theorie des Preisindex. *Demonstratio Math.*, **6**, 561–73. *MR*: **50**, #1702 (1975).

(*b*) Characterization of the CES production functions by quasilinearity. In *Production theory* (*Proc. Internat. Sem. Univ. Karlsruhe, Karlsruhe, 1973*). (Lecture Notes in Econom. and Math. Systems, Vol. 99). Springer, Berlin, 1974, pp. 21–33. *MR*: **52**, #4976 (1976).

Eichhorn, W. and Kolm, S.-C. Technical progress, neutral inventions, and Cobb-Douglas. In *Production theory* (*Proc. Internat. Sem. Univ. Karlsruhe, Karlsruhe, 1973*). (Lecture Notes in Econom. and Math. Systems, Vol. 99). Springer, Berlin, 1974, pp. 35–45. *MR*: **53**, #2302 (1977).

Färe, R. A characterization of Hicks neutral technical process. In *Production theory* (*Proc. Internat. Sem. Univ. Karlsruhe, Karsruhe, 1973*). (Lecture Notes in Econom. and Math. Systems, Vol. 99). Springer, Berlin, 1974, pp. 47–51. *MR*: **52**, #16539 (1976).

Forte, B. Why Shannon's entropy. In *Convegno di Informatica Teorica* (*INDAM, Roma, 1973*). (Symposia Math., Vol. 15). Academic Press, London, 1975, pp. 137–52. *MR*: **53**, #7610 (1977).

Forte, B. and Ng, C.T. On a characterization of the entropies of degree β. *Utilitas Math.*, **4**, 193–205. *MR*: **48**, #10667 (1974).

Garsia, A.M. An exposé of the Mullin-Rota theory of polynomials of binomial type. *Linear and Multilinear Algebra*, **1**, 47–65. *MR*: **47**, #9321a (1974).

Ger, R. Thin sets and convex functions. *Bull. Acad. Polon. Sci. Sér. Sci. Math. Astronom. Phys.*, **21**, 413–16. *MR*: **48**, #4232 (1974).

Gołąb, S. Theorie der Quasigruppen. In *Grundlagen der Geometrie und algebraische Methoden* (*Internat. Kolloq., Pädagog. Hochsch. "Karl Liebknecht", Potsdam, 1973*). (Potsdamer Forschungen, Reihe B, Heft 3). Pädagog. Hochsch. "Karl Liebknecht", Potsdam, 1974, pp. 137–51. *MR*: **51**, #778 (1976).

Haruki, H. On the equivalence of Hille's and Robinson's functional equations. *Ann. Polon. Math.*, **28**, 261–4. *MR*: **49**, #7430 (1975).

Haruki, Sh.

(*a*) A note on a pentomino functional equation. *Ann. Polon. Math.*, **27**, 129–31. *MR*: **47**, #670 (1974).

(*b*) Multiple integrals evaluated by functional equations. *Ann. Polon. Math.*, **27**, 197–9. *MR*: **47**, #671 (1974).

(*c*) A note on a square type functional equation. *Canad. Math. Bull.*, **16**, 443–5. *MR*: **49**, #909 (1975).

Haruki, Sh. (*cont.*)
(*d*) On the functional equation $\sum_{i=1}^{3}(X_i^t - 1)\cdot f = 0$ and two related equations. *Utilitas Math.*, **4**, 3–7. *MR*: **49**, #5619 (1975).

Hille, E. On a class of adjoint functional equations. *Acta. Sci. Math. (Szeged)*, **34**, 141–61. *MR*: **48**, #2611 (1974).

Howroyd, T.D. Cancellative medial groupoids and arithmetic means. *Bull. Austral. Math. Soc.*, **8**, 17–21. *MR*: **47**, #6927 (1974).

Jou, S.S. Convex functions on topological groups. *Glas. Mat. Ser. III*, **8(28)**, 175–8. *MR*: **51**, #6205 (1976).

Kagan, A.M., Linnik, Ju.V. and Rao, C.R. *Characterization problems in mathematical statistics.* Wiley & Sons, New York-London-Sydney, 1973. *MR*: **49**, #11689 (1975).

Kampé, de Fériet, J. Une équation fonctionnelle de la théorie de l'information généralisant l'équation de Cauchy. *Demonstratio Math.*, **6**, 665–85. *MR*: **50**, #1770 (1975).

Kannappan, Pl. and Ng, C.T. Measurable solutions of functional equations related to information theory. *Proc. Amer. Math. Soc.*, **38**, 303–10. *MR*: **47**, #672 (1974).

Kopp, P.E., Strauss, D., and Yeadon, F.J. Positive Reynolds operators on Lebesgue spaces. *J. Math. Anal. Appl.*, **44**, 350–65. *MR*: **48**, #4813 (1974).

Kuczma, M.
(*a*) On some set classes occurring in the theory of convex functions. *Comment. Math. Prace Mat.*, **17**, 127–35. *MR*: **48**, #4224 (1974).
(*b*) Cauchy's functional equation on a restricted domain. *Colloq. Math.*, **28**, 313–15. *MR*: **49**, #3366 (1975).

Kurepa, S.
(*a*) On bimorphisms and quadratic forms on groups. *Aequationes Math.*, **9**, 30–45. *MR*: **48**, #4565 (1974).
(*b*) A weakly measurable selfadjoint cosine function. *Glas. Mat. Ser. III*, **8(28)**, 73–9. *MR*: **48**, #12147 (1974).

Lehmann, I. Einige Ergebnisse zur Theorie der bisymmetrischen Quasigruppen. In *Grundlagen der Geometrie und algebraische Methoden (Internat. Kolloq., Pädagog. Hochsch. "Karl Liebknecht", Potsdam, 1973)*. (Potsdamer Forschungen, Reihe B, Heft 3). Pädagog. Hochsch. "Karl Liebknecht", Potsdam, 1974, pp. 154–6. *MR*: **51**, #10489 (1976).

Lin, S.-Y.T. On a functional equation arising from the Monteiro-Botelho-Teixeira axioms for a topological space. *Aequationes Math.*, **9**, 281–3. *MR*: **48**, #6750 (1974).

Lorch, E.R. *Precalculus.* Norton, New York, 1973.

Luchter, J.
(*a*) The determination of non-homogeneous linear geometric objects of the first class for which the number of coordinates is not greater than the dimension of the space. *Demonstratio Math.*, **5**, 231–43. *MR*: **49**, #6060 (1975).
(*b*) On some multiplicative functional equation defined on a differential group. *Demonstratio Math.*, **5**, 245–60. *MR*: **49**, #5616 (1975).
(*c*) Non-existence of linear differential geometric objects of class $s \geqslant 3$, whose component number is not greater than the dimension of the space. *Demonstratio Math.*, **6**, 211–22. *MR*: **52**, #1542 (1976).

Miller, J.B. Analytic structure and higher derivations on commutative Banach algebras. *Aequationes Math.*, **9**, 171–83. *MR*: **51**, #13696 (1976).

Morita, K. Nomography in Japan (Russian). In *Nomographic collection, No. 9*, Vychisl. Centr Akad. Nauk SSSR, Moscow, 1973, pp. 3–8, 190. *MR*: **50**, #6759 (1975).

Moszner, Z.
(*a*) Structure de l'automate plein réduit et inversible. *Aequationes Math.*, **9**, 46–59. *MR*: **47**, #3585 (1974).
(*b*) The translation equation and its application. *Demonstratio Math.*, **6**, 309–27. *MR*: **49**, #9461 (1975).

Ng, C.T.
(*a*) On the functional equation $f(x) + \sum_{i=1}^{n} g_i(y_i) = h(T(x, y_1, y_2, \ldots, y_n))$. *Ann. Polon. Math.*, **27**, 329–36. *MR*: **48**, #6752 (1974).
(*b*) Local boundedness and continuity for a functional equation on topological spaces. *Proc. Amer. Math. Soc.*, **39**, 525–9. *MR*: **47**, #7265 (1974).

Paganoni Marzegalli, S. Existence theorems for a general class of functional equations (Italian). *Istit. Lombardo Accad. Sci. Lett. Rend. A*, **107**, 34–43. *MR*: **49**, #11099 (1975).

Piquet, C. Opérateurs multiplicativement liés. In *Séminaire Choquet, 13e année (1973/74)*. (Initiation à l'analyse, Exp. No. 4). Secrétariat Mathématique, Paris, 1975. *MR*: **57**, #17379 (1979).

Rota, G.-C., Kahaner, D. and Odlyzko, A. On the foundation of combinatorial theory, VIII. Finite operator calculus. *J. Math. Anal. Appl.*, **42**, 684–760. *MR*: **49**, (1974), #10556.

Rudin, W. *Functional analysis*. McGraw-Hill, New York-Düsseldorf-Johannesburg, 1973. *MR*: **51**, #1315 (1976).

Shih, K.S. On the functional equations $P_n(x+y) = \sum_{k=0}^{n} \binom{n}{k} P_{n-k}(x) P_k(y)$. *Chinese J. Math.*, **1**, 113–17. *MR*: **52**, #14726 (1976).

Simone, J.N. On number theoretic functions which satisfy $f(x+y) = f(f(x)+f(y))$. *Math. Mag.*, **46**, 213–15. *MR*: **48**, #2045 (1974).

Stehling, F. Neutral inventions and CES production functions. In *Production theory (Proc. Internat. Sem. Univ. Karlsruhe, Karlsruhe, 1973)*. (Lecture Notes in Econom. and Math. Systems, Vol. 99). Springer, Berlin, 1974, pp. 65–94. *MR*: **52**, #2548 (1976).

Strecker, R. Über entropische Gruppoide. In *Grundlagen der Geometrie und algebraische Methoden (Internat. Kolloq., Pädagog. Hochsch. "Karl Liebknecht" Potsdam, 1973)*. (Potsdamer Forschungen, Reihe B, Heft 3). Pädagog. Hochsch. "Karl Liebknecht", Potsdam, 1974, pp. 153–4. *MR*: **50**, #7384 (1975).

Szymiczek, K. Note on semigroup homomorphisms. *Uniw. Śląski w Katowicach Prace Nauk.—Prace Mat.*, **3**, 75–8. *MR*: **47**, #8735 (1974).

Targonski, G. Charakterisierung von Operatoren durch Funktionalgleichungen. *Demonstratio Math.*, **6**, 861–70. *MR*: **50**, #13952 (1975).

Täuber, J. Über die Lösungen einer von A.D. Alexandrov verallgemeinerten Cauchyschen Funktionalgleichung. *An. Univ. Timisoara Ser. Sti. Mat.*, **11**, 175–82. *MR*: **52**, #14724 (1976).

Taylor, M.A.
(*a*) Certain functional equations on groupoids weaker than quasigroups. *Aequationes Math.*, **9**, 23–9. *NR*: **51**, #13507 (1976).
(*b*) Cartesian nets and groupoids. *Canad. Math. Bull.*, **16**, 347–62. *MR*: **49**, #444 (1975).

Vrbová, P. Quadratic functionals and bilinear forms. *Časopis Pěst. Mat.*, **98**, 159–61. *MR*: **49**, #11082 (1976).

1974

Aczél, J.
(*a*) Determination of all additive quasiarithmetic mean codeword lengths. *Z. Wahrsch. Verw. Gebiete* **29**, 351–360. *MR*: **50**, #16076 (1975).
(*b*) General solution of a functional equation connected with a characterization of statistical distributions. In *Statistical distributions in scientific work, Vol. 3. (Proc. NATO Advanced Study Institute, University of Calgary, Calgary, 1974)*. Reidel, Dordrecht-Boston, 1975, pp. 47–55.

Aczél, J. and Eichhorn, W. Systems of functional equations determining price and productivity indices. *Utilitas Math.*, **5**, 213–26. *MR*: **50**, #13944 (1975).

Aczél, J., Forte, B., and Ng, C.T.
(*a*) On a triangular functional equation and some applications, in particular to the generalized theory of information. *Aequationes Math.*, **11**, 11–30. *MR*: **50**, #5246 (1975).
(*b*) Why the Shannon and Hartley entropies are 'natural'. *Adv. in Appl. Probab.*, **6**, 131–46. *MR*: **49**, #4658 (1975).

Baker, J.A. On the functional equation $f(x)g(y) = \sum_{i=1}^{n} h_i(a_i x + b_i y)$. *Aequationes Math.*, **11**, 154–62. *MR*: **51**, #10932 (1976).

Belousov, V.D. and Livshic, E.S. Functional equation of general associativity on binary quasigroups (Russian). *Mat. Issled.*, **9**, No. 4 (34), 5–22. *MR*: **54**, #10463 (1977).

Carroll, T.B. A characterization of completely multiplicative arithmetic functions. *Amer. Math. Monthly*, **81**, 993–5. *MR*: **50**, #4462 (1975).

Cheban, A.M. Functional equations of Bol-Moufang type (Russian). *Mat. Issled.*, **9**, 4 (34), 150–60. *MR*: **54**, #10464 (1977).

Chinda, K.P. The equation $F(F(x,y), F(z,y)) = F(x,z)$. *Aequationes Math.*, **11**, 196–8. *MR*: **52**, #11395 (1976)

Crombez, G. and Six, G. On topological *n*-groups. *Abh. Math. Sem. Univ. Hamburg*, **41**, 115–24. *MR*: **52**, #14143 (1976).

Csiszár, I. Information measures: a critical survey. In *Transactions of the Seventh Prague Conference on Information Theory, Statistical Decision Functions, Random Processes and of the Eighth European Meeting of Statisticians (Tech. Univ. Prague, Prague, 1974)*, Vol. B. Academia, Prague, 1978, pp. 73–86.

Daróczy, Z. Remarks on the entropy equation. In *Symposium en quasigroupes et équations fonctionnelles (Belgrade-Novi Sad, 1974)*. Math. Inst., Belgrade, 1976, pp. 31–4. *MR*: **54**, #9848 (1977).

Davison, T.M.K. On the functional equation $f(m+n-mn) + f(mn) = f(m) + f(n)$. *Aequationes Math.*, **10**, 206–11. *MR*: **49**, #11086 (1975).

Donnell, W. A note on entropic groupoids.

Donnellg W. (*cont.*)
Portugal. Math., **33**, 77–8. *MR*: **48**, #11357 (1974).

Ecsedi, I.
(*a*) A certain class of discontinuous solutions of the functional equation $f(ax+by)g(cx+dy)=h(x)k(y)$ (Hungarian). *Magyar Tud. Akad. Mat. Fiz. Oszt. Közl.*, **22**, 3–10. *MR*: **51**, #6203 (1976).
(*b*) On a functional equation which contains a real-valued function of vector variable. *Period. Math. Hungar.*, **5**, 333–42. *MR*: **51**, #6204 (1976).

Eichhorn, W. (compil.) Elfte internationale Tagung über Funktionalgleichungen, Oberwolfach vom 14 bis 20. December 1973. *Aequationes Math.* **11**, 277–314.

Etigson, L. Equivalence of 'cube' and 'octahedron' functional equations. *Aequationes Math.*, **10**, 50–6. *MR*: **49**, #912 (1975).

Forte, B. and Ng, C.T. Entropies with the branching property. *Ann. Mat. Pura Appl.* (4), **101**, 355–73. *MR*: **51**, #15188 (1976).

Galambos, J. Characterizations of probability distributions by properties of order statistics, I. In *Statistical Distributions in Scientific Work, Vol. 3 (Proc. NATO Advanced Study Institute, University of Calgary, Calgary. 1974)*. Reidel, Dordrecht-Boston, 1975, pp. 71–88.

Godini, G. Uniqueness theorems for a class of functional equations. *Rev. Roumaine Math. Pures Appl.*, **19**, 1013–20. *MR*: **50**, #7871 (1975).

Gołąb, S.
(*a*) Sur un système d'équations fonctionnelles lié au rapport anharmonique. *Ann. Polon. Math.*, **29**, 273–80. *MR*: **50**, #813 (1975).
(*b*) *Tensor Calculus.* Elsevier, Amsterdam-London-New York, 1974. (Polish edn. in 1956). *MR*: **49**, #11411 (1975).

Haruki, H. A famous definite integral. *Math. Notae*, **24**, 23–5. *MR*: **53**, #8700 (1977).

Hollocou, Y. Sur l'équation fonctionnelle de Cauchy. *C.R. Acad. Sci. Paris Sér. A*, **278**, 649–51. *MR*: **49**, #11083 (1975).

Hosszú, M. and Csikós, M. Normenquadrat über Gruppen. In *Symposium en quasigroupes et équations fonctionnelles (Belgrade-Novi Sad, 1974)*. Math. Inst., Belgrade, 1976, pp. 35–9. *MR*: **55**, #915 (1978).

Hsu, I.C. On a cubic functional equation defined on groups. *Elem. Math.*, **29**, 112–17.

Kannappan, Pl. and Ng, C.T.
(*a*) A functional equation and its application to information theory. *Ann. Polon. Math.*, **30**, 105–12. *MR*: **49**, #5621 (1975).
(*b*) On functional equations connected with directed divergence, inaccuracy and generalized directed divergence. *Pacific J. Math.*, **54**, 157–67. *MR*: **51**, #10939 (1976).

Kotz, S. Characterizations of statistical distributions: a supplement to recent surveys. *Internat. Statist. Rev.*, **42**, 39–65. *MR*: **50**, #6015 (1975).

Kucharzewski, M. and Zajtz, A. Über die multiplikative Funktionalgleichung für stochastische Matrizen. *Aequationes Math.*, **11**, 128–37. *MR*: **51**, #3737 (1976).

Lajkó, K.
(*a*) Applications of extensions of additive functions. *Aequationes Math.*, **11**, 68–76. *MR*: **49**, #7637 (1975).
(*b*) Special multiplicative deviations. *Publ. Math. Debrecen*, **21**, 39–45. *MR*: **51**, #1186 (1976).

Maksa, Gy. A functional equation with differences. In *Symposium en quasigroupes et equations fonctionnelles (Belgrade-Novi Sad, 1974)*. Math. Inst., Belgrade, 1976, pp. 49–52. *MR*: **52**, #912 (1978).

Mills, T.M. An algebraic functional equation. *Portugal, Math.*, **33**, 51–6. *MR*: **48**, #11837 (1974).

Mokanski, J.P. Extensions of functions satisfying Cauchy and Pexider type equations defined on arbitrary groups. *Mathematica (Cluj)*, **16**, 99–108. *MR*: **53**, #13908 (1977).

Morgan, C.L. Addition formulae for field-valued continuous functions on topological groups. *Aequationes Math.*, **11**, 77–96. *MR*: **49**, #8999 (1975).

Nagy, B.
(*a*) On a generalization of the Cauchy equation. *Aequationes Math.* **10**, 165–71. *MR*: **50**, #5912 (1975).
(*b*) On the generators of cosine operator functions. *Publ. Math. Debrecen*, **21**, 151–4. *MR*: **50**, #14368 (1975).
(*c*) On cosine operator functions in Banach spaces. *Acta Sci. Math. (Szeged)*, **36**, 281–9. *MR*: **51**, #11191 (1976).

Ng. C.T.
(*a*) Representation for measures of information with the branching property. *Inform. and Control*, **25**, 45–56. *MR*: **49**, #7019 (1975).
(*b*) On the measurable solutions of the functional equation $\Sigma_{i=1}^{2}\Sigma_{j=1}^{3}$

$F_{i,j}(p_i q_j) = \Sigma_{i=1}^{2} G_i(p_i) + \Sigma_{j=1}^{3} H_j(q_j)$. *Acta Math. Acad. Sci. Hungar.*, **25**, 249–54. *MR*: **50**, #7875 (1975).

Paganoni, L.

(*a*) On the uniqueness of the solutions of a certain class of functional equations (Italian). *Rend. Istit. Mat. Univ. Trieste*, **6**, 77–88. *MR*: **50**, #5252 (1976).

(*b*) An extension of a theorem of Steinhaus (Italian). *Istit. Lombardo Accad. Sci. Lett. Rend. A*, **108**, 262–73. *MR*: **50**, #7457 (1975).

(*c*) On the equivalence of measurability and continuity for the solutions of a class of functional equations (Italian). *Riv. Mat. Univ. Parma*, **3**, 175–88. *MR*: **56**, #6174 (1978).

Peljuh, G.P. and Sharkovskiĭ, A.N. *Introduction to the theory of functional equations* (Russian). Naukova Dumka, Kiev, 1974. *MR*: **51**, #3735 (1976).

Rao, C.R.

(*a*) Functional equations and characterization of probability distributions. In *Proc. Internat. Congress of Mathematicians.* (*Vancouver, 1974*), *Vol. 2.* Canadian Mathematical Congress, 1975, pp. 163–8. *MR*: **54**, #8945 (1977).

(*b*) Inaugural Linnik memorial lecture – Some problems in the characterization of the multivariate normal distribution. In *Statistical distributions in scientific work , Vol. 3* (*Proc. NATO Advanced Study Institute, University of Calgary, Calgary, 1974*). Reidel, Dordrecht-Boston, 1975, pp. 1–13.

Rimán, J. On an extension of Pexider's equation. In *Symposium en quasigroupes et équations fonctionnelles* (*Belgrade-Novi Sad, 1974*). Math. Inst., Belgrade, 1976, pp. 65–72. *MR*: **54**, #8064 (1977).

Rukhin, A.L. A remark on the d'Alembert functional equation (Russian). *Zap. Nauchn. Sem. Leningrad. Otdel. Mat. Inst. Steklov.*, **47**, 182–3, 190, 195. *MR*: **52**, #8708 (1976).

Samuelson, H. To e via convexity. *Amer. Math. Monthly*, **81**, 1012–14.

Schmidt, H. Über das Additionstheorem der Binomialkoeffizienten, *Aequationes Math.* **10**, 302–6.

Schwarz, W. and Spilker, J. Über zahlentheoretische Funktionen, die $f(x+y) = f(f(x)+f(y))$ erfüllen. *Mitt. Math. Sem. Giessen*, **111**, 80–6. *MR*: **51**, #8014 (1976).

Strambach, K. Distributive quasigroups. In *Foundations of geometry* (*Proc. Conf. Univ. Toronto, Toronto, 1974*). Univ. Toronto Press, Toronto, 1976, pp. 251–76. *MR*: **54**, #2861 (1977).

Strecker, R. Über entropische Gruppoide. *Math. Nachr.*, **64**, 363–71. *MR*: **50**, #13323 (1975).

Ušan, J.

(*a*) The precision with which the group and the substitutions are determined in the solution of the system of functional equations $\Lambda_{j\varepsilon\{2,\ldots,n\}} X_1[X_2(a_1,\ldots,a_m), a_{m+1},\ldots, a_{m+n-1}] = X_{2j-1}[a_1,\ldots,a_{j-1}, X_{2j}(a_j,\ldots,a_{j+m-1}), a_{j+m},\ldots,a_{m+n-1}]$, $m=n$ (Russian). *Publ. Inst. Math.* (*Beogard*) (*N.S.*), **17(31)**, 173–82. *MR*: **52**, #11397 (1976).

(*b*) Some systems of functional equations of general associativity, and their connection with the functional equations of general associativity on an algebra of quasigroups (Russian). In *Symposium en quasigroupes et équations fonctionnelles* (*Belgrade-Novi Sad, 1974*). Math. Inst., Belgrade, 1976, pp. 73–80. *MR*: **55**, #552 (1978).

Van der Mark, J. On the functional equation of Cauchy. *Aequationes Math.*, **10**, 57–77. *MR*: **48**, #11832 (1974).

Volkmann, P. Eine Charakterisierung der positiv definiten quadratischen Formen. *Aequationes Math.*, **11**, 174–82. *MR*: **50**, #10776 (1975).

1975

Aczél, J. On a system of functional equations determining price and productivity indices. *Utilitas Math.*, **7**, 345–62. *MR*: **51**, #10940 (1976).

Aczél, J. and Benz, W. Über das harmonische Produkt und eine korrespondierende Funktionalgleichung. *Abh. Math. Sem. Univ. Hamburg*, **43**, 1–10. *MR*: **52**, #8704 (1976).

Aczél, J. and Daróczy, Z. *On Measures of Information and Their Characterizations.* (Mathematics in Science and Engineering; Vol. 115). Academic Press, New York-San Francisco-London. *MR*: **58**, #33509 (1979).

Baumgartner, E. Zur Einführung der Logarithmus- und Exponentialfunktionen in der Sekundarstufe, II. *Didaktik Math.*, **3**, 1–28.

de Bragança, S.L. Finite dimensional Baxter algebras. *Stud. Appl. Math.*, **54**, 75–89. *MR*: **56**, #6464 (1978).

Buche, A.B. On an exponential-cosine operator-valued functional equation.

Buche, A.B. (*cont.*)
Aequationes Math., **13**, 233–41. *MR*: **53**, #3789 (1977).

Chernoff, P.R. Solution of elementary problem E2479. *Amer. Math. Monthly*, **82**, 668–9.

Dhombres, J.
(*a*) Itération linéaire d'ordre 2. *C.R. Acad. Sci. Paris Sér. A*, **280**, 275–7. *MR*: **51**, #13506 (1976).
(*b*) A functional characterization of Markovian linear exaves. *Bull. Amer. Math. Soc.*, **81**, 703–6. *MR*: **51**, #13751 (1976).

Diderrich, G.T. The role of boundedness in characterizing Shannon entropy. *Inform. and Control*, **29**, 149–61. *MR*: **54**, #9849 (1977).

Dionin, V. Solution of the system of functional equations $\Lambda X_1[X_2(a_1^n), a_{n+1}^{2n+d-1}] = X_{2j-1}[a_1^{i-1}, X_2(a_j^{i+n-1}), a_{j+n}^{2n+d-1}]$, $j\varepsilon\{2, 3, \ldots, n+d\}$ on an algebra of quasigroups (Russian). *Mat. Vesnik*, **12** (**27**), 135–42. *MR*: **52**, #8313 (1976).

Đorđević, R.Ž. and Mitrović, Ž.M. On a function related to Binet-Cauchy's functional equation. *Univ. Beograd. Publ. Elektrotehn. Fak. Ser. Mat. Fiz.*, **1975**, No. 498–541, 125–8. *MR*: **52**, #3770 (1976).

Fearnley-Sander, D. A characterization of the determinant. *Amer. Math. Monthly*, **82**, 838–40. *MR*: **53**, #2977 (1977).

Forte, B. and Ng, C.T. Derivation of a class of entropies including those of degree β. *Inform. and Control*, **28**, 335–51. *MR*: **51**, #9962 (1976).

Ger, R. On some functional equations with a restricted domain. *Fund. Math.*, **89**, 131–49. *MR*: **52**, #8706 (1976).

Ger, R. and Kuczma, M. On inverse additive functions. *Boll. Un. Mat. Ital.* (4), **11**, 490–5. *MR*: **52**, #6234 (1976).

Gołąb, S. Über das Carnotsche Skalarprodukt in schwach normierten Vektorräumen. *Aequationes Math.*, **13**, 9–13. *MR*: **53**, #3664 (1977).

Gudder, S. and Strawther, D. Orthogonally additive and orthogonally increasing functions on vector spaces. *Pacific J. Math.*, **58**, 427–36. *MR*: **52**, #11542 (1976).

Haruki, H. A characteristic property of orthogonal pencils of coaxal circles from the standpoint of conformal mapping. *Ann. Polon. Math.*, **31**, 171–7. *MR*: **52**, #14295 (1976).

Jean, R. *Mesure* et *intégration*. Les Presses de l'Université du Québec, Montréal, 1975. *MR*: **57**, #6341 (1979).

Kannan, D. and Kannappan, Pl. On a characterization of Gaussian measures in a Hilbert space. *Ann. Inst. H. Poincaré Sect. B.*, **11**, 397–404. *MR*: **52**, #15586 (1976).

Kucharzewski, M. and Szociński, B. Über Homomorphismen einer Gruppe von Matrizen. *Ann. Polon. Math.*, **30**, 237–42. *MR*: **55**, #8068 (1978).

Livshic, E.S. Functional equations of the second kind over binary quasigroups (Russian). *Mat. Issled.* **10**, No. 2(36), 168–81, 285. *MR*: **53**, #3177 (1977).

Lundberg, A. and Ng, C.T. Uniqueness theorems for the representation $\phi(f(x)g(y) + h(y))$. *Aequationes Math.*, **13**, 77–87. *MR*: **52**, #11396 (1976).

Marsaglia, G. and Tubilla, A. A note on the 'lack of memory' property of the exponential distribution. *Ann. Probab.*, **3**, 353–4. *MR*: **51**, #2073 (1975).

Moszner, Z. Sur les solutions d'une équation fonctionnelle, II. *Aequationes Math.* **13**, 269–73. *MR*: **54**, #773 (1977).

Moszner, Z. and Tabor, J. L'équation de translation sur une structure avec zéro. *Ann. Polon. Math.* **31**, 255–64. *MR*: **53**, #13910 (1977).

Nashed, M.Z. A functional equation which characterizes polynomial operators with applications to uniqueness. *Applicable Anal.*, **5**, 55–63. *MR*: **52**, #15156 (1976).

Neuman, F. (compil.) Twelfth International Symposium on Functional Equations (Waterloo and Victoria Harbour, Ont., August 30–September 9, 1974). *Aequationes Math.* **12**, 262–310. *MR*: **51**, #10923 (1976).

Olkin, I. Problem **P128**. *Aequationes Math.*, **12**, 290–2.

Paganoni, L. Cauchy, Ostrowski and Steinhaus measures on the real axis (Italian). *Atti Accad. Sci. Torino Cl. Sci. Fis. Mat. Natur.*, **109**, 145–55. *MR*: **53**, #5823 (1977).

Strambach, K. Rechtsdistributive Quasigruppen auf l-Mannigfaltigkeiten. *Math. Z.*, **145**, 63–8. *MR*: **52**, #3425 (1976).

Tabor, J. Solution of Cauchy's functional equation on a restricted domain. *Colloq. Math.*, **33**, 204–8. *MR*: **52**, #8701 (1976).

1976

Aczél, J.
(*a*) Picardus ab omni naevo liberatus (On the axiomatics of vector addition). In *General Inequalities 1* (*Proc. First Internat. Conf. on General Inequalities, Oberwolfach, 1976*). Birkhäuser, Basel-Stuttgart, 1978, pp. 249–53. *MR*: **80**, #51017 (1980).

(*b*) Some recent applications of functional equations to combinatorics, probability distributions, information measures and to the theory of index numbers in mathematical economics. In *Theory and Applications of Economic Indices* (*Proc. Internat. Sympos., Univ. Karlsruhe, 1976*). Physica, Würzburg, 1978, pp. 565–90. *MR*: **58**, #23220 (1979).

Babescu, G.H. On the D'Alembert's functional equation in the Banach algebras. *Bul. Şti. Tehn. Inst. Politehn. 'Traian Vuia' Timişoara*, **21**(**35**), 136–8. *MR*: **55**, #13110 (1978).

Baker, J.A. On the functional equation $f(x)g(y) = p(x + y)q(x/y)$. *Aequationes Math.*, **14**, 493–506. *MR*: **55**, #3581 (1978).

Belousov, V.D. and Livshic, E.S. Balanced functional equations on quasigroups of arbitrary arity (Russian). *Quasigroups and combinatorics. Mat. Issled.* No. **43**, 9–29, 203. *MR*: **58**, #29536 (1979).

Benz, W. On characterizing Lorentz transformations. In *General Inequalities 1* (*Proc. First Internat. Conf. on General Inequalities, Oberwolfach, 1976*). Birkhäuser, Basel-Stuttgart, 1978, p. 319.

Berg, L. Functional equations in a field. *Math. Nachr.*, **74**, 309–12. *MR*: **54**, #10912 (1977).

Bilinski, S. Die linearadditiven Zweiindizesfunktionen. *Aequationes Math.*, **14**, 95–104. *MR*: **54**, #1065 (1977).

Buche, A.B. and Vasudeva, H.L. The generalized Cauchy equation for operator-valued functions. *Aequationes Math.*, **14**, 387–90. *MR*: **54**, #8355 (1977).

Burk, R. and Gehrig, W. Indices of income inequality and societal income. An axiomatic approach. In *Theory and Applications of Economic Indices.* (*Proc. Internat. Sympos., Univ. Karlsruhe, 1976*). Physica, Würzburg, 1978, pp. 309–56. *MR*: **57**, #15319 (1979).

Corovei, I. The extensions of the functions satisfying the cosine functional equation. *Mathematica* (*Cluj*), **18**(**41**), 131–6. *MR*: **57**, #6927 (1979).

Dhombres, J.G.

(*a*) Interpolation linéaire et équations fonctionnelles. *Ann. Polon. Math.* **32**, 287–302. *MR*: **55**, #3590 (1978).

(*b*) Functional equations, averaging operators, interpolation operators, and linear extension operators (Chinese). *Nanta Math.*, **9**, 109–16. *MR*: **57**, #6930 (1979).

(*c*) Associativity on the real axis. *Glas. Mat. Ser. III*, **11**(**31**), 37–40. *MR*: **54**, #2745 (1977).

Dhombres, J. and Spilker, J. Über die Funktionalgleichung $f(f(x)\cdot f(y)) = f(x\cdot y)$. *Manuscripta Math.*, **18**, 371–90. *MR*: **53**, #13905 (1977).

Eichhorn, W.

(*a*) Fisher's tests revisited. *Econometrica*, **44**, 247–56. *MR*: **56**, #10853 (1978).

(*b*) Inequalities and functional equations in the theory of the price index. In *General Inequalities 1* (*Proc. First Internat. Conf. on General Inequalities, Oberwolfach, 1976*). Birkhäuser, Basel-Stuttgart, 1978, pp. 23–8.

(*c*) What is an economic index? An attempt of an answer. In *Theory and applications of economic indices* (*Proc. Internat. Sympos., Karlsruhe, 1976*). Physica, Würzburg, 1978, pp. 3–42. *MR*: **58**, #4228 (1979).

Eichhorn, W. and Voeller, J. *Theory of the Price Index. Fisher's Test Approach and Generalizations.* (Lecture Notes in Econom, and Math. Systems, Vol. 140). Springer, Berlin-New York, 1976. *MR*: **56**, #16479 (1978).

Ger, R.

(*a*) On a method of solving of conditional Cauchy equations. *Univ. Beograd. Publ. Electrotehn. Fak. Ser. Mat. Fiz.*, **1976**, No. 554–76, 159–65. *MR*: **56**, #6179 (1978).

(*b*) On some functional equations with a restricted domain. *Bull. Acad. Polon. Sci. Sér. Sci. Math. Astronom. Phys.*, **24**, 429–32. *MR*: **53**, #13907 (1977).

(*c*) Certain functional equations with a restricted domain (Polish). *Uniw. Śląski w Katowicach Prace Nauk*, **132**, 1–36. *MR*: **55**, #13116 (1978).

Haruki, H. A functional equation connected with the sine addition theorem. *Portugal. Math.*, **35**, 75–79. *MR*: **56**, #909 (1978).

Hemdan, H.T. Solution of the 'cube' functional equation in terms of 'trilinear coefficients'. *Canad. Math. Bull.*, **19**, 181–91. *MR*: **54**, #13361 (1977).

Hofmann, K.H. Topological semigroups. History, theory, applications. *Jber. Deutsch. Math.-Verein.*, **78**, 9–59.

Jung, C.F.K., Boonyasombat, V., Barbançon, G., and Jung, F.R. On the functional equation $f(x + f(y)) = f(x)f(y)$. *Aequationes Math.* **14**, 41–8. *MR*: **54**, #774 (1977).

Kareńska, Z. On some functional equations occurring in the theory of geometric objects. *Demonstratio Math.*, **9**, 379–91. *MR*: **54**, #10911 (1977).

Kato, T. *Perturbation Theory for Linear Operators.* Second edn. (Grundlehren Math. Wiss., Vol. 132). Springer, Berlin-Heidelberg-New York, 1976. *MR*: **53**, #11389 (1977).

Kuczma, M. (compil.) Dreizehnte internationale Tagung über Funktionalgleichungen, Oberwolfach vom 6.7 bis 12.7. 1975. *Aequationes Math.* **14**, 197–236. *MR*: **53**, #8695 (1977).

Kuczma, M. and Smítal, J. On measures connected with the Cauchy equation. *Aequationes Math.* **14**, 421–8. *MR*: **53**, #13502 (1977).

Kurepa, S. Decomposition of weakly measurable semigroups and cosine operator functions. *Glas. Mat. Ser. III*, **11**(31), 91–5. *MR*: **53**, #11431 (1977).

Lambek, J. *Lectures on Rings and Modules.* Second edn. Chelsea, New York, 1976. *MR*: **54**, #7514 (1977).

Miron, R. and Radó, F. On the definitions of a module and a vector space by independent axioms. *Mathematica* (*Cluj*), **18(41)**, 179–86. *MR*: **58**, #28017 (1979).

Myers, D.E. On the exponential function and Pólya's proof. *Portugal. Math.*, **35**, 35–9. *MR*: **55**, #3588 (1978).

Nagy, B. Cosine operator functions and the abstract Cauchy problem. *Period. Math. Hungar.*, **7**, 213–17. *MR*: **56**, #9023 (1978).

Narens, L. and Luce, R.D. The algebra of measurement. *J. Pure Appl. Algebra*, **8**, 197–233. *MR*: **57**, #5203 (1979).

Nguyen Huu Bong. Some apparent connection between Baxter and averaging operators. *J. Math. Anal. Appl.* **56**, 330–45. *MR*: **55**, #1130 (1978).

Paganoni, L. Uniqueness structure and parametrization for a class of a functional equation's solutions. *Riv. Mat. Univ. Parma* (4), **2**, 337–45. *MR*: **56**, #9127 (1978).

Paganoni Marzegalli, S. Boundedness and continuity of the solutions of a class of functional equations (Italian). *Rend. Istit. Mat. Univ. Trieste*, **8**, 73–83. *MR*: **55**, #3585 (1978).

Papp, F.J. The functional equation $f(xy) = ff(x)xyff(y)$ in a partially ordered monoid. *Aequationes Math.*, **14**, 33–5. *MR*: **53**, #10681 (1977).

Rätz, J.

(*a*) Quadratic functionals satisfying a subsidiary inequality. In *General Inequalities 1* (*Proc. First Internat. Conf. on General Inequalities, Oberwolfach. 1976*). Birkhäuser, Basel-Stuttgart, 1978, pp. 261–70.

(*b*) On the homogeneity of additive mappings. *Aequationes Math.*, **14**, 67–71 *MR*: **53**, #13320 (1977).

Robinson, D.A. Concerning functional equations of the generalized Bol-Moufang type. *Aequationes Math.*, **14**, 429–34. *MR*: **54**, #2860 (1977).

Sander, W. Verallgemeinerungen eines Satzes von H. Steinhaus in Fund. Math. 1 (1920), 93–104. *Manuscripta Math.*, **18**, 25–42. *MR*: **52**, #14210 (1976).

Schwerdtfeger, H. Invariants of a class of transformation groups. *Aequationes Math.*, **14**, 105–10. *MR*: **53**, #14291 (1977).

Seneta, E. *Regularly Varying Functions.* (Lecture Notes in Math., Vol. 508). Springer, Berlin-New York, 1976. *MR*: **56**, #12189 (1978).

Sharir, M. A counterexample on extreme operators. *Israel J. Math.*, **24**, 320–37. *MR*: **55**, #11096 (1978).

Smítal, J.

(*a*) A necessary and sufficient condition for continuity of additive functions. *Czechoslovak Math. J.*, **26(101)**, 171–3. *MR*: **53**, #6138 (1977).

(*b*) On convex functions bounded below. *Aequationes Math.*, **14**, 345–50. *MR*: **55**, #3189 (1978).

Sokolov, E.I.

(*a*) The Gluskin-Hosszú theorem for Dörnte *n*-groups (Russian). Nets and quasigroups. *Mat. Issled.*, no. **39**, 187–9, 235. *MR*: **56**, #523 (1978).

(*b*) Solution of the functional equation of generalized transitivity (Russian). *Mat. Issled.*, no. **43**, 160–1, 210. *MR*: **58**, #29543 (1979).

Świątak, H. The regularity of the locally integrable and continuous solutions of nonlinear functional equations. *Trans. Amer. Math. Soc.*, **221**, 97–118. *MR*: **55**, #3583 (1978).

Tretiak, O.J. and Eisenstein, B.A. Separator functions for holomorphic filtering. *IEEE Trans. Acoust. Speech Signal process*, ASSP-**24**, 359–64. *MR*: **55**, #122253 (1978).

1977

Aczél, J.

(*a*) Functions of binomial type mapping groupoids into rings. *Math. Z.*, **154**, 115–24. *MR*: **55**, #13115 (1978).

(*b*) General solution of a system of functional equations satisfied by the sums of powers. *Mitt. Math. Sem. Giessen*, **123**, 121–8. *MR*: **58**, #12054 (1979).

Aljančic, S. and Arandelović, D. O-regularly varying functions. *Publ. Inst. Math. (Beograd) (N.S.)*, **22(36)**, 5–22. *MR*: **57**, #6317 (1979).

Alo, R.A., De Korvin, A., and Roberts, C. Averaging operators on normed Köthe spaces. *Ann. Mat. Pura Appl.* (4), **112**, 33–48. *MR*: **55**, #8871 (1978).

Barcz, E., Kania, M., and Moszner, Z. Sur les automates commutatifs. *Zeszyty Nauk. Uniw. Jagielloń. Prace Nauk.—Prace Mat.*, **19**, 195–9. *MR*: **58**, #33592 (1979).

Benz, W. The functional equation of distance preservance in spaces over rings. *Aequationes Math.*, **16**, 303–7. *MR*: **57**, #17064 (1979).

Corovei, I.
(*a*) The cosine functional equation for nilpotent groups. *Aequationes Math.*, **15**, 99–106. *MR*: **55**, #8603 (1978).
(*b*) The extensions of functions satisfying the functional equation $f(xy) + f(xy^{-1}) = 2f(x)g(y)$. *Mathematica (Cluj)*, **19(42)**, 33–39. *MR*: **58**, #29537 (1979).

Daróczy, Z., Lajkó, K, and Székelyhidi, L. Functional equations on ordered fields. *Publ. Math. Debrecen*, **24**, 173–9. *MR*: **56**, #6178 (1978).

Daróczy, Z. and Maksa, Gy. Nonnegative information functions. In *Analytic Function Methods in Probability Theory (Proc. Colloq. Methods of Complex Anal. in the Theory of Probab. and Statist., Kossuth L. Univ., Debrecen, 1977)*. (Colloq. Math. Soc. János Bolyai, Vol. 21). North-Holland, Amsterdam, 1980, pp. 67–78. *MR*: **81h**, #39008.

Dhombres, J.G.
(*a*) Solution générale sur un groupe abélien de l'équation fonctionnelle $f(x*f(y)) = f(f(x)*y)$. *Aequationes Math.*, **15**, 173–93. *MR*: **57**, #10302 (1979).
(*b*) Itération linéaire d'ordre deux. *Publ. Math. Debrecen*, **24**, 277–87. *MR*: **58**, #6798 (1979).
(*c*) On some extensions of homomorphisms. *Nanta Math.*, **10**, 135–41. *MR*: **80e**, #39003.

Forte, B. Non-symmetric entropies in information theory and statistical mechanics. In *Théorie de l'information, développements récents et applications (Cachan, France, 1977)*. (Colloq. Internat. CNRS, Vol. 276). CNRS, Paris, 1978, pp. 69–76. *MR*: **81f**, #94008.

Forti, G.L. On the functional equation $f(L + x) = f(L) + f(x)$. *Istit. Lombardo Accad. Sci. Lett. Rend. A*, **111**, 296–302. *MR*: **80h**, #39009.

Ger, R.
(*a*) On an alternative functional equation. *Aequationes Math.*, **15**, 145–62. *MR*: **58**, #1801 (1979).
(*b*) Almost additive functions on semigroups. *Aequationes Math.*, **15**, 291.
(*c*) Functional equations with a restricted domain. *Rend. Sem. Mat. Fiz. Milano*, **47**, 175–84. *MR*: **80j**, #39010.

Grząślewicz, A. On the solutions of the generalizing equation of homomorphism. *Wyż. Szkol. Ped. Kraków. Rocznik Nauk.-Dydakt. Prace Mat.*, **61**, 31–60. *MR*: **58**, #29539 (1979).

Haruki, Sh. Four different unknown functions satisfying the triangle mean value property for harmonic polynomials. *Ann. Polon. Math.*, **33**, 219–21. *MR*: **55**, #3276 (1978).

Kannappan, Pl. On Shannon's entropy and a functional equation of Abel. *Period. Math. Hungar.*, **8**, 41–44. *MR*: **56**, #16190 (1978).

Kannappan, Pl. and Taylor, M.A. On closure conditions. *Notices Amer. Math. Soc.*, **24**, A-64, #742-20-5. *MR*: **56**, #16190 (1978).

Kuczma, M. On mappings preserving linear dependence and independence. *Uniw. Śląski w Katowicach Prace Nauk. Prace Mat.*, **158**, 29–49. *MR*: **55**, #913 (1978).

Lester, J.A. and McKiernan, M.A. On null cone preserving mappings. *Math. Proc. Cambridge Philos. Soc.*, **81**, 455–62. *MR*: **57**, #15192 (1979).

Lundberg, A. On the functional equation $f(\lambda(x) + g(y)) = \mu(x) + h(x + y)$. *Aequationes Math.*, **16**, 21–30. *MR*: **58**, #29529 (1979).

Luneburg, H. and Plaumann, P. Die Funktionalgleichung von Gołąb und Schinzel in Galoisfeldern. *Arch. Math. (Basel)*, **28**, 55–59. *MR*: **58**, #29542 (1979).

Lutz, D. Which operators generate cosine operator functions? *Atti Accad. Naz. Lincei Rend. Cl. Sci. Fis. Mat. Natur.* (8), **63**, 314–17. *MR*: **80k**, #47042.

Maksa, Gy. On the functional equation $f(x + y) + g(xy) = h(x) + h(y)$. *Publ. Math. Debrecen* **24**, 25–9. *MR*: **56**, #6177 (1978).

Mangolini, M. On some properties of the solutions of the generalized Pexider functional equations (Italian). *Istit. Lombardo Accad. Sci. Lett. Rend. A*, **111**, 194–202. *MR*: **80m**, #39008.

Mangolini, M. and Paganoni, L. Regularity

Mangolini, M. and Paganoni, L. (*cont.*) theorems for a class of functional equations (Italian). *Rend. Sem. Mat. Univ. Padova*, **57**, 93–105. *MR*: **80h**, #39008.

McKennon, K. and Dearden, B. Functional equations for polynomials. *Proc. Amer. Math. Soc.*, **63**, 23–28. *MR*: **55**, #10490 (1978).

McKiernan, M.A.
(*a*) The matrix equation $a(x\circ y) = a(x) + a(x)a(y) + a(y)$. *Aequationes Math.*, **15**, 213–23. *MR*: **58**, #1802 (1979).
(*b*) Equations of the form $H(x\circ y) = \sum_i f_i(x)g_i(y)$. *Aequationes Math.*, **16**, 51–8. *MR*: **58**, #1803 (1979).

Nagy, B.
(*a*) Approximation theorems for cosine operator functions. *Acta Math. Acad. Sci. Hungar.*, **29**, 69–76. *MR*: **55**, #13294 (1978).
(*b*) A sine functional equation in Banach algebras. *Publ. Math. Debrecen*, **24**, 77–90. *MR*: **57**, #943 (1979).
(*c*) On some functional equations in Banach algebras. *Publ. Math. Debrecen* **24**, 257–61. *MR*: **57**, #3682 (1979).

Ng, C.T. Representation for measures of information having the branching property over a simple graph. In *Théorie de l'information, développements récents et applications (Cachan, France, 1977)*. (Colloq. Internat. CNRS, Vol. 276). CNRS, Paris, 1978, pp. 121–4.

O'Connor, T.A.
(*a*) A solution of the functional equation $\phi(x-y) = \sum_1^n a_j(x)\overline{a_j(y)}$ on a locally compact abelian group. *J. Math. Anal. Appl.* **60**, 120–2. *MR*: **56**, #12695 (1978).
(*b*) A solution of D'Alembert's functional equation on a locally compact abelian group. *Aequationes Math.*, **15**, 235–8. *MR*: **56**, #12694 (1978).

Paganoni Marzegalli, S. Cauchy's equation on a restricted domain. *Boll. Un. Mat. Ital. A* (5), **14**, 398–408. *MR*: **56**, #12690 (1978).

Reis, C.M. and Shih, H.J. A note on a problem of Hua. *Semigroup Forum*, **14**, 7–13. *MR*: **57**, #9875 (1979).

Röhmel, J. Eine Charakterisierung quadratischer Formen durch eine Funktionalgleichung. *Aequationes Math.*, **15**, 163–8. *MR*: **58**, #22130 (1979).

Sathyabhama, V. Transitivity, associativity and bisymmetry equations on GD-groupoids. *Publ. Inst. Math. (Beograd) (N.S.)*, **21(35)**, 175–83. *MR*: **56**, #12148 (1978).

Schwerdtfeger, H. Invariants à cinq points dans le plan projectif. *C.R. Acad. Sci. Paris Sér. A*, **285**, 127–8. *MR*: **57**, #10576 (1979).

Sempi, C. (compil.) The Fourteenth International Symposium on Functional Equations (Lecce and Castro Marina, Italy, May 21–28, 1976). *Aequationes Math.* **15**, 265–300.

Sharir, M. A non-nice extreme operator. *Israel J. Math*, **26**, 306–12. *MR*: **56**, #12970 (1978).

Strambach, K. Kommutative distributive Quasigruppen. *Math.-Phys. Semesterber.*, **24**, 71–83. *MR*: **56**, #519 (1978).

Światak, H.
(*a*) On alternative functional equations. *Aequationes Math.*, **15**, 35–47. *MR*: **55**, #13117 (1978).
(*b*) Existence and regularity problems for non-linear functional equations. *Acta Math. Acad. Sci. Hungar.*, **30**, 21–31. *MR*: **57**, #942 (1979).

Vitushkin, A.G. On representation of functions by means of superpositions and related topics. *Enseign. Math.* (2), **23**, 255–320. *MR*: **58**, #17008 (1979).

Yellott, J.I., Jr. The relationship between Luce's choice axiom, Thurstone's theory of comparative judgement, and the double exponential distribution. *J. Math. Psych.*, **15**, 109–44. *MR*: **56**, #8096 (1978).

1978

Aczél, J. Some recent results on characterizations of measures of information related to coding. *IEEE Trans. Inform. Theory* IT-**24**, 592–5, *MR*: **58**, #26526 (1979).

Artin, E. *Algèbre géometrique*. Traduit par M. Lazard. Gauthier-Villars, Paris, 1978.

Anumann, G. Der abbildungstheoretische Zugang zur Topologie, II *Bayer. Akad. Wiss. Math.-Natur. Kl. Sitzungsber* L., **1978**, 85–93. *MR*: **81a**, #54002.

Blackorby, C., Primont, D., and Russell, R.R. *Duality, Separability and Functional Structure: Theory and Economic Applications*. North Holland, New York-Oxford-Shannon, 1978.

Dhombres, J. and Ger, R. Conditional Cauchy equations. *Glas. Mat. Ser. III*, **13** **(33)**, 39–62. *MR*: **58**, #17636 (1979).

Diderrich, G. Local boundedness and the Shannon entropy. *Inform. and Control*, **36**, 292–308. *MR*: **58**, #4631 (1979).

Eichhorn, W. *Functional Equations in Economics*. (Applied Math. and Computation, No. 11). Addison–Wesley, Reading, Mass 1978. *MR*: **80e**, #90034.

Elliott, P.D.T.A. On certain asymptotic functional equations. *Aequationes Math.*, **17**, 44–52. *MR*: **81e**, #39003.
Forti, G.L. Bounded solutions with zeros of the functional equation $f(L+x)=f(L)+f(x)$. *Boll. Un. Mat. Ital. A* (5), **15**, 248–56. *MR*: **57**, #17061 (1979).
Galambos, J. and Kotz, S. *Characterization of Probability Distributions. A Unified Approach with an Emphasis on Exponential and Related Models.* (Lecture Notes in Math., Vol. 675). Springer, Berlin-New York, 1978. *MR*: **80a**, #62022.
Ger, R.
(*a*) Note on almost additive functions. *Aequationes Math.* **17**, 73–6. *MR*: **57**, #13279 (1979).
(*b*) On some functional equations with a restricted domain, II. *Fund. Math.*, **98**, 250–72. *MR*: **57**, #17065 (1979).
(*c*) Homogeneity sets for Jensen-convex functions. In *General Inequalities 2 (Proc. Second Internat. Conf. on General Inequalities, Oberwolfach, 1978)*. Birkhäuser, Basel-Boston-Stuttgart, 1980, pp. 193–201. *MR*: **82d**, #39007.
Grząślewicz, A.
(*a*) On extensions of homomorphisms. *Aequationes Math.*, **17**, 199–207. *MR*: **58**, #6799 (1979).
(*b*) Some remarks to additive functions. *Math. Japon.*, **23**, 573–8. *MR*: **80b**, #39004.
Grząślewicz, A., Powązka, Z., and Tabor, J. On Cauchy's nucleus. *Publ. Math. Debrecen*, **25**, 47–51. *MR*: **58**, #23229 (1979).
Janardan, K.G. A new functional equation analogous to Cauchy–Pexider functional equation and its application. *Biometrical J.*, **20**, 323–8. *MR*: **80f**, #39009.
Kátai, I. On additive functions. *Publ. Math. Debrecen*, **25**, 251–7. *MR*: **80b**, #10005.
Kirsch, A. Beziehungen zwischen der Additivität und der Homogenität von Vektorraum-Abbildungen. *Math.-Phys. Semesterber.*, **25**, 207–10. *MR*: **80b**, #15005.
Krapež, A. On solving a system of balanced functional equations on quasigroups, I. *Publ. Inst. Math. (Beograd) (N.S.)*, **23** **(37)**, 117–27. *MR*: **80a**, #39007.
Kuczma, M.
(*a*) On some alternative functional equations. *Aequationes Math.*, **17**, 182–98. *MR*: **80k**, #39003.
(*b*) Functional equations on restricted domains. *Aequationes Math.*, **18**, 1–34. *MR*: **58**, #29532 (1979).
Lajkó, K. Generalized difference-property for functions of several variables. *Publ. Math. Debrecen* **25**, 169–76. *MR*: **58**, #29540 (1979).
Moszner, Z. On the equivalence of Thurstone models. *J. Math. Psych.*, **17**, 263–5. *MR*: **58**, #4467 (1979).
Neuman, F. (compil.) Fifteenth International Symposium on Functional Equations (Oberwolfach, Germany, May 22–28, 1977). *Aequationes Math.* **17**, 348–78.
Ng. C.T. Inverse systems and the translation equation on topological spaces. *Acta. Math. Acad. Sci. Hungar.*, **31**, 227–32. *MR*: **57**, #13280 (1979).
Paganoni, L. On uniqueness structures: some conditions and applications. *Rend. Circ. Mat. Palermo* (2), **27**, 41–52. *MR*: **80h**, #39016.
Pokropp, F. The functional equation of aggregation. In *Functional equations in Economics*, by W. Eichhorn. Addison-Wesley, Reading, Mass., pp. 122–39. *MR*: **80e**, #90034.
Rätz, J. On approximately additive mappings. In *General Inequalities 2 (Proc. Second Internat. Conf. on General Inequalities, Oberwolfach, 1978)*. Birkhäuser, Basel, 1980, pp. 233–51. *MR*: **83e**, #90131.
Sander, W.
(*a*) Verallgemeinerte Cauchy–Funktionalgleichungen. *Aequationes Math.*, **18**, 357–69. *MR*: **80h**, #39017.
(*b*) Regularitätseigenschaften von Funktionalgleichungen. *Glas. Mat. Ser. III*, **13(33)**, 237–47. *MR*: **80h**, #39018.
Schwerdtfeger, H.
(*a*) Invariants of a class of transformation groups, II. *Aequationes Math.*, **17**, 292–4. *MR*: **58**, #12720 (1979).
(*b*) On $(n+1)$-point invariants of the group $GL(n, F)$, II. *Linear and Multilinear Algebra*, **6**, 63–4. *MR*: **58**, #919 (1979).
Speiser, J.M., Whitehouse, H.J., and Berg, N.J. Signal processing architectures using convolutional technology. *Real Time Signal Processing, SPIE* **154**, 66–80.
Stehling, F. The functional equation $f(yx) = g(y)f(x) + b(y)$ on a restricted domain. *Aequationes Math.*, **17**, 360–1.
Täuber, J. and Neuhaus, N.
(*a*) Monotone Lösungen zweier Komplexfunktionalgleichungen, die aus der Jensenfunktionalgleichung hervorgehen. *Rev. Roumaine Math. Pures Appl.*, **23**, 299–307. *MR*: **58**, #12059 (1979).

Täuber, J. and Neuhaus, N. (*cont.*)
(*b*) Über die monotonen Lösungen der auf Mengen verallgemeinerten Jensenfunktionalgleichung. *Rev. Roumaine Math. Pures Appl.*, **23**, 309–12. *MR*: **58**, #12060 (1979).
Taylor, M.A. On the generalized equations of associativity and bisymmetry. *Aequationes Math.*, **17**, 154–63. *MR*: **58**, #931 (1979).
Vincze, E.
(*a*) Über die Einführung der komplexen Zahlen mittels Funktionalgleichungen. *Publ. Math. Debrecen*, **25**, 303–9. *MR*: **80f**, #39010.
(*b*) Über ein Funktionalgleichungsproblem von I. Olkin. *Publ. Techn. Univ. Heavy Industry (Miskolc) Ser. D*, **33**, 113–24.

1979

Baker, J., Lawrence, J., and Zorzitto, F. The stability of the equation $f(x+y) = f(x)f(y)$. *Proc. Amer. Math. Soc.*, **74**, 242–6. *MR*: **80d**, #39009.
Benz, W.
(*a*) On a conjecture of I. Fenyö. *C.R. Math. Rep. Acad. Sci. Canada*, **1**, 249–52. *MR*: **80i**, #39006.
(*b*) Über eine Funktionsgleichung von G. Aumann. *Bayer. Akad. Wiss. Math.-Natur. Kl. Sitzungsber.*, **1979**, 7–13. *MR*: **81m**, #39013.
Borelli Forti, C. and Forti, G.L. Vector-valued solutions of a functional equation. *Boll. Un. Mat. Ital. B*(5), **16**, 266–77. *MR*: **80h**, #39012.
Brillouët, N. *Opérateurs extrémaux et opérateurs sympathiques.* (Thèse Doctorale 3-ème Cycle). Université de Nantes, Nantes, 1979.
Buchs, A.B. The generalized Cauchy equation for selfadjoint operators on a Hilbert space. *Indian J. Pure Appl. Math.*, **10**, 166–70. *MR*: **80e**, #39006.
Daróczy, Z. and Járai, A. On the measurable solution of a functional equation arising in information theory. *Acta Math. Acad. Sci. Hungar.*, **34**, 105–16. *MR*: **80i**, #39008.
Dhombres, J.G. *Some Aspects of Functional Equations.* Chulalongkorn University, Department of Mathematics, Bangkok, 1979. *MR*: **80j**, #39001.
Eichhorn, W. Wirtschaftliche Kennzahlen. In *Quantitative Wirtschafts- und Unternehmensforschung* (*Ergebnisband des St. Galler Symposiums, 1979*). Springer, Berlin-Heidelberg-New York, 1980, pp. 143–77.
Forti, G.L. The general solution of the functional equation $\{cf(x+y) - af(x) - bf(y) - d\}\{f(x+y) - f(x) - f(y)\} = 0$ (Italian). *Matematiche (Catania)*, **34**, 219–42.
Ger, R. Almost additive functions on semigroups and a functional equation. *Publ. Math. Debrecen*, **26**, 219–28. *MR*: **81e**, #39010.
Głowacki, E. and Kuczma, M. Some remarks on Hosszú's functional equation on integers. *Uniw. Śląski w Katowicach Prace Nauk.—Prace Mat.*, **9**, 53–63. *MR*: **81m**, #39014.
Grząślewicz, A.
(*a*) On the solution of the system of functional equations related to quadratic functionals. *Glas. Mat. Ser. III* **14(34)**, 77–82. *MR*: **80b**, #39004.
(*b*) On some homomorphisms in product Brandt groupoids. *Wyż Szkoł. Ped. Kraków. Rocznik Nauk.-Dydakt. Prace Mat.*, **9**, 67–72. *MR*: **81i**, #20111.
Grząślewicz, A. and Sikorski, P. On some homomorphisms in Ehresmann groupoids. *Wyż Szkol. Ped. Kraków. Rocznik Nauk. Dydakt. Prace Mat.*, **9**, 55–66. *MR*: **81i**, #20110.
Hardy, G.H. and Wright, E.M. *An Introduction to the Theory of Numbers.* Fifth edn. Oxford University Press, New York, 1979. *MR*: **81i**, #10002.
Haruki, H. On a generalization of Hille's functional equation. *Ann. Polon. Math.*, **36**, 205–11. *MR*: **80i**, #30040.
Haruki, Sh. A property of quadratic polynomials. *Amer. Math. Monthly*, **86**, 577–9. *MR*: **80g**, #26010.
Járai, A. On measurable solutions of functional equations. *Publ. Math. Debrecen*, **26**, 17–35. *MR*: **80f**, #39008.
Johnson, R.W. Axiomatic characterization of the directed divergences and their linear combinations. *IEEE Trans. Inform. Theory*, IT- **25**, 709–16. *MR*: **80m**, #94027.
Jou, S.S. On the functional equation $f[x + yf(x)] = f(x)f(y)$. *Kyungpook Math. J.* **19**, 115–18. *MR*: **80j**, #39007.
Kannappan, Pl.
(*a*) An application of a differential equation in information theory. *Glas. Mat. Ser. III*, **14(34)**, 269–74. *MR*: **83f**, #39006.
(*b*) Solution 2 of Problem 177 (**P177S2**). *Aequationes Math.*, **19**, 114–15.
Krapež, A.
(*a*) On solving a system of balanced functional equations on quasigroups, II. *Publ. Inst. Math.* (*Beograd*) (*N.S.*),

25 (**39**), 70–8. *MR*: **81a**, #39006.
(*b*) On solving a system of balanced functional equations on quasigroups, III. *Publ. Inst. Math.* (*Beograd*) (*N.S.*), **26** (**40**), 145–56. *MR*: **81h**, #39012.

Laird, P.G. On characterizations of exponential polynomials. *Pacific J. Math.*, **80**, 503–7. *MR*: **80h**, #43003.

Lajkó, K. Remark to a paper by J.A. Baker. *Aequationes Math.*, **19**, 227–31. *MR*: **83a**, #39008.

Lawrence, J., Mess, G., and Zorzitto, F. Near derivations and information functions. *Proc. Amer. Math. Soc.* **76**, 117–22. *MR*: **80d**, #39005.

Lester, J.A. A characterization of non-Euclidean non-Minkowskian inner product space isometries. *Utilitas Math.*, **16**, 101–9. *MR*: **81f**, #51014.

McKiernan, M.A.
(*a*) General Pexider equations, I. Existence of injective solutions. *Pacific J. Math.*, **82**, 499–502. *MR*: **81f**, #39009a.
(*b*) General Pexider equations, II. An application of the theory of webs. *Pacific J. Math.*, **82**, 503–14. *MR*: **81f**, #39009b.

Moszner, Z. and Waśko, M. Sur les solutions d'une équation fonctionnelle, III. *Aequationes Math.*, **19**, 79–88. *MR*: **80g**, #20074.

Nagy, B. On cosine operator functions. *Publ. Math. Debrecen*, **26**, 133–8. *MR*: **81f**, #47042.

Natale, E. Medial idempotent commutative n-quasigroups. *Rend. Accad. Sci. Fis. Mat. Napoli*, (4), **46**, 221–9. *MR*: **82a**, #20083.

Pavlović, V. The Moufang functional equation $B_1(x,B_2(y,B_3(x,z))) = B_4(B_5(B_6(x,y),x),z)$ on GD-groupoids. *Aequationes Math.*, **19**, 37–47. *MR*: **80h**, #20112.

Penney, R.C. and Rukhin, A.L. D'Alembert's functional equation on groups. *Proc. Amer. Math. Soc.*, **77**, 73–80. *MR*: **80h**, #39013.

Reich, L. and Schwaiger, J. Analytische und fraktionelle Iteration formal-biholomorpher Abbildungen. In *Jahrbuch Überblicke Mathematik, 1979*. Bibliographisches Inst., Mannheim, 1979, pp. 123–44. *MR*: **81h**, #58056.

Roberts, F.S. *Measurement Theory. With applications to Decision Making, Utility and the Social Sciences.* (Encyclopedia of Mathematics and its Applications, Vol. 7). Addison-Wesley, Reading, Mass., 1979. *MR*: **81b**, #90003.

Ruzsa, I.Z. Additive functions with bounded difference. *Period. Math. Hungar.*, **10**, 67–70. *MR*: **58**, #10795 (1979).

Schwaiger, J. (compil.) Sechzehntes Internationales Symposium über Funktionalgleichungen (Schloss Retzhof, Österreich, 3–10. September 1978). *Aequationes Math.* **19**, 245–99.

Schwerdtfeger, H. *Geometry of Complex Numbers*. Second edn. Dover, New York, 1979. *MR*: **82g**, #51032.

Sen, D.K. The special-relativistic velocity-addition law and a related functional equation. *Internat. J. Theoret. Phys.*, **18**, 179–84. *MR*: **81a**, #83004.

Shimizu, R. On a lack of memory property of the exponential distribution. *Ann. Inst. Statist. Math.*, **31**, 309–13. *MR*: **81f**, #62026.

Székelyhidi, L. Remark on a paper of M.A. McKiernan: 'On vanishing n-th ordered differences and Hamel-bases'. *Ann. Polon. Math.*, **36**, 245–7. *MR*: **80h**, #39011.

Ušan, J. *Quasigroups* (Serbo-Croatian). Inst. Mat., Novi Sad, 1979. *MR*: **81g**, #20133.

Wirsing, E. Additive and completely additive functions with restricted growth. In *Recent Progress in Analytic Number Theory* (*Durham. 1979*), Vol. 2, Academic Press, London, 1982, pp. 231–80. *MR*: **83a**, #10096.

Zdun, M.C. *Continuous and Differentiable Iteration Semigroups.* (Uniw. Sląski w Katowicach Prace Nauk., No. 308). Uniw. Śląski, Katowice, 1979. *MR*: **81e**, #39006.

1980

Aczél, J.
(*a*) General solution of a system of functional equations satisfied by sums of powers on arithmetic progressions. *Aequationes Math.*, **21**, 39–43. *MR*: **81m**, #39009.
(*b*) Information functions on open domain, III. *C.R. Math. Rep. Acad. Sci. Canada*, **2**, 281–5. *MR*: **81m**, #39012.
(*c*) Some good and bad characters I have known and where they led. (Harmonic analysis and functional equations). In *1980 Seminar on Harmonic Analysis.* (Canad. Math. Soc. Conf. Proc., Vol. 1). Amer. Math. Soc., Providence, R.I., 1981, pp. 177–87. *MR*: **84j**, #39009.

Aczél, J. and Wagner, C. A characterization of weighted arithmetic means. *SIAM J. Algebraic Discrete Methods*, **1**, 259–60. *MR*: **81m**, #39012.

Bähl, E. and Rüthing, D. Die Funktionalgleichung der

Bähl, E. and Rüthing, D. (*cont.*) Exponentialfunktion. *Praxis Math.*, **22**, 37–40. *MR*: **81a**, #39003.

Baker, J.A.
(*a*) On a theorem of Daróczy, Lajkó and Székelyhidi. *Publ. Math. Debrecen* **27**, 283–8. *MR*: **82h**, #39007.
(*b*) The stability of the cosine equation. *Proc. Amer. Math. Soc.*, **80**, 411–16. *MR*: **81m**, #39015.

Bedingfield, N.K. An initial value problem for a functional equation. *Aequationes Math.*, **20**, 90–6. *MR*: **81g**, #39004.

Benz, W.
(*a*) Eine Beckman-Quarles-Charakterisierung der Lorentztransformation des R^n. *Arch. Math.* (*Basel*), **34**, 550–9. *MR*: **82h**, #51022.
(*b*) On a functional equation arising from hyperbolic geometry. *Aequationes Math.*, **21**, 113–9. *MR*: **82b**, #51024.

Borelli Forti, C. Some classes of solutions of the functional equation $f(L+x)=f(L)+f(x)$. *Riv. Mat. Univ. Parma* (4) **6**, 37–45. *MR*: **82m**, #39008.

Borelli Forti, C. and Fenyö, I. On the difference equations with variable increments (Italian). *Stochastica*, **4**, 93–101. *MR*: **82b**, #39002.

Chung, J.K. A generalization of a linear functional equation, II. *Univ. Beograd. Publ. Elektrotehn. Fak. Ser. Mat. Fiz.*, **1980**, no. 678–715, 183–190. *MR*: **83h**, #39012.

Corovei, I. The functional equation $f(xy)+f(xy^{-1})=2f(x)g(y)$ for nilpotent groups. *Mathematica* (*Cluj*), **22(45)**, 33–41. *MR*: **82j**, #39007.

Cowell, F.A. On the structure of additive inequality measures. *Rev. Econom. Stud.*, **47**, 521–31.

Daróczy, Z. Über die stetigen Lösungen der Aczél-Benz'schen Funktionalgleichung. *Abh. Math. Sem. Univ. Hamburg*, **50**, 210–8. *MR*: **81k**, #39005.

Davison, T.M.K. A functional equation for quadratic forms. *Aequationes Math.*, **21**, 1–7. *MR*: **82a**, #39004.

Davison, T.M.K. and Redlin, L. Hosszú's functional equation over rings generated by their units. *Aequationes Math.*, **21**, 121–8. *MR*: **82h**, #39004.

Fenyö, I. Observations on some theorems of D.H. Hyers (Italian). *Istit. Lombardo Accad. Sci. Lett. Rend. A*, **114**, 235–42. *MR*: **85a**, #39008.

Forti, G.L. An existence and stability theorem for a class of functional equations. *Stochastica*, **4**, 23–30. *MR*: **81i**, #39006.

Gehrig, W. On generalized quasilinear functions and their applications in economic theory. *Methods Oper. Res.*, **38**, 103–14.

Golovinskiĭ, I.A. Early history of analytic iteration and functional equations (Russian). *Istor.-Mat. Issled.*, **25**, 25–31. *MR*: **82f**, #01040.

Haruki, H. On a theorem of inverse geometry. *Ann. Polon. Math.*, **38**, 163–8. *MR*: **82b**, #51020.

Haruki, Sh. Note on the equation $(\Delta_{x,t}^2-\Delta_{y,t}^2)f=0$. *Aequationes Math.*, **21**, 49–52. *MR*: **82f**, #39008.

Kairies, H.H. (compil.) Siebzehnte internationale Tagung über Funktionalgleichungen, Oberwolfach vom 17.6. bis 23.6.1979. *Aequationes Math.* **20**, 286–315.

Kannappan, Pl. On directed divergence, inaccuracy and a functional equation connected with Abel, II. *Period. Math. Hungar.*, **11**, 213–19. *MR*: **82c**, #39008.

Kannappan, Pl. and Ng, C.T.
(*a*) On functional equations and measures of information, II. *J. Appl. Probab.*, **17**, 271–7. *MR*: **83h**, #39015.
(*b*) Measurable solutions of the functional equation $f(q,u)-g(pq,uv)-h(q-pq,uw)=A(p,q,v,w)$. *Nanta Math.*, **13**, 69–77.

Krapež, A.
(*a*) Generalized associativity on groupoids. *Publ. Inst. Math.* (*Beograd*) (*N.S.*), **28** (**42**), 105–12. *MR*: **83c**, #20084.
(*b*) Almost trivial groupoids. *Publ. Inst. Math.* (*Beograd*) (*N.S.*) **28(42)**, 113–24. *MR*: **83c**, #20085.

Laczkovich, M. Functions with measurable differences. *Acta Math. Acad. Sci. Hungar.* **35**, 217–35. *MR*: **82a**, #39002.

Lajkó, K.
(*a*) Some general functional equations. *Publ. Math. Debrecen*, **27**, 177.
(*b*) On the functional equation $f(x)g(y)=h(ax+by)k(cx+dy)$. *Period. Math. Hungar.*, **11**, 187–95. *MR*: **81k**, #39007.

Lester, J.A. The solution of a system of quadratic functional equations. *Ann. Polon. Math.*, **37**, 113–17. *MR*: **81g**, #39010.

Lutz, D.
(*a*) Compactness properties of operator cosine functions. *C.R. Math. Rep. Acad. Sci. Canada*, **2**, 277–80. *MR*: **82f**, #47047.
(*b*) Über operatorwertige Lösungen der Funktionalgleichung des Cosinus. *Math. Z.*, **171**, 233–45. *MR*: **81i**, #47043.

Maksa, Gy. Bounded symmetric information

functions. *C.R. Math. Rep. Acad. Sci. Canada*, **2**, 247–52. *MR*: **82a**, #94032.

Marchionna, C. Continuity of the solutions of a class of functional equations (Italian). *Boll. Un. Mat. Ital. B* (5), **17**, 727–39. *MR*: **81i**, #39008.

Midura, S. Sur les solutions de l'équation fonctionnelle $f[x+3f^2(y)-3f(x)f(y)-y]=f(x)-f(y)$. *Aequationes Math.*, **20**, 302–3.

Moszner, Z. Sur l'équation $f(x)f(y)f(-x-y)=g(x)g(y)g(-x-y)$. *Aequationes Math.*, **21**, 20–32. *MR*: **82a**, #39003.

Ng, C.T.
(*a*) Information functions on open domains. *C.R. Math. Rep. Acad. Sci. Canada*, **2**, 119–23. *MR*: **83a**, #39010a.
(*b*) Information functions on open domains, II. *C.R. Math. Rep. Acad. Sci. Canada*, **2**, 155–8. *MR*: **83a**, #39010b.

Paganoni, L. Solutions of a functional equation on restricted domain (Italian). *Boll. Un. Mat. Ital. B* (5), **17**, 979–93.

Paganoni, L. and Paganoni Marzegalli, S. Cauchy's functional equation on semigroups. *Fund. Math.*, **110**, 63–74. *MR*: **82c**, #39009.

Radó, F. On the characterization of plane affine isometries. *Resultate Math.*, **3**, 70–3. *MR*: **82e**, #51015.

Rădulescu, M. On a supra-additive and supra-multiplicative operator of $C(X)$. *Bull. Math. Soc. Sci. Math. R.S. Roumanie* (*N.S.*), **24(72)**, 303–5. *MR*: **82g**, #47024.

Redlin, L. (compil.) *Proceedings of the Eighteenth International Symposium on Functional Equations (Waterloo and Scarborough, Ontario, Canada, 26 August–6 September, 1980)*. Centre for Information Theory, University of Waterloo, Waterloo, Ont., 1980. *MR*: **83f**, #39001.

Reich, L. Neuere Ergebnisse über kontinuierliche Iteration formaler Potenzreihen. *Bonner Math. Schriften*, **121**, 62–79. *MR*: **83h**, #39011.

Sokolov, E.I. Investigation of the functional equation $f(\sqrt[n]{x^n+y^n})=F(g(x),h(y),l(x),k(y))$ over a ring of continuous functions over R (Russian). In *General Algebra and Discrete Geometry*. Shtiintsa, Kishinev, 1980, pp. 99–107, 163. *MR*: **83h**, #39011.

Świątak, H. On an equation of information theory. *Aequationes Math.*, **20**, 18–22. *MR*: **81d**, #39004.

Székelyhidi, L. Almost periodic functions and functional equations. *Acta Sci. Math. (Szeged)*, **42**, 165–9. *MR*: **81f**, #39010.

Taylor, M.A. A Pexider equation for functions defined on a semigroup. *Acta Math. Acad. Sci. Hungar.*, **36**, 211–13. *MR*: **83d**, #39012.

Truesdell, C. and Muncaster, R.G. *Fundamentals of Maxwell's Kinetic Theory of a Simple Monatomic Gas Treated as a Branch of Rational Mechanics*. Academic Press, New York-London-Toronto-Sydney-San Francisco, 1980. *MR*: **81f**, #80001.

Wirsing, E. Additive functions with restricted growth on the numbers of the form $p+1$. *Acta Arith.*, **37**, 345–57. *MR*: **82b**, #10056.

Yadrenko, M.I. *Spectral Theory of Random Fields*. (Russian) Kiev State Univ., Kiev, 1980. (English transl. Springer, Berlin-Heidelberg-New York; Optimization Software, New York, 1983). *MR*: **82e**, #60081.

1981

Aczél, J.
(*a*) Notes on generalized information functions. *Aequationes Math.*, **22**, 97–107. *MR*: **83e**, #94020.
(*b*) Second solution to problem 206 (**P206S2**). *Aequationes Math.*, **22**, 310–11.
(*c*) On the Thomsen condition on webs. *J. Geom.*, **17**, 155–60. *MR*: **83e**, #20081.

Aczél, J., Kannappan, Pl., Ng, C.T., and Wagner, C. Functional equations and inequalities in 'rational group decision making'. In *General Inequalities 3 (Proc. Third Internat. Conf. on General Inequalities, Oberwolfach, 1981)*. Birkhäuser, Basel-Boston-Stuttgart, 1983, pp. 239–43. *MR*: **86j**, #39009.

Aczél, J. and Ng, C.T. On general information functions. *Utilitas Math.*, **19**, 157–70. *MR*: **82m**, #39011.

Aczél, J. and Wagner, C. Rational group decision making generalized: the case of several unknown functions. *C.R. Math. Rep. Acad. Sci. Canada*, **3**, 139–42.

Baker, J.A. A generalized Pexider equation. *Publ. Math. Debrecen*, **28**, 265–70. *MR*: **84c**, #39003.

Baker, J.A. and Davidson, K.R. Cosine exponential and quadratic functions. *Glas. Mat. Ser. III*, **16 (36)**, 269–74. *MR*: **83e**, #39010.

Benz, W.
(*a*) Über eine Charakterisierung des hyperbolischen Flächeninhaltes und eine Funktionalgleichung. *Publ. Math. Debrecen*, **28**, 259–63. *MR*: **83k**, #53016.
(*b*) Die Heavisidefunktion als spezielle

Benz, W. (*cont.*)
Lösung ihrer Funktionalgleichung. *Aequationes Math.*, **23**, 151–5. *MR*: **84d**, #39005.

Bergman, G.M. Solution to problem 178 (**P178S1**). *Aequationes Math.*, **23**, 311–2.

Borelli Forti, C. and Forti, G.L. On the structure of semigroups occurring in the representation of the solutions of a functional equation. *Rend. Sem. Mat. Univ. Politec. Torino*, **39**, 99–106. *MR*: **83m**, #39007.

Cowell, F.A. and Kuga, K. Additivity and the entropy concept: Axiomatic approach to inequality measurement. *J. Econom. Theory*, **25**, 131–43.

Davidson, K.R. and Ng, C.T. Information measures and cohomology. *Utilitas Math.*, **20**, 27–34. *MR*: **83e**, #94021.

Davison, T.M.K. On Gleason's definition of quadratic forms. *Canad. Math. Bull.*, **24**, 233–6. *MR*: **82h**, #15030.

Ebanks, B.R. A characterization of separable utility functions. *Kybernetika* (*Prague*), **17**, 244–55. *MR*: **82m**, #90024.

Falmagne, J.C. On a recurrent misuse of a classical functional equation result. *J. Math. Psych.* **23**, 190–3. *MR*: **83d**, #26004.

Fenyö, I. and Paganoni, L. On a system of functional equations. *C.R. Math. Rep. Acad. Sci. Canada*, **3**, 109–12. *MR*: **82g**, #39007.

Fischer, P. and Mokanski, J.P. A class of symmetric biadditive functionals. *Aequationes Math.*, **23**, 169–74. *MR*: **84d**, #15025.

Flatto, L. and Jacobson, D. Functions satisfying a discrete mean value property. *Aequationes Math.*, **22**, 173–93. *MR*: **84g**, #31004.

Forti, G.L. and Paganoni, L. A method for solving a conditional Cauchy equation on abelian groups. *Ann. Mat. Pura Appl.* (4), **127**, 79–99. *MR*: **84d**, #39009.

Ger, R. Almost approximately additive mappings. In *General Inequalities 3* (*Proc. Third Internat. Conf. on General Inequalities, Oberwolfach, 1981*). Birkhäuser, Basel-Boston-Stuttgart, 1983, pp. 263–76. *MR*: **86k**, #39008.

Haruki, Sh. On the theorem of S. Kakutani-M. Nagumo and J.L. Walsh for the mean value property of harmonic and complex polynomials. *Pacific J. Math.*, **94**, 113–23. *MR*: **83d**, #30025.

Heuvers, K.J. Functional equations which characterize *n*-forms and homogeneous functions of degree *n*. *Aequationes Math.*, **22**, 223–48. *MR*: **83h**, #39013.

Humblet, P.A. Generalization of Huffman coding to minimize the probability of buffer overflow. *IEEE Trans. Inform. Theory*, **IT-27**, 230–2.

Janssen, A.J.E.M. Note on a paper by M. Laczkovich: "Functions with measurable differences" [Acta Math. Acad. Sci. Hungar. **35** (1980), no. 1–2, 217–235]. *Nederl. Akad. Wetensch. Indag. Math.*, **43**, 309–13. *MR*: **83i**, #26005.

Jurkat, W.B. and Shawyer, B.L.R. On continuous maps of sequence spaces. *Analysis*, **1**, 209–10. *MR*: **83g**, #40005.

Jussila, O. History of the functional equation $f(x+y)=f(x)+f(y)$ (Finnish). *Arkhimedes*, **33**, 193–203. *MR*: **83d**, #01033.

Kominek, Z. On the continuity of Q-convex functions and additive functions. *Aequationes Math.*, **23**, 146–50. *MR*: **84e**, #26012.

Krapež, A. Functional equations of generalized associativity, bisymmetry, transitivity and distributivity. *Publ. Inst. Math.* (*Beograd*) (*N.S.*) **30**, 81–87. *MR*: **84d**, #39007.

Laczkovich, M. A generalization of Kemperman's inequality $2f(x)\leqslant f(x+h)+f(x+2h)$. In *General Inequalities 3* (*Proc. Third Internat. Conf. on General Inequalities, Oberwolfach*, 1981). Birkhäuser, Basel-Boston-Stuttgart, 1983, pp. 281–93. *MR*: **86h**, #39021.

Lajkó, K. On the general solution of rectangle-type functional equations. *Publ. Math. Debrecen*, **28**, 137–43. *MR*: **82i**, #39007.

Lawrence, J.
(*a*) The cocyle equation in division rings. *Aequationes Math.*, **22**, 70–2. *MR*: **82h**, #39005.
(*b*) The Shannon kernel of a non-negative information function. *Aequationes Math.*, **23**, 233–5.

Lehrer, K. and Wagner, C. *Rational Consensus in Science and Society. A Philosophical and Mathematical Study.* (Philosophical Studies Series in Philosophy, Vol. 24). Reidel, Dordrecht-Boston-London, 1981. *MR*: **84k**, #03018.

Lewin, L. *Polylogarithms and Associated Functions.* North Holland, New York-Amsterdam, 1981. *MR*: **83b**, #33019.

Ljubenova, E.T. On d'Alembert's functional equation on an abelian group. *Aequationes Math.*, **22**, 54–5. *MR*: **82k**, #39006.

Losonczi, L. and Maksa, Gy. The general solution of a functional equation of information theory. *Glas. Mat. Ser. III*,

16(36), 261–8. *MR*: **83h**, #39010.
Lutz, D. Strongly continuous operator cosine functions. In *Functional Analysis (Dubrovnik, 1981)*. (Lecture Notes in Math., Vol. 948). Springer, Berlin, 1982, pp. 73–97. *MR*: **84f**, #47044.
Maksa, Gy.
(*a*) On near-derivations. *Proc. Amer. Math. Soc.*, **81**, 406–8. *MR*: **81m**, #39010.
(*b*) On the bounded solutions of a functional equation. *Acta Math. Acad. Sci. Hungar.*, **37**, 445–50. *MR*: **82g**, #94008.
(*c*) The general solution of a functional equation related to the mixed theory of information. *Aequationes Math.*, **22**, 90–6. *MR*: **82i**, #39006.
(*d*) First solution to problem 206 (**P206S1**). *Aequationes Math.*, **22**, 311.
Nikodem, K., On additive set-valued functions. *Rev. Roumaine Math. Pures Appl.*, **6**, 1005–13. *MR*: **82m**, #28005.
Pokropp, F. Macroeconomic production correspondence and consistent aggregation. *Methods Oper. Res.*, **39**, 145–60.
Polonijo, M. Associativity and n-ary cyclic bisymmetry. *Rostock. Math. Kolloq.* **16**, 81–6. *MR*: **83d**, #39010.
Powązka, Z. On the differentiability of generalized subadditive functions. *Zeszyty Nauk. Uniw. Jagielloń. Prace Mat.*, **22**, 189–94. *MR*: **82m**, #39012.
Rätz, J.
(*a*) Zur Charakterisierung des Skalarproduktes. *Elem. Math.*, **36**, 94–7. *MR*: **82i**, #15026.
(*b*) On light-cone-preserving mappings of the plane. In *General Inequalities 3 (Proc. Third Internat. Conf. on General Inequalities, Oberwolfach, 1981)*. Birkhäuser, Basel-Boston-Stuttgart, 1983, pp. 349–67. *MR*: **84h**, #00007.
Redlin, L. (compil.) *Proceedings of the Nineteenth International Symposium on Functional Equations (Nantes and LaTurbaille, France, 3–13 May, 1981)*. Centre for Information Theory, University of Waterloo, Waterloo, Ont., 1981. *MR*: **83b**, #39001.
Reich, L. Über die allgemeine Lösung der Translationsgleichung in Potenzreihenringen. *Ber. Math.-Statist. Sekt. Forsch. Graz*, **159**, 1–22. *MR*: **83k**, #39008.
Sander, W. Pexider equations. *Glas. Mat. Ser. III*, **16 (36)**, 275–85. *MR*: **83f**, #39007.
Sweet, L. On the generalized cube and octahedron equations. *Aequationes Math.* **22**, 29–38. *MR*: **83m**, #49050.
Świątak, H. On alternatives equivalent to the Cauchy functional equation and a related equation *Aequationes Math.*, **23**, 66–75. *MR*: **84a**, #39007.
Székelyhidi, L.
(*a*) The stability of linear functional equations. *C.R. Math. Rep. Acad. Sci. Canada*, **3**, 63–7. *MR*: **82e**, #39007.
(*b*) On a stability theorem. *C.R. Math. Rep. Acad. Sci. Canada*, **3**, 253–5. *MR*: **83a**, #39012.
(*c*) Functional equations on abelian groups. *Acta Math. Acad. Sci. Hungar.* **37**, 235–43. *MR*: **82f**, #39010.
(*d*) An extension theorem for a functional equation. *Publ. Math. Debrecen*, **28**, 275–9. *MR*: **82k**, #39005.
Targonski, G. *Topics in Iteration Theory.* (Studia Math., Skr. 6). Vandenhoeck Ruprecht, Göttingen-Zürich, 1981. *MR*: **82j**, #39001.
Valenti, S. and Rodono, L.G. Sur une suite liée à l'équation fonctionnelle $f(xy) = g(y)f(x) + g(x)f(y)$. *Atti Accad. Sci. Lett. Arti Palermo Parte I* (4) **38**, 87–97. *MR*: **83k**, #39009.
Volkmann, P. Existenz einer zwischen zwei Funktionen v,w gelegenen Lösung von Funktionalgleichungen der Form $u(\Phi(x_1,\ldots,x_n)) = \phi(u(x_1),\ldots,u(x_n))$, wenn v, w entsprechenden Funktionalungleichungen genügen. In *General Inequalities 3 (Proc. Third Internat. Conf. on General Inequalities, Oberwolfach, 1981)*. Birkhäuser, Basel-Boston-Stuttgart, 1983, pp. 531–2.
Volkmann, P. and Weigel, H. Systeme von Funktionalgleichungen. *Arch. Math. (Basel)*, **37**, 443–9. *MR*: **83e**, #39014.
Žižović, M.R. Topological medial n-quasigroups. In *Algebraic Conference (Novi Sad. 1981)*. Inst. Mat. Novi Sad, 1982, pp. 61–6. *MR*: **84i**, #22004.

1982

Aczél, J. and Chung, J.K. Integrable solutions of functional equations of a general type. *Studia Sci. Math. Hungar.*, **17**, 51–67. *MR*: **85i**, #39008.
Albert, M. and Baker, J.A. Bounded solutions of a functional inequality. *Canad. Math. Bull.*, **25**, 491–5. *MR*: **84k**, #39018.
Bagyinszki, J. The solution of the Hosszú equation over finite fields. *Közl. MTA Számitástech. Automat. Kutató Int. Budapest*, **25**, 25–33. *MR*: **84d**, #39008.

Blanton, G. and Baker, J.A. Iteration groups generated by C^n functions. *Arch. Math.* (*Brno*), **18**, 121–7. *MR*: **85e**, #26004.

Borisov, I.S. A criterion for Gaussian random processes to be Markov processes (Russian). *Teor. Veroyatnost. i Primenen.*, **27**, 802–5. *MR*: **84e**, #60054.

Brillouët, N. *Équations fonctionnelles et théorie des groupes.* Publ. Math. Univ., Nantes, 1982.

Brunner, N. A note on the difference property. *Collect. Math.*, **33**, 221–3. *MR*: **85j**, #39001.

Campbell, L.L. Information submartingales. In *Transactions of the Ninth Prague Conference on Information Theory, Statistical Decision Functions, Random Processes, Prague. 1982.* Academia, Prague, 1983, pp. 171–6. *MR*: **85i**, #00018a.

Dankiewicz, K. and Moszner, Z. Prolongements des homomorphismes et des solutions de l'équation de translation. *Wyż. Szkoła Ped. Krakow. Rocznik Nauk.-Dydakt. Prace Mat.*, **10**, 27–44.

Daróczy, Z. and Páles, Zs. On comparison of mean values. *Publ. Math. Debrecen*, **29**, 107–15. *MR*: **84e**, #39007.

Dhombres, J.
(*a*) Recent applications of functional equations. In *Proc. Second World Conf. Math. at the Service of Man, Las Palmas, 1982*, Vol. 2. Univ. Politec., Barcelona, 1982.
(*b*) *Moyennes.* Publ. Math. Univ., Nantes, 1982.
(*c*) Remark 7. *Aequationes Math.*, **24**, 289.

Dyiak, C. On a characterization of trigonometrical and hyperbolic functions by functional equations. *Publ. Inst. Math.* (*Beograd*) (*N.S.*) **32(46)**, 45–8. *MR*: **84i**, #39005.

Ebanks, B.R.
(*a*) A note on symmetric utility rules for gambles. *J. Math. Econom.*, **9**, 107–11. *MR*: **83c**, #90018.
(*b*) The Twentieth International Symposium on Functional Equations, August 1-August 7, 1982, Oberwolfach, Germany. *Aequationes Math.*, **24**, 261–97.

Eichhorn, W. and Gehrig, W. Measurement of inequality in economics. In *Modern Applied Mathematics – Optimization and Operations Research.* North Holland, Amsterdam-New York, 1982, pp. 657–93.

Fishburn, P.C.
(*a*) A note on linear utility. *J. Econom. Theory*, **27**, 444–6. *MR*: **83k**, #90016.
(*b*) Nontransitive measurable utility, *J. Math. Psych.*, **26**, 31–67. *MR*: **84g**, #90020.
(*c*) *The Foundations of Expected Utility.* (Theory and Decision Library, Vol. 31). Reidel, Dordrecht-Boston-London, 1982. *MR*: **85k**, #90021.

Forti, G.L.
(*a*) *On some Cauchy Equations on Thin Sets.* (Univ. Studi Milano, No. 16/S(II)). Ist. Mat. F. Enriques, Milano, 1982.
(*b*) On an alternative functional equation related to the Cauchy equation. *Aequationes Math.*, **24**, 195–206. *MR*: **85c**, #39011.

Galambos, J. The role of functional equations in stochastic model building. *Aequationes Math.*, **25**, 21–41. *MR*: **85g**, #62028.

Glazek, K. and Gleichgewicht, B. Bibliography of *n*-groups (polyadic groups) and some group-like *n*-ary structures. In *Proc. Symp. n-ary Structures* (*Skopje, 1982*). Makedonska Akad. Nauk Umetnost., Skopje, 1982, pp. 77–86. *MR*: **85j**, #20078.

Hildebrand, A. A characterization of the logarithm as an additive arithmetic function. *Arch. Math.* (*Basel*), **38**, 535–9. *MR*: **84c**, #10049.

Járai, A. Regularity properties of functional equations. *Aequationes Math.*, **25**, 52–66. *MR*: **84k**, #39015.

Kannappan, Pl. Information functions on open domain. IV. *C.R. Math. Rep. Acad. Sci. Canada*, **4**, 207–12. *MR*: **84b**, #94017.

Kovács, K. On the characterization of additive and multiplicative functions. *Studia Sci. Math. Hungar.*, **18**, 1–11. *MR*: **86m**, #11004.

Kuwagaki, A. Sur l'équation fonctionnelle des vecteurs dans R^3: $f([x,y]) = [f(x), f(y)]$. *Studia Humana Natur.* (*Kyoto*), **16**, 1–6.

Losonczi, L. and Maksa, Gy. On some functional equations of the information theory. *Acta Math. Acad. Sci. Hungar.*, **39**, 73–82. *MR*: **85b**, #39003.

Lundberg, A. Generalized distributivity for real, continuous functions. I: Structure theorems and surjective solutions. *Aequationes Math.*, **24**, 74–96. *MR*: **85d**, #39010.

Maksa, Gy. Solution on the open triangle of the generalized fundamental equation of information with four unknown functions. *Utilitas Math.*, **21**, 267–82. *MR*: **83k**, #94015.

Moszner, Z. Quelques remarques sur les solutions continues de l'équation de

translation sur un groupe. *Demonstratio Math.*, **15**, 279–84. *MR*: **84k**, #39017.
Neuman, F.
(*a*) Functions of the form $\sum_{i=1}^{N} f_i(x)g_i(t)$ in L_2. *Arch. Math. (Brno)*, **18**, 19–22. *MR*: **85a**, #26026.
(*b*) Factorizations of matrices and functions of two variables. *Czechoslovak Math. J.*, **32 (107)**, 582–8. *MR*: **84d**, #15017.
Páles, Zs. Characterization of quasideviation means. *Acta Math. Acad. Sci. Hungar.* **40**, 243–60. *MR*: **84j**, #26018.
Parsloe, A.J. An asymptotic form of Cauchy's functional equation. *Aequationes Math.*, **25**, 194–208. *MR*: **85c**, #39007.
Rukhin, A.L. The solution of the functional equation of d'Alembert's type for commutative groups. *Internat. J. Math. Sci.*, **5**, 315–35. *MR*: **84g**, #39006.
Sander, W. Boundedness properties for functional inequalities. *Manuscripta Math.* **39**, 271–6. *MR*: **84d**, #39013.
Székelyhidi, L.
(*a*) On a theorem of Baker, Lawrence and Zorzitto. *Proc. Amer. Math. Soc.*, **84**, 95–6. *MR*: **83a**, #39011.
(*b*) The stability of d'Alembert-type functional equations. *Acta Sci. Math. (Szeged)*, **44**, 313–20. *MR*: **84m**, #39020.
(*c*) Note on a stability theorem. *Canad. Math. Bull.*, **25**, 500–1. *MR*: **83m**, #39002.
(*d*) On the zeros of exponential polynomials. *C.R. Math. Rep. Acad. Sci. Canada* **4**, 189–94. *MR*: **83k**, #22012.
(*e*) On a class of linear functional equations. *Publ. Math. Debrecen* **29**, 19–28.
Targonski, G. Unsolved problems in iteration theory. In *Théorie de l'Itération et ses Applications* (*Toulouse, 1982*). (Colloq. Internat. CNRS, Vol. 332). CNRS, Paris, 1982. *MR*: **86m**, #580789.
Tsutsumi, A. and Haruki, Sh. Functional equations and hypoellipticity. *Proc. Japan Acad. Ser. A Math. Sci.*, **58**, 105–8. *MR*: **83j**, #39008.
Turdza, E. Stability of Cauchy equations. *Wyż. Szkola Ped. Krakow. Rocznik Nauk.-Dydakt. Prace Mat.*, **10**, 141–6.
Volkmann, P. Sur un système d'inéquations fonctionnelles. *C.R. Math. Rep. Acad. Sci. Canada*, **4**, 155–8. *MR*: **83k**, #39010.
Volkmann, P. and Weigel, H. Sur la solution de certaines équations fonctionnelles. *Uniw. Śląski w Katowicach Prace Nauk. —Prace Mat.*, **12**, 22–9. *MR*: **83g**, #39005.
Wagner, C. Allocation, Lehrer models, and the consensus of probabilities. *Theory and Decision*, **14**, 207–20. *MR*: **83m**, #90002.
Wołodźko, S. A modification of a construction of Z. Moszner. *Wyż. Szkoła Ped. Krakow. Rocznik Nauk.-Dydakt.*, **10**, 165–9.

1983

Aczél, J.
(*a*) On locally bounded maps of sequence spaces. *Real Anal. Exchange*, **9**, 111–15.
(*b*) Some unsolved problems in the theory of functional equations, II. *Aequationes Math.*, **26**, 255–60.
(*c*) Diamonds are not the Cauchy extensionist's best friend. *C.R. Math. Rep. Acad. Sci. Canada*, **5**, 259–64. *MR*: **85a**, #39005.
Aczél, J. and Ng, C.T. Determination of all semisymmetric recursive information measures of multiplicative type on *n* positive discrete probability distributions. *Linear Algebra Appl.*, **52/53**, 1–30. *MR*: **84j**, #94016.
Aczél, J. and Saaty, T.L. Procedures for synthesizing ratio judgements. *J. Math. Psych.*, **27**, 93–102. *MR*: **85a**, #90002.
Albert, M. and Baker, J.A. Functions with bounded *n*-th differences. *Ann. Polon. Math.*, **43**, 93–103.
Benz, W. On a functional equation in connection with derivations. *Math. Z.*, **183**, 495–501. *MR*: **84m**, #39019.
Blanton, G. Smoothness in disjoint groups of real functions under composition. *C.R. Math. Rep. Acad. Sci. Canada*, **5**, 169–72. *MR*: **85a**, #26006.
Chew, S.H. A generalization of the quasilinear mean with applications to the measurement of income inequality and decision theory resolving the Allais paradox. *Econometrica*, **51**, 1065–92. *MR*: **85g**, #90019.
Cholewa, P.W. The stability of the sine equation. *Proc. Amer. Math. Soc.*, **88**, 631–4. *MR*: **84m**, #39016.
Christian, R.R. Another way to introduce natural logarithms and *e*. *Two-Year College Math. J.* **14**, 424–6.
Corovei, I. The sine functional equation for groups. *Mathematica (Cluj)*, **25 (48)**, 11–19. *MR*: **85d**, #39012.
Darsow, W.F. and Frank, M.J. Associative functions and Abel–Schröder systems.

Publ. Math. Debrecen, **30**, 253–72. *MR*: **85f**, #39007.

Dhombres, J. Sur les fonctions simultanément suradditives et surmultiplicatives. *C.R. Math. Rep. Acad. Sci. Canada*, **5**, 207–10. *MR*: **84k**, #16053.

Diminnie, C.R., Freese, R.W., and Andalafte, E.Z. An extension of Pythagorean and isosceles orthogonality and a characterization of inner-product spaces. *J. Approx. Theory*, **39**, 295–8. *MR*: **85d**, #46028.

Forti, G.L. On some conditional Cauchy equations on thin sets. *Bol. Un. Mat. Ital. B* (6), **2**, 391–402. *MR*: **84h**, #39007.

Foster, J.E. An axiomatic characterization of the Theil measure of income inequality. *J. Econom. Theory*, **31**, 105–21. *MR*: **85a**, #90074.

Gehrig, W. On a characterization of the Shannon concentration measure. *Utilitas Math.* **24**, 67–85. *MR*: **85c**, #94010.

Grzegorczyk, A. and Tabor, J. Monotonic solution of the translation equation. *Ann. Polon. Math.*, **43**, 253–60.

Haruki, H. On a functional equation for Jacobi's elliptic function $\mathrm{cn}(z;k)$. *Abstracts Amer. Math. Soc.*, **4**, 464, #808–39–75.

Haruki, Sh. A truncated cube functional equation. *Proc. Japan Acad. Ser. A Math. Sci.*, **59**, 372–4. *MR*: **85e**, #39005.

Hyers, D.H. The stability of homomorphisms and related topics. In *Global Analysis–Analysis of Manifolds*. (Teubner-Texte zur Math., No. 57). Teubner, Leipzig, 1983, pp. 140–53. *MR*: **86a**, #39004.

Ježek, J. and Kepka, T. Medial groupoids. *Rozpravy Československé Akad. Věd. Řada Mat. Přírod. Věd.*, **93**, no. 2, 1–93. *MR*: **85k**, #20165.

Kannappan, Pl. and Ng. C.T. On a generalized fundamental equation of information. *Canad. J. Math.*, **35**, 862–72.

Kestelman, H. Advanced Problem 6436. *Amer. Math. Monthly*, **90**, 485.

Kovács, K. On the characterization of additive and multiplicative functions. *Studia Sci. Math. Hungar.*, **18**, 1–11. *MR*: **86m**, #11004.

Krause, G.M. Interior idempotents and nonrepresentability of groupoids. *Stochastica* **7**, 5–10. *MR*: **86a**, #26015.

Kuhn, M. A note on t-convex functions. In *General Inequalities 4* (*Proc. Fourth Internat. Conf. on General Inequalities, Oberwolfach, 1983*). Birkhäuser, Basel-Boston-Stuttgart, 1984, pp. 269–76.

Laird, P.G. and Mills, J. On systems of linear functional equations. *Aequationes Math.*, **26**, 64–73. *MR*: **85c**, #39010.

Lajkó, K. Some general functional equations. *Publ. Math. Debrecen*, **30**, 225–33. *MR*: **86d**, #39011.

Luce, R.D. and Cohen, M. Factorizable automorphisms in solvable conjoint structures. I. *J. Pure Appl. Algebra*, **27**, 225–61. *MR*: **84d**, #92048.

Moszner, Z. Sur un problème concernant des 1-paramétriques Lie-groupes. *Tensor* (*N.S.*) **40**, 32–4. *MR*: **87f**, #39013.

Paganoni, L. and Rusconi, D. A characterization of some classes of functions F of the form $F(x,y) = g(\alpha f(x) + \beta f(y) + \gamma)$ or $F(x,y) = \phi(h(x) + k(y))$. *Aequationes Math.*, **26**, 138–62. *MR*: **85f**, #39008.

Przeworska-Rolewicz, D. Non-Leibniz algebras. *Studia Math.*, **77**, 69–79. *MR*: **85c**, #47003.

Sander, W. Some functional inequalities. *Monatsh. Math.*, **95**, 149–57. *MR*: **84h**, #39011.

Schweizer, B. and Sklar, A. *Probabilistic Metric Spaces*. North-Holland, New York-Amsterdam, 1983. *MR*: **86g**, #54045.

Skof. F.
(*a*) Local properties and approximation of operators (Italian). *Rend. Sem. Mat. Fis. Milano* **53**, 113–29. *MR*: **87m**, #39009.
(*b*) On the approximation of locally δ-additive mappings. *Atti Accad. Sci. Torino Cl. Sci. Fis. Mat. Natur.*, **117**, 377–89.

Sprecher, D.A. On functional complexity and superpositions of functions. *Real Anal. Exchange*, **9**, 412–31. *MR*: **86b**, #26001.

Székelyhidi, L.
(*a*) Local polynomials and functional equations. *Publ. Math. Debrecen*, **30**, 283–90. *MR*: **85i**, #39010.
(*b*) Almost periodicity and functional equations. *Aequationes Math.*, **26**, 163–75. *MR*: **85h**, #39003.

Voxman, W. and Goetschel, R. A note on the characterization of the max and min operator. *Inform. Sci.*, **30**, 5–10. *MR*: **84j**, #04001.

Vukman, J. Bilinear quadratic functionals (Slovenian). *Obzornik Mat. Fiz.*, **30**, no. 3, 80–5. *MR*: **85f**, #15022.

Zorzitto, F. The Twenty-first International Symposium on Functional Equations, August 6-August 13, 1983, Konolfingen, Switzerland. *Aequationes Math.*, **26**, 225–94.

1984

Aczél, J.
(*a*) On history, applications and theory of functional equations. In *Functional Equations: History, Applications and Theory*. Reidel, Dordrecht-Boston-Lancaster, 1984, pp. 3–12.
(*b*) Some recent results on information measures, a new generalization and some 'real life' interpretations of the 'old' and new measures. In *Functional Equations: History, Applications and Theory*. Reidel, Dordrecht-Boston-Lancaster, 1984, pp. 175–89.
(*c*) Related functional equations applied to Korovkin approximation and to the characterization of Rényi entropies – links to uniqueness theory. *C.R. Math. Rep. Acad. Sci. Canada*, **6**, 319–36. *MR*: **86f**, #39003.
(*d*) On weighted synthesis of judgements. *Aequationes Math.*, **27**, 288–306. *MR*: **86g**, #90008.

Aczél, J. and Alsina, C. Characterization of some classes of quasilinear functions with applications to triangular norms and to synthesizing judgements. In *Contributions to Production Theory, Natural Resources, Economic Indices and Related Topics.* (Methods of Operations Research, Vol. 487). Athenäum-Hain-Hanstein-Königstein, 1984, pp. 3–12.

Aczél, J., Ng, C.T., and Wagner, C. Aggregation theorems for allocation problems. *SIAM J. Algebraic Discrete Methods*, **5**, 1–8. *MR*: **85a**, #90139.

Adamaszek, I. Almost trigonometric functions. *Glas. Mat. Ser. III*, **19(39)**, 83–104. *MR*: **87m**, #39012.

Birsan, T. Une application d'un théorème de J. Aczél. *Publ. Math. Debrecen*, **31**, 113–18. *MR*: **86f**, #54047.

Brockett, P.L. General bivariate Makeham laws. *Scand. Actuar. J.*, 1984, 150–6. *MR*: **86k**, #62178.

Cholewa, P.W.
(*a*) Remarks on the stability of functional equations. *Aequationes Math.* **27**, 76–86. *MR*: **86d**, #39016.
(*b*) The stability for a generalized Cauchy type functional equation. *Rev. Roumaine Math. Pures Appl.*, **29**, 457–60. *MR*: **86c**, #39013.

Daróczy, Z. and Kotora, E. Die Aczél-Benzsche Funktionalgleichung auf der additiven Gruppe der ganzen Zahlen, *Publ. Math. Debrecen*, **31**, 221–7. *MR*: **86d**, #39008.

Dhombres, J.
(*a*) On the historical role of functional equations. In *Functional Equations: History, Applications and Theory.* Reidel, Dordrecht-Boston-Lancaster, 1984, pp. 17–31.
(*b*) Some recent applications of functional equations. In *Functional Equations: History, Applications and Theory.* Reidel, Dordrecht-Boston-Lancaster, 1984, pp. 67–91.
(*c*) Autour de $f(x)f(y) - f(xy)$. *Aequationes Math.*, **27**, 231–5. *MR*: **86b**, #39008.
(*d*) Utilisation des équations fonctionnelles pour la caractérisation des espaces prèhilbertiens. *Rend. Sem. Mat. Fis. Milano*, **54**, 159–86.

Drljević, H. On the stability of the functional quadratic on *A*-orthogonal vectors. *Publ. Inst. Math.* (*Beograd*) (*N.S.*), **36(50)**, 111–18. *MR*: **87c**, #47084.

Ebanks, B.R. Kurepa's functional equation on Gaussian semigroups. In *Functional Equations: History, Applications and Theory*. Reidel, Dordrecht-Boston-Lancaster, 1984, pp. 167–73.

Elliott, P.D.T.A. Cauchy's functional equation in the mean. *Adv. in Math.*, **51**, 253–7. *MR*: **86d**, #39009.

Forte, B., Ng, C.T. and Lo Schiavo, M. Additive and subadditive entropies for discrete random variables. *J. Combin. Inform. System Sci.*, **9**, 207–16.

Gajda, Z.
(*a*) Additive and convex functions in linear topological spaces. *Aequationes Math.*, **27**, 214–9.
(*b*) On some properties of Hamel bases connected with the continuity of polynomial functions. *Aequationes Math.*, **27**, 57–75. *MR*: **86e**, #39006.

Gehrig, W.
(*a*) Functional equation methods applied to economic problems: some examples. In *Functional Equations: History, Applications and Theory.* Reidel, Dordrecht-Boston-Lancaster, 1984, pp. 33–52.
(*b*) On a characterization of the Shannon concentration measure. In *Functional Equations: History, Applications and Theory*. Reidel, Dordrecht-Boston-Lancaster, 1984, pp. 191–205.

Haruki, H.
(*a*) An improvements of the Nevalinna-Pólya theorem. In *Functional Equations: History, Applications and Theory*. Reidel, Dordrecht-Boston-Lancaster, 1984, pp. 113–26.
(*b*) On a conformal-mapping property. *Ann. Polon. Math.* **44**, 39–44. *MR*: **86a**, #30013.

(*c*) On a functional equation for Jacobi's elliptic function cn(z; k). *Aequationes Math.*, **27**, 317–21. *MR*: **85m**, #33003.

Jáhad, A. A remark to a paper of J. Aczél and J.K. Chung. *Studia Sci. Math. Hung.*, **19**, 273–4. *MR*: **87m**, #39006.

Kovács, K. The characterization of complex-valued additive functions. *Studia Soc. Math. Hungar.*, **19**, 159–61. *MR*: **86f**, #26003.

Krapež, A.

(*a*) Functional equations on groupoids and related structures. In *Functional Equations: History, Applications and Theory.* Reidel, Dordrecht-Boston-Lancaster, 1984, pp. 57–64.

(*b*) Groupoids with Δ-kernels. In *Functional Equations: History, Applications and Theory.* Reidel, Dordrecht-Boston-Lancaster, 1984, pp. 93–7.

Krtcha, M. A characterization of the Edgeworth-Marshall index. In *Contributions to Production Theory, Natural Resources, Economic Indices and Related Topics.* (Methods of Operation Research, Vol. 48). Athenäum-Hain-Hanstein, Königstein, 1984, pp. 127–35.

Kuwagaki, A. Sur l'équation fonctionnelle du type Pexider pour des 3-vecteurs: $f([x, y]) = [g(x), h(y)]$. *Mem. Konan Univ., Sci. Ser.*, **31**, 113–18.

Laczkovich, M.

(*a*) On Kemperman's inequality $2f(x) \leq f(x+h) + f(x+2h)$. *Colloq. Math.*, **49**, 109–15. *MR*: **86g**, #26028.

(*b*) On the difference property of the class of pointwise discontinuous functions and some related classes. *Canad. J. Math.*, **36**, 756–68. *MR*: **85j**, #26003.

Matkowski, J. Cauchy functional equation on a restricted domain and commuting functions. In *Iteration Theory and Its Functional Equations (Proc. Schloss Hofen, 1984).* (Lecture Notes in Math., Vol. 1163). Springer, Berlin-Heidelberg-New York-Tokyo, 1985, pp. 101–6. *MR*: **87a**, #39001.

Mitchell, T.M. A functional equations approach to the theory of production, technical change and invariance. *Z. Nationalökonom.*, **44**, 177–87. *MR*: **86i**, #90014.

Páles, Zs. On the characterization of means defined on a linear space. *Publ. Math. Debrecen* **31**, 19–27. *MR*: **86d**, #26024.

Papp, F.J. Some generalizations of the functional equation satisfied by Russell's 'wage income' and 'optical consumption plan' functions. *Internat. J. Math. Ed. Sci. Tech.*, **15**, 107–15. *MR*: **85j**, #39007.

Pokropp, F. The functional equation of aggregation with weakly monotonically increasing functions. In *Contributions to Production Theory, Natural Resources, Economic Indices and Related Topics.* (Methods of Operations Research, Vol. 48). Athenäum-Hain-Hanstein, Königstein, 1984, pp. 63–81. *MR*: **86h**, #39009.

Reich, L. and Schwaiger, J. On polynomials in additive and multiplicative functions. In *Functional Equations: History, Applications and Theory.* Reidel, Dordrecht-Boston-Lancaster, 1984, pp. 127–60.

Sander, W. Applications of certain 𝔎-families. *Fund. Math.*, **122**, 77–84. *MR*: **85i**, #28002.

Targonski, G. New directions and open problems in iteration theory. *Ber. Math.-Statist. Sekt. Forsch. Graz.*, **229**, 1–51. *MR*: **86k**, #39009.

Tsutsumi, A. and Haruki, Sh. The regularity of solutions of functional equations and hypoellipticity. In *Functional equations: History, Applications and Theory.* Reidel, Dordrecht-Boston-Lancaster, 1984, pp. 99–112.

Vilks, A. Zur Klein-Nataf-Aggregation bei limitationalen Produktionsfunktionen. In *Contributions to Production Theory, Natural Resources, Economic Indices and Related Topics*, (Methods of Operations Research, Vol. 48). Athenäum-Hain-Hanstein, Königstein, 1984, pp. 285–93.

Volkmann, P. Une équation fonctionnelle pour les différences divisées. *Mathematica (Cluj)*, **26** (49), 175–81. *MR*: **87a**, #39016.

Volkmann, P. and Weigel, H. Über ein Problem von Fenyö. *Aequationes Math.*, **27**, 135–49. *MR*: **85h**, #39005.

Yanushkyavichene, O.L. Estimate of the stability of a characterization of an exponential law (Russian). *Teor. Veroyatnost. i Primenen.*, **29**, 279–88—*Theory Probab. Appl.*, **29**, 281–92. *MR*: **86a**, #60022.

1985

Aczél, J.

(*a*) A mean value property of the derivative of quadratic polynomials – without mean values and derivatives. *Math. Mag.*, **58**, 42–5. *MR*: **86c**, #39012.

(*b*) Solutions of a system of functional-differential equations arising from an

optimization problem. *Linear Algebra Appl.*, **71**, 9–13. *MR*: **87j**, #39018.
Aczél, J. and Paganoni, L.A. generalization of affine functions. *C.R. Math. Rep. Acad. Sci. Canada*, **7**, 257–62. *MR*: **86k**, #39011.
Alsina, C. and Ger, R. Associative operations close to a given one. *C.R. Math. Rep. Acad. Sci. Canada*, **7**, 207–10. *MR*: **87f**, #39014.
Baker, J.A. A note on iteration groups. *Aequationes Math.*, **28**, 129–31. *MR*: **86g**, #26003.
Balk, B.M. A simple characterization of Fisher's price index. *Statist. Hefte (N.F.)*. **26**, 59–63.
Baron, K. A remark on the stability of the Cauchy equation. *Wyż. Szkoła Ped. Krakow. Rocznik Nauk.-Dydakt. Prace Mat.*, **11**, 7–12.
Benz, W.
(*a*) On mappings preserving a single Lorentz-Minkowski-distance in Hjelmslev's classical plane. *Aequationes Math.*, **28**, 80–7. *MR*: **86f**, #51010.
(*b*) Isometrien in normierten Räumen. *Aequationes Math.*, **29**, 204–9.
Buchwalter, H. Formes quadratiques sur un semi-groupe involutif. *Math. Ann.* **271**, 619–35.
Cenzer, D. The stability problem: new results and counterexamples. *Lett. Math. Phys.*, **10**, 155–63. *MR*: **87e**, #39015.
Chung, J.K., Kannappan, Pl., and Ng, C.T. A generalization of the cosine-sine functional equation on groups. *Linear Algebra Appl.*, **66**, 259–77. *MR*: **87c**, #39008.
Corovei, I. The functional equation $f(xy) + f(yx) = f(x)f(y) + f(y)f(x)$ on a semigroup into a ring. *Mathematica (Cluj)*, **27(50)**, 97–9. *MR*: **87i**, #39019.
Daróczy, Z. and Kátai, I. On additive number-theoretical functions with values in a compact abelian group. *Aequationes Math.*, **28**, 288–92. *MR*: **86l**, #39010.
Davison, T.M.K. (compil.) *Proceedings of the Twenty-third International Symposium on Functional Equations (Gargnano, Italy, June 2-June 11, 1985)*. Centre for Information Theory, University of Waterloo, Waterloo, Ont., 1985.
Ebanks, B.R.
(*a*) Measurable solutions of functional equations connected with information measures on open domains. *Utilitas Math.*, **27**, 217–23. *MR*: **87b**, #94020.
(*b*) (compil.) The Twenty-second International Symposium on Functional Equations, Dec. 16–22, 1984, Oberwolfach, Germany. *Aequations Math.*, **29**, 62–111.
Escalone, E.F. Problem No. 296. *College Math. J.*, **16**, 153.
Fenyö, I. On the general solution of a system of functional equations. *Aequationes Math.*, **28**, 233–5. *MR*: **86m**, #39018.
Forti, G.L. A note on a general difference-functional equation. *Stochastica*, **9**, 19–24. *MR*: **87j**, #39004.
Foster, J.E. Inequality measurement. In *Fair Allocation (AMS Short Course, Anaheim, Cal., Jan. 7–8, 1985)*. Amer. Math. Soc., Providence, RI, 1985, pp. 31–68. *MR*: **87g**, #90016.
Ger, R. A consistency equation for three geometries. *Aequationes Math.*, **29**, 50–5. *MR*: **87b**, #39007.
Grabmüller, H. Cosine families of unbounded operators. *Appl. Anal.*, **19**, 1–38. *MR*: **86h**, #47060.
Haruki, H. A proof of Euler's identity by a functional equation. *Aequationes Math.*, **28**, 138–43. *MR*: **86c**, #30045.
Haruki, Sh. A nonsymmetric partial difference functional equation analogous to the wave equation. *Proc. Japan Acad. Ser. A Math. Sci.*, **61**, 302–4. *MR*: **87d**, #39003.
Heuvers, K.J. A family of symmetric biadditive nonbilinear functions. *Aequationes Math.*, **29**, 14–8. *MR*: **87j**, #15049.
Kannappan, Pl. and Sahoo, P.K. On a functional equation connected to sum form nonadditive information measures on an open domain. *C.R. Math. Rep. Acad. Sci. Canada*, **7**, 45–50. *MR*: **86c**, #39011.
Krapež, A. and Taylor, M.A. On the Pexider equation. *Aequationes Math.*, **28**, 170–89. *MR*: **86e**, #39010.
Kuczma, M.
(*a*) *An introduction to the theory of functional equations and inequalities. Cauchy's equation and Jensen's inequality.* Uniw. Ślask—P.W.N., Warszawa-Kraków-Katowice, 1985. *MR*: **86i**, #39008.
(*b*) On some analogies between measure and category and their applications in the theory of additive functions. *Ann. Math. Sil.* **13**, 155–62.
Kuwagaki, A. Sur l'équation fonctionnelle de type de Pexider pour des matrices carrées: $F([X, Y]) = [G(X), H(Y)]$. *Mem. Konan Univ., Sci. Ser.* **32**, 43–9.
Lawrence, J.
(*a*) The stability of multiplicative

Lawrence, J. (*cont.*)
semigroup homomorphisms to real normed algebras, I. *Aequationes Math.*, **28**, 94–101. *MR*: **86d**, #46041.
(*b*) Orthogonality and additive functions on normed linear spaces. *Colloq. Math.*, **49**, 253–5. *MR*: **87g**, #46028.

Losonczi, L.
(*a*) Sum form equations on an open domain, I. *C.R. Math. Rep. Acad. Sci. Canada*, **7**, 85–90. *MR*: **86i**, #39009.
(*b*) An extension theorem. *Aequationes Math.*, **28**, 293–9. *MR*: **86i**, #39005.

Luce, R.D. and Narens, L. Classification of concatenation measurement structures according to scale types. *J. Math. Psych.*, **29**, 1–72. *MR*: **87j**, #92044.

Lundberg, A. Generalized distributivity for real continuous functions. II. Local solutions in the continuous case. *Aequationes Math.*, **28**, 236–51. *MR*: **86k**, #39012.

Moszner, Z. Sur la stabilité de l'équation d'homomorphisme. *Aequationes Math.*, **29**, 290–306. *MR*: **87b**, #39009.

Narens, L. *Abstract Measurement Theory.* MIT Press, Cambridge, MA-London, 1985.

Neuwirth, E. Latent additivity and a differential equation. *Publ. Math. Debrecen*, **32**, 129–31. *MR*: **87b**, #92034.

Ng, C.T. On a functional equation related to income inequality measures. *Aequationes Math.*, **28**, 161–9. *MR*: **86d**, #39013.

Paganoni, L. On an alternative Cauchy equation. *Aequationes Math.*, **29**, 214–21.

Papp, F.J. The d'Alembert functional equation. *Amer. Math. Monthly*, **92**, 273–5. *MR*: **86g**, #39003.

Rašković, M. An application of nonstandard analysis to functional equations. *Publ. Inst. Math.* (*Beograd*) (*N.S.*) **37(51)**, 23–4. *MR*: **87a**, #39013.

Rätz, J. The orthogonally additive mappings. *Aequationes Math.*, **28**, 35–49. *MR*: **87b**, #39012.

Reich, L. Über Iteration ohne Regularitätsbedingungen im Ring der formalen Potenzreihen in einer Variablen. *Aequationes Math.*, **29**, 44–9. *MR*: **87d**, #13018.

Russell, R.R. A note on decomposable inequality measures. *Rev. Economic Stud.*, **52**, 347–52. *MR*: **87a**, #90015.

Sablik, M. Note on a Cauchy conditional equation. *Rad. Mat.*, **1**, 241–5. *MR*: **87e**, #39015.

Sander, W. About solutions of a functional equation. *Aequationes Math.*, **29**, 135–44. *MR*: **87d**, #39027.

Schroth, P. Note on initial topologies in rational vector spaces induced by real valued linear mappings. *Aequationes Math.*, **28**, 252–4. *MR*: **86g**, #39010.

Smítal, J. On a problem of Aczél and Erdös concerning Hamel bases. *Aequationes Math.* **28**, 135–7. *MR*: **86e**, #04004.

Stettler, R. Eine axiomatische Herleitung der 2-dimensionalen Lorentztransformation. *Z. Angew. Math. Phys.*, **36**, 538–48. *MR*: **87a**, #83005.

Székelyhidi, L.
(*a*) Regularity properties of polynomials on groups. *Acta Math. Hungar.*, **45**, 15–19. *MR*: **86h**, #39018a.
(*b*) Regularity properties of exponential polynomials on groups. *Acta Math. Hungar.*, **45**, 21–6. *MR*: **86h**, #39018b.

Targonski, G. Iterationstheorie zwischen Funktionalgleichungen und Dynamik. *Jbuch. Überblicke Math.*, **1985**, 9–28.

Tissier, A. Solution of Advanced Problem 6436. *Amer. Math. Monthly*, **92**, 291.

Volkmann, P. Caractérisation de la fonction $f(x)=x$ par un système de deux inéquations fonctionnelles. *Wyż. Szkoła Ped. Krakow. Rocznik Nauk.-Dydakt. Prace Mat.*, **97**, 177–82.

Vukman, J. Some results concerning the Cauchy functional equation in certain Banach algebras. *Bull. Austral. Math. Soc.*, **31**, 137–44. *MR*: **86f**, #39005.

1986

Aczél, J. Characterizing information measures: approaching the end of an era. In *Uncertainty in Knowledge-Based Systems* (*Int. Conf. Inf. Proc. Manag. Uncert. Knowl.-Based Systems, Paris, June–July, 1986*). (Lecture Notes in Comp. Sci., Vol. 286). Springer, Berlin-Heidelberg-New York-London-Paris-Tokyo, 1987, pp. 359–84. *MR*: **88d**, #39013.

Aczél, J. and Alsina, C. On synthesis of judgements. *Socio-Econom. Plann. Sci.*, **20**, 333–9.

Aczél, J., Roberts, F.S. and Rosenbaum, Z. On scientific laws without dimensional constants. *J. Math. Anal. Appl.*, **119**, 389–416. *MR*: **88b**, #00019.

Amir, D. *Characterizations of Inner Product Spaces.* (Operations Theory, Advances and Applications, Vol. 20). Birkhäuser, Basel-Boston-Stuttgart, 1986.

Bacchelli, B. Representation of continuous associative functions. *Stochastica*, **10**, 13–28. *MR*: **88e**, #39014.

Bauer, H.
(*a*) A class of means and related

inequalities. *Manuscripta Math.*, **55**, 199–211. *MR*: **87j**, #26027.
(*b*) Mittelwerte und Funktionalgleichungen. *Bayer. Akad. Wiss. Math.-Natur. Kl. Sitzungsber.*, **1986**, 1–9.

Benz, W. A class of nowhere continuous solutions of Baxter's equation. *Aequationes Math.*, **31**, 169–172. *MR*: **88d**, #39015.

Brillouët, N. and Dhombres, J. Équations fonctionnelles et recherche de sous-groupes. *Aequationes Math.*, **31**, 253–93.

Cheng, L. The impact of changes in relative weights on the optimal solution of a maximization problem. *J. Math. Econom.*, **15**, 143–50.

Daróczy, Z. On the general solution of a system of functional equations on a commutative semigroup. *Period. Math. Hungar.*, **17**, 27–35. *MR*: **87h**, #39006.

Dhombres, J. Quelques aspects de l'histoire des équations fonctionnelles liés à l'évolution du concept de fonction. *Arch. Hist. Exact Sci.*, **36**, 91–181. *MR*: **88a**, #01019.

Diderrich, G.T. Boundedness on a set of positive measure and the fundamental equation of information. *Publ. Math. Debrecen*, **33**, 1–7. *MR*: **87i**, #39014.

Drljević, H. On a functional which is quadratic on *A*-orthogonal vectors. *Publ. Inst. Math. (Beograd) (N.S.)*, **40(54)**, 63–71.

Ebanks, B.R. The equation $F(x) + M(x)F(x^{-1}) = 0$ for additive F and multiplicative M on the positive cone of R^n. *C.R. Math. Rep. Acad. Sci. Canada*, **8**, 247–52. *MR*: **87m**: #39011.

Ebanks, B.R. and Maksa, Gy. Measures of inset information on the open domain-I. Inset entropies and information functions of all degrees. *Aequationes Math.*, **30**, 187–201.

Eberl, W., Jr. A functional equation for the Mellin-Stieltjes transforms of gamma distributions. *J. Math. Anal. Appl.*, **116**, 538–52. *MR*: **87h**: #44007.

Fenyö, I. On an inequality of P.W. Cholewa. In *General Inequalities 5 (Proc. 5th Int. Conf. General Inequalities, Oberwolfach, May 4–10, 1986)*. Birkhäuser, Basel-Boston, 1987, pp. 277–80.

Fifer, Z. Set-valued Jensen functional equation. *Rev. Roumaine Math. Pures Appl.*, **31**, 297–302.

Gajda, Z. Christensen measurable solutions of generalized Cauchy functional equations. *Aequationes Math.*, **31**, 147–58. *MR*: **88d** #3902.

Ger, J. Convex functions in metric spaces. *Radovi Mat.*, **2**, 217–36.

Gueraggio, A. The functional equations in the foundations of financial mathematics (Italian). *Riv. Mat. Sci. Econom. Social.*, **9**, 33–52.

Ito, T. and Nara, C. Quasi-arithmetic means of continuous functions. *J. Math. Soc. Japan*, **38**, 697–720. *MR*: **87k**, #39003.

Járai, A. On regular solutions of functional equations. *Aequationes Math.*, **30**, 21–54. *MR*: **87g**, #39010.

Johnson, B.E. Approximately multiplicative functionals. *J. London Math. Soc.* (2), **34**, 489–510. *MR*: **87k**, #46105.

Kovács, K. On the characterization of additive functions with monotonic norms. *J. Number Theory*, **24**, 298–304.

Kumari, K., Parnami, J.C. and Vasudeva, H.L. On the functional equation $\Delta_y^n f(x) - n!\ f(y) = 0$. *Math. Today*, **4**, 11–20. *MR*: **88d**, #39022.

Luce, R. D. Uniqueness and homogeneity of ordered rational structures. *J. Math. Psych.*, **30**, 391–415. *MR*: **88h**, #06027.

Losonczi, L. Sum form equations on an open domain. II. *Utilitas Math.*, **29**, 125–32. *MR*: **88d**, 94006.

Maehara, R. On a connected dense proper subgroup of R^2 whose complement is connected. *Proc. Amer. Math. Soc.*, **97**, 556–8. *MR*: **87i**, #54042.

Maksa, Gy. and Ng, C.T. The fundamental equation of information on open domain. *Publ. Math. Debrecen*, **33**, 9–11. *MR*: **87j**, #39015.

McDougall, J.A. and Ozeki, A. The octahedron equation implies the cube equation. *Aequationes Math.*, **31**, 243–6.

McKiernan, M.A.
(*a*) On iteration groups and related functional equations. *Aequationes Math.*, **31**, 37–46. *MR*: **87i**, #39012.
(*b*) Pexider type functional equations and Cartan equivalence problems. *Utilitas Math.*, **30**, 57–62. *MR*: **87m**, #39010.

Morsisyan, Yu.M. *Introduction of the theory of Algebras with Hyperidentities* (Russian). Univ. of Erevan Press, Erevan, 1986. *MR*: **88f**, #08001.

Pannenberg, M. A discrete integral representation for polynomials of fixed maximal degree and their universal Korovkin closure. *J. Reine Agnew. Math.*, **370**, 74–82. *MR*: **87j**, #41083.

Perelli, A. and Zannier, U. On a functional equation associated with the homomorphisms of a group into a field (Italian). *Boll. Un. Mat. Ital. B* (6), **5**,

Perelli, A. (*cont.*)
235–45. *MR*: **87h**, #39008.

Sablik, M. On extending functions to solutions of some functional equations. *Aequationes Math.*, **31**, 142–6.

Šemrl, P. On quadratic and sesquilinear functionals. *Aequationes Math.*, **31**, 184–90. *MR*: **88h**, #39007.

Szabó, Gy. On mappings, orthogonally additive in the Birkhoff-James sense. *Aequationes Math.*, **30**, 93–105. *MR*: **87k**, #46040.

Székelyhidi, L.
(*a*) Note on Hyers's theorem. *C.R. Math. Rep. Acad. Sci. Canada*, **8**, 127–9. *MR*: **87g**, #39019.
(*b*) The Fourier transform of mean periodic functions. *Utilitas Math.*, **29**, 43–8. *MR*: **88c**, #42021.
(*c*) The Fourier transform of exponential polynomials. *Publ. Math. Debrecen*, **33**, 13–20. *MR*: **87i**, #35082.

Szigeti, J. Homological methods in the solution of certain functional equations. *Aequationes Math.*, **31**, 310–14.

Talwalker, S. Functional equations in characterizations of discrete distributions by Rao-Rubin conditions and its variants. *Commun. Statist.—Theor. Math.*, **15**, 961–79. *MR*: **87f**, #62028.

Volkmann, L. Über ein System von zwei Funktionalgleichungen. *Bull. Math. Soc. Sci. Math. R.S. Roumanie* (*N.S.*), **30(78)**, 79–87. *MR*: **88d**, #39017.

Vukman, J. An application of S. Kurepa's extension of Jordan-Neumann characterization of pre-Hilbert space. *Glas. Mat. Ser. III*, **21(41)**, 349–52.

1987

Aczél, J.
(*a*) *A Short Course on Functional Equations Based upon Recent Applications to the Social and Behavioral Sciences.* Reidel, Dordrecht-Boston-Lancaster-Tokyo, 1987.
(*b*) Scale-invariant equal sacrifice in taxation and conditional functional equations. *Aequationes Math.*, **32**, 336–49.

Alsina, C. and Garcia-Roig, J.L. On a functional equation related to the Ptolemaic inequality. *Aequationes Math.*, **34**, 298–303.

Alsina, C. and Sklar, A. A characterization of continuous associative operations whose graphs are ruled surfaces. *Aequationes Math.*, **33**, 114–19.

Bell, D.E. Multilinear representation for ordinal utility functions. *J. Math. Psych.*, **31**, 44–59.

Benz, W.
(*a*) The cardinality of the set of discontinuous solutions of a class of functional equations. *Aequationes Math.*, **32**, 58–62.
(*b*) Ein Beitrag zu einem Problem von Herrn Fenyö. *Abh. Math. Sem. Univ. Hamburg*, **57**, 21–5.
(*c*) Ästhetische Rechteckmasse die massstabunabhängig sind. *Math. Methods Appl. Sci.*, **9**, 53–8. *MR*: **88b**, #51037.

Choczewski, B. (compil.) Proceedings of the II International Conference on Functional Equations and Inequalities, June 22–27, 1987, Sczawnica (Poland). *Wyż. Szkoła Ped. Krakow. Rocznik Nauk.-Dydakt. Prace. Mat.*, **12**, 163–234.

Christensen, J.P.R. and Fischer, P. Small sets and a class of general functional equations. *Aequationes Math.*, **33**, 18–23.

Davison, T.M.K. Röhmel's equation for quadratic forms. *Aequationes Math.*, **34**, 78–81.

Dhombres, J. Moyennes. In *Espaces de Marcinkiewicz – corrélations – mesures – systèmes dynamiques.* Masson, Paris-New York-Barcelona-Milan-Mexico-São Paulo, 1987, pp. 197–246. *MR*: **88g**, #11049.

Dhombres, J., Dahan-Dalmedico, A., Bkouche, R. and Guillemot, M. *Mathématiques au fil des âges.* Gauthier-Villars, Paris, 1987. *MR*: **88e**, #10007.

Ebanks, B.R., Kannappan, Pl., and Ng, C.T. Generalized fundamental equation of information of multiplicative type. *Aequationes Math.*, **32**, 19–31. *MR*: **88d**, 39018.

Fenyö, I. On the hyperbolic linear functional equations. *Aequationes Math.*, **32**, 252–58.

Filipescu, R. A simple characterisation of the exponential distribution. *Bull. Math. Soc. Sci. Math. R.S. Roumaine* (*N.S.*), **31(79)**, 23–4.

Forti, G.L. The stability of homomorphisms and amenability with applications to functional equations. *Abh. Math. Sem. Univ. Hamburg*, **57**, 215–26.

Gajda, Z. A solution of a problem of J. Schwaiger. *Aequationes Math.*, **32**, 38–44. *MR*: **88f**, #39005.

Gracia, J.M. On Pólya's functional equation. *Internat. J. Math. Ed. Sci. Tech.*, **19**, 403–11. *MR*: **88h**, μ→←←∅.

Gronau, D. Johannes Kepler und die Logarithmen. *Ber. Math.-Statist. Sekt. Forsch. Graz*, **284**, 1–41.

Gzyl, H. Characterization of vector valued, gaussian stationary Markov processes.

Statist. Probab. Lett., **6**, 17–19.
Iverson, G.J. Thurstonian psychophysics: Case III. *J. Math. Psych.*, **31**, 219–47.
Kovács, K. A note on a theorem of P. Erdös. *Studia Sci. Math. Hungar.*, **22**, 203–6.
Kurepa, S.
(*a*) Quadratic and sesquilinear forms. In *Functional Analysis, II.* (Lecture Notes in Math., Vol. 1242). Springer, Berlin-Heidelberg-New York-London-Paris-Tokyo, 1987, pp. 43–76. *MR*: **88e**, #46001.
(*b*) On quadratic forms. *Aequationes Math.*, **34**, 125–38.
(*c*) On the definition of a quadratic norm. *Publ. Inst. Math. (Beograd) (N.S.)*, **42**, 35–41.
Maher, P.J. What makes an operator linear. *Internat. J. Math. Ed. Sci. Tech.*, **18**, 177–9.
Mak, K.-T. Coherent continuous systems and the generalized functional equation of associativity. *Math. Oper. Res.*, **12**, 597–625.
Maksa, Gy. A characterization of the signed hyperbolic distance. *C.R. Math. Rep. Acad. Sci. Canada*, **9**, 21–4. *MR*: **88h**, #51018.
Mehring, G.H. Der Hauptsatz über Iteration im Ring der formalen Potenzreihen *Aequationes Math.*, **32**, 274–96. *MR*: **88h**, #13014.
Ng, C.T. The equation $F(x)+M(x)G(1/x)=0$ and homogeneous biadditive forms. *Linear Algebra Appl.*, **93**, 255–79. *MR*: **88g**, #39006.
Nikodem, K. Midpoint convex functions majorized by midpoint concave functions. *Aequationes Math.*, **32**, 45–51.
Paganoni, L. On a functional equation concerning affine transformations. *J. Math. Anal. Appl.*, **127**, 475–91.
Páles, Zs. On the characterization of quasiarithmetic means with weight function. *Aequationes Math.*, **32**, 171–94. *MR*: **88h**, #26007.
Radó, F. and Baker, J.A., Pexider's equation and aggregation of allocations. *Aequationes Math.*, **32**, 227–39.
Rudin, W. *Real and Complex Analysis.* Third edn. McGraw-Hill, New York-St. Louis-San Francisco-Auckland-Bogotá-Hamburg-Johannesburg-London-Madrid-Mexico-Milan-Montreal-New Delhi-Panama-Paris-São Paulo-Singapore-Sydney-Tokyo-Toronto, 1987.
Saaty, T.L. and Vargas, L.G. Stimulus-response with reciprocal kernels. The rise and fall of sensation. *J. Math. Psych.*, **31**, 83–92.
Sablik, M.
(*a*) (compil.) The Twenty-fourth International Symposium on Functional Equations. August 12—August 20, 1986, South Hadley, MA. *Aequationes Math.*, **32**, 96–150.
(*b*) Generating solutions of some Cauchy and cosine functional equations. *Aequationes Math.*, **32**, 216–26. *MR*: **88f**, #39009.
Strambach, K. Multiplikationen mit grossen Automorphismengruppen. *Aequationes Math.*, **32**, 312–26. *MR*: **88g**, #51041.
Sundberg, C. and Wagner, C. A functional equation arising in multi-agent statistical decision theory. *Aequationes Math.*, **32**, 32–7. *MR*: **88d**, #39021.
Székelyhidi, L.
(*a*) On addition theorems. *C.R. Math. Rep. Acad. Sci. Canada*, **9**, 139–41.
(*b*) Remarks on Hyers's theorem. *Publ. Math. Debrecen*, **34**, 131–5. *MR*: **88g**, #39005.
Tabor, J. On mappings preserving the stability of the Cauchy functional equation. *Wyż. Szkoła Ped. Krakow. Rocznik Nauk.-Dydakt. Prace Mat.*, **12**, 139–47.
Vukman, J.
(*a*) Some functional equations in Banach algebras and an application. *Proc. Amer. Math. Soc.*, **100**, 133–6. *MR*: **88d**, #46093.
(*b*) A note on additive mappings in noncommutative fields. *Bull. Austral. Math. Soc.*, **36**, 499–502.
Young, H.P. Progressive taxation and the equal sacrifice principle. *J. Public Econom.*, **32**, 203–14.

1988

Aczél, J. 'Cheaper by the dozen': Twelve functional equations and their applications to the 'laws of science' and to measurement in economics. In *Measurement in Economics. Theory and Applications of Economic Indices.* Physica, Heidelberg, 1988, pp. 3–17.
Blanton, G. Smoothness in disjoint groups of real functions under composition. *Aequationes Math.*, **35**, 1–16.
Cooke, W.P. L'Hôpital's rule in Poisson derivation. *Amer. Math. Monthly*, **95**, 253–4.
Dhombres, J. Relations de dependance entre les équations fonctionnelles de Cauchy. *Aequationes Math.*, **35**, 186–212.
Ebanks, B.R., Kannappan, Pl. and Ng, C.T. Recursive inset entropies of multiplicative type on open domains. *Aequationes Math.*, **36**, 268–93.

Eichhorn, W. and Gleissner, W.
(*a*) The equation of measurement. In *Measurement in Economics. Theory and Applications of Economic Indices.* Physica, Heidelberg, 1988, pp. 19–27.
(*b*) The solutions of important special cases of the equation of measurement. In *Measurement in Economics. Theory and Applications of Economic Indices.* Physica, Heidelberg, 1988, pp. 29–37.

Farnsworth, D. and Orr, R. Transformation of power means and a new class of means. *J. Math. Anal. Appl.*, **129**, 394–400.

Gajda, Z. On stability of the Cauchy equation on semigroups. *Aequationes Math.*, **36**, 76–9.

Galambos, J. *Advanced Probability Theory.* (Probability: Pure and Appl. Ser., Vol. 3). Marcel Dekker, New York-Basel, 1988.

Ger, R.
(*a*) (compil.) The Twenty-fifth International Symposium on Functional Equations, August 16–22, 1987, Hamburg-Rissen, Germany. *Aequationes Math.*, **35**, 82–121.
(*b*) Almost approximately convex functions. *Math. Slovaca*, **38**, 61–78.

Gronau, D. and Reich, L. (compil.) Seminar on functional equations Katowice-Graz, May 22–23, 1986. *Aequationes Math.*, **35**, 281–98.

Johnson, B.E. Approximately multiplicative maps between Banach algebras. *J. London Math. Soc.*, (2), **37**, 294–316.

Kulkarni, R.D. D'Alembert's functional equation on products of topological groups. *Indian J. Pure Appl. Math.*, **19**, 539–48.

Mak, K.-T. and Sigmon, K. Standard threads and distributivity. *Aequationes Math.*, **36**, 251–67.

Paganoni, L. On a functional equation with applications to measurement in economics. In *Measurement in Economics. Theory and Applications of Economic Indices.* Physica, Heidelberg, 1988, pp. 39–45.

Páles, Zs.
(*a*) On a Pexider-type functional equation for quasideviation means. *Acta Math. Hungar.*, **51**, 205–24.
(*b*) On homogeneous quasideviation means. *Aequationes Math.*, **36**, 132–52.

Šemrl, P. On quadratic functionals. *Bull. Austral. Math. Soc.*, **37**, 27–8.

Smital, J. *On Functions and Functional Equations.* Adam Hilger, Bristol–Philadelphia, 1988.

Stam, A.J. Polynomials of binomial type and compound Poisson processes. *J. Math. Anal. Appl.*, **130**, 493–500.

Tabor, J. On functions behaving like additive functions. *Aequationes Math.*, **35**, 164–85.

Vukman, J. Some remarks on derivations in Banach algebras and related results. *Aequationes Math.*, **36**, 165–75.

Young, H.P. Distributive justice in taxation. *J. Econom.* Theory, **44**, 321–35.

AUTHOR INDEX

Numbers in italics refer to the Bibliography

SUBJECT INDEX

Numbers in italics indicate the pages where the subject is defined (not used when only one page is listed)